Extremophiles as Astrobiological Models

Scrivener Publishing
100 Cummings Center, Suite 541J
Beverly MA, 01915-6106

Astrobiology Perspectives on Life of the Universe

Series Editors: Richard Gordon and Joseph Seckbach

In his 1687 book *Principia*, Isaac Newton showed how a body launched atop a tall mountain parallel to the ground would circle the Earth. Many of us are old enough to have witnessed the realization of this dream in the launch of Sputnik in 1957. Since then our ability to enter, view and understand the Universe has increased dramatically. A great race is on to discover real extraterrestrial life, and to understand our origins, whether on Earth or elsewhere. We take part of the title for this new series of books from the pioneering thoughts of Svante Arrhenius, who reviewed this quest in his 1909 book *The Life of the Universe as Conceived by Man from the Earliest Ages to the Present Time*. The volumes in ***Astrobiology Perspectives on Life of the Universe*** will each delve into an aspect of this adventure, with chapters by those who are involved in it, as well as careful observers and assessors of our progress. Guest editors are invited from time to time, and all chapters are peer-reviewed.

Publishers at Scrivener
Martin Scrivener (martin@scrivenerpublishing.com)
Phillip Carmical (pcarmical@scrivenerpublishing.com)

Extremophiles as Astrobiological Models

Edited by
Joseph Seckbach and Helga Stan-Lotter

Scrivener Publishing

WILEY

This edition first published 2021 by John Wiley & Sons, Inc., 111 River Street, Hoboken, NJ 07030, USA and Scrivener Publishing LLC, 100 Cummings Center, Suite 541J, Beverly, MA 01915, USA

For more information about Scrivener publications please visit www.scrivenerpublishing.com.

Wiley Global Headquarters
111 River Street, Hoboken, NJ 07030, USA

For details of our global editorial offices, customer services, and more information about Wiley prod-ucts visit us at www.wiley.com.

Library of Congress Cataloging-in-Publication Data

ISBN 978-1-119-59168-9

Cover image: Courtesy of the editors
Cover design and illustrations by Russell Richardson

Set in size of 11pt and Minion Pro by Manila Typesetting Company, Makati, Philippines

Printed in the USA

Contents

Preface

Foreword

Life on Earth is ubiquitous, with most organisms living in so-called "normal" environments that we consider ambient habitats. Many microorganisms are known to tolerate and resist harsh and extreme external conditions; they are called *extremophiles* [6; 7]. Among them are Archaea, Bacteria, fungi, plants and microscopical animals, like the tardigrades. Some of these microorganisms are living at elevated pressure, such as in the depths of the oceans; some are able to tolerate extreme temperatures and pH values, desiccation or strong UV radiation. Some extremophiles may be under the stress of more than one factor, and we refer to them as *polyextremophiles* [8]. The severe environments most probably resemble the conditions on early Earth. Some of the extremophiles may thus be considered as "living fossils" since their environments resemble the conditions that have existed during the time when life is thought to have arisen on Earth, more than 3.8 billion years ago.

Are there any life forms outside the terrestrial regions? Why should such extremophilic organisms not also live in extraterrestrial places? If these organisms can thrive in such harsh conditions on Earth, they should be able to exist on celestial bodies with similar conditions.

The origin of life on Earth is still unknown. Life has been suggested to evolve from hot springs, geysers, ocean depths, volcanoes, etc. Some scientists think that life originated in cold environments, and suggest connections to bacteria in extremely cold territories like Antarctica or the North Pole.

Who and what is in this new book?

In 1995 the first extrasolar planet was discovered by Mayor and Queloz [4]; now that number is 4,301 [5; July 2020] and the question of whether there is life—possibly intelligent life—in space is more timely than ever.

Research is hampered by the fact that no other life-bearing planet has been found which could serve as a comparison. Or has it? The question of whether there is—or was—life on Mars, at least in the form of simple microbes, has not yet been solved unequivocally [1]. The Jovian moon Europa and the Saturnian moons Titan and Enceladus are also considered promising candidates for worlds with life [3].

This book deals with the description of extremophilic microorganisms which live in environments with similarities to those known from several planets and moons in the solar system. Also, on Earth, environmental conditions occur which are lethal or at least harmful to many organisms, but specialists nevertheless survive or even thrive under these conditions.

The first chapters (Part I) consider extremophiles which are found in environments with possible or very likely similarities to the conditions on extrasolar planets or other celestial bodies. For example, the pH can be very low, such as 3.0, and temperatures can be high (> 90 °C), such as in volcanic steam vents or fumaroles (Bizzoco and Kelley). Using a special collector for the steam, the authors discovered a great diversity of thermophiles, which mostly belong to the Archaea.

Acidophiles are of special interest because their chemolithotropic metabolism obtains energy from reduced minerals, thus creating the extreme acidic conditions in which they thrive. An extensive geomicrobiological characterization of the Rio Tinto basin in Spain has proven the prominent role of the iron cycle in the ecosystem (Amils and Fernández-Remolar). The identification of iron sulfates and oxides on Mars, analogous to those generated in the Tinto basin by microbial metabolism, has made Rio Tinto one of the best geochemical and mineralogical terrestrial Mars analogues.

Recent findings suggest that microbiomes which are found in brackish, marine and hypersaline modern sapropels ('rotten mud') include yet uncultured Archaea that may be close to the evolutionary roots of eukaryotes and life itself (Andrei *et al.*). The extreme geochemistry of certain sapropels, as well as their relevance in the preservation of biomarkers, might qualify them as analogs for early Earth habitats or for the exploration of habitable extraterrestrial milieus.

Expansive evaporite mineral deposits on Mars are evidence of ancient lacustrine systems (Bayles *et al.*). As the surface water dried up, hypersaline lakes would have filled the ancient lake basins. Halite and gypsum contain fluid inclusions where microorganisms may be entombed over geologic time. Haloarchaea are also resistant to other extremes, such as high radiation doses. These properties make them excellent analogues for life that could have existed in the hypersaline lakes on Mars and perhaps remained preserved in the evaporitic minerals there.

Historical observations of NASA's activities towards Mars (Viking experiments) are presented by Oremland, including the early enthusiasm for astrobiology (although that name was not yet coined in the 1970s). His research focuses on soda lakes, which are alkaline (pH ≥ 9.5), often hypersaline (salinity > 35 g/L), mineral-rich water bodies, and their amazingly intense microbial populations, such as haloalkaliphilic arsenotrophs, which are capable of using As(V), Fe(III), or S(0) as electron acceptors. The possibility of similar environments on Mars or planetoids (Enceladus, Titan) is considered.

Antarctica is a rich source of extremophiles, not just psychrophiles, but polyextremophiles which are exposed to unusually high level of UV radiation for Earth, hyperarid conditions, hypersaline conditions, and extremely low nutrients (Abbott and Pearce). Notable similarities exist to conditions known to occur on Mars, or to what is known of the icy moons of Jupiter and Saturn.

Today, Earth harbors vast oceanic ecosystems—most of them only barely explored—which support life (psychrophilic, barophilic and chemosynthetic) adapted to conditions that may occur in other planets in the universe (George). Climate change-induced processes may slow down major ocean currents, such as the Antarctic Bottom Water (ABW) flowing into the lower hadal or ultra-abyssal zone at a depth greater than 8,000 meters, with potential creation of hypoxic hadal zones at extreme ocean depths and its consequences.

In Part II, experiments with extremophiles in space, e.g., on the International Space Station, are presented. Bacteria as well as Archaea, lichens, fungi, algae and tiny animals (tardigrades) are now being investigated for their tolerance to extreme conditions in simulated or real space environments. Experimental results from exposure studies on the International Space Station and on space probes for up to 1.5 years are presented.

Anhydrobiotic cyanobacteria of the genus *Chroococcidiopsis* possess a remarkable resistance to desiccation and radiation. These cyanobacteria were exposed to laboratory simulations that mimic planetary conditions, such as dryness, UV and gamma radiation, and also to real space conditions on the International Space Station (Billi). The resistances found are discussed with respect to lithopanspermia.

Lichen species, which are composite organisms consisting of algae or cyanobacteria in a mutualistic relationship with filaments of various fungi, are well-known survivors of the most stressful environments. The first space experiments were carried out with *Rhizocarpon geographicum* and *Xanthoria elegans* in the Foton-M2 satellite in 2005 (de la Torre Noetzel and Sancho). Space simulation and space exposure experiments have focused on these and other lichen species, demonstrating their high capacities of survival and recovery.

The halophilic archaeon *Halococcus morrhuae* and the biofilm-forming bacterium *Halomonas muralis* were exposed to space conditions during the EXPOSE-R2 mission. *Hlm. muralis* was much less resistant to extreme conditions than *Hcc. morrhuae* (Leuko *et al.*). Exposure to outer space had a strong detrimental effect on the survival and genomic stability of *Hcc. morrhuae*; however, *Hcc. morrhuae* could be re-cultivated from samples exposed on the ISS for up to 534 days. These results add to our understanding that life may be able to survive the travel through space.

The cryptoendolithic endemic black fungus *Cryomyces antarcticus* was isolated from sandstone collected in the McMurdo Dry Valleys in Antarctica, a place which is considered the coldest hyperarid desert on Earth and one of the best terrestrial analogues for Mars (Onofri *et al.*). *C. antarcticus* is able to survive intense ionizing radiation, probably due to the presence of a highly melanized thick cell wall.

In Part III, several authors provide reviews on specific topics, such as Jönsson on the properties of tardigrades (also called water bears), which were used in the Foton-M3 mission and survived a combined exposure to space vacuum, cosmic radiation and UV radiation. The evidence for their metabolic arrest (cryptobiosis) and multiple resistances to environmental extremes make these tiny animals (size ca. 1 mm or less) particularly useful astrobiological models.

Nicholson reviews bacterial endospores, the very first model of cells used for astrobiological purposes, and demonstrates their connection to and impact on the history of origin-of-life studies as well as the concept of lithopanspermia. Modern experimental testing of this concept involves using ballistic devices to simulate the interplanetary exchange of rocks containing bacterial spores.

Paul and Mormile consider the very basic problem of available sources of energy for microbial life on Mars. They analyzed the utilization of several available compounds—hydrogen and methane as key molecules, but also iron, sulfur and others—by extremophilic microorganisms, providing numerous examples of a wide range of well-described biochemical reactions. A cautious look at the possible connection to the energy demands of potential human settlements is also included.

Part IV contains articles on theory and hypotheses. The search for extraterrestrial life is closely connected with the question of the origin of life on Earth and its early evolution.

Newer proposals emphasize the importance of environmental oscillations and suggest experiments for testing this on a laboratory scale (Kompanichenko and Levchenko).

Several theories exist on the existence of a prebiotic world, which are outlined by Jheeta. The implications of the vast—and probably still underestimated—horizontal gene transfer and the potential role of viruses and RNA are presented. One or several transitional (T)-LUCAs (last universal common ancestor) are postulated as leading to the emergence of the first cells.

Recent evidence for an unprecedented archaeal diversity (uncultivated strains) and novel bacterial phyla, due to significantly improved sampling of the subsurface and whole genome sampling, has greatly extended our views on the LUCA of all life, including viruses and their consideration as ancient forms of life (Lineweaver).

Instrumentation for the search of sulfur isotopes as biomarkers for potential habitats for extremophiles in the Solar System is proposed by Chela-Flores, together with a discussion of the pros and cons for possible life on the celestial bodies close to Earth (Moon, Mars, Europa, Titan, and the icy moons).

Conclusions

The search for extraterrestrial life has been declared as a goal for the 21st century by NASA, ESA and other space agencies. For meaningful missions, careful planning of sites to be selected and knowledge of their properties is essential. The study of extremophiles on Earth has already provided rich information about the physico-chemical limits of life; these studies need to be extended with a focus on the specific technical requirements of future space missions. For Mars, a recent report suggested four potential sites for extremophiles—caves, deep subsurface, ices, and salts—and outlined measurement techniques and detection methods [2].

Acknowledgments

We thank all authors for their contributions and the reviewers of chapters for their work. Thanks also go to Martin Scrivener and Richard Gordon for their helpful suggestions and excellent cooperation.

References

1. American Museum of Natural History. (https://www.amnh.org/learn-teach/curriculum-collections/cosmic-horizons-book/fossil-microbes-mars), accessed May 29, 2020.
2. Carrier, B.L., Beaty, D.W., Meyer, M.A., Blank, J.G., DasSarma, D.J., Des Marais, D.J., Eigenbrode, J.L. *et al.*, Mars extant life: What's next? Conference Report. *Astrobiology*, 20, 6, 785–824, 2020.
3. Lunine, J.I., Ocean worlds exploration. *Acta Astronaut.*, 131, 123–130, 2017.
4. Mayor, M. and Queloz, D., A Jupiter-mass companion to a solar-type star. *Nature*, 378, 6555, 355–359, 1995.
5. *The Extrasolar Planets Encyclopaedia*, (http://exoplanet.eu; accessed July 31, 2020).

6. Rothschild, L.J. and Mancinelli, R.L., Life in extreme environments. *Nature*, 409, 1092–1101, 2001.
7. Seckbach, J. (Ed.), *Enigmatic Microorganisms and Life in Extreme Environments*, Kluwer Academic Publishers, Dordrecht/Boston/London, 1999.
8. Seckbach, J., Oren, A., Stan-Lotter, H. (Eds.), *Polyextremophiles: Life under Multiple Forms of Stress*, Springer Dordrecht, Heidelberg, New York, London, 2013.

Joseph Seckbach and Helga Stan-Lotter

Part I

EXTREMOPHILES IN ENVIRONMENTS ON EARTH WITH SIMILARITY TO SPACE CONDITIONS

1

Volcanic Steam Vents: Life at Low pH and High Temperature

Richard L. Weiss Bizzoco* **and Scott T. Kelley**

Department of Biology, San Diego State University, San Diego, California, USA

Abstract

Following meteoric recharge, fundamental components of volcanic magma (heat, gases and ions) are carried by rising steam from the subsurface through porous and fractured passageways until the heated steam exits at the surface in caves and vents. Deposits accumulate as steam condenses on cooler cave ceilings and vent walls, leaving evaporites, precipitates, sublimates, and particulates in solid form along with vapor condensates. Over time these become available in sufficient quantities in matrix material to support growth of microorganisms. This study describes and characterizes some of the chemistry of steam vent habitats, the means developed to sample steam and steam deposits aseptically, using a steam condenser and a simple sample collection device. We describe and evaluate controls producing contamination-free steam and steam deposit samples for analysis. The chemistry of different types of vents studied so far includes nonsulfur, sulfur and iron. Recent iron vent and salt cave studies will be considered. These steam bathed sites harbor both known and unknown organisms, typically but not exclusively, Archaea, that have been grown or are present in quantities sufficient to identify by cloning, and then isolate and describe some of their physical features as well as the means used for enrichment and isolation. We report two recent new isolates from an extreme nonsulfur cave and an iron vent (~pH 3; 85.5 °C) and describe revised equipment used to capture steam samples from low pH high-temperature cave/vent habitats. Overall, we report on steam cave/vent habitats, sampling, enrichment, isolation and identification, as well as structural features of isolated organisms found at low pH and near maximum temperatures for these high-altitude geothermal habitats. Our understanding of the features of life in one of the most extreme and abundant habitats that provide the limits of existence will help produce a model for life beyond Earth's extreme boundaries.

Keywords: Acid, Archaea, fumarole, iron, nonsulfur, salt, sulfur, thermophile

1.1 Introduction

A prominent feature of many, if not most geothermal fields is the presence of abundant fumaroles (i.e., steam vents). These are especially visible early in the day when the cool air

**Corresponding author*: rbizzoco@sdsu.edu
Richard L. Weiss Bizzoco: https://www.researchgate.net/profile/Richard_Bizzoco
Scott T. Kelley: https://scholar.google.com/citations?user=oHXnsCEAAAAJ&hl=en

Joseph Seckbach and Helga Stan-Lotter (eds.) Extremophiles as Astrobiological Models, (3–20)

condenses the rising water vapor into smoke-like clouds. These geothermal steam vents differ chemically from each other and can be grouped into nonsulfur, sulfur and iron fumaroles [1.1]. A number of different chemical processes cause steam deposits to form and give individual vents their main chemical identities. Fumaroles can issue a greater or lesser volume of steam depending on the meteoric water supply, temperature, size, and type of vent. In turn, steam deposits are caused by condensation of steam on cooler cave or vent walls or surfaces. This allows their collection by sampling of steam cave or vent surfaces at the point of contact between steam and cave surface and yields distinctive chemical-microbiological samples. These samples result from one of the following processes: 1) evaporation of soluble ions leaving deposits of salts or evaporites; 2) oxidation of soluble Fe(II) ions to insoluble precipitates such as iron oxides, hydroxides, or oxyhydroxides; 3) gases like H_2S rise with steam, becoming less soluble and form into solid sublimates that are oxidized into sulfur deposits; 4) steam and particulates travel over cave surfaces leading to adhesion or entrapment within the surface matrix; and 5) ammonia forms as vapors rising with steam [1.11] [1.13] and concentrates at the steam deposit site [1.1] comparable to the way H_2S acts in acidic systems.

In this study we describe the use of a portable steam collector designed for sterile microbiological sampling to capture volcanic steam and microbial steam vent life forms. Steam exits fumaroles in a continuous flow or in bursts that are both hot and diffuse. This often makes collecting a sufficient quantity of steam vapors for microbiological analysis a challenging prospect. Our collector uses a difference in temperature to condense steam, resulting in hot water guided into a sterile polypropylene collection tube (Figure 1.1). The condenser captures up to 2–3 mL min^{-1} from vents above 90 °C, making on-site collection a practical procedure. Once we collected steam water samples, we examined the condensed water samples with DNA staining (using DAPI) for microbial life and also established cultures to grow the steam vent microorganisms from locations in Hawai'i, California, New Mexico, Wyoming and Russia. By determining the number of organisms in condensed water samples and the steam flow rate, we were able to estimate the overall fumarole dispersion, as steam vapors exited different types of fumaroles. We concentrated our efforts

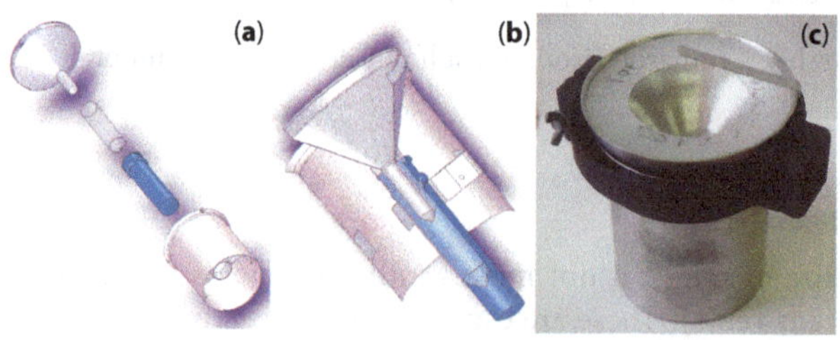

Figure 1.1 Steam condenser: (a) Collector body is a stainless cylinder, 1 or 2 liters with overlying funnel condenser and silicone plugged guide rod. Central collection tube holder lies near cylinder bottom. (b) Cutaway diagram showing interior of condenser parts. (c) Condenser has a fitted splash shield to prevent water loss and contamination. Mounting feature (black) allows extension pole or tripod attachment. Operation: Ascending steam contacts underside of cooler cone funnel, transfers heat to water reservoir above, steam condenses to water that drains down funnel to guide rod, and into 50 mL sterile screw cap collection tube. Following collection, sterile tube is capped. (Image credit: the authors).

on the Archaea, a group of organisms known to live at pH and temperature extremes. In some extreme fumaroles, we found these to be the only organisms present. Finally, we sampled both steam and steam cave/vent deposits and found that the deposit matrix formed at the steam-cave surface contact site served as a unique concentrating device to entrap nutrients critical for the survival of specific steam vent organisms. Hawai'i has so far presented the most chemically diverse types of steam vents, making this an ideal site for examining extremophiles and Archaea as a model for life beyond Earth's boundaries. Other collection sites provide unique steam cave/vent features that increase our understanding of fumarolic microbial life.

1.2 Steam Cave and Vent Sites

Samples of steam were collected from fumaroles in four permanently protected geothermal areas, Lassen Volcanic National Park, Hawai'i Volcanoes National Park, Valles Caldera National Preserve, and Solfatara Crater, Italy (Table 1.1). Our Hawai'i Volcanoes National Park sites are coded, as part of our collection agreement with the park. These sites have a wide and interesting variety of chemical steam caves and vents not seen in other geothermal areas. Our choice of Hawai'i as a site with an active volcano proved to be interesting. In many ways, its basaltic lava flows resemble those of our nearby planet Mars, both chemically and physically providing a match for using Earth's extremophile life as a model for life on other planets. Mars is one of those planets close enough for experimental approaches and applications.

In active steam caves, meteoric waters characteristically descend through porous and fractured surface lava and eventually meet upward convection of heat and rising volcanic gases. Heat generates abundant steam that rises as a vapor-gas mixture through fractured lava crevices and fissures towards the surface. Near the surface, passageways—both horizontal and vertical—allow the vapors to condense on the cooler cave ceilings and walls, depositing characteristic chemical signatures and forming a matrix-like material rich in nutrients for extreme microorganisms. The steam exits the cave or vent opening and provides a diffused smoke-like flow known as a fumarole. Venting steam commonly proceeds in an artesian or burst flow. With large volumes of steam, visibility of the cave interior or vent opening can be obscured. Steam cave/vent openings can be small on the order of a few centimeters or large, one to three meters or larger for wide steam vents or vertical caves. Our selection of steam caves was based on the idea that while we and others [1.7] [1.10] [1.12] [1.16] had already investigated and identified organisms, especially Archaea in steam vents, there was not much information on organisms from those steam caves/vents regarding their *in-situ* appearance and some of the sites where they have previously been isolated left gaps in our information on the physical appearance of steam vent microorganisms at the structural level.

1.3 Steam Cave and Vent Sample Collection

The steam cave and vent sites were selected for their chemical properties. These were initially recognized in flowing hot springs as iron and sulfur. Here, deposition of iron occurred

Table 1.1 Fumarole sampling locations and physical properties.

Location	pH	Temp (°C)	Type	Chemistry
Hawai'i Volcanoes National Park[1]				
Hawai'i (1)	5.5	65.0	Cave	Nonsulfur
Hawai'i (2)	5.2	68.0	Cave	Nonsulfur
Hawai'i (3)	3.0	82.0	Cave	Sulfur
Hawai'i (4)	3.0	82.0	Cave	Iron/sulfur
Hawai'i (5)	5.0	68.0	Cave	Salt
Hawai'i (6)	4.5	66.0	Cave	Salt/sulfur
Lassen Volcanic National Park[2]				
Sulphur Works, lower (1)	4.5	87.0	Cave	Nonsulfur
Sulphur Works, lower (2)	3.2	93.0	Vent	Sulfur
Sulphur Works, lower (3)	4.0	91.5	Cave	Sulfur
Sulphur Works, upper (4)	3.0	85.5	Vent	Iron
Sulphur Works, upper (5)	2.5	81.2	Cave	Iron
Valles Caldera National Preserve[3]				
Sulphur Springs, New Mexico	2.0	89.0	Cave	Nonsulfur
Pozzuoli, Italy[4]				
Solfatara Crater	3.0	94.0	Vent	Sulfur

[1] June 2016 sampling
[2] Oct 2018 sampling
[3, 4] included for comparison

in the anoxic zone of the spring and at the point of disappearance of iron at the anoxic/oxic interface, sulfur deposition took place and finally iron deposits formed on the existing sulfur in the anoxic/oxic zone. The same phenomenon can be seen in very small sulfur caves where the interior has no sulfur, the cave opening is small and sulfur deposition occurs only at the anoxic/oxic interface or on the outer oxic surface, sometimes along with iron deposits over sulfur. The other categories, nonsulfur and salt (the latter so far only seen in Hawai'i), represent a departure from the flowing hot spring analogy. Nonsulfur steam caves have not yet been fully categorized, although unusual thin filamentous organisms have been enriched from the H5 salt cave, and from sulfur caves such as SW 3 in Lassen. Nonsulfur caves farther below are considered for their solid surface chemistry by means of energy dispersive X-ray microanalysis (EDX). Sampling of steam cave or vent ceilings depends on the nature of the surface. With highly acidic solfatara areas, cave ceilings can be delicate and sometimes only careful hand sampling is essential. With deeper nonsulfur caves sometimes encountered in Hawai'i, this is less of a problem because the cave ceilings are formed of hard lava and only shallow samples are obtained. However, sampling through a small opening into a deep cave interior is not without its challenges, principally resulting from the high level of heat encountered, with or without leather gloves.

1.3.1 Steam Collection

Samples of steam were collected from fumaroles in four permanently protected geothermal areas, Lassen Volcanic National Park, Hawai'i Volcanoes National Park, Valles Caldera National Preserve, and Solfatara Crater, Italy. Our steam collector (Figure 1.1) consists of a stainless shroud with a central holder for the sample tube. The condenser with a splash shield fits into the shroud and is slightly raised at 3 points, allowing excess steam to bypass the edges of the condenser during collection. The central lower guide rod directs condensed steam water into a sterile 50 mL polypropylene tube. During steam collection the open tube rests in a holder attached near the bottom of the shroud. The steam condenser can be hand-held, but is also used attached to a tripod by an external mount, which by design serves to protect samples from aerial contaminants. During collection, rising steam contacts the water-cooled condenser and the guide rod drains the condensed steam water directly into the collection tube. After collection, the tube is immediately capped. Overall, the unit is portable, durable, corrosion resistant, easily autoclaved and produces contamination-free steam condensed samples and controls in a relatively short time period. In an average collection a 20–25 mL sample is collected in 15 min. Faster collections occur with a high steam flow rate and elevated temperature. These range between 1–3 mL min^{-1} in extreme fumaroles near or above 90 °C; examples of those are found in Lassen Volcanic National Park [1.8].

We found that collected steam water carried with it organisms that were detected microscopically when stained with DAPI (a dsDNA stain) and examined by fluorescence microscopy. Ellis *et al.* [1.8] found organisms in steam from volcanic fumaroles collected at several widely distant geothermal regions: Kamchatka, Russia, Lassen Volcanic National Park, Yellowstone National Park, Sulphur Springs, New Mexico, Valles Caldera National Preserve and Hawai'i Volcanoes National Park. The steam collector was small enough to transport sterile collectors by airplane. The efficiency of collection allowed the collector to be easily used under most field conditions. We were able to estimate cell numbers and found the efficiency of the collector improved with collection from high temperature, low pH steam vents. DAPI staining and phase contrast microscopy showed that steam from all vents examined contained cells. We were able to estimate cell concentrations in steam water, extract DNA for PCR and cloning and isolate Archaea from enrichments and from subcultures. Concentrations ranged from a low value of 150 cells ml^{-1} in Kamchatka, Russia, to a high number of 1100 cells ml^{-1} in Lassen Volcanic National Park SW2, a sulfur vent. Microscopic observations revealed that Hawaiian fumaroles contained several morphological types and the greatest diversity of phylotypes of any fumaroles examined. By applying lab collection rates in controlled experiments, we found that the collector efficiency was almost 14%, and the flow rate of one Yellowstone fumarole was determined to be about 15 mL min^{-1}. This allowed us to estimate a dispersal rate of 3×10^3 cells min^{-1} or 4×10^6 cells day^{-1}. Hawai'i steam samples contained the greatest diversity of halophiles, isolated initially from steam vent water and later from steam deposits, and suggested that halophiles may be present in these habitats though they do not appear able to grow at elevated temperatures [1.8]. This seems to be consistent with other areas enriched with halophiles that may also experience elevated temperatures by insolation.

1.3.2 Steam Deposit Collection

For sampling steam deposit samples when steam caves are too hot for handheld sampling tools, or samples are collected deep within a cave interior, an extension pole is used with a sterile 50 mL polypropylene tube angled up to 45° towards the cave opening. With this approach, only the upper trailing edge of the open tube contacts the deposit matrix. Thus, the steam deposit material removed by the upper collection tube edge falls directly into the open tube, which is capped immediately upon collection. In most fumaroles, collected material is visible and only a few mm (2–4 mm) at the steam-cave surface deposit is removed. Most collection sites have either thick or easily removed deposits of sulfur, salt, iron or other matrix material, making this procedure a relatively easy task. In cases where the steam contact site cannot be seen, the surface is lava, so only a light sampling is carried out, just sufficient to remove the adherent steam deposit material.

Hawai'i presented the greatest variety of chemical types of steam vents. Three of the vents, two coded and one noncoded, were nonsulfur vents. This type of vent is recharged by meteoric input. Rainwater descends through porous ground, contacts rising heat and is converted to steam. Ammonia released by degassing magma creates an ascending mixture of ionized NH_4^+ and NH_3 gas traveling upward through fissures and fractured lava, reaching passageways that lead to horizontal or vertical caves and vents to the atmosphere as fumaroles. In Hawai'i these clouds of rising steam vapors containing volcanic gases and ionized elements support the growth of ammonia oxidizing archaea (AOA) located below the vent opening and adhering to the cave ceilings and walls. Ammonia oxidizing archaea gain their nutrients from entrapped particles, condensing steam and concentrating volcanic gases such as ammonia that rise from closely positioned magma that degases as it rises towards the surface. Altogether these events in the Hawaiian nonsulfur caves support an unusual habitat that thrives on the concentrating effects of surfaces with matrix material and contact between the steam-gas mixture. Here, the cooler cave ceiling or surface condenses steam from the continuous flow of warm thermal vapors and nutrients. Figure 1.2 shows two nonsulfur sites, one in Hawai'i and one in Lassen Volcanic National Park along with a sulfur cave and an iron vent in Lassen, and a salt cave site in Hawai'i Volcanoes National Park. We have documented the presence of AOA in three nonsulfur steam vents in Hawai'i and have recently isolated Archaea-like cultures from the nonsulfur cave and Archaea from the iron vent in Lassen Park. We have not yet determined whether the SW1 nonsulfur cave Archaea-like spheres belong to the *Thaumarchaeota*, a newly proposed phylum for the ammonia oxidizing archaea [1.4] [1.14] [1.15]. The Hawai'i H1 steam seems similar to the steam at the nonsulfur vent in Lassen Volcanic National Park (Table 1.2) and there is sufficient ammonia at both sites for AOA, and clones were recovered at the Hawai'i H1 site [1.1].

We recognized that a majority of microorganisms growing in natural habitats attach to surfaces. Since there is no sediment in steam caves, organisms tend to attach to the type of surface deposit created by the steam flow and chemical identity of the steam vent or cave (Table 1.1). The chemical features of solid surfaces can be characterized by X-ray microanalysis. As a result, steam deposit samples from caves and vents can be analyzed directly for their chemical composition. In our case, we used an X-max 50 mm2 X-ray detector

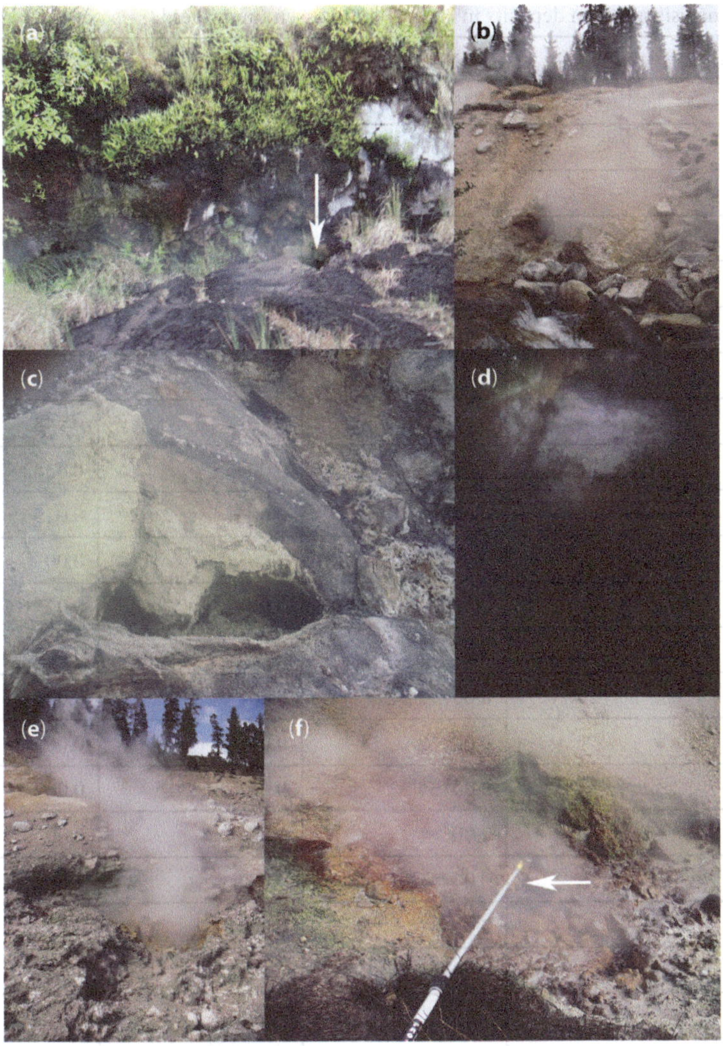

Figure 1.2 Steam deposit sampling sites: (a) Nonsulfur steam cave site Hawai'i 1, rising steam vapors fill the air to the left of the cave opening (arrow). Ammonia oxidizing archaea (AOA) clones recovered from H1. (b) Nonsulfur steam cave site SW1 Lassen, unknown spherical cell culture recovered and subcultured. (c) Sulfur steam cave SW3 Lassen, steam vapor clouds mix with air and deposit yellow S° inside cave and on outside rock surfaces. *Sulfolobus* cultures recovered from SW3 and subcultured, pH 4.5, 85 °C. (d) Salt cave site Hawai'i 5, rainwater-lava interaction result in salt cave deposits (white) on the ceiling, wall and floor deep within Hawai'i 5. Mixed culture enrichments, pH 4.5, 55 °C, include abundant unknown thin filaments. (e) Iron steam vent site SW4 Lassen, unknown archaeon recovered and subcultured. (f) Measuring temperature in iron vent SW4 with mercury maximum recording thermometer (arrow). B, E Cultures isolated and subcultured at pH 4.5, 80 °C; SW4 isolate was also subcultured at pH 3, 80 °C. (Image credit: the authors).

Table 1.2 Analysis of steam deposits in Hawai'i and Lassen fumaroles.[1]

Analyte (mg L^{-1})	H1NS°C[2]	SW1NS°C	SW4FeV[3]
Na	3.454	3.457	3.453
Ca	2.285	2.119	2.870
Al	1.024	1.258	2.006
Fe (total)	0.975	0.974	0.997
Si	18.579	21.932	25.706
B	0.697	0.723	0.760
K	2.864	3.965	3.401
Mg	0.119	0.138	0.945
Zn	0.0147	0.00455	0.0362
Mn	0.00604	0.00618	0.0256
Mo	0.00	0.00	0.00
Se	0.0289	0.0242	0.0251
Ni	0.00	0.00	0.00
Pb	0.00	0.00	0.00
Cr	0.00	0.00	0.00
Cd	0.00164	0.00178	0.00236
Cu	0.00	0.00	0.00
Hg	0.00	0.00	0.00
As (total)	0.0242	0.0201	0.0209
S	0.0312	5.143	17.833
Sr	0.0832	0.00596	0.0187
NO_2/NO_3, N-NO_3 (μM)	0.00	0.00	3.453
NH_4,N-NH_4 (μM)	19.550	79.023	5.780
PO_4,P-PO_4 (μM)	0.00	27.372	0.00
SO4 (μM)	0.00	0.00	222.0
Conductivity (μScm^{-1})	13.9	167.2	516

[1]Analysis by inductively coupled plasma optical emission spectrometry and wet chemistry.
[2]Column 1 Data from Bizzoco and Kelley [1.3].
[3]Column 3 data from Bizzoco and Kelley [1.2].
H1NS°C, Hawai'i 1 nonsulfur cave, SW1NS°C Sulphur Works 1 nonsulfur cave, SW4FeV, Sulphur Works 4 iron vent.

and a Quanta 450 FEI-SEM operated at 20 kV with Oxford Inca software. Samples were attached with an adhesive-conductive carbon tab to a stub and analyzed. We collected spectra (Figure 1.3) from three different types of caves/vents that supported growth:, (a) Hawai'i H1 nonsulfur cave, (b) Lassen SW1 nonsulfur cave, (c) Lassen SW4 iron vent and Hawai'i H5 salt cave. The salt cave sample remains to be analyzed. The Hawaiian sites H1 and H5 were 65–68 °C and the Lassen sites SW1 and SW4 were 85.5–87 °C.

The Hawai'i H1 sample (Figure 1.3a) was collected from a hard lava cave ceiling (Figure 1.2a) and contained strong peaks for aluminum (Al), silicon (Si) and iron (Fe). The aluminum peak was also formed by a smaller peak of iridium (Ir). Although characterized as a nonsulfur cave, a small sulfur (S) peak was present in the collected sample and was a characteristic of this cave site. Also observed was a strong peak for oxygen (O) and moderate peaks for titanium (Ti) and several biologically important elements, potassium (K), sodium (Na), magnesium (Mg) and calcium (Ca); each of these peak identities was further documented by analyzing several spectra (data not shown). Titanium has been found along with silicon coated on rock surfaces in the Ka'u desert on the Big Island [1.6] and in most Hawai'i steam cave/vent samples, but titanium has no known role in metabolism. The Lassen nonsulfur cave site SW1 was a shallow cave (Figure 1.2b) composed of soft backing material and the immediate sampling was guided by visual access. There was a distinctive red gelatinous material in the collected sample that proved to be noncellular upon microscopic examination. The nonsulfur SW1 spectrum (Figure 1.3b) had far fewer peaks than the Hawaiian H1 counterpart and contained as its main peaks silicon (Si) and oxygen (O) with smaller peaks for titanium (Ti) and carbon (C). Lassen iron vent SW4 was sampled as a thin (~2–3 mm) soft iron (III) oxide deposit collected from a hard rounded rock surface within the vent (Figure 1.2e, f). The spectrum (Figure 1.3c) contained strong peaks for iron (Fe), silicon (Si), oxygen (O) and aluminum (Al) with a moderate sulfur (S) peak and a smaller carbon (C) peak.

1.3.3 Steam and Steam Deposit Collection: Control Methods

Steam Controls

To control for the presence of foreign DNA in our sample analysis of DNA from steam-condensed samples, we used Millipore filtered (0.22 μm membrane) autoclaved Barnstead purified water. The water was converted to steam next to a cave or vent sampled and collected in identical fashion to steam samples from fumaroles. After collection, samples were analyzed with cave or vent samples by (1) PCR, (2) filtration of 10–15 mL samples on 13 mm black Millipore membranes (0.22 μm pore size) and DAPI microscopy, and (3) for thermophile culture pH 3 and 4.5 at 55, 70, 80, and 85 °C. Control samples proved to be negative.

Steam Deposit Controls

Steam deposit samples from Hawai'i H1 nonsulfur cave were collected by scraping lightly along the inside of the steam cave ceiling. Lava presents a hard surface that cannot be penetrated with a slight 50 mL polypropylene collection tube scraping able to collect steam

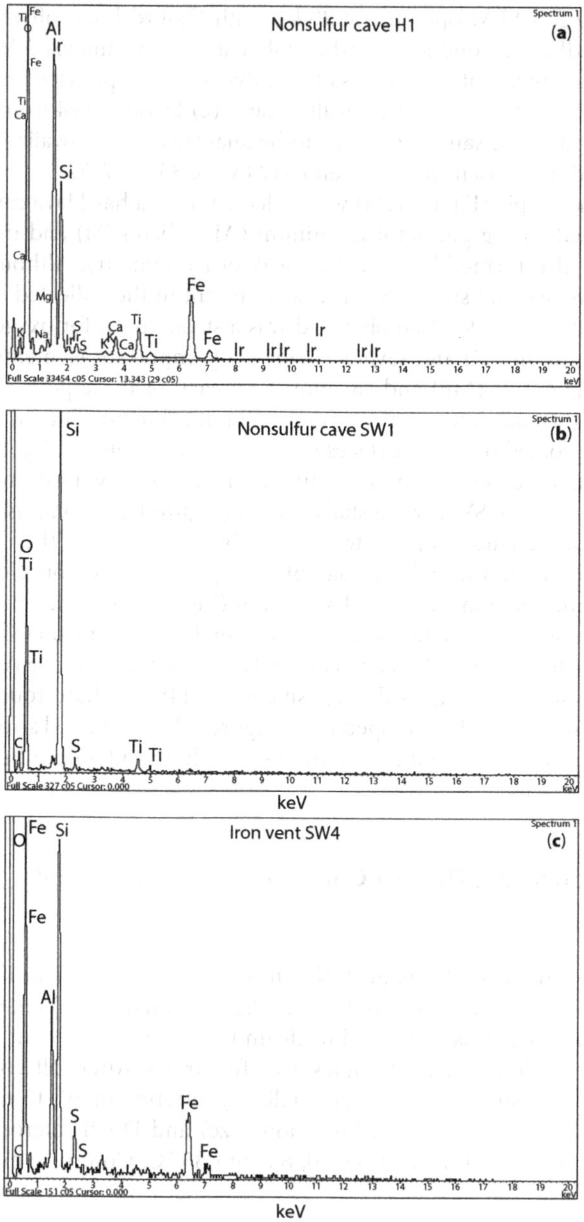

Figure 1.3 Steam caves/vents spectra: (a) Hawai'i H1 nonsulfur cave. (b) Lassen nonsulfur cave SW1. (c) Lassen iron vent SW4. (Image credit: the authors).

deposits. We used an adjacent nonfumarolic, ambient temperature lava cave and performed an identical collection procedure with scraping-tube collection used for the steam cave. The other soft surfaces of Lassen nonsulfur cave SW1 and SW2 sulfur cave and Hawai'i H5 salt cave were shallow samplings only a few mm of thick deposits and did not include material other than from the nonsulfur steam deposits, salt deposits or sulfur sublimates. Steam deposit controls were analyzed and evaluated along with the collected steam cave/vent samples, as described above for PCR, DAPI microscopy and thermophile culture and were negative. We are currently expanding our controls in several ways to evaluate the contribution growing cells might make to the chemical profile from EDX analysis.

1.4 Culture Isolation

In some of the earliest work on low pH, high temperature organisms, the concept was advanced [1.5] that steam vents were lacking in microbial life and there was little encouragement to study any of those sites, since there were thousands of pools and springs available and already known. In steam caves/vents there is no mixing, settling or sediments as there might be in hot spring pools or flowing springs. Surfaces become an important part of the habitat since organisms must remain in place or be swept away from the habitat. As steam sweeps over these niche-habitats a continuous, though perhaps changing supply of nutrients passes through, some of which become concentrated or deposited as a nutrient source. Thus, the underside of ledges, ceilings of caves, or sides of vents all become important areas for colonization and consequently, for collection. In this regard, we collected from a steam cave ledge, but only on the underside where steam rose up and hit the ledge deep inside a cave, or on the salt cave shelf where pure white salt crystals existed (Figure 1.2d) and could be collected. When we collected from narrow caves on the ceiling often it was simply a hard ceiling so that no soil or other features existed beyond the hard lava ceiling and the steam deposits on the ceiling structure, sometimes several meters inside a horizontal cave.

Typically, steam deposit samples collected from caves and vents were enriched in vials in mineral salts medium [1.8]. A small amount of a natural sample was aseptically added to a vial containing 10 mL liquid medium at temperatures in stepwise increases. At lower temperatures (55–75 °C), Teflon caps were used; at higher temperatures removable screw cap butyl stoppers replaced these. We established enrichments in the present study with three high temperature sites in Lassen Volcanic National Park, a nonsulfur cave SW1, a sulfur cave SW3 and an iron vent SW4. Cultures set at pH 4.5, 85 °C were positive in 10–15 days. All of the cultures that were positive were slightly below the habitat temperature. When cultures were examined microscopically there was commonly a limited variety of organisms ranging from irregular spheres to very thin rods and irregular spheres in a mixed culture usually with thin filaments. Examples of DAPI stained cultures appear in Figures 1.4 to 1.6. All the SW1 and SW4 cultures were initially isolated at pH 4.5, 80 °C, but with subculture one of them, SW4, grew more densely at pH 3, 80 °C, its habitat pH, 6.50×10^8 cells mL^{-1} compared to 1.48×10^8 cells mL^{-1} at pH 4.5, 80 °C. Large cell groups consisting of hundreds of cells were seen throughout the iron vent cultures growing at pH 4.5 and 80 °C (Figure 1.6).

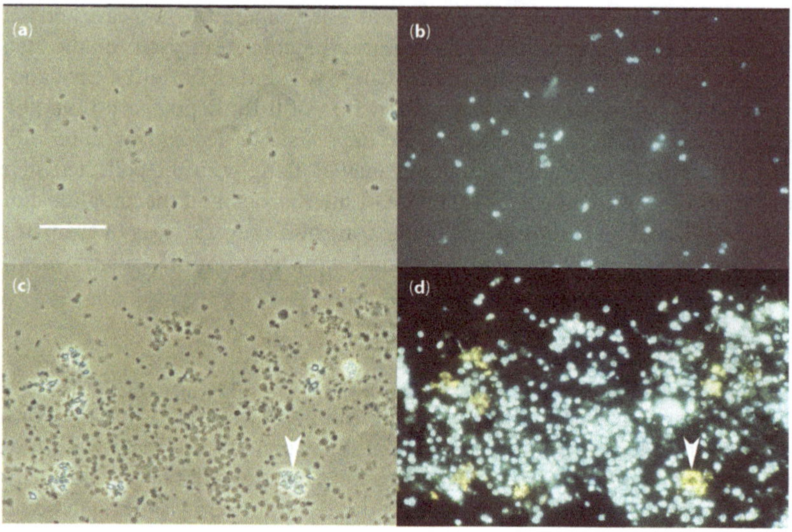

Figure 1.4 Sulphur Works nonsulfur cave SW1 culture pH 4.5, 80 °C. (a) Phase contrast, numerous cell pairs and single cells appear phase dense. (b) DAPI stain. Cells stained by DAPI appear blue. Cell pairs often appear as a result of cell division. (c) Phase contrast. Large cell groups appear in culture associated with reddish tinged cave deposit material (arrowhead). Phase dense cells appear spherical. (d) DAPI stain. Reddish tinge minerals (arrowhead) and associated cells at the surface adhering to mineral particles. Bar, A–D, 10 μm. (Image credit: the authors).

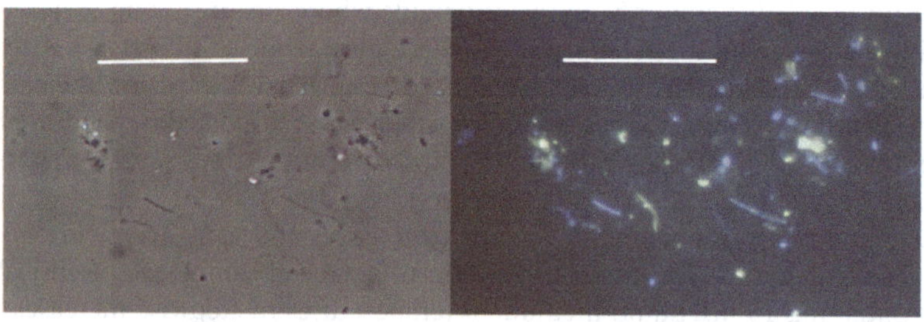

Figure 1.5 Sulphur Works sulfur cave SW3 culture pH 4.5, 85 °C. (a) Phase contrast (b) DAPI stain. Mixed culture; spheres identified by PCR as *Sulfolobus acidocaldarius* and thin filaments, identity unknown. Microscopic field with approximately equal numbers of spheres and thin filaments. Bar A–B, 10 μm. (Image credit: the authors).

The appearance of the irregular spheres isolated from SW1 and SW4 resembled Archaea, but more information was essential to determine their taxonomic status. In some instances, particularly where heavy metals may be detrimental to DNA integrity, it may not be possible to obtain amplifiable DNA, or if DNA is obtained it may not sequence properly. If it is not possible to obtain DNA for identification from cultures, fluorescence in situ hybridization (FISH) is used to identify the taxonomic group. An example of this approach is given in Figure 1.7. The controls include DAPI stain, to identify Bacteria and Archaea, a nonsense sequence probe (NON338) and no probe added. Both samples and controls were

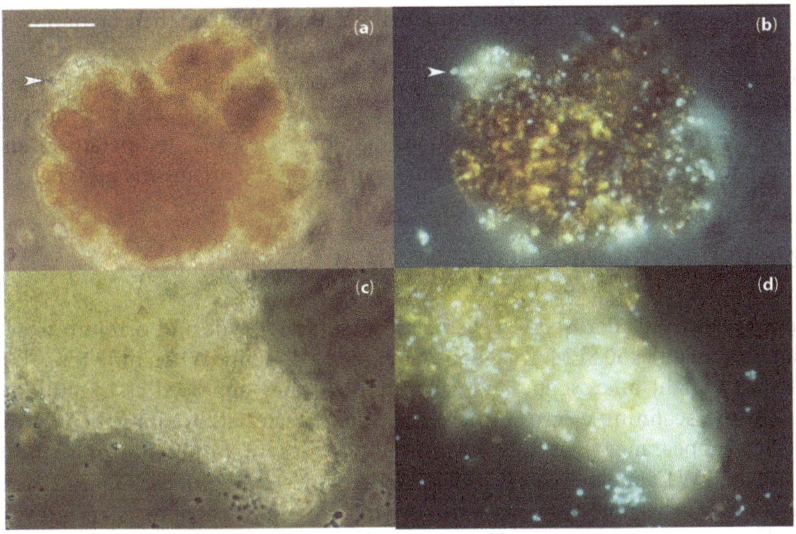

Figure 1.6 Sulphur Works iron vent SW4 culture, pH 4.5, 80 °C. (a) Phase contrast. Silica, iron and minerals form a heterogeneous base for cell growth in culture. Arrowheads (a,b) indicate same cell location. (b) DAPI stain. Cells appear as blue fluorescent spheres mainly surrounding edges of steam deposits. (c) Phase contrast. Faint red tinged siliceous particle with few cells seen, other than at lower right edge. (d) DAPI stain. Same small cell group evident at lower right edge in phase contrast and DAPI images; blue DAPI stained cells, not seen by phase contrast, appear dispersed throughout the deposit particle. Bar A–D, 5 μm. (Image credit: the authors).

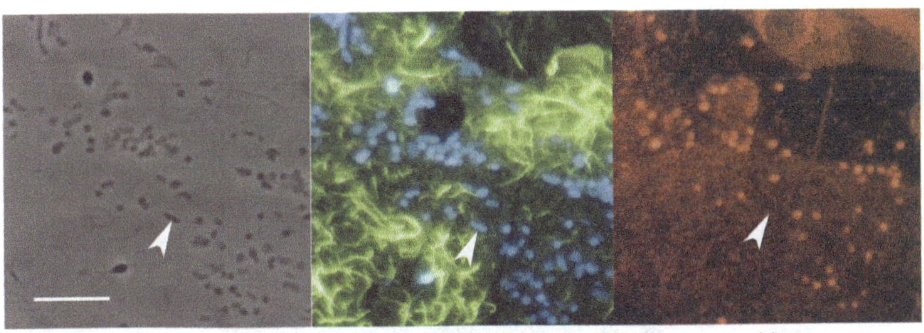

Figure 1.7 FISH labeled culture pH 4.5, 80 °C, isolated from iron vent, Sulphur Works SW4. (a) Phase contrast. (b) DAPI stain. (c) Fluorescent image of cells labeled with Archaea-specific probe Arch915-CY3 probe. Labeled cells were all bright orange archaeal spheres. None of the rod-shaped bacterial cells (A,B, arrowhead) were labeled by the probe (c). Only spherical Archaea were fluorescent. Arrowheads (a-c) mark location of a typical bacterial rod nonlabeled in C. Bar A–C, 5 μm. (Image credit: the authors).

processed identically and at the same time. Negative results were obtained with the controls and indicated that if the Archaea-specific Arch915 probe labeled cells, the labeling when positive indicated the cells were *Archaea*. In this sample from a culture of mixed rods and spheres only the spheres were labeled by the probe, indicating that they were members of the taxonomic group *Archaea*, although no further information on their taxonomy is available from this direct labeling technique.

A salt cave liquid enrichment, Hawai'i H5, was obtained at pH 4.5, 55 °C. The culture was a mixture of thin filaments, wide rods about 0.5 µm in diameter, and pleomorphic cells. Further work with more selective conditions (ionic species, pH, temperature) and solid surface isolations may assist in obtaining pure cultures of the organisms present.

1.5 Cell Structure of Isolates

For scanning electron microscopy, cells fixed in glutaraldehyde and osmium were examined in an FEI Quanta 450 SEM. Images were recorded digitally. When individual cells were observed in the light microscope and cells tumbled, they appeared as irregular spheres. Similarly, in the scanning electron microscope cell isolates showed a generally irregular spherical SEM appearance (Figure 1.8). Most cells were less than 1 µm in size and the nonsulfur cave isolate SW1 appeared as large (Figure 1.8a) and small groups (Figure 1.8a, insets). Paired and single cells more typically were seen in light microscope images (Figure 1.4a, b). The SW4 iron vent cells grown at pH 3 were highly concentrated and readily seen in SEM images as paired or single cells (Figure 1.8b). The cells appeared to be irregular spheres. At higher magnification, irregular ridges and depressions were prominent (Figure 1.8b, inset). This can be typical of later stages of growth when competition for food, and increased surface to volume ratios exist. The origin of ridges and depressions remains uncertain and requires further study to eliminate the possibility of artifact resulting from sample preparation. In the field of view shown here, the iron vent SW4 isolate had a few cell pairs that

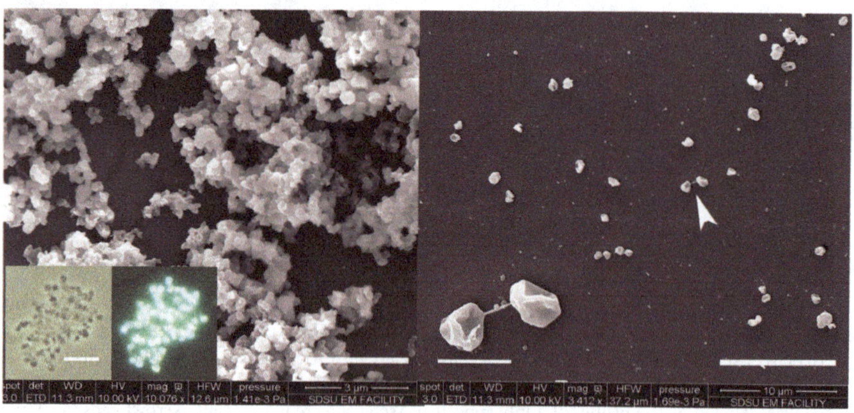

(a) (b)

Figure 1.8 Scanning electron microscope images of steam vent isolates. (a) Cell cluster from nonsulfur steam cave SW1 with irregular spheres grown at pH 4.5, 80 °C. Left inset, SW1 phase contrast; right inset, SW1 DAPI stain. (b) Archaea grown from iron vent SW4 appear in the SEM mainly as pairs and single cells; arrowhead marks inset cell pair. Inset, ridges, depressions, and thin cell-to-cell connections appear. Cells grown at pH 3, 80 °C. Bar A, 3 µm. Bar B, 10 µm. Bar insets A (phase-DAPI stain), 5 µm. Bar inset B, 1 µm. (Image credit: the authors).

appeared to be held together by thin cell-to-cell bridges, presumably resulting from cell division. Though somewhat difficult to discern, one example marked by an arrowhead, is more readily visible in the inset. The thin cell bridge measured ~0.07 × 0.5 μm and tapered at both ends where it connected to cells. We did not determine that cells were motile as no flagella were seen, though some Archaea with flagella are known to be motile [1.9].

It seemed unusual that the two new Lassen isolates grew only at 80 °C, and not above. Yet, the habitat temperatures for nonsulfur cave (SW1) and iron vent (SW4) were 87 and 85.5 °C, respectively. Furthermore, in prior years the habitat temperatures were both above 90 °C. Temperature drops off quickly even at short distances away from the cave or vent opening. Thus, locations providing gases, nutrients or energy sources may not lie directly in the path of venting steam, possibly explaining the discrepancy between the higher environmental temperature and the lower isolation temperature. It may also be the case, as reported for hot springs [1.5], that different temperature strains exist within natural populations bridging a broader range of temperatures than that of the isolates obtained in our study.

1.6 Environmental Models

Models of the Lassen SW4 iron-oxidizing environment and the Hawai'i H5 salt cave environment are presented below (Figure 1.9 and 1.10) to provide an overview of these habitats.

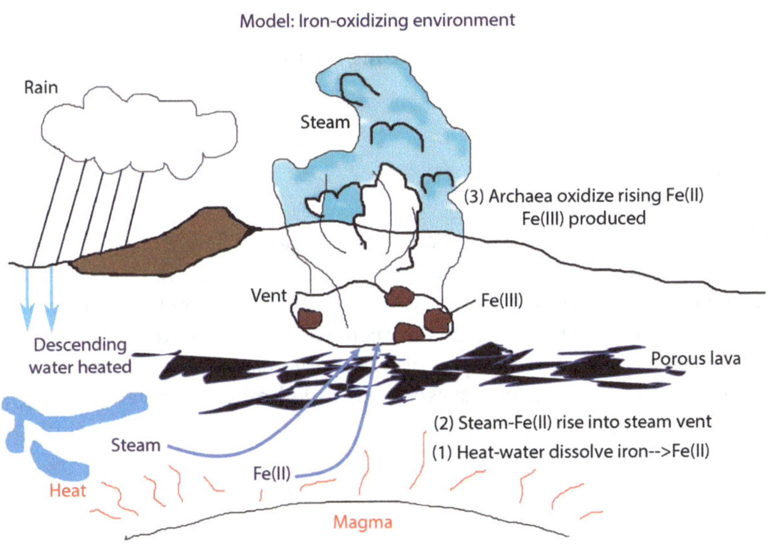

Figure 1.9 Iron-oxidizing environment. Heated rainwater from magmatic heat convection produces steam that contacts lava and extracts soluble Fe(II). (1) Fe(II) is solubilized in the steam water and (2) rises with steam through porous lava fissures and cracks and reaches the iron vent openings of SW4, as the Fe(II) contacts air (3) Archaea utilize the Fe(II) in their growth by oxidizing the iron to Fe (III). Air oxidation also takes place and adds Fe(III) to the iron-rich layer formed on the solid rocks within the iron vent. (Image credit: the authors).

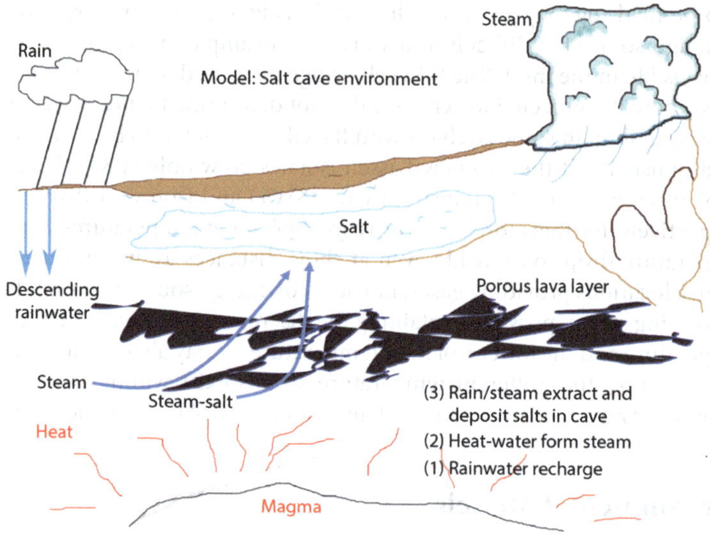

Figure 1.10 Salt cave environment. (1) Rainwater recharge of the Hawaiian ground water passes slowly through the porous lava rich ground. Fractured lava is heated by the underlying magma and as (2) rainwater descends through crevices and fissures it extracts soluble salts that are present. At greater depths the descending water boils and provides steam and heat so that (3) salts continue to slowly dissolve, rise with steam and deposit as evaporites on a cooler nearby wall, ceiling or as in this model, a shelf. (Image credit: the authors).

Models of the sulfur and ammonia oxidizing environments are available in a previous study [1.3].

1.7 Conclusions

The worldwide abundance of fumaroles compared to hot springs make this an ideal choice as a habitat for early development of life on Earth. These primitive steam habitats characterized by two extreme physical features, low pH and high temperature, typically flourish with Archaea. The fact that these organisms seem to have persisted from the earliest times argues that they may represent the type of living organism that could have evolved elsewhere beyond planet Earth. Steam caves/vents, while limited in diversity [1.7], represent and remain a rich and potentially useful and reliable source for isolation of unknown and unexplored life forms.

Acknowledgments

We thank the following National Park Service members for assistance with access to collections; Michael Magnuson, Wildlife Biologist, Lassen Volcanic National Park; Rhonda Loh, Chief of Natural Resources Management and Keola Awong, Cultural Specialist and Native Hawaiian Liaison for facilitating our collections of fumaroles and supporting our research efforts in Hawai'i Volcanoes National Park. Jeff Sutton and Aaron Pietruszka, US Geological Survey, offered guidance on the Geology/Geochemistry of our collection areas. Benno Spingler prepared Figure 1.1; Xzayla Zabiti and Anthony Correy provided ideas and participated in the design and construction of the steam collector. We thank our graduate students, Dean Ellis, Courtney Benson, Kate Wall, and Jenny Cornell, and undergraduates, Christa Anderson and Wendy Gutierrez. The authors acknowledge the assistance of Steve Barlow, and use of equipment at the San Diego State University Electron Microscopy Facility acquired by NSF Instrumentation grant DBI-0959908, SB Barlow. Ingrid Niesman, Director of the SDSU Electron Microscope Laboratory facilitated the SEM work. Generous contributions of Schering-Plough Biopharma assisted this study.

References

1.1. Benson, C.A., Bizzoco, R.W., Lipson, D.A., Kelley, S.T., Microbial diversity in nonsulfur, sulfur and iron geothermal steam vents. *FEMS Microbiol. Ecol.*, 76, 1, 74–88, 2011.

1.2. Bizzoco, R.L.W. and Kelley, S.T., Microbial diversity in acidic high-temperature steam vents [Chapter 30], in: *Polyextremophiles: Life under Multiple Forms of Stress [Volume 27 in the series: Cellular Origin, Life in Extreme Habitats and Astrobiology, series editor: Joseph Seckbach]*, J. Seckbach, A. Oren, H. Stan-Lotter (Eds.), pp. 315–332, Springer, Dordrecht, 2013.

1.3. Bizzoco, R.L.W. and Kelley, S.T., Geothermal steam vents of Hawai'i [Chapter 2], in: *Model Ecosystems in Extreme Environments [MEET]*, Volume 2 in the series: Astrobiology: Exploring Life on Earth and Beyond, P. Rampelotto, J. Seckbach, R. Gordon, J. Seckbach, P. Rampelotto (Eds.), pp. 23–40, Imperial College Press, UK, 2019.

1.4. Brochier-Armanet, C., Boussau, B., Gribaldo, S., Forterre, P., Mesophilic crenarchaeota: Proposal for a third archaeal phylum, the Thaumarchaeota. *Nat. Rev. Microbiol.*, 6, 3, 245–252, 2008.

1.5. Brock, T.D., *Thermophilic Microorganisms and Life at High Temperatures*, Springer, New York, 1978.

1.6. Chemtob, S.M., Jolliff, B.L., Arvidson, R.E., Si- and Ti-rich surface coatings on Hawaiian basalt and implications for remote sensing on Mars, in: *Lunar Planetary Science 37th Conference*, 2006.

1.7. Cockell, C.S., Harrison, J.P., Stevens, A.H., Payler, S.J., Hughes, S.S., Kobs Nawotniak, S.E., Brady, A.L., Elphic, R.C., Haberle, C.W. *et al.*, A low-diversity microbiota inhabits extreme terrestrial basaltic terrains and their fumaroles: Implications for the exploration of Mars. *Astrobiology*, 19, 3, 284–299, 2019.

1.8. Ellis, D.G., Bizzoco, R.W., Kelley, S.T., Halophilic *Archaea* determined from geothermal steam vent aerosols. *Environ. Microbiol.*, 10, 6, 1582–1590, 2008.

1.9. Grogan, D.W., Phenotypic characterization of the Archaebacterial genus *Sulfolobus*: Comparison of five wild-type strains. *J. Bacteriol.*, 171, 12, 6710–6719, 1989.

1.10. Hughes, S.S., Haberle, C.W., Kobs Nawotniak, S.E., Sehlke, A., Garry, W.B., Elphic, R.C., Payler, S.J., Stevens, A.H., Cockell, C.S., Brady, A.L. *et al.*, Basaltic terrains in Idaho and Hawai'i as planetary analogs for Mars geology and astrobiology. *Astrobiology*, 19, 3, 260–283, 2019.

1.11. Jones, M.E., Ammonia equilibrium between vapor and liquid aqueous phases at elevated temperatures. *J. Phys. Chem.*, 67, 5, 1113–1115, 1963.

1.12. King, G.M., Chemolithotrophic bacteria: Distributions, functions and significance in volcanic environments. *Microbes Environ.*, 22, 4, 309–319, 2007.
1.13. Nordstrom, D.K., Ball, J.W., McCleskey, R.B., Ground water to surface water: Chemistry of thermal outflows in Yellowstone National Park, in: *Geothermal Biology and Geochemistry in Yellowstone National Park*, W.P. Inskeep and T.R. McDermott (Eds.), pp. 73–94, Montana State University Publications, Bozeman, MT, 2005.
1.14. Stahl, D.A. and de la Torre, J.R., Physiology and diversity of ammonia-oxidizing Archaea. *Annu. Rev. Microbiol.*, 66, 83–101, 2012.
1.15. Stieglmeier, M., Alves, R.J.E., Schleper, C., The Phylum Thaumarchaeota, in: *The Prokaryotes*, E. Rosenberg, E.F. DeLong, S. Lory, E. Stackebrandt, F. Thompson (Eds.), pp. 347–362, Springer-Verlag Berlin, Heidelberg, 2014.
1.16. Wall, K., Cornell, J., Bizzoco, R.W., Kelley, S.T., Biodiversity hot spot on a hot spot: Novel extremophile diversity in Hawaiian fumaroles. *MicrobiologyOpen*, 4, 2, 267–281, 2015.

2

Rio Tinto: An Extreme Acidic Environmental Model of Astrobiological Interest

Ricardo Amils[1,2]* and David Fernández-Remolar[3,4]

[1]*Centro de Astrobiología (CAB, CSIC-INTA), Torrejón de Ardoz, Spain*
[2]*Centro de Biología Molecular Severo Ochoa (CBMSO, CSIC-UAM), Universidad Autónoma de Madrid, Madrid, Spain*
[3]*State Key of Lunar and Planetary Sciences, Macau University of Science and Technology, Macau SAR, People's Republic of China*
[4]*CNSA Macau Center for Space Exploration and Science, Macau SAR, People´s Republic of China*

Abstract

Among extremophiles, acidophiles are of special interest because their chemolithotrophic metabolism obtains energy from reduced minerals, thus creating the extreme acidic conditions in which they thrive. Rio Tinto is a 92 km long extreme acidic environment, which is the product of the metabolic activity of chemolithotrophic microorganisms thriving in the high concentration of metal sulfidic minerals existing in the Iberian Pyrite Belt. An extensive geomicrobiological characterization of the Tinto basin has proven the prominent role of the iron cycle in the ecosystem. The identification of iron sulfates and oxides on Mars, analogous to those generated in the Tinto basin by microbial metabolism, has made Rio Tinto one of the best geochemical and mineralogical terrestrial Mars analogues.

Keywords: Acidophiles, chemolithotrophy, pyrite, iron cycle, subsurface geomicrobiology, Mars, Meridiani Planum, methane

2.1 Introduction

Although Darwin's observation of samples from an Argentinian salt pond led him to predict the existence of life associated with diverse extreme environments [2.26], almost two hundred years had to pass before extremophiles gained serious scientific consideration. At the beginning of the XXth century the interest in preventing the spoiling of salt-preserved codfish by halophilic microorganisms, with its damaging economic and social consequences, led to a need to understand how these microorganisms overcome the osmotic

**Corresponding author*: ramils@cbm.csic.es
Ricardo Amils: https://www.researchgate.net/profile/Ricardo_Amils, http://www.cbm.uam.es/ramils, https://orcid.org/0000-0002-7560-1033
David Fernández-Remolar: https://www.researchgate.net/profile/David_Fernandez-Remolar, https://scholar.google.es/citations?user=NDpdhrIAAAAJ&hl=es, https://cab.academia.edu/DavidFernandezRemolar

Joseph Seckbach and Helga Stan-Lotter (eds.) Extremophiles as Astrobiological Models, (21–44)

pressure created by the saturated salt preservation conditions. Unfortunately, the interest in halophiles decreased after an effective solution for fish preservation using salt from continental mines that contained fewer viable halophilic organisms, was put into practice. In the forties, the interest in extremophiles re-emerged after the discovery that the massive oxidation of metal components in metal mining activities was mainly due to the metabolic activity of chemolithoautotrophic microorganisms that obtained energy from reduced mineral sources, and was not the result of chemical oxidation promoted by the atmospheric oxygen, as had been sustained for a long period of time [2.23]. Extremophile studies grew as a result of increased biotechnological interest in thermophiles based on the application of the Arrhenius equation [2.107]. Finally, the discovery of a new domain of life, Archaea, by Carl R. Woese (originally named Archaebacteria), in which many known extremophiles were included (halophiles, thermophiles, acidophiles, strict anaerobes), led to the explosion of microbial ecology's search for new extremophilic champions, in an attempt to establish the limits of life [2.108] [2.109].

The systematic exploration of extreme environments has led to the discovery of habitats that had been considered uninhabitable only a few years ago. As a consequence, interest in extreme environments has grown exponentially not only to answer fundamental questions about the mechanisms used by extremophiles to grow in different extreme environments but to explore their biotechnological potential (e.g., biomining, catalysis, energy generation, bioremediation, etc.).

The exploration of extreme environments has also played an important role in the development of Astrobiology. As specified in the NASA Astrobiology roadmap (https://astrobiology.nasa.gov/research/astrobiology-at-nasa/astrobiology-strategy) and the European AstRoMap (http://astromap.esf.org/astromap-roadmap.html) one of the main goals of this transdisciplinary area of research is to characterize extreme environments, the microorganisms developing in them and the mechanisms used to cope with the extreme conditions of the environment, to estimate the possible existence of life outside of planet Earth. The experiments performed by the Viking mission (https://www.nasa.gov/mission_pages/viking), the first astrobiological mission aimed at searching for signs of life on Mars, determined that there was little chance that life could have developed on its surface given the extreme conditions detected there [2.76]. In the last fifty years important advances in microbiology have confronted this rather negative vision. Thanks to the exploration of extreme habitats here on Earth we now know that life is extremely robust and can adapt to a wide range of extreme conditions, thus raising the probability of finding life on other planetary bodies.

In this chapter we will review our current knowledge of Rio Tinto as an acidic geomicrobiological model system and address its astrobiological implications. The main interest of acidophiles is that unlike the rest of the extremophiles, which are simple adaptations to different geophysical constrains (radiation, pressure, temperature, ionic strength, water activity, chaotropicity, etc.), the chemolithotrophic metabolisms of these microorganisms generate the extreme acidic conditions of pH and the associated high concentration of heavy metals in which they dwell.

2.2 Acidic Chemolithotrophy

Natural acidic environments are generally associated with volcanic and mining activities. In the first case sulfur chemolithotrophic microorganisms are the main players while metal and coal mining activities expose sulfidic minerals to aerobic chemolithotrophic microorganisms, facilitating their development and causing the so-called acid mine drainage (AMD) or acid rock drainage (ARD), both of which lead to serious environmental problems due to their high content of soluble heavy toxic metals [2.61] [2.62].

The mechanism by which chemolithotrophic microorganisms conserve energy using reduced sulfidic minerals has been under debate for many years [2.30]. The demonstration that ferric iron generated by iron oxidizing microorganisms is responsible for the chemical oxidation of metal sulfides has conclusively clarified this controversial issue [2.98]. We now know that the differences observed over the years in enrichment cultures using diverse metal sulfides depends on the chemical oxidation mechanism used, which is determined by the crystallographic structure of the mineral substrates. Three metal sulfides—pyrite, tungstenite and molybdenite—can be oxidized by ferric iron through the so-called thiosulfate mechanism, generating sulfuric acid [2.99]. The rest of the metal sulfides can experience ferric iron oxidation through the polysulfide mechanism, producing, in this case, elemental sulfur as final product [2.99], and requiring the additional activity of sulfur oxidizing microorganisms to generate sulfuric acid. The main players in these reactions are the iron-oxidizing microorganisms responsible for maintaining a high concentration of the chemical oxidizing agent, ferric iron.

The acidophilic strict chemolithoautotroph *Acidithiobacillus ferrooxidans* (formerly known as *Thiobacillus ferrooxidans*) was isolated for the first time in a coal mine in the middle of the last century [2.23]. Although *At. ferrooxidans* can obtain energy by oxidizing both reduced sulfur and iron, bioenergetic considerations ignored the role of reduced iron as an important source of energy for chemolithotrophic organisms for many years [2.90] [2.7] [2.30]. But the isolation and characterization of the strict chemolithotroph *Leptospirillum ferrooxidans*, which can only grow using ferrous iron as its source of energy, and the evaluation of its important role in biohydrometallurgical operations, has finally changed this point of view [2.91] [2.61] [2.48] [2.30]. In addition, it is now well-known that iron can also be oxidized in anaerobic conditions through anoxygenic photosynthesis using reduced iron as reducing power [2.106] or anaerobic respirations using nitrate as an electron acceptor [2.16], although the mechanism in this case is still very controversial [2.64] [2.20] [2.65] [2.110].

The current demonstration that subsurface chemolithotrophic microorganisms participate very actively in the dark biosphere, already predicted by Darwin almost two hundred years ago, has opened interesting perspectives not only in microbial ecology but also in astrobiology [2.52] [2.15] [2.86] [2.21] [2.111] [2.9] [2.10] [2.12] [2.87] [2.43]. There is an increasing list of alternative sources of chemolithotrophic energy (H_2, Fe^{2+}, S^{2-}, S^o, As^{3+}, Mn^{2+}, NH_4^+ ...) and possible electron acceptors for anaerobic respiration (SO_4^{2-}, NO_3^-, NO_2^-, Fe^{3+}, CO_2, AsO_4^{3-}, Mn^{4+}...), amplifying the metabolic versatility of this energy conservation system. Moreover, microbial iron- and sulfur-oxidizers coupled with the activity of the correspondent microbial reducers play a critical role in the operation of two critical biogeochemical cycles, the iron and sulfur cycles.

Extreme acidic environments differ significantly in their characteristics and as a consequence in their microbial ecology. High temperatures generated by biological activity facilitate the development of thermophilic and thermotolerant acidophiles. Acidic ecosystems associated with metal mining activities are, at the geological scale, rather young.

Nonetheless, some mining activities, such as those in the Tinto area, which have been in exploitation for more than 5.000 years, have a somewhat relatively long history [2.68].

2.3 Rio Tinto Basin

Size (92 km long), pH (mean pH 2.3), high concentration of heavy metals (Fe, Cu, Zn, As...) and a high level of microbial diversity, mainly eukaryotic, make Rio Tinto an uncommon extreme acidic environment [2.72] [2.5] [2.6] [2.57] [2.2] [2.3] [2.13] (Figure 2.1). Rio Tinto rises in Peña de Hierro, in the core of the Iberian Pyrite Belt (IPB) and merges with the Atlantic Ocean at Huelva. The IPB is a geological unit 250 km long and 60 km wide located in the northernmost zone of the Variscan Iberian Massif. The IPB is one of the largest sulfidic deposits on Earth. Massive bodies of pyrite and chalcopyrite, as well as minor quantities of lead and zinc sulfides, constitute the main mineral ores of the IPB [2.17] [2.69]. Hydrothermalism during the Hercynian orogenesis was responsible for its generation [2.17] [2.70]. In the 1700 km^2 river basin three zones can be clearly defined according to their chemical and geological characteristics: the northern (from Peña de Hierro to Niebla), the transitional (from Niebla to San Juan del Puerto) and the estuarine (from San Juan to the Atlantic Ocean). The last zone can be considered to have the most extreme conditions detected in the Tinto basin, because the microorganisms thriving in this part of the river have to face drastic changes in pH, from around 3 to neutrality, as a consequence of the twice daily tidal influence of the Atlantic Ocean [2.103]. Climographs exhibit characteristic basin bimodality, with humid and temperate periods (fall, winter and early spring), alternating with dry and warm periods (late spring and summer).

Figure 2.1 View of colorful filamentous algae in the red waters of the origin of Rio Tinto. (Image credit: the authors).

Oxygen content varies from saturation to strict anaerobic conditions, which agree with the range, -280 to +650 mV of redox potentials measured. The comparison with other local rivers indicates that the acidity and the concentration of iron in the Tinto basin are at least one order of magnitude higher than the acidic Odiel and Agrio rivers, both also associated with mining activities [2.72].

A peculiar characteristic of the Tinto ecosystem is its rather constant pH, which is the consequence of the buffer capacity of ferric iron.

$$Fe^{3+} + 3H_2O \leftrightarrow Fe(OH)_3 + 3H^+ \tag{2.1}$$

When the river is diluted by neutral tributaries or rain, hydrolysis of ferric iron occurs, precipitating ferric hydroxides and generating protons. In the summer, intense heat evaporates the water in the river and protons are consumed, dissolving the ferric hydroxide precipitates. Due to this buffering capacity a pH of around 2.3 remains constant along the river course with the exception of the estuarine zone. Its dimensions and relatively easy access make the Tinto basin an excellent model for the study of microbial ecology associated with an extreme acidic environment [2.12].

2.4 Biodiversity in the Tinto Basin

Combining conventional and molecular microbial ecology methods allowed the most characteristic organisms associated with the Tinto basin to be identified [2.72] [2.57] [2.6] [2.49] [2.94] [2.96] [2.3]. Remarkably, over eighty per cent of the water column diversity corresponds to microorganisms belonging to only three bacterial genera: *Leptospirillum* spp. (strict aerobic iron oxidizers), *Acidiphilium* spp. (iron reducers), and *Acidithiobacillus ferrooxidans* (an aerobic iron oxidizer and an anaerobic iron reducer), all of them well-known members of the iron cycle [2.57].

Although other bacterial and archaeal iron-oxidizers (members of *Ferrimicrobium*, "*Ferrovum*," *Ferrimicrobium* and *Thermoplasma* genera) or bacterial iron-reducers (members of *Ferrimicrobium*, *Metallibacterium* and *Acidobacterium* genera) have been identified in the Tinto ecosystem [2.57] [2.47] [2.49]. Their low numbers, as detected by fluorescence in situ hybridization, suggest that they have a less important role in the water column.

Regarding the sulfur cycle, only *At. ferrooxidans* was detected in significant numbers in the water column. Sulfate-reducing microorganisms were also detected in the sediments at different locations along the river [2.75] [2.49] [2.94] [2.95] [2.96] [2.44] [2.45], thus a subsidiary sulfur cycle is also operative along the course of the river.

It is assumed that the toxicity of high concentrations of heavy metals existing in extreme acidic environments impairs the development of eukaryotic microorganisms in these ecosystems [2.59]. Nevertheless, multicolored algal biofilms can be easily observed covering important areas along the river (Figure 2.1). A careful estimation of this phenomenon allowed López-Archilla and collaborators to conclude that more than sixty per cent of the primary production in the river is due to photosynthetic eukaryotic organisms [2.72]. Members of the *Chlamydomonas*, *Chlorella* and *Euglena* genera, all belonging to the Chlorophyta phylum, are the dominant eukaryotic microorganisms detected in the Tinto basin [2.5] [2.6] [2.2] [2.3]. Also, filamentous algae of the *Klebsormidium* and *Zygnemopsis* genera were identified in the river. Species of the *Dunaliella* and *Cyanidium* genera can be found colonizing the most extreme area in the origin of the river. Members of these genera are well known for their tolerance to toxic heavy metals [2.1] [2.3]. Large brown biofilms made of pennate diatoms of the genus *Pinnularia* can be detected along the course of the Tinto basin [2.2] [2.3] [2.4].

Unicellular and filamentous fungi are the most diverse eukaryotes detected in the Tinto basin [2.72] [2.73]. A recent characterization of Rio Tinto fungal diversity from samples obtained along the river has rendered three hundred and fifty isolates, which is a small representation of the fungal diversity existing in the Tinto basin. Internal transcribed spacer (ITS) sequence analysis showed that Ascomycetes is the most abundant phylum, followed by Basidiomycetes and Zygomycetes. Among the Ascomycetes, more than fifty per cent cluster within the Eurotiomycetes class, followed by Dothideomycetes, Sordariomycetes, Helotiales and Mucorales. A systematic survey of fungal metal tolerance showed that members of the Sordariomycetes and the Eurotiomycetes classes are the most resistant to high concentration of toxic heavy metals, in some cases several orders of magnitude higher than the concentration measured in the water column. Interestingly enough, many different fungal isolates from the Tinto basin can grow in the presence of high concentrations of individual toxic metals, such as As, Ni, Co, Cr, Cu, but their tolerance decreases severely in the presence of combinations of these toxic heavy metals. In addition, some metals with low toxicological profile, such as Fe and Al, can increase the fungal tolerance of individual or combined toxic metals.

Heterotrophic protists are also found along the river. The mixotrophic community is dominated by cercomonads and stramenopiles belonging to the *Ochromonas, Bodo* and *Cercomonas* genera. The protistan consumer community is characterized by two different ciliates identified as members of the *Euplotes* and *Oxytricha* genera. Amoebas related to the *Vahlkampfia* and *Naegleria* genera are found in the most acidic part of the river. Members of the heliozoan belonging to the *Actinophyros* genus are the most characteristic predators of the benthic food chain [2.5] [2.2] [2.3]. The bdelloid rotifer of the genus *Rotifera* is the only member of the animal phylum detected so far in Rio Tinto [2.5].

Due to the interest of aerobic chemolithotrophs in biohydrometallurgical processes, the characterization of the anoxic sediments from acidic environments had been neglected in the past, with few exceptions [2.74]. A comparative analysis of the microbial diversity associated with sediments and the correspondent water column showed a significantly higher level of diversity in the sediments [2.49]. Most of the microorganisms identified in this study were somehow related to the iron cycle. Most of them have been isolated or identified in biomining operations or AMD sites [2.60] [2.61] [2.57] [2.58] [2.92] [2.47]. Nonetheless, some bacteria, such as members of Firmicutes, Acidobacteria, Actinobacteria, Planctomycetes and Chloroflexi phyla, were identified for the first time in the Tinto basin [2.49].

A thorough analysis of two closely located anoxic sediments from the origin of the river has recently shown major phylogenetic differences. In one of the sediments, members of the Firmicutes followed by Acidobacteria phyla corresponded to the most numerous group of Bacteria. In the other sediment, members of the Proteobacteria followed by members of the Actinobacteria phyla were the most represented [2.94]. Members of Firmicutes and Acidobacteria were also present but at rather low numbers. Sulfate-reducing bacteria have always been identified in the strict anoxic conditions existing in the lower part of the sediments, indicating that this energy conservation system is a frequent metabolic activity in the anaerobic part of the ecosystem, probably due to the high concentration of sulfates detected [2.75] [2.94] [2.11].

2.5 Tinto Basin Sedimentary Geomicrobiology

As previously mentioned, most of the biomass from the Tinto ecosystem forms colorful biofilms on the riverbed and on its rock surfaces. Iron minerals precipitate on the surface of these biofilms according to the hydrological cycles, consolidating into distinct fluvial terraces [2.35] [2.37].

During the dry season the evaporation of river water drives precipitation of schwertmannite and hydronium jarosite, whereas gypsum, coquimbite, copiapite and other sulfate minerals are transported to the surface by capillary action, generating effluorescences [2.37]. During the rainy season, hydrolysis of sulfates in addition to the iron hydrolysis results in the precipitation of amorphous iron oxyhydroxides and nanophase goethite. In the oldest river terraces, increasing goethite crystallinity and its replacement by hematite can be observed (Figure 2.2).

Biosignatures, such as textures of coated microbial streamers, surface blisters generated by biogenic gas and microfossils preserved in iron oxides, can be detected in the sedimentary rocks, which could be of help in designing strategies for future astrobiological exploration missions [2.37] [2.39] [2.8]. Until recently, the extreme conditions of pH and high concentration of toxic heavy metals identified in the Tinto basin were considered the consequence of over 5000 years of mining activities [2.51] [2.68] [2.27]. However, recent geological, geophysical and hydrogeological information does not support this assumption. The recharge area of Peña de Hierro aquifer was detected northwest of the pit lake at a depth ranging from 100 to 400 mbs [2.56]. Its groundwater moves southwards along the fracture network and when it reaches the massive sulfide bodies it activates the metabolism of chemolithotrophic microorganisms, generating acidic waters with high concentrations of iron and sulfate [2.38] [2.56]. This acidic groundwater is pumped along open strike-slip normal faults and eventually reaches the surface, where it sources the acidic springs that feed the Rio Tinto headwaters. Thus, the acidic waters with high content of iron and sulfate are the consequence of the subsurface oxidation of metal sulfides, mainly pyrite, by chemolithotrophic microorganisms operating in strict anaerobic conditions. Consequently, mining activities that brought sulfidic minerals to the surface are not the main cause of the distinctive acidity and high concentration of heavy metals detected in the river. This hypothesis is strongly supported by the sedimentary record of the ancient terraces deposited along the Tinto basin [2.37]. The oldest, containing finely laminated and massive ironstones, has been dated 2.1 (million years ago) Ma. This means that several millions of years before the oldest recorded mining activity, the natural bioleaching of native unexposed sulfidic minerals (subterranean and, as a consequence, in anoxic conditions) gave rise to the ionic conditions that generated the old terraces [2.38]. Furthermore, the oldest gossan deposits in the Tinto basin, representing the remaining materials of the *in-situ* bioleaching of the sulfidic deposits, were dated older than 7 Ma [2.32], underscoring the natural origin of the river.

Recently, the remarkable biomineralization of hydronium jarosite in non-permissive ionic conditions by *Purpureocillium lilacinum*, a filamentous fungi isolated from Rio Tinto

Figure 2.2 Comparison between ferruginous deposits of the Burns Formation (a) cropping up in the Endurance Crater and those formed by acidic fluvial activity (b) in Rio Tinto. In the fluvial deposits of Rio Tinto, jarosite and hematite were detected using different techniques such as X-ray diffraction (c), which form as a consequence of the low pH. The same mineral association has been found on Mars. (Image credits: image A, PIA07110, NASA/JPL/Cornell; images B & C, the authors).

[2.82], has been described. In addition, siderite was identified in the modern sediments of the river [2.42], which was also produced in cultures of the acidophilic iron reducer *Acidiphilium* sp., one of the major microorganisms identified in the Tinto basin [2.97]. These results clearly show that the presence of microorganisms can generate nucleation sites which could radically modify the expected mineral precipitation schemes predicted by the global physicochemical conditions of the environment in which they develop [2.82] [2.83] [2.84]. Furthermore, the detection of organic molecules in the oldest Rio Tinto terraces indicates that even in the extreme acidic and oxidant conditions of the system this ferruginous environment can preserve the molecular information of organisms that have inhabited this extreme environment for long periods of time [2.22] [2.89].

Considering the geomicrobiological characteristics of the Tinto ecosystem we propose that Rio Tinto is mainly under the control of iron [2.7]. Iron oxidizing microorganisms are responsible for the solubilization of the massive metal sulfides of the IPB and the associated high concentrations of iron and sulfate measured in the river. Iron has diverse properties of biological interest, which make the Rio Tinto basin an attractive astrobiological precedent [2.11]. Reduced iron is a good electron donor for respiration (both aerobic and anaerobic). Oxidized iron is a useful electron acceptor for anaerobic respiration. Oxidized iron is also a good buffer for controlling the pH of the system. Moreover, it has been shown that soluble ferric iron can protect sensitive organisms from damaging UV radiation [2.54] [2.55].

This iron-controlled environment appears to be adequate for the chemolithotrophic microorganisms detected in the Tinto basin. Nonetheless, taking into consideration that eukaryotic diversity seems to be much greater than prokaryotic [2.72] [2.5] [2.3], and that most of the primary production in the system depends on eukaryotic photosynthesis, how do eukaryotes benefit from adapting to an extreme acidic environment with high concentration of toxic heavy metals? One possible answer to this question might be related to free access to an indispensable element for life, iron. Iron is difficult to obtain at neutral pH, due to its meager solubility in this condition [2.14] [2.77] [2.19] [2.18]. A reliable advantage for the eukaryotes growing in the Tinto basin is an unrestricted iron supply provided by the chemolithotrophy promoted by the extremely high concentration of IPB iron sulfides [2.53] [2.7] [2.11].

2.6 The Iberian Pyrite Belt Dark Biosphere

As discussed above, we favor the hypothesis that the extreme conditions of pH and the high concentration of heavy metals detected along the Tinto basin is the result of an underground bioreactor in which interaction of metal sulfides, underground water and chemolithoautotrophic microorganisms generates the metabolic products, mainly iron and sulfuric acid, detected in the river. To demonstrate this hypothesis two devoted drilling projects to intersect this subsurface bioreactor and obtain information on the oxidation of metal sulfides in anaerobic conditions were carried out.

The central objective of the first, the Mars Astrobiology Research and Technology Experiment (MARTE project), a joint effort between the Centro de Astrobiología (CAB, CSIC-INTA) and the NASA Astrobiology Institute (NAI), was to gather information about the microbial activity operating in the subsurface of the IPB. Peña de Hierro (Iron Mountain, a recurrent name associated with mining operations), on the north flank of Rio Tinto anticline, was selected for drilling. Complex massive sulfide lenses or stockwork veins of pyrite and quartz generated by hydrothermal activity can be found at the upper part of the IPB volcanic sequence [2.69]. Faults intersect the Early Carboniferous volcanic tuff-hosted pyrite bodies.

Three boreholes, BH1, BH4 and BH8, were continuously cored by rotary diamond-bit drilling using a wireline system that produced 60 mm diameter cores, recovered inside a plastic liner to prevent excess contact between the core and the drilling fluid. Tap water with

NaBr as a contamination tracer was used as drilling fluid to refrigerate the bit. Retrieved cores were placed in plastic bags, flushed with N_2 to remove oxygen, sealed and transported to a nearby laboratory in the Museo Minero de Riotinto. Cores were then placed in an anaerobic chamber and samples were obtained aseptically from the interior of the selected cores with a modified hand drill operated at low speed (Figure 2.3).

Water from upslope springs was used to characterize the groundwater before contacting the ore body. The water from these springs had a neutral pH, a low ionic content and was saturated with O_2. The environmental conditions within the ore body were obtained from the analysis of the core samples of two drilling boreholes, BH4 and BH8, at a depth of 165 mbs, separated by a distance of 10 m. The water table was detected at 90 mbs in both boreholes.

Wells were cased with PVC tubes to avoid collapse of the borehole walls and to install multilevel diffusion samplers (MLDS) at different depths for the analysis of ions and gases of the underground aquifer [2.40] [2.9]. Ionic chromatography analyses of the core samples leachates were used to evaluate the potential resources for microbial metabolisms and detect drilling contamination. Samples with Br concentrations above the background level were not analyzed further. Sulfate, a good indicator of the bio-oxidation of metal sulfides, was abundant in most samples. Both reduced and oxidized iron were extracted from several

Figure 2.3 Processing the selected cores for the generation of samples in an anaerobic chamber in the Museo Minero laboratory. (Image credit: the authors).

core samples evidencing an active iron cycle. Nitrate and nitrite were also detected in high concentrations in different samples [2.40] [2.9].

Twice a year the MLDS were analyzed to follow the evolution of fluid formation in the boreholes. Similar patterns were observed for both boreholes. The average pH was ca. 3.5 and remained acidic for two years after drilling. Sulfate concentrations were lower than in rock leachates. The dissolved, oxidized and reduced iron ratios were variable along the length of the borehole, underlying the functional activity of the iron cycle. The highest concentration of dissolved H_2 was found in the upper part of the water table and dissolved methane was detected in many samples, indicating methanogenic activity in the subsurface of the IPB [2.40] [2.9].

Core samples from both boreholes were examined with an antibody microarray (LDCHIP200) [2.85] and an oligonucleotide hybridization microarray [2.50], giving positive signals for sulfur and iron oxidizers, methanogenic archaea, sulfate reducers and Gram-positive bacteria. Denitrifying and hydrogenotrophic bacteria were identified by 16S rRNA cloning and sequencing. Enrichment cultures showed the presence of aerobic pyrite and iron oxidizers, anaerobic respiration of thiosulfate using nitrate, sulfate reducers and methanogens at different depths [2.87] [2.11].

The environment down-gradient from the metal sulfide ore body was sampled by drilling borehole BH1. Compared to BH4 and BH8 boreholes, BH1 showed lower iron and sulfate concentrations in the leachates while sulfate concentrations in the MLDS were much higher, indicating that the groundwater had a strong interaction with the ore. Dissolved hydrogen had lower concentrations and dissolved methane had higher concentration in BH1 than in BH4 and BH8. Enrichment cultures with samples from this borehole showed mainly methanogenic and sulfate-reducing activities [2.40] [2.9].

To further investigate the characteristics of the subsurface geomicrobiology of the IPB, researchers from the Centro de Astrobiología were granted an Advance ERC project, Iberian Pyrite Belt Subsurface Life Detection (IPBSL, 2011–2015) [2.10]. Two geophysical techniques, Time-domain Electromagnetic Sounding and Electric Resistivity Tomography, were used to obtain more precise information on subsurface areas most likely intersecting the underground bioreactor. Two boreholes, BH10 (620 mbs) and BH11 (340 mbs) were selected for drilling [2.56] (Figure 2.4).

Drilling was performed in similar conditions to those described previously for the MARTE project. Rock leachates obtained from samples at regular intervals were analyzed overnight by ion chromatography to determine the concentration of water-soluble anions, facilitating the selection of cores for further analysis using complementary methodologies. Chromatograms showed the presence of oxidized inorganic in ions, such as nitrate and sulfate, as well as reduced organic acids such as acetate. Proteins and sugars were also detected in different samples, demonstrating the existence of microorganisms at different depths.

Cores were logged at the drilling site and samples from selected cores were taken for further petrographic, mineralogical (XRD), elemental (ICP-MS) and stable isotopic analysis. Pyrite and its alteration products, hematite and magnetite, were identified mineralogically in samples from both boreholes. Iron and other metals were identified in leachates from these samples.

Gas chromatography of core samples from both boreholes detected H_2, CO_2 and CH_4. Samples from the BH10 borehole were analyzed with the immunosensor LDChip300, a new generation of antibody microarray containing three hundred antibodies with diverse

Figure 2.4 Borehole BH11 drilling in Peña de Hierro. (Image credit: the authors).

and complementary specificity. Positive immunological reactions were detected using specific antibodies against sulfate-reducing bacteria and methanogenic archaea, which agree with the results obtained using other techniques.

DNA and RNA were efficiently extracted from core samples from both boreholes and most of them gave positive amplifications for bacterial 16S rRNA genes, which are currently under analysis by cloning and pyrosequencing. Samples from the two boreholes were fixed and stained for examination using fluorescence in situ hybridization (CARD-FISH). So far, the results show positive hybridization signals for both Bacteria and Archaea [2.31] and some specific phyla probes, including α-, β- and γ-Proteobacteria, and Gram-positive Bacteroidetes and Euryarchaea, at different depths (Figure 2.5). Further hybridizations with

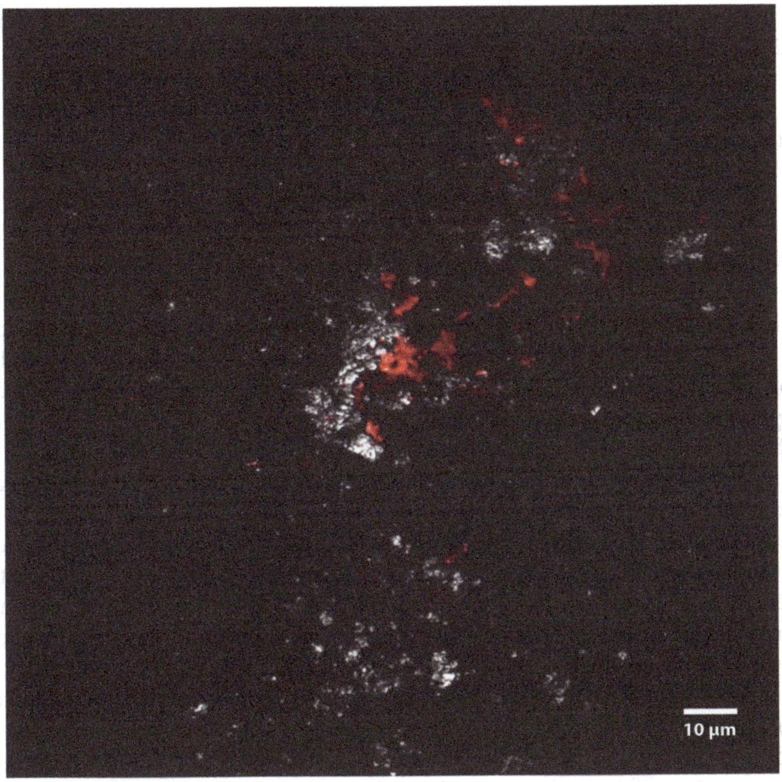

Figure 2.5 Bacteria detected using the CARD-FISH probe EUB338-1 at 420 mbs. (Image credit: the authors).

probes selected or designed after identification by sequence analysis are under development. Biofilms were detected in core samples using fluorescence in situ hybridization and sugar specific fluorescent probes [2.31]. This is an interesting observation because these results challenge the concept that under strict oligotrophic conditions, like those existing in the deep subsurface, microorganisms cannot afford to waste energy producing biofilms. This means that subsurface lifestyle in a solid matrix makes use of biofilms not only to control the bio-reactions in the micro-niches but also to efficiently interconnect them [2.31]. Different electron donors and acceptors have been used to prepare anaerobic enrichment cultures. The following activities have been detected using samples from both boreholes: iron and sulfur oxidizers, iron and sulfate reducers, methanogens, acetogens, methanotrophs and denitrifiers.

The results obtained so far in the MARTE and IPBSL drilling projects give us the following scenario: as groundwater encounters the volcanogenic-hosted massive sulfide (VHMS) system, both biotic and abiotic processes are triggered. Electron donors available for microbial metabolism include metal sulfides, ferrous iron, NH_4^+, hydrogen, nitrite and reduced organic matter. The list of possible electron acceptors includes ferric iron, sulfate, nitrate and CO_2. These compounds can support different anaerobic respiration metabolisms. This is not a complete list; further analysis of the enrichment cultures allowed us to isolate microorganisms developing in the strict anaerobic conditions detected in the subsurface [2.88] [2.67] and their physiological and genomic characterization should provide new insight into the array of metabolisms operating in the subsurface of the IPB. These results confirm the hypothesis that microorganisms are active in the subsurface of the IPB and are responsible for the extreme conditions detected in the river. This observation has important astrobiological implications [2.11] [2.43].

2.7 Methanogenesis in Non-Methanogenic Conditions

Although methane can be produced abiotically, more than eighty per cent of the methane existing in the Earth atmosphere is generated by methanogenic archaea. With only a few exceptions, methanogenic activities are normally detected in environments with circumneutral pH and negative (reduced) redox potentials [2.66] [2.102]. These conditions are very different from those detected in the Tinto basin (acidic and high (oxidized) redox potentials). For a long time, methanogenic activities were not considered to be operating in the Tinto basin for this reason.

After the detection of methane in the borehole fluids of the MARTE project and in the Martian atmosphere [2.46] [2.81], regular inspections for methanogenic activity were implemented in the study of the anoxic sediment of the Tinto basin. The first sampling station in which methane generation was detected was Campo de Galdérias, at the origin of the river [2.100]. Sediments from this station exhibited specific locations with negative reduced redox potential, surrounded by the high positive oxidation redox potential characteristic of the river. Pressure applied to the ground around this sampling site released gases occluded in the sediments. Microcosms using these reduced sediments showed very active methane generation after reaching negative redox potentials and a significant pH increase, following the spike with different methanogenic substrates. The highest methane production was obtained after addition of methanol [2.100].

A second site, JL Dam, was selected for further analysis. Cores from this sampling site exhibited distinctive black bands with negative reduced redox potential and circumneutral pH among acidic reddish-brown sediments with positive oxidized redox potentials, similar to those detected in the sediments collected along the course of the river. The sequence of amplified 16S rRNA gene from the blackish bands corresponded to *Methanosaeta concilii*. Enrichment cultures using different methanogenic substrates allowed the identification of *M. concilii* using organic substrates, *Methanobacterium bryantii* using H_2 and *Methanosarcina barkeri* using methanol [2.100].

How can we explain the development of methanogenic activities in an ecosystem in which the characteristic pH and redox potential are the opposite of the required conditions? This apparent contradiction is resolved when we analyze the results at the microscopic level. The generation of micro-niches in a semi-solid matrix, such as sediments, or even more, in a solid matrix within a deep subsurface rock, could facilitate the growth of microorganisms generating environmental conditions quite different from the restrictive ones existing in the ecosystem [2.31]. As mentioned above, the use of fluorescence in situ hybridization allowed the detection of these micro-niches in the porous rocks of the deep subsurface of the IPB, many of which included biofilms with different types of microorganisms, sharing space and metabolisms [2.31].

The presence of methanogens in an environment controlled by iron has important astrobiological implications since it could be a model for the biotic generation of Martian atmospheric methane [2.46] [2.81] [2.105]. The recurrent argument that the environmental conditions of Mars are not suitable for generating biological methane could be contradicted by the methanogenic activities detected in the sediments of Rio Tinto and the subsurface of the IPB, a geochemical and mineralogical terrestrial analogue of Mars [2.36] [2.37] (Figure 2.2).

2.8 Rio Tinto: A Geochemical and Mineralogical Terrestrial Analog of Mars

In 2000, Christensen *et al.* [2.24] [2.25] reported the occurrence of extensive outcrops of hematite in the region of Terra Meridiani. This mineral is abundant in Rio Tinto as the result of the maturation of ferric acidic minerals. This enigmatic enrichment led to the selection of Meridiani as the landing site for the Opportunity rover in 2004. During the exploration of this area, Opportunity detected hematite occurring in spheroidal concretions that were named blueberries. However, the most intriguing and revealing finding was the abundance of sulfates [2.93] [2.101] [2.80] where the main mineral phase was jarosite, as detected by Mössbauer spectroscopy on board the Opportunity [2.63]. Later on, it was observed that the acidic sulfates were very extensive in different Mars regions, supporting the idea that Mars had been exposed to a low-pH episode lasting hundreds of millions of years [2.79] [2.33] [2.29], whose origin has been related to the weathering of metallic sulfides [2.34] [2.112]. Very recently, the formation of iron sulfides associated with hydrothermal materials that are similar to the hydrothermal rocks of the Rio Tinto basement was reported [2.78]. This suggests that the biogeochemical cycles of the putative biosphere that could have emerged on Mars might have been dominated by S and Fe metabolism, as is observed in Rio Tinto.

In this context, Rio Tinto becomes an excellent terrestrial reference in two aspects. On one hand, it provides direct information about the physicochemical and biological processes that are involved in the formation of the iron-rich acidic minerals such as iron oxides and sulfates [2.57] [2.58] [2.37]. On the other hand, the evolution of the acidic system over the last 30 million years is recorded in the form of alteration materials and iron-rich terraces that show the diagenetic pathways followed by the iron-rich minerals over time [2.35] [2.37] [2.41]. The association of modern and ancient deposits at Rio Tinto can provide key information to understanding the origin and maturation of the iron-rich deposits that are found in vast areas of Mars like Aram Chaos, Meridiani Planum, Valles Marineris, Mawrth Vallis and Syrtis Major [2.63] [2.79] [2.33] [2.29] [2.71]. From an astrobiological perspective, Rio Tinto shows us how biological activity is produced and recorded under extreme acidic conditions over millions of years. The modern environment and its ancient acidic materials of surface and subsurface areas show a wide diversity of biological traces, including microstructures, minerals, stable isotope signatures and molecular compounds, suggesting that these biosignatures survived the diagenesis and maturation processes over 30 million of years of geological evolution [2.43]. As discussed before, the micron-sized crystals of Fe-bearing carbonates likely produced by *Acidiphilium* sp. are found in the acidic sediments going from modern to the oldest ferruginous materials of over 30 million years [2.42]. Although carbonate minerals are highly unstable under the acidic conditions of Rio Tinto, mineral precipitation is favored by metabolic activity in micron-size cellular scale sites that allow net preservation in the oldest acidic materials of the basin. Furthermore, the occurrence of microbial-derived minerals is also observed associated with microbial structures in the deep subsurface of the Rio Tinto crust [2.43]. Over the course of the last 30 million years, iron and sulfur biogeochemical cycling in the basement has been recorded as carbonate-bearing microstructures that formed through metabolic processing of hydrothermal minerals by using sulfur as electron acceptors or donors (e.g., barite and pyrite), depending on the redox conditions. In the Rio Tinto subsurface, some of the reactions that followed the microbes to obtain energy mimic or are the reverse of the biochemical processes that would have appeared in a S and Fe world prebiotic scenario for an emergent metabolism [2.43] [2.104]. Therefore, there is a close connection between a substrate that promotes the geochemical cycles of S and Fe and the microbial metabolism that couples them to obtain energy in the most favorable ways. If life arose on Mars, the dominance of S and Fe geochemical processes rooted in an early metabolism very likely would have produced the same metabolic minerals in the interior of the red planet [2.43].

It is often claimed that acidic environments are inhospitable for the preservation of molecular traces of life since the highly oxidizing and aggressive pH would rapidly destroy the biological and organic compounds. The microscopic study of modern and ancient Rio Tinto materials [2.39] has shown that the high mineralization rates of the acidic solutions encourages the preservation of biological structures in great detail (Figure 2.6a, b1, b2). Paradoxically, in the molecular analysis of samples, the preservation of organic compounds was not considered under such low pH and high Eh conditions. Later on, the sample analysis of the different terrace and gossan levels showed that fragments of biomolecules and complex organic compounds are found in all deposits from the most modern sediments to the oldest materials of the acidic system. Using peptide extraction techniques on ancient acidic materials from the Rio Tinto terraces, Colín-García *et al.* [2.22] showed preservation of protein fragments during the last 2.1 million years. Such peptidic sequences provided qualitative information about the organisms in the past of the acidic system, which coincide with some microorganisms living in the modern system. Very recently, the sample analysis of surface and subsurface using TOF-SIMS detected diverse organic compounds that are associated with microbial structures (Figure 2.6a, c1 to c3). The preservation of molecular traces of life in acidic conditions is even more extraordinary as these rocks have been exposed to intense diagenesis that has changed the composition of the mineral matrix from sulfates to iron oxides under varying pH and Eh conditions. The occurrence of biomolecules and organic compounds of biological origin in the Rio Tinto rocks demonstrates the preservation of molecular traces of life in materials formed under extreme acidic conditions and despite exposure to drastic diagenesis.

The preservation of different biosignatures in the ancient acidic materials of Rio Tinto strongly supports that, if life emerged on Mars, traces of its activity in acidic deposits are just as likely to remain as they are in materials that formed under mildly neutral conditions in the red planet [2.28]. Given that acidic environments were abundant in the Late Noachian to Hesperian ages (more than 3.5 billion years ago), the next astrobiological missions should increase the chances of finding traces of life on the red planet by seeking them in acidic deposits.

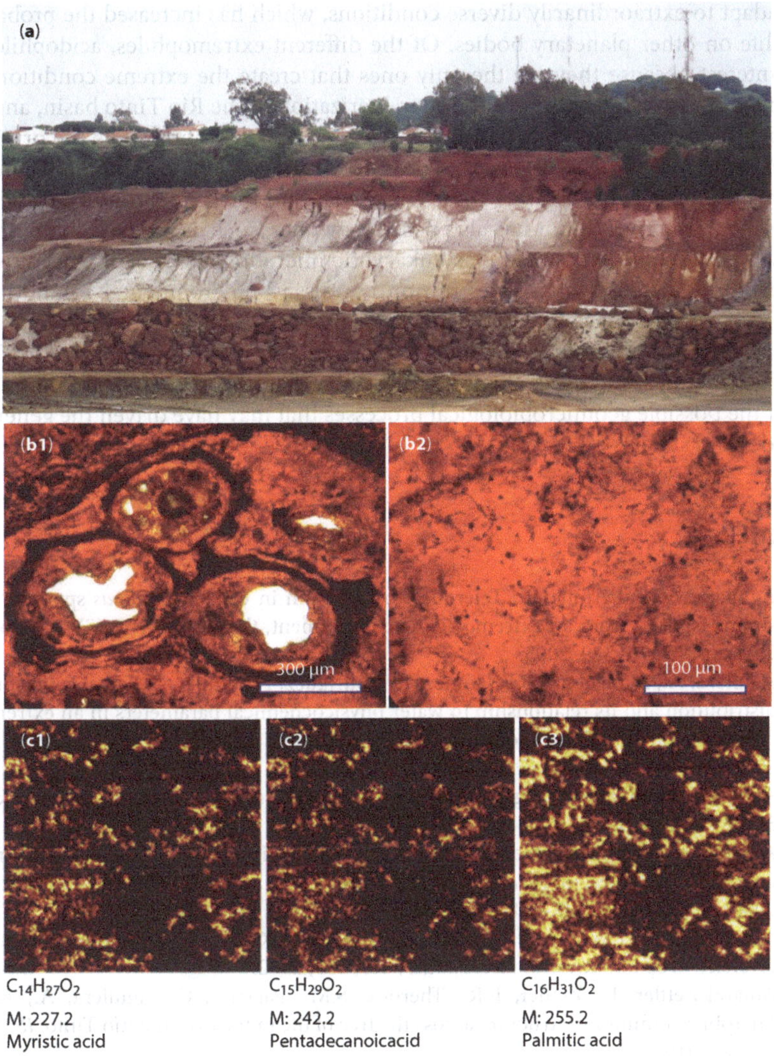

Figure 2.6 Preservation of different traces of life in the oldest Rio Tinto terrace (a) that has been dated as 7 to 2.1 million years old [2.32, 2.37]. b1 and b2 show preservation of plants and micron-sized bacterial filaments, respectively. The analysis of the terrace samples by TOF-SIMS shows the detection of filament structures outlined by the distribution of the negative fragments of different lipids biomarkers like Myristic (c1), Pentadecanoic (c2), and Palmitic acids (c3). (Image credit: the authors).

2.9 Conclusions

In the last fifty years, important advances in microbiology have challenged the view generated by the Viking Missions, which were very pessimistic about the existence of life on Mars. Thanks to research on extremophiles, we now know that life is extremely robust and able to adapt to extraordinarily diverse conditions, which has increased the probability of finding life on other planetary bodies. Of the different extremophiles, acidophiles are of special interest because they are the only ones that create the extreme conditions of the environment in which they thrive. The characterization of the Rio Tinto basin, an extreme acidic environment 92 km long, has answered some basic questions, like the origin of the extreme acidic conditions of the ecosystem, the identification and isolation of the microorganisms responsible for creating these conditions and the generation of iron secondary minerals along the river, but it has also posed new ones, such as the paradox of the unexpectedly high level of diverse complex eukaryotic microorganisms or the need to explain the detection of functional activities in non-permissive environmental conditions, all of which are of astrobiological interest. The discovery of Noachian iron lithological units on Mars, similar to those produced biologically in the Tinto basin, have given Rio Tinto the status of geochemical and mineralogical terrestrial analogue that can lead to a more detailed vision of the possible geomicrobiological processes that may have driven the generation of iron oxides and sulfates on Mars.

References

2.1. Aguilera, A. and Amils, R., Tolerance to cadmium in *Chlamydomonas* sp. (Chlorophyta) strains isolated from an extreme acidic environment, the Tinto River (SW, Spain). *Aquat. Toxicol.*, 75, 316–329, 2005.

2.2. Aguilera, A., Manrubia, S.C., Gómez, F., Rodríguez, N., Amils, R., Eukaryotic community distribution and its relationship to water physicochemical parameters in an extreme acidic environment, Río Tinto (Southwestern Spain). *Appl. Environ. Microbiol.*, 72, 5325–5330, 2006.

2.3. Aguilera, A., Eukaryotic organisms in extreme acidic environments, the Río Tinto case. *Life*, 3, 363–374, 2013.

2.4. Aguilera, A., Olsson, S., Puente-Sánchez, F., Physiological and phylogenetic diversity in acidophilic eukaryotes, in: *Acidophiles: Life in Extremely Acidic Conditions*, B. Johnson and R. Quatrini (Eds.), pp. 107–118, Caister Academic Press, UK, 2016.

2.5. Amaral-Zettler, L.A., Gómez, F., Zettler, E., Keenan, B.G., Amils, R., Sogin, M.L., Eukaryotic diversity in Spain's River of Fire. *Nature*, 417, 137, 2002.

2.6. Amaral-Zettler, L., Zettler, E.R., Theroux, S.M., Palacios, C., Aguilera, A., Amils, R., Microbial community structure across the tree of life in the extreme Río Tinto. *ISME J.*, 5, 1, 42–50, 2010.

2.7. Amils, R., González-Toril, E., Gómez, F., Fernández-Remolar, D., Rodríguez, N., Malki, M., Zuluaga, J., Aguilera, A., Amaral-Zettler, L.A., Importance of chemolithotrophy for early life on Earth: The Tinto River (Iberian Pyritic Belt) case, in: *Origins*, J. Seckbach (Ed.), pp. 463–480, Kluwer Academic Publishers, Amsterdam, NL, 2004.

2.8. Amils, R., González-Toril, E., Fernández-Remolar, D., Gómez, F., Aguilera, A., Rodríguez, N., Malki, M., García-Moyano, A., González-Fairén, A., de la Fuente, V., Sanz, J.L., Extreme environments as Mars terrestrial analogs: The Río Tinto case. *Planet. Space Sci.*, 55, 370–381, 2007.

2.9. Amils, R., Fernández-Remolar, D., Gómez, F., González-Toril, E., Rodríguez, N., Briones, C., Prieto-Ballesteros, O., Sanz, J.L., Díaz, E., Stevens, T.O. *et al.*, Subsurface geomicrobiology of the Iberian Pyritic Belt, in: *Microbiology of Extreme Soils [Volume 13 in the series: Soil Biology]*, P. Dion and C.S. Nautiyal (Eds.), pp. 205–223, Springer-Verlag, Berlin, GE, 2008.

2.10. Amils, R., Fernández-Remolar, D., Parro, V., Rodríguez-Manfredi, J.A., Timmis, K., Oggerin, M., Sánchez-Román, M., López, F.J., Fernández, J.P., Puente, F. *et al.*, Iberian Pyrite Belt Subsurface Life (IPBSL), a drilling Project of biohydrometallurgical interest. *Adv. Mat. Res.*, 825, 15–18, 2013.

2.11. Amils, R., Fernández-Remolar, D., the IPBSL Team, Río Tinto: A geochemical and mineralogical terrestrial analogue of Mars. *Life*, 4, 511–534, 2014.

2.12. Amils, R., Lessons learned from thirty years of geomicrobiological studies of Río Tinto. *Res. Microbiol.*, 167, 7, 539–545, 2016.

2.13. Amils, R. and Fernández-Remolar, D., Acidophiles and Astrobiology, in: *Acidophiles, Life in Extremely Acidic Conditions*, B. Johnson and R. Quatrini (Eds.), pp. 285–300, Caister Academic Press, UK, 2016.

2.14. Archibald, F., *Lactobacillus plantarum*, an organism not requiring iron. *FEMS Microbiol. Lett.*, 19, 29–32, 1983.

2.15. Bachofen, R., Ferloni, P., Flynn, L., Review: Microorganisms in the subsurface. *Microbiol. Res.*, 153, 1–22, 1998.

2.16. Benz, M., Brune, A., Schink, B., Anaerobic and aerobic oxidation of ferrous iron at neutral pH by chemoheterotrophic nitrate-reducing bacteria. *Arch. Microbiol.*, 169, 159–165, 1998.

2.17. Boulter, C.A., Did both extensional tectonics and magmas act as major drivers of convection cells during the formation of the Iberian Pyrite Belt massive sulphide deposits? *J. Geol. Soc. London*, 153, 181–184, 1996.

2.18. Boyd, W.P., Watson, A.J., Law, C.S., Abraham, E.R., Trull, T., Murdoch, R., Bakker, D.C., Bowie, A.R., Buesseler, K.O., Chang, H. *et al.*, A mesoscale phytoplankton bloom in the polar Southern Ocean stimulated by iron fertilization. *Nature*, 407, 695–702, 2005.

2.19. Braun, V. and Killmann, H., Bacterial solution to the iron.supply problems. *Trends Biochem. Sci.*, 24, 104–109, 1999.

2.20. Carlson, H.K., Clark, I.C., Blazewicz, S.J., Iavarone, A.T., Coates, J.D., Fe(II) oxidation is an innate capability of nitrate-reducing bacteria that involves abiotic and biotic reactions. *J. Bacteriol.*, 195, 3260–3268, 2013.

2.21. Chapelle, F.H., O´Nelly, K., Bradley, P.M., Methé, B.A., Ciufo, S.A., Knobel, L.L., Lovley, D.R., A hydrogen- based subsurface microbial community dominated by methanogens. *Nature*, 415, 312–314, 2002.

2.22. Colín-García, M., Kanawati, B., Harir, M., Schmidt-Kopplin, P., Amils, R., Parro, V., García, M., Fernández-Remolar, D., Detection of peptidic sequences in the ancient acidic sediments of Río Tinto, Spain. *Orig. Life Evol. Biosph.*, 41, 523–527, 2011.

2.23. Colmer, A.R., Temple, K.L., Hinkle, H.E., An iron-oxidizing bacterium from the acid drainage of some bituminous coal mines. *J. Bacteriol.*, 59, 317–328, 1950.

2.24. Christensen, P.R., Bandfield, J.L., Clark, R.N., Edgett, K.S., Hamilton, V.E., Hoefen, T., Kieffer, H.H., Kuzmin, R.O., Lane, M.D., Malin, M.C. *et al.*, Detection of cristaline hematite mineralization on Mars by the thermal emission spectrometer: Evidence for near-surface water. *J. Geophys. Res.*, 104, 9623–9642, 2000.

2.25. Christensen, P.R., Morris, R.V., Lane, M.D., Banfield, J.L., Malin, M.C., Global mapping of martian hematite mineral deposits: Remnants of water-driven processes on early Mars. *J. Geophys. Res.*, 106, 23873–23885, 2001.

2.26. Darwin, C., Voyages of the Adventure and Beagle, in: *Volume III, Journal and remarks*, pp. 1832–1836, Henry Colburn, London, UK, 1839.

2.27. Davis, Welty, A.T., Borrego, J., Morales, J.A., Pendon, J.G., Ryan, J.G., Río Tinto estuary (Spain): 5000 years of pollution. *Environ. Geol.*, 39, 1107–1116, 2000.

2.28. Ehlmann, B.L., Mustard, J.F., Murchie, S.L., Poulet, F., Bishop, J.L., Brown, A.J., Calvin, W.M., Clark, R.N., Marais, D.J.D., Milliken, R.E. *et al.*, Orbital Identification of Carbonate-Bearing Rocks on Mars. *Science*, 322, 1828–1832, 2008.

2.29. Ehlmann, B.L. and Mustard, J.F., An in-situ record of major environmental transitions on early Mars at Northeast Syrtis Major. *Geophys. Res. Lett.*, 39, L11292, 2012.

2.30. Ehrlich, H.L., Newman, D.K., Kappler, A., *Ehrlich's Geomicrobiology*, 6th edition, CRC, Boca Ratón, USA, 2015.

2.31. Escudero, C., Vera, M., Oggerin, M., Amils, R., Active microbial biofilms in deep continental subsurface poor porous rock samples from the Iberian Pyrite Belt. *Sci. Rep.*, 8, e1538, 2018.

2.32. Essalhi, M., Sizaret, S., Barbanson, L., Chen, Y., Lagroix, F., Demory, F., Nieto, J., Sáez, R., Capitán, M.A., A case study of the internal structures of gossans and weathering processes in the Iberian Pyrite Belt using magnetic fabrics and paleomagnetic dating. *Miner. Deposita*, 46, 981–999, 2011.

2.33. Farrand, W.H., Glotch, T.D., Rice, J.W., Jr., Hurowitz, J.A., Swayze, G.A., Discovery of jarosite within Mawrth Vallis region of Mars: Implications for the geologicalhistory of the region. *Icarus*, 204, 478–488, 2009.

2.34. Fairen, A.G., Fernández-Remolar, D., Dohm, J.M., Baker, V.R., Amils, R., Inhibition of carbonate synthesis in acidic oceans from Mars. *Nature*, 431, 423–426, 2004.

2.35. Fernández-Remolar, D.C., Rodríguez, N., Gómez, F., Amils, R., Geological record of an acidic environment driven by iron hydrochemistry: The Tinto River system. *J. Geophys. Res.*, 108, E7, 5080, 2003.

2.36. Fernández-Remolar, D.C., Gómez-Elvira, J., Gómez, F., Sebastián, E., Martín, J., Manfredi, J.A., Torres, J., González Kesler, C., Amils, R., The Tinto River, an extreme acidic environment as an analogue of the Terra Meridiani hematite site of Mars. *Planet. Space Sci.*, 52, 239–248, 2004.

2.37. Fernández-Remolar, D.C., Morris, R.V., Gruener, J.E., Amils, R., Knoll, A.H., The Rio Tinto Basin, Spain: Mineralogy, sedimentary geobiology and implications for interpretation of outcrop rocks of Meridiani Planum, Mars. *Earth Planet. Sci. Lett.*, 240, 149–167, 2005.

2.38. Fernández-Remolar, D., Gómez, F., Prieto-Ballesteros, O., Schelble, R.T., Rodríguez, N., Amils, R., Some ecological mechanisms to generate habitability in planetary subsurfaces areas by chemolithotrophic communities: The Río Tinto subsurface ecosystem as a model system. *Astrobiology*, 8, 157–173, 2008.

2.39. Fernández-Remolar, D.C. and Knoll, A.H., Fossilization potential of iron-bearing minerals in acidic environments of Rio Tinto, Spain: Implications for Mars exploration. *Icarus*, 194, 72–85, 2008.

2.40. Fernández-Remolar, D., Prieto-Ballesteros, O., Rodríguez, N., Gómez, F., Amils, R., Gomez-Elvira, J., Stoker, C., Underground habitats found in the Río Tinto Basin: A model for subsurface life habitats on Mars. *Astrobiology*, 8, 1023–1046, 2008.

2.41. Fernández-Remolar, D., Prieto-Ballesteros, O., Gómez-Ortiz, D., Fernández-Sampedro, M., Sarrazin, P., Gailhanou, M., Amils, R., Río Tinto sedimentary mineral assemblages: A terrestrial perspective that suggests some formation pathway of phyllosilicates on Mar. *Icarus*, 211, 114–138, 2011.

2.42. Fernández-Remolar, D.C., Preston, L.J., Sánchez-Román, M., Izawa, M.R.M., Huang, L., Southam, G., Banerjee, N.R., Osinski, G.R., Flemming, R., Gómez-Ortíz, D. *et al.*, Carbonate precipitation under bulk acidic conditions as a potential biosignature for searching life on Mars. *Earth Planet. Sci. Lett.*, 351, 13–26, 2012.

2.43. Fernández-Remolar, D., Banerjee, N., Gómez-Ortiz, D., Izawwa, M., Amils, R., A mineralogical archive of the biogeochemical sulfur cycle preserved in the subsurface of the Río Tinto system. *Am. Mineral.*, 103, 394–411, 2018.

2.44. Florentino, A.P., Brienza, C., Stams, A.J.M., Sánchez-Andrea, I., *Desulfurella amilsii* sp.nov., a novel acidotolerant sulfur-respiring bacterium isolated from acidic river sediments. *Int. J. Syst. Evol. Microbiol.*, 66, 1249–1253, 2016.

2.45. Florentino, A.P., Stams, A.J.M., Sánchez-Andrea, I., Genome sequence of *Desulfurella amilsii* strain TR1 and comparative genomics of Desulfurellaceae family. *Front. Microbiol.*, 8, e222, 2017.
2.46. Formisano, V., Atreya, S., Encrenaz, T., Ignatiev, N., Giuranna, M., Detection of methane in the atmosphere of Mars. *Science*, 306, 1758–1761, 2004.
2.47. García-Moyano, A., González-Toril, E., Aguilera, A., Amils, R., Prokaryotic community composition and ecology of macroscopic floating filaments from an extreme acidic environment, Río Tinto, (SW, Spain). *Syst. Appl. Microbiol.*, 30, 601–614, 2007.
2.48. Garcia-Moyano, A., González-Toril, E., Moreno-Paz, M., Parro, V., Amils, R., Evaluation of *Leptospirillum* spp. in Rio Tinto, a model of interest to biohydrometallurgy. *Hydrometallurgy*, 94, 155–161, 2008.
2.49. García Moyano, A., González-Toril, E., Aguilera, A., Amils, R., Comparative microbial ecology study of the sediments and the water column of the Río Tinto, an extreme acidic environment. *FEMS Microbiol. Ecol.*, 81, 303–314, 2012.
2.50. Garrido, P., González-Toril, E., García-Moyano, A., Moreno-Paz, M., Amils, R., Parro, V., An oligonucleotide prokaryotic microarray (PAM): Its validation and its use to monitor seasonal variations in extreme acidic environments with total environmental RNA. *Environ. Microbiol.*, 10, 836–850, 2008.
2.51. Geen, A., Adkins, J.F., Boyle, E.A., Nelson, C.H., Palanques, A., A 120-year record of widespread contamination from mining of the Iberian Pyrite Belt. *Geology*, 25, 291–294, 1997.
2.52. Gold, T., The deep hot biosphere. *Proc. Natl. Acad. Sci. USA*, 89, 6045–6049, 1992.
2.53. Gómez, F., Fernández-Remolar, D., González-Toril, E., Amils, R., The Tinto River, an extreme Gaian environment, in: *Gaia 2000*, L. Margulis, J. Miller, P. Boston, S. Schneider, C. Crist (Eds.), pp. 321–333, MIT Press, Boston, USA, 2003.
2.54. Gómez, F., Aguilera, A., Amils, R., Soluble ferric iron as an efective protective agent against UV radiation: Implications for early life. *Icarus*, 191, 352–359, 2007.
2.55. Gómez, F., Mateo-Martí, E., Prieto.Ballesteros, O., Martín-Gago, J., Amils, R., Protection of chemolithotrophic bacteria exposed to Mars environmental conditions. *Icarus*, 209, 2, 482–487, 2010.
2.56. Gómez-Ortiz, D., Fernández-Remolar, D., Granda, A., Quesada, C., Granda, T., Prieto-Ballesteros, O., Molina, A., Amils, R., Identification of the subsurface sulfide bodies responsible for acidity in Río Tinto source water. Spain. *Earth Planet. Sci. Lett.*, 391, 36–41, 2014.
2.57. González-Toril, E., Llobet-Brosa, E., Casamayor, E.O., Amann, R., Amils, R., Microbial ecology of an extreme acidic environment, the Tinto River. *Appl. Environ. Microbiol.*, 69, 4853–4865, 2003.
2.58. González-Toril, E., Aguilera, A., Rodríguez, N., Fernández-Remolar, D., Gómez, F., Díaz, E., García-Moyano, A., Sanz, J.L., Amils, R., Microbial ecology of Río Tinto, a natural extreme acidic environment. *Hydrometallurgy*, 10, 329–333, 2010.
2.59. Gross, W., Ecophysiology of algae living in highly acidic environments. *Hydrobiology*, 433, 31–37, 2000.
2.60. Hallberg, K.B. and Johnson, D.B., Biodiversity of acidophilic prokaryotes. *Adv. Appl. Microbiol.*, 49, 37–84, 2001.
2.61. Johnson, D.B. and Hallberg, K.B., The microbiology of acidic mine waters. *Res. Microbiol.*, 154, 466–473, 2003.
2.62. Johnson, D.B. and Hallberg, K.B., Acid mine drainage remediation options: A review. *Sci. Total Environ.*, 338, 3–14, 2005.
2.63. Klingelhöfer, G., Morris, R.V., Bernhardt, B., Schröder, C., Rodionov, D.S., de Souza, P.A., Jr., Yen, A., Gellert, R., Evlanov, E.N., Zubkov, B. *et al.*, Jarosite and hematite at Meridiani Planum from the Mössbauer spectrometer on the Opportunity rover. *Science*, 306, 1740–1745, 2005.

2.64. Klueglein, N. and Kappler, A., Abiotic oxidation of F(II) by reactive nitrogen species in cultures of the nitrate-reducing Fe(II) oxidizer *Acidovorax* sp. BoFeN1 questioning the existence of enzymatic Fe(II) oxidation. *Geobiology*, 11, 180–190, 2013.

2.65. Klueglein, N., Zeitvogel, F., Stierhof, Y.D., Floetenmeyer, M., Konhauser, K.O., Kappler, A., Potential role of nitrite for abiotic Fe(II) oxidation and cell encrustation during nitrate reduction by denitrifying bacteria. *Appl. Environ. Microbiol.*, 80, 1051–1061, 2014.

2.66. Kotsyurbenko, O.R., Friedrich, M.W., Simankova, M.V., Nozhenvnikova, A.N., Golyshin, P.N., Timmis, K.N., Conrad, R., Shift from acetoclastic to H_2 dependent methanogenesis in a West Siberian peat bog at low pH values and isolation of an acidophilic *Methanobacterium* strain. *Appl. Environ. Microbiol.*, 73, 2344–2348, 2007.

2.67. Leandro, T., da Costa, M.S., Sanz, J.L., Amils, R., Complete genome of *Tessaracoccus* sp. strain T2.5-30 isolated from 139.5 m deep on the subsurface of the Iberian Pyrite Belt. *Genome Announc. J.*, 5, 17, #e00238–17, 2017.

2.68. Leblanc, M., Morales, J.A., Borrego, J., Elbaz-Poulichet, F., A 4500-year-old mining pollution in Southwestern Spain: Long-term implications for modern mining pollution. *Econ. Geol.*, 95, 655–662, 2000.

2.69. Leistel, J.M., Marcoux, E., Theiblemont, D., Quesada, C., Sánchez, A., Almodóvar, G.R., Pascual, E., Saez, R., The volcanic-hosted massive sulphide deposits of the Iberian Pyrite Belt. *Miner. Deposita*, 33, 2–30, 1998.

2.70. Lescuyer, J.L., Leistel, J.M., Mrcoux, E., Milési, J.P., Thiéblemont, D., Late Devonian-Early Carboniferous peak sulphide mineralization in the Western Hercynides. *Miner. Deposita*, 33, 208–220, 1998.

2.71. Lichtenberg, K.A., Arvidson, R.E., Morris, R.V., Murchie, S.L., Bishop, J.L., Fernandez Remolar, D., Glotch, T.D., Noe Dobrea, E., Mustard, J.F., Andrews-Hanna, J. *et al.*, Stratigraphy of hydrated sulfates in the sedimentary deposits of Aram Chaos, Mars. *J. Geophys. Res.: Planets*, 115, ED00D17, 2010.

2.72. López-Archilla, A.I., Marín, I., Amils, R., Microbial community composition and ecology of an acidic aquatic environment: The Tinto River, Spain. *Microb. Ecol.*, 41, 20–35, 2001.

2.73. López- Archilla, A.I., González, A.E., Terrón, M.C., Amils, R., Diversity and ecological relationships of the fungal populations of an acidic river of Southwestern Spain: The Tinto River. *Can. J. Microbiol.*, 50, 923–934, 2005.

2.74. Lu, S., Gischkat, S., Reiche, M., Akob, D.M., Hallberg, K.B., Küsel, K., Ecophysiology of Fe-cycling bacteria in acidic sediments. *Appl. Environ. Microbiol.*, 76, 8174–8183, 2010.

2.75. Malki, M., González-Toril, E., Sanz, J.L., Gómez, F., Rodríguez, N., Amils, R., Importance of the iron cycle in biohydrometallurgy. *Hydrometallurgy*, 83, 223–228, 2006.

2.76. Margulis, L., Mazur, P., Barghoorn, E.S., Halvorson, H.O., Jukes, T.H.J., Kaplan, I.R., The Viking Mission: Implications for life in the Vallis Marineris area. *Science*, 305, 78–81, 1979.

2.77. Martin, J.H., Glacial-interglacial CO_2 change: The iron hypothesis. *Paleooceanography*, 5, 1–13, 1990.

2.78. Michalski, J.R., Dobrea, E.Z.N., Niles, P.B., Cuadros, J., Ancient hydrothermal seafloor deposits in Eridania basin on Mars. *Nat. Commun.*, 8, e15978, 2017.

2.79. Milliken, R.E., Swayze, G.A., Arvidson, R.E., Bishop, J.L., Clark, R.N., Ehlmann, B.L., Green, R.O., Grotzinger, J.P., Morris, R.V., Murchie, S.L. *et al.*, Opaline silica in young deposits on Mars. *Geology*, 36, 847–850, 2008.

2.80. McLennan, S.M., Bell, J.F., III, Calvin, W.M., Christensen, P.R., Clark, B.C., de Souza, P.A., Farmer, J., Farrand, W.H., Fike, D.A., Gellert, R. *et al.*, Provenance and diagenesis of the Burns formation, Meridiani Planum, Mars. *Earth Planet. Sci. Lett.*, 240, 95–121, 2005.

2.81. Mumma, M.J., Villanueva, G.L., Novak, R.E., Hewagama, T., Bonev, B.P., DiSanti, M.A., Mandell, A., Smith, M.D., Strong release of methane on Mars in Northen Summer 2003. *Science*, 323, 1041–1045, 2009.
2.82. Oggerin, M., Tornos, F., Rodríguez, N., del Moral, C., Sánchez-Román, M., Amils, R., Specific jarosite biomineralization by *Purpureocillium lilacinum*, an acidophilic fungi isolated from Río Tinto. *Environ. Microbiol.*, 15, 2228–2237, 2013.
2.83. Oggerin, M., Rodríguez, N., del Moral, C., Amils, R., Fungal jarosite biomineralization in Río Tinto, a process of biohydrometallurgical interest. *Res. Microbiol.*, 165, 719–725, 2014.
2.84. Oggerin, M., Tornos, F., Rodríguez, N., Amils, R., Fungal iron biomineralization in Río Tinto. *Minerals*, 6, 2, 37, 2016.
2.85. Parro, V., Fernández-Remolar, D., Rodríguez-Manfredi, J.A., Cruz-Gil, P., Rivas, L.A., Ruiz-Bermejo, M., Moreno-Paz, M., García-Villadangos, M., Gómez-Ortiz, D., Blanco-López, Y. *et al.*, Classification of modern and old Río Tinto sedimentary deposits through the biomolecular record using a Life Marker Biochip: Implications for detecting Life on Mars. *Astrobiology*, 11, 29–44, 2011.
2.86. Pedersen, K., Exploration of deep intraterrestrial microbial life: Current perspectives. *FEMS Microbiol. Lett.*, 185, 9–16, 2000.
2.87. Puente-Sánchez, F., Moreno-Paz, M., Rivas, L.A., Cruz-Gil, P., García-Villadangos, M., Gómez, M.J., Postigo, M., Garrido, P., González-Toril, E., Briones, C. *et al.*, Deep subsurface sulfate reduction and methanogenesis in the Iberian Pyrite Belt revealed through geochemistry and molecular biomarkers. *Geobiology*, 12, 34–47, 2014.
2.88. Puente-Sánchez, F., Sánchez-Román, M., Amils, R., Parro, V., *Tessaracoccus lapidicaptus* sp. nov., a novel actinobacterium isolated from the deep subsurface of the Iberian Pyrite Belt (Huelva, Spain). *Int. J. Syst. Evol. Microbiol.*, 64, 3546–3552, 2014.
2.89. Preston, L., Shuster, J., Fernández-Remolar, D., Banerjee, N., Osinski, G.R., Southam, G., The preservation and degradation of filamentous bacteria and biomolecules within iron oxide deposits at Rio Tinto, Spain. *Geobiology*, 9, 233–249, 2011.
2.90. Pronk, J.T., Bruyn, J.C., Bos, P., Kuenen, J.G., Anaerobic growth of *Thiobacillus ferrooxidans*. *Appl. Environ. Microbiol.*, 58, 2227–2230, 1992.
2.91. Rawlings, D.E., Heavy metal mining using microbes. *Annu. Rev. Microbiol.*, 56, 65–91, 2002.
2.92. Rawlings, D.E., Characteristics and adptability of iron- and sulfur-oxidizing microorganisms used for the recovery of metals from minerals and their concentrates. *Microb. Cell Fact.*, 4, 13–28, 2005.
2.93. Rieder, R., Gellert, R., Anderson, R.C., Brückner, J., Clark, B.C., Dreibus, G., Economou, T., Klingelhöfer, G., Lugmair, G.W., Ming, D.W. *et al.*, Chemistry of rocks and soils at Meridiani Planum from the alpha particle X-ray spectrometer. *Science*, 306, 1746–1749, 2004.
2.94. Sánchez-Andrea, I., Rodríguez, N., Amils, R., Sanz, J.L., Microbial diversity in anaerobic sediments at Río Tinto, a naturally acidic environment with a high heavy metal content. *Appl. Environ. Microbiol.*, 77, 17, 6085–6093, 2011.
2.95. Sánchez-Andrea, I., Rojas-Ojeda, P., Amils, R., Sanz, J.L., Screening of anaerobic activities in sediments of an acidic environment: Tinto River. *Extremophiles*, 16, 829–839, 2012.
2.96. Sánchez-Andrea, I., Stams, A.J.M., Amils, R., Sanz, J.L., Enrichment and isolation of acidophilic sulfate-reducing bacteria from Tinto River sediments. *Environ. Microbiol. Rep.*, 5, 5, 672–678, 2013.
2.97. Sánchez-Román, M., Fernández-Remolar, D., Amils, R., Sánchez-Navas, A., Schmid, T., San Martíon-Uriz, P., Rodríguez, N., McKenzie, J.A., Microbial mediated formation of Fe-carbonate minerals under extreme acidic conditions. *Sci. Rep.*, 4, e4757, 2014.

2.98. Sand, W., Gehrke, T., Hallmann, R., Schippers, A., Sulfur chemistry, biofilm and the (in) direct attack mechanisms. A critical evaluation of bacterial leaching. *Appl. Microbiol. Biotechnol.*, 43, 961–966, 1995.
2.99. Sand, W., Gehrke, T., Jozsa, P.G., Schippers, A., Biochemistry of bacterial leaching. Direct vs indirect bioleaching. *Hydrometallurgy*, 59, 159–175, 2001.
2.100. Sanz, J.L., Rodríguez, N., Díaz, E., Amils, R., Methanogenesis in the sediments of Río Tinto, an extreme acidic environment. *Environ. Microbiol.*, 13, 8, 2336–2341, 2011.
2.101. Squyres, S.W., Grotzinger, J.P., Arvidson, R.E., Bell, J.F., III, Calvin, W., Christensen, P.R., Clark, B.C., Crisp, J.A., Farrand, W.H., Herkenhoff, K.E. *et al.*, In situ evidence for an ancient aqueous environment in Meridiani Planum, Mars. *Science*, 306, 1709–1714, 2004.
2.102. Taconi, K.A., Zappi, M.E., French, W.T., Brown, L.R., Methanogenesis under acidic pH conditions in a semy-continuous reactor system. *Bioresour. Technol.*, 99, 8075–8081, 2008.
2.103. Vizzioli, C., Bacterial diversity associated to the tidal area of Río Tinto. PhD thesis, Universidad Autónoma de Madrid, Spain, 2017.
2.104. Wächtershäuser, G., Groundworks for an evolutionary biochemistry: The iron Sulphur world. *Prog. Biophys. Mol. Biol.*, 58, 85–201, 1992.
2.105. Webster, C.R., Mahaffy, P.R., Atreya, S.K., Flesch, G.J., Mischna, M.A., Meslin, P.-Y., Farley, K.A., Conrad, P.G., Christensen, L.E., Pavlov, A.A. *et al.*, Mars methane detection and variability at Gale crater. *Science*, 347, 415–417, 2015.
2.106. Widdel, F., Schnell, S., Heising, S., Ehrenreich, A., Assmus, B., Schink, B., Ferrous iron oxidation by anoxygenic phototrophic bacteria. *Nature*, 162, 834–836, 1993.
2.107. Wiegel, J., Ljungdahl, L.G., Demain, A.L., The importance of thermophilic bacteria in biotechnology. *Crit. Rev. Biotechnol.*, 3, 39–108, 1985.
2.108. Woese, C.R. and Fox, G.E., Phylogenetic structure of the prokaryotic domain: The primary kingdoms. *Proc. Natl. Acad. Sci. USA*, 77, 5088–5090, 1977.
2.109. Woese, C.R., Kandler, O., Wheelis, M.L., Toward a natural system of organisms: Proposal for the domains archaea, bacteria and eukarya. *Proc. Natl. Acad. Sci. USA*, 87, 4576–4579, 1990.
2.110. Yan, R., Kappler, A., Peiffer, S., Interference of nitrite with pyrite under acidic conditions: Implications for studies of chemolithotrophic denitrification. *Environ. Sci. Technol.*, 49, 11403–11410, 2015.
2.111. Zhang, G., Dong, H., Xu, Z., Zhao, D., Zhang, C., Microbial diversity in ultra-high-pressure rocks and fluids from the Chinese Continental Scientific Drilling Project in China. *Appl. Environ. Microbiol.*, 71, 3213–3227, 2005.
2.112. Zolotov, M. and Shock, E., Formation of jarosite-bearing deposits through aqueous oxidation of pyrite at the Meridiani Planum, Mars. *Geophys. Res. Lett.*, 32, L21203, 2005.

3

Blossoms of Rot: Microbial Life in Saline Organic-Rich Sediments

Adrian-Ştefan Andrei[1†], Paul-Adrian Bulzu[2†] and Horia Leonard Banciu[2*†]

[1]*Department of Aquatic Microbial Ecology, Institute of Hydrobiology, Biology Centre CAS, České Budějovice, Czech Republic*
[2]*Department of Molecular Biology and Biotechnology, Faculty of Biology and Geology, Babeş-Bolyai University of Cluj-Napoca, Cluj-Napoca, Romania*

Abstract
The organic-rich sediments (or "sapropels") are complex organo-mineral and microoxic/anoxic milieus underlying stagnant aquatic basins with high input of organic matter in addition to persistent salinity and density gradients. Sapropels of brackish, marine and hypersaline aquatic systems were found to harbor one of the most diverse microbiomes reported to date, and their study may contribute to elucidating the phylogenetic branching of the tree of life. In the present chapter, we explore the links between the sapropels' chemistry and innate biological diversity and postulate that highly connected systems favor phylogenetic and metabolic specialization. In these salty organic-rich sediments, the fermentative and sulfate-reducing bacteria alongside methanogenic archaea and a plethora of bacterial and archaeal candidate lineages are proposed to cooperate in the final steps of organic matter degradation. Moreover, recent findings suggest that brackish, marine and hypersaline modern sapropels-associated microbiomes include yet-uncultured *Archaea* that may lay close to the evolutionary roots of eukaryotes and life itself. Future investigations of both ancient and modern organic-rich sediments may have deep implications on: i) shedding light on the past events that shaped the evolutionary history of life and ii) formulating and testing hypotheses on the mechanisms that microbes may use to cope with thermodynamic (e.g., use of low Gibbs free energy metabolic conversions) and environmental (e.g., microoxic/anoxic and saline) constraints that could be considered analogous to extraterestrial conditions.

Keywords: Biogeochemical cycling, density stratification, energetics, evolution of life, metagenomics, salinity, sapropels, uncultured diversity

**Corresponding author*: horia.banciu@ubbcluj.ro
†*These authors contributed equally to this work.*
Adrian-Ştefan Andrei: https://www.researchgate.net/profile/Adrian_Stefan_Andrei
Paul-Adrian Bulzu: https://www.researchgate.net/profile/Paul_Bulzu
Horia Leonard Banciu: https://www.researchgate.net/profile/Horia_Banciu, https://scholar.google.com/citations?user=XW6_3FcAAAAJ&hl=en

Joseph Seckbach and Helga Stan-Lotter (eds.) Extremophiles as Astrobiological Models, (45–82)

3.1 Introduction

The organic carbon-rich sediments or "sapropels" (from ancient Greek "*sapros*" - rotten or decayed and "*pelos*" - mud) are generically defined as dark-colored, fine-grained (i.e., <0.1 mm), geologically young (i.e., from the Pliocene epoch up to the present) and unconsolidated sediments with elevated contents of organic carbon (>2% by weight) [71] [127]. Historically, the German palaeobotanist Henry Potonié (1904) coined the term "sapropel" as the scientific international equivalent of the German word "*Faulschlamm*" (literally meaning "rotten mud") used to designate "*mehr minder dunkelen, grau-braun-grünlichen Schlamm, der sich am Boden stagnirender Gewässer absetzt, entstanden aus den sich zersetzenden (faulenden) Organismen, die im Wasser gelebt haben, und ihren Ausscheidungen.*" (Personal translation, "the more or less darker, greyish-brownish-greenish mud settling on the bottom of stagnant waters arisen from the decomposing (rotting) organisms that lived in the water and their excrements") [3.194]. Although originally used to denote freshwater lake sediments enriched in completely decomposed plant (or algal)-derived biomass of recent [3.248] [3.257] or fossil origin [3.192], since the 1950s the term "sapropel" has been widely used in sedimentology, petrology and marine geology specifically referring to the fossil organic-rich sediment sequences formed in the eastern Mediterranean Sea [3.127] [3.173] [3.202], Black Sea [3.36] or Baltic Sea [3.23] during the late Pliocene to early Holocene epochs.

The coastal zones are presently considered the most productive organic matter depositional settings [3.102], surpassing by area and volume the lacustrine basins with documented high sapropel productivity [3.250]. Consequently, the genesis, geochemistry and microbial diversity and function in modern organic-rich marine sediments have been studied in far more detail than in their saline lacustrine counterparts. In an early attempt to summarize the ecology of hypersaline lakes, Larsen [3.138] made a general statement that "the sediment of the bottom floor of the hypersaline lakes and ponds is often characteristically black and contains H_2S, indicating activity of sulfate reducing bacteria," This claim was lately reiterated by Jones *et al.* [3.119], who drew a similar conclusion on the black and anoxic lacustrine sediments in shallow Eastern African soda lakes. Present-day saline organic-rich sediments ("black muds" or sapropels *sensu lato* [3.71]) were also reported in coastal saline Tekirghiol Lake in Southeast Romania (Ţuculescu, 1965; cited by [3.13]), circum-Mediterranean crystallizer ponds [3.106] and sebkhas [3.45], a Canadian salt playa [3.139], tropical marine-derived shallow lakes [3.99] [3.110], the Dead Sea [3.174], shallow hypersaline lakes Mar Chiquita in Argentina [3.161] and Kyzylkak in Kazakhstan [3.131], soda lakes [3.80] [3.251] [3.252] and meromictic saline lakes Mahoney [3.26] [3.50] [3.96], Ursu and Fără Fund in central Romania [3.5] (Figure 3.1).

In this chapter, we present an overview of the genesis and geochemical characteristics of recent organic carbon-rich sediments in a range of saline depositional environments. Furthermore, the diversity of microbes inhabiting these habitats is briefly discussed with an emphasis on their possible functional roles in early diagenetic processes and the biogeochemical cycling of carbon, sulfur and nitrogen. In the end, we highlight the factors constraining life within saline sapropels and the relevance of sapropel-associated biomarkers and biosignatures to the science of astrobiology.

Figure 3.1 Macroscopic aspect of black mud freshly collected from hypersaline meromictic Ursu Lake (Transylvanian Basin, central Romania) in 2015. (Image credit: the authors).

3.2 Overview of Saline Aquatic Systems

Saline aquatic environments comprise water bodies where salts are concentrated by fast evaporation rate that outbalance the freshwater inflow, resulting in excessive salinity values (i.e., >3% w/v of total salts) [3.116]. Natural and man-made saline aquatic systems are distributed on all continents and latitudes with a total volume of about 44% of all lakes on Earth [3.166]. The saline water bodies are very diverse in their geochemistry, hydrology, and morphometry, spanning from marine to continental origin, from fluctuating volume and salinity (e.g., ephemeral saline lakes and marine-derived, lagoonal or coastal lakes) to permanent salt-saturated brines; from small-surfaced shallow ponds (e.g., circum-Mediterranean solar evaporation ponds and Andean salars) to large and deep basins (e.g., Caspian and Aral seas, Dead Sea, Great Salt Lake); from cold (e.g., high Arctic and Antarctic salt lakes) to hot (e.g., alkaline-saline lakes of East African Rift Valley); and from acidic (e.g., temporary acid brines in Southern Australia) to circumneutral (chloride-based brines) and alkaline water (e.g., carbonate-dominated lakes in Northern America, Eastern Africa, Southern Siberia and Inner Mongolia) [3.98] [3.116] [3.118]. Besides their cultural, aesthetic or economic benefits [3.263], saline basins of all hydro-geochemical categories are regarded as ecosystems of yet-to-be explored dynamics and complexity [3.19] [3.143] deserving dedicated conservation management [3.190] [3.209] [3.268]. At the very least, the extreme geochemistry of certain types of brines as well as their relevance in the preservation of biomarkers [3.21] [3.58] [3.148] and biosignatures [3.18] [3.210] [3.254], and their associated peculiar biology [3.143], may qualify these environments to serve as analogs for early Earth habitats [3.32] or for the exploration of habitable extraterrestrial milieus [3.47] [3.79] [3.193] [3.197].

3.3 Prerequisites of Organic Carbon-Rich Sediment Genesis in Saline Lakes

Saline waters are presently viewed as highly productive ecosystems [3.97] [3.119] [3.143]. For example, the polymictic shallow alkaline-saline lakes of the East African Rift Valley were shown to produce up to 9.6 g $C{\cdot}m^{-2}{\cdot}d^{-1}$ (Aleem and Samaan, 1969, cited by [3.97]), whereas maximum daily rates of primary production in the photic zone of meromictic (permanently stratified) saline Mahoney [3.93], Big Soda [3.48] and Mono lakes [3.114] were estimated at 0.47, 2.8 and 10 g $C{\cdot}m^{-2}$, respectively. The main primary producers in saline lakes are the oxygenic phototrophs such as cyanobacteria (e.g., *Spirulina-Cyanospira-Arthrospira* group in soda lakes, halotolerant *Aphanothece-Cyanothece* group in many pH-neutral saline lakes), unicellular algae (diatoms in brackish, marine or low-salt lakes or green algae *Dunaliella* spp. in highly saline pH-neutral or alkaline inland lakes), or anoxygenic phototrophic sulfur bacteria at the oxic-anoxic transition zone of permanently stratified saline lakes [3.93] [3.97] [3.119] [3.182] [3.216]. In productive saline systems, the photosynthetic biomass feeds abundant grazing zooplankton consisting of anostracans (*Artemia* spp.), copepods (*Cletocamptus* spp., *Diaptomus* spp.), cladocerans (*Moina* spp.) and rotifers (*Brachionus* spp.) [3.2] [3.3] [3.91] that additionally contribute to the regeneration of nutrients [3.123]. During episodes of massive organic matter inflow, the water basins turn to an eutrophic state [3.206]. Such settings favor an accelerated sinking rate and rapid burial of decaying biomass with subsequent oxygen depletion [3.151]. The mineralization of entrapped organic matter is thus further impeded, triggering the formation of sediments enriched in organic carbon [3.1] [3.108] [3.237]. Furthermore, in semi-stagnant (i.e., monomictic) or stagnant (i.e., meromictic) saline lakes lacking vertical water mixing, the ventilation is hampered, leading to seasonal or permanent bottom-water anoxia and onset of a strong reducing milieu by the biological production and release of sulfides, ammonium and/or methane [3.3] [3.22] [3.63] [3.112] [3.129] [3.140] [3.222] [3.258]. Concerted high-surface productivity and strong density gradient augmented by accelerated organic matter deposition under high redox conditions in bottom water are postulated to control the formation of organic-rich sediments in estuarine [3.73], marine [3.72] [3.204] and lacustrine environments [3.82] [3.123] [3.212].

3.4 Chemistry of Recent Organic Carbon-Rich Sediments in Saline Water Bodies

The chemistry of sapropels is puzzlingly complex. It is largely composed of organic and mineral fractions with composition and proportions that are mainly influenced by age and environmental settings such as ionic balance of water and local geological characteristics. A significant (~90%) proportion of organic matter appears to be strongly adsorbed on the surface of mineral grains and thus cannot be physically separated from its mineral matrix [3.102]. Generally, the organic fraction of modern sapropels includes a mixture of hydrolyzable (labile) and non-hydrolyzable (refractory) macromolecules. The hydrolyzable organic molecules are seemingly more abundant in young or upper sapropel layers (e.g., humic acids of terrigenous sources, proteinaceous material, cellulose and hemicelluloses), while the refractory biopolymers, including resistant polysaccharides (e.g., chitinous exoskeletal fragments), polysaccharide derivatives (e.g., humins), highly aliphatic (e.g., *n*-alkanes, branched alkanes such as pristane, phytane and algaenans) or aromatic biomacromolecules (e.g., hopanes from bacteria; gammaceranes from ciliated protozoa; and sterol-derived steranes from algae or higher plants; and diagenetic products of photosynthetic pigments such as chlorobactene, isorenieratane, β-carotane, etc.) from decayed organisms, accumulate in aged or deep sapropel layers [3.54] [3.59] [3.99] [3.130] [3.250]. The pentacyclic triterpanes, such as hopanes, are the most frequently detected class of aromatic hydrocarbons in organic-rich sediments [3.99] [3.188]. Low pristane-to-phytane (Pr/Ph) ratio [3.241], abundant gammacerane [3.55], C35 homohopane [3.191], β-carotane [3.29] [3.162] or *n*-docosane alkanes [3.17] were proposed as proxies for euxinic (i.e., anoxic and sulfidic) and/or evaporitic (hypersaline) conditions during deposition of oil source rocks (see Table 3.1 for relevant chemistry features of selected saline sapropels with varying salinities). The common component of organic-rich sediments, the oil-precursor kerogen (lipid-derived, insoluble, high molecular weight organic substances [3.67]), is presumed to develop in the early stages of diagenesis [3.100] [3.142] [3.146] and demonstrated to originate from resistant biopolymers [3.60] [3.250].

The organic matter is interspersed in a mineral matrix of clastic (e.g., silicates such as clays and quartz, calcite and magnesium-rich calcite, pyrite, etc.) and biogenic origin (e.g., biogenic silicates, aragonite-type carbonates, etc.). In saline evaporitic conditions, for example, a dominance of dolomite ($CaMg(CO_3)_2$) over calcite was frequently recorded [3.57] [3.125] [3.159] following a yet-unclear process controlled by complicated interplay between abiotic (temperature, Mg/Ca ionic ratio, alkalinity) [3.118] [3.265] and biotic (e.g., sulfate-reducing bacteria, accumulation of carboxyl-bearing biomolecules) factors [3.61] [3.200].

The distinctive dark (blackish, grayish or olive-brownish) coloration of sapropels is presumably attributed to enhanced organic carbon and metal sulfides accumulation [3.11]. The concurrent accumulation of metals (especially Fe and Mn) and preservation of organics in sapropels is influenced directly by chemical binding of metals and sulfur with the hetero-atomic functional groups of organic substances and/or indirectly by anaerobic sulfate and metals reduction or, alternatively, by anoxic diagenesis and subsequent precipitation of metallic sulfides and elemental metals [3.99] [3.196].

As stated above, both organic and mineral fractions are largely variable depending on the local geochemical and limnological settings. Additionally, the identification of organic compounds is heavily dependent on the extraction and fractionation approaches, making our knowledge of fossil and recent sapropels incomplete and their classification problematic [3.95] [3.196] [3.208].

3.5 Microbial Life in Saline Sapropels

The ecological principles underpinning the microbial diversity and interactions within saline organic-rich sediments can be summarized as follows:

i) The microbial diversity associated with saline and anoxic sapropels is shaped by concerted environmental (salinity/water activity) and thermodynamic (low Gibbs free energy reactions) constraints, and
ii) The polydisperse, multicomponent nature of sapropels results in a large number of possible microbial interactions.

Table 3.1 Composition of recent sapropels from selected, well-investigated brackish, marine and saline environments.

Type and origin of sediment	Composition of: Organic fraction	Composition of: Mineral fraction	References
Recent surficial sapropel from meromictic, saline (~8.5% salinity) and permanently euxinic Mahoney Lake, Canada	3.4–20.5% wt C_{org} Fatty acids (by GC-IR-MS): C16:0, C18:0, and C18:1 dominating among short-chained fatty acids and significant abundance of long-chained - ≥ C20 - fatty acids mostly of higher plants origin; Low sterols (C27 and C29), and n-alcohols contents mostly of higher plants origin; Dominant pigments and their derivatives (by UHPLC-QQQ): bacteriopheophytin *a* and okenone (of probable purple sulfur bacteria origin), pheophytin *a*, pyrobacteriopheophytin *a*.	Laminated sediments with carbonate contents: aragonite and Mg-rich calcite probably from upper-layer precipitation by high phototrophic activity; high contents of Fe and Mn.	[3.26] [3.94] [3.156] [3.189]
Recent sapropels from estuarine, brackish (2.9–3.2% salinity), seasonally hypoxic Grevelingen Lake, The Netherlands	1.8–4.4% wt C_{org} Fatty acids (by GC-IR-MS): phytoplankton-derived C14:0, C16:0, C16:1, and C18:0; C15:0 probably of sulfate-reducing bacteria origin; polyunsaturated C20:5 of algal origin. Dominant pigments (by spectrophotometry): functional chlorophyll *a* and pheopigments of algal origin	Carbonates: calcite, dolomite, aragonite of probable biogenic sources; high contents of Fe and Mn.	[3.92] [3.103] [3.124] [3.150] [3.172]

(*Continued*)

Table 3.1 Composition of recent sapropels from selected, well-investigated brackish, marine and saline environments. (*Continued*)

Type and origin of sediment	Composition of: Organic fraction	Composition of: Mineral fraction	References
Recent sapropel of shallow, marine (3.2% salinity), tropical, coastal Mangrove Lake, Bermuda Islands.	24–52% dry wt C_{org} Organic matter is dominated by humin (proto-kerogen) of mostly algal origin and containing carbohydrates and proteins well preserved in surface of sediments but significantly hydrolyzed in deeper layers; hydrocarbons (GC-MS) are enriched in algal-derived nC22 and plant-derived nC29 and nC31 alkanes. Fatty acids (by GC-MS): microbially or algal-derived C16:0; C26 to C30:0 probably of vascular plant wax origin; the bacterial biomarkers homohopane (C31) and trisnorhopane (C27) (indicator of euxinic conditions) were major terpanes in the saline fraction of sediment.	Carbonates: aragonite; low contents of silicates and metals (Fe and Mn), sulfate-rich.	[3.99] [3.130]

(*Continued*)

Table 3.1 Composition of recent sapropels from selected, well-investigated brackish, marine and saline environments. (*Continued*)

Type and origin of sediment	Composition of		References
	Organic fraction	**Mineral fraction**	
Recent marine sapropel in deep sea brine (25–26% salinity), Kebrit Deep, Red Sea	0.6–4.2% dry wt C_{org} Organic matter mostly of marine plankton (dinoflagellates) origin. Fatty acids (by GC-MS): plankton-derived n-C16 and n-C18 fatty acids; dominance of short-chain n-alkanes (nC15 to nC25) of algal and microbial origins; high dinosterol contents and low pristane/phytane ratio ≤1 suggests carbon input by either high dinoflagellate productivity or halophilic microbial communities; dominant hopanoid is trisnorhopane (C27) suggesting recent bacterial organic-matter degradation; type II kerogen was suggested.	Carbonate-rich, fine-grained, dark-grey laminated, anoxic and sulfidic sediments with high Fe (pyrite), Zn and Pb contents	[3.25]
Recent sapropel in coastal evaporitic sebkha (~16% salinity), Moknine, Tunisia	0.33–1.55% dry wt C_{org} Organic matter mostly derived from terrestrial halophyte (*Zygophillum* sp., *Atriplex* sp., *Salicornia* sp.) biomass; Hydrocarbons (by GC-MS): predominance of nC18 and nC22 alkanes (of probable microalgal origin), nC25 and nC31 alkanes (from higher plants); C29β and C30β hopanes dominate among terpanes. Gammacerane was detected.	Clay, silicate-based; calcite, halite and gypsum.	[3.45] [3.46]

Abbreviations: C_{org} – organic carbon.

These statements are briefly explained below.

i) The main environmental factor constraining the microbial diversity (i.e., structure and composition) in saline habitats is the low water activity, causing elevation of osmotic pressure [3.87] [3.143]. The increased osmotic pressure naturally selects for organisms capable of balancing the energy yield of catabolic reactions with expenses required both by salt adaptation (maintenance) and growth. Adaptation to high salinity is an energy-demanding process [3.180] and, therefore, the organisms using the cost-effective "salt-in" strategy such as members of the class Halobacteria (Euryarchaeota phylum) and the genus *Salinibacter* (Bacteroidetes) among aerobes, and of the order Halanaerobiales (Clostridia class, Firmicutes phylum) among anaerobes, are apparently prevalent at extreme salt concentrations [3.181]. Additionally, the low oxygen/highly reduced conditions typically occurring in organic-rich sediments stimulate diversification of heterotrophic anaerobic specialists with respiratory or fermenting, slow-growing metabolism making use of energetically weak yet very diverse electron acceptors (e.g., SO_4^{2-}, Fe^{3+}, NO_2^-, N_2O, CO, CO_2, etc.) and implicitly of reactions yielding low Gibbs free energy [3.68] [3.85] [3.89] [3.147].
ii) The intricate organomineral matrix of sapropels consists of numerous solid phases (e.g., minerals, organometallics, organosulfuric compounds, low and high molecular weight biomolecules, etc.) mixed with saline pore water, presumably promoting a broad spectrum of microbial interactions and strategies related to substrate use (e.g., degradation of complex carbohydrates, proteins and fatty acids, uptake of precursors of compatible solutes, etc.), electron transfer (e.g., syntrophy), alternative energy-conservation systems (e.g., light-sensitive rhodopsins, V-type ATPase, etc.), and motility (e.g., flagellar mobility, substrate adhesion and gliding) [3.10] [3.147]. Ultimately, the *in-situ* conditions (e.g., pH, temperature, particle surface chemistry, redox state) might impact the thermodynamic favorability of cell metabolism [3.68].

3.5.1 Higher-Order Microbial Taxonomy Detected in Modern Saline Sapropels

Sediment-associated microbial consortia mostly include microaerophilic and anaerobic Bacteria and Archaea that participate in sequential conversion of settled organic compounds. In contrast to aerobic organic matter oxidation that is driven by aerobic-respiring heterotrophs, the anaerobic degradation leading to complete organic matter mineralization is predominantly achieved by mutualistic and syntrophic cooperation among anaerobic consortia [3.133]. A general picture can be drawn from the available information on microbial diversity associated with saline organic-rich sediments. Thus, in most if not all available reporting, the Bacteria (70–80% of total relative abundance of recovered environmental DNA) appear to dominate over Archaea (10–20%) in organic-rich sediments regardless of the salinity or alkalinity of the overlying water [3.5] [3.201] [3.247] [3.251]. The high-level taxonomic composition of microbial communities most frequently reported in saline organic-rich sediments includes Actinobacteria (Actinobacteria and Nitriliruptoria classes), Bacteroidetes (Rhodothermia and Balneolia classes). Firmicutes (Clostridia class), Proteobacteria (Gammaproteobacteria and Deltaproteobacteria classes), and Euryarchaeota (Halobacteria, Thermoplasmata and Methanomicrobia classes) that act as core bacterial and archaeal cultured lineages, respectively [3.5] [3.201] [3.247] [3.251] (Figure 3.2). A dramatic expansion of our knowledge on microbial diversity residing in the aquatic sediments has been brought by the culture-independent approaches that have unveiled novel bacterial and archaeal lineages. In moderately saline and hypersaline sapropels, commonly detected bacterial and archaeal candidate lineages are "*Candidatus* Parcubacteria" phylum ("*Ca.* Patescibacteria" superphylum), "Parvarchaeota," "Nanohaloarchaeota," and "Woesearchaeota" candidate phyla (DPANN superphylum), whereas "*Ca.* Asgardarchaeota" and "*Ca.* Bathyarchaeota" were mostly associated with low-saline (brackish and marine) sediments. Their proportions within the microbial assemblages, however, vary with the salinity but are also dependent on the sampling sites and employed molecular approaches (i.e., marker gene clone library, small-subunit (SSU) rRNA gene amplicon-based or metagenomic shotgun sequencing) [3.5] [3.96] [3.201] [3.247] [3.251].

3.5.2 Anaerobic Organic Matter Degradation in Saline Organic Carbon-Rich Sediments

The particulate organic matter that settle on the bottom of aquatic systems undergoes sequential mineralization carried out by functionally distinct microbial communities. Chemically labile high-molecular weight organic compounds, such as proteins, nucleic acids, lipids, or polysaccharides (e.g., cellulose, hemicelluloses, chitin, etc.), are aerobically hydrolyzed to smaller and more biologically accessible molecules (e.g., oligo-, di- and monomers of sugars and nucleotides, peptides, amino acids, fatty acids, glycerol, etc.) mainly by fungi and actinobacteria. Dimers and monomers are further used either by anaerobic-respiring heterotrophs or fermented to low-molecular weight organics (e.g., short-chain volatile fatty acids –VFAs, such as formate, acetate, propionate, butyrate; alcohols such as ethanol). The latter are then anaerobically converted by terminal metabolism to end-products such as CO_2, H_2, H_2S, CH_4, etc. [3.9] [3.37] [3.133] [3.261] (Figure 3.3). The anaerobic steps of organic matter mineralization comprising fermentation, anaerobic respiration (denitrification, metal oxides and sulfate reduction), and methanogenesis are key processes in the early diagenesis of all aquatic sediments [3.37] [3.120] [3.136]. Below, we outline the main steps of anaerobic organic-matter degradation as inferred in saline organic-rich sediments.

3.5.2.1 Hydrolysis of Biomacromolecules

Biopolymer breakdown is readily favored under oxic conditions at varied salinities (i.e., from brackish/marine to hypersaline) and at low-salt (Cl-based) or haloalkaline microoxic/anoxic conditions, whereas concomitant high-salt (Cl-based) and anoxic settings are postulated as inhibiting for the hydrolysis of polymeric substances [3.105] [3.152]. As hydrolysis is the rate-limiting step of anaerobic organic matter digestion [3.133] [3.176], its potential inhibition at high salinity at least partially explains massive organic matter accumulation under such conditions. For example, in oxygenated sediments of different salinities, concerted

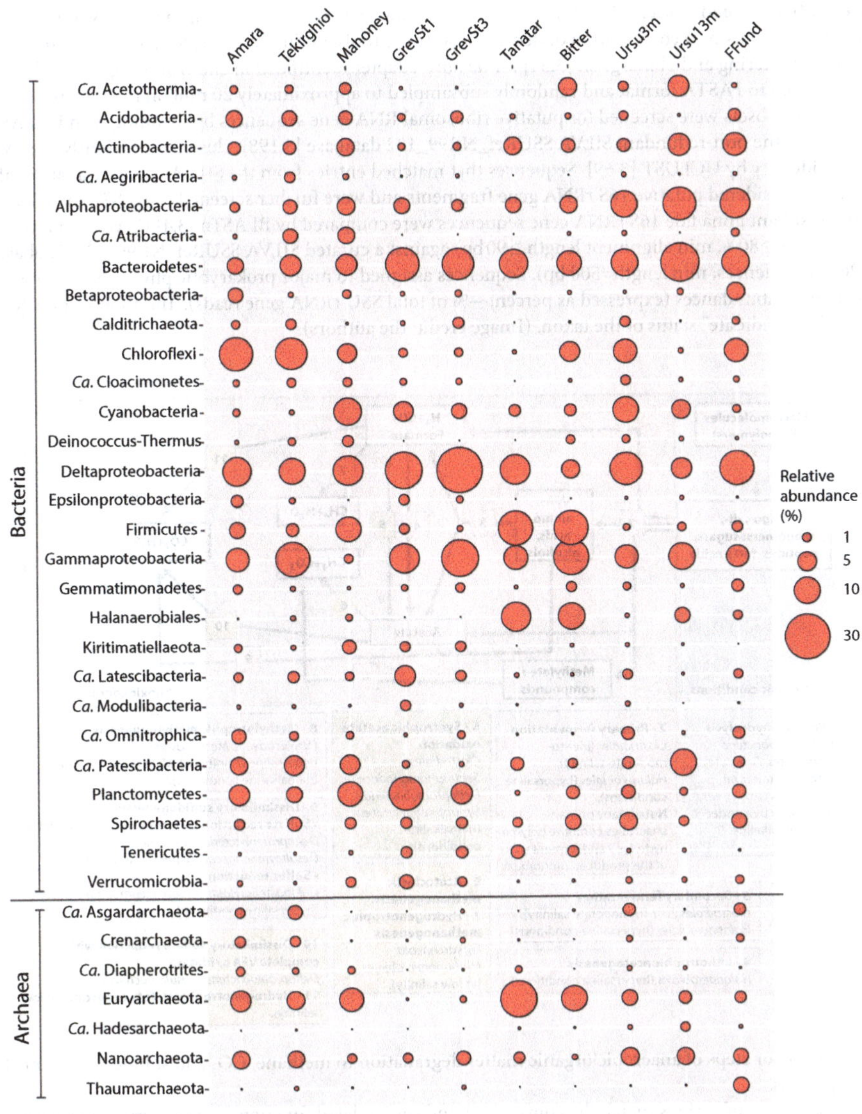

Figure 3.2 Relative abundance estimation for major taxa in organic-rich sediments from eight selected brackish-to-hypersaline aquatic systems.

To obtain a snapshot of the microbial diversity inherent in recently formed sapropels sampled worldwide, we chose 10 representative next-generation sequencing datasets (including both amplicon and Whole-genome shotgun data) from eight water systems as follows: inland brackish sulfated Amara and coastal saline Tekirghiol lakes [3.31]; inland meromictic and euxinic saline Mahoney Lake [3.96]; marine Grevelingen Lake stations 1 and 3 [3.150]; inland Tanatar and Bitter soda lakes [3.251] [3.252]; inland meromictic, pH-neutral hypersaline lakes Ursu (3-m and 13-m depth sediments) and Fără Fund [3.5] [3.6]. We have performed taxonomic classification and abundance estimations based on 16S rRNA gene fragments as described previously [3.7] [3.31].

(Continued)

Figure 3.2 (Continued) Briefly, FASTQ files containing raw sequencing data of sapropel DNA were quality filtered by using a combination of dedicated software tools: bbduk.sh [3.33] for adapter removal and contaminant filtering and bbmerge.sh [3.34] for *de novo* adapter identification and trimming. Filtered datasets were converted to FASTA format and randomly subsampled to approximately 20 million reads by reformat.sh [3.33]. All subsets were screened for putative ribosomal RNA gene sequences by scanning with UBLAST [3.69] against the non-redundant SILVA SSURef_NR99_132 database [3.198], clustered previously at 85% sequence identity by UCLUST [3.69]. Sequences that matched entries from the SILVA database at an E value <1e-5 were considered putative 16S rRNA gene fragments and were further screened using SSU-ALIGN [3.170]. Resultant bona fide 16S rRNA gene sequences were compared by BLASTn [3.4] (E-value cutoff <1e-5, identity ≥80%, min alignment length ≥ 90 bp) against a curated SILVA SSURef_NR99_132 database (n=626874 sequences, min length=500 bp). Sequences assigned to major prokaryotic phyla were used to calculate their abundances (expressed as percent—% of total SSU rRNA gene reads). The abbreviation "*Ca.*" denotes the "candidate" status of the taxon. (Image credit: the authors).

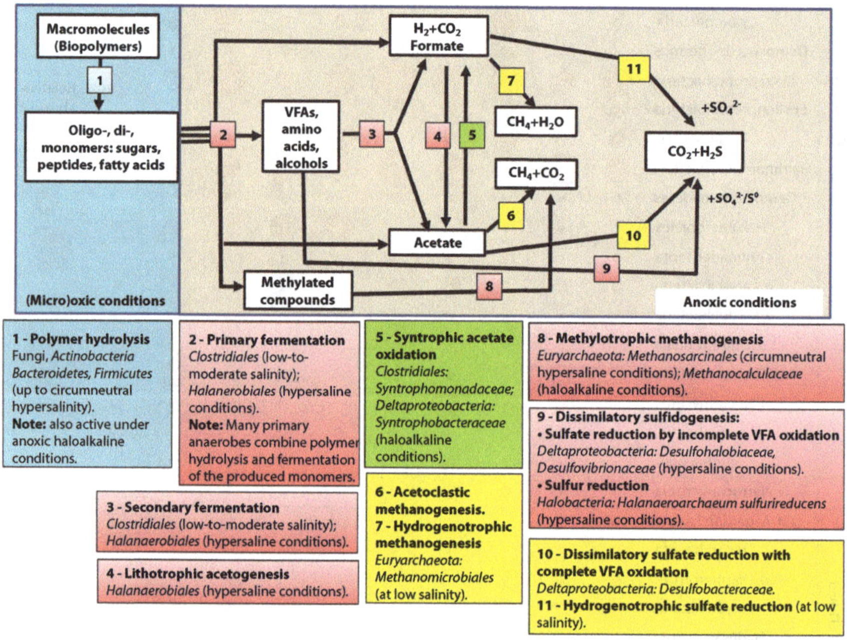

Figure 3.3 Major steps of anaerobic organic matter degradation to methane, CO_2 and sulfide in present-day saline sapropels.

The occurrence of various metabolic conversions and their related functional types of microorganisms are denoted by color marks as follows: blue – functional types demonstrated in a broad range of salinity and (micro)oxic conditions; red – present in broad range of salinity up to hypersaline settings under anoxic conditions; green – present in haloalkaline conditions up to near-salt saturation but absent at circumneutral hypersaline conditions; yellow – present under anoxic conditions only at low salinity and lacking in hypersaline settings. Abbreviation used: VFAs – volatile fatty acids. (Image credit: the authors).

hydrolysis of biomacromolecules mostly derived from plants (e.g., celluloses, hemicelluloses, lignins) was proposed to be achieved by active halotolerant or halophilic communities of fungi (e.g., unicellular, actinomycetous black yeasts or filamentous fungi of Eurotiomycetes) [3.19] [3.271] in addition to bacteria of orders Actinomycetales within the Actinobacteria phylum [3.12] [3.235] [3.266] and Bacillales (Firmicutes) [3.90]. NGS approaches applied on environmental DNA from O_2-depleted, organic matter-enriched saline sediments pointed to the occurrence of consortia with putative hydrolytic abilities comprising members of classes Clostridia (Firmicutes), Bacteroidia (Bacteroidetes), and Gammaproteobacteria (Proteobacteria) [3.164] [3.201] [3.270] [our unpublished results] (see Figure 3.2). In spite of isolation of anaerobic strains with cellulolytic (*Clostridium alkalicellum*) [3.272], pectinolytic (*Natronoflexus pectinivorans, Natranaerovirga hydrolytica*) [3.227] [3.231], chitinolytic (*Halanaerobacter chitinivorans, Chitinivibrio alkaliphilus*) [3.149] [3.232] or proteolytic (*Proteinivorax tanatarense, Anoxynatronum buryatiense*) [3.126] [3.199] capacities from moderately saline alkaline or hypersaline sediments, the anaerobic degradation of polymers under highly saline conditions remain poorly understood. Most of the above-mentioned polymer-degrading bacteria and some facultatively anaerobic members of the Halobacteria class, such as *Halorhabdus* spp. [3.8], might be involved in subsequent catabolization of sugar monomers, amino acids and/or long-chain fatty acids either through anaerobic respiration or primary fermentation (Figure 3.3). The latter route for carbon flow is expectedly dominant under anoxic and hypersaline conditions, as nitrate-dependent anaerobic respiration is hampered by the absence of autotrophic nitrification at salinities above 15% w/v NaCl [3.184]. Anaerobic respiration by Fe^{3+} reduction was suggested to be active in microoxic upper layers of sulfate-rich hypersaline sediments [3.74], a process that in turn seems hindered in euxinic (hyper)saline conditions where metal (Fe^{3+}) oxides are precipitated as pyrite (FeS_2) [3.104].

3.5.2.2 *Primary and Secondary Fermentation of Low Molecular Weight Molecules*

Typical primary fermentative bacteria flourishing in anoxic saline sediments are members of orders Clostridiales at low to moderate salinities, Halanaerobiales in chloride-based hypersaline conditions, and Natranaerobiales at soda-based, haloalkaline conditions (see reviews by [3.216] [3.218], and systematic descriptions in [3.185] [3.186]). The extremely halophilic members of Halanaerobiaceae, Halobacteroidaceae and Natranaerobiaceae families (Halanaerobiales) share the energetically favorable "salt-in" osmoadaptative strategy, thus explaining their wide distribution in anoxic hypersaline conditions [3.181] [3.184]. The products of primary fermentation, including formate and other VFAs, alcohols, methylated amines, CO_2, and H_2, are further used by secondary fermentative (i.e., acetogenic) bacteria, sulfate-reducing bacteria and methanogenic archaea (Figure 3.3). Several halotolerant and halophilic sulfate-reducing bacteria are also capable of fermentation with formation of acetate. So far, only a few salt-adapted acetogens have been isolated. Moderately haloalkaliphilic *Natronincola* and *Tindallia* genera (Clostridiales), and extremely haloalkaliphilic *Natroniella* sp. and *Fuchsiella* sp. (Halanaerobiales) were recovered from sediments of alkaline saline lakes [3.275] [3.276] (also see the reviews by [3.62] [3.216]). *Acetohalobium arabaticum* (Halanaerobiales) is an extremely halophilic homoacetogen isolated from a hypersaline cyanobacterial mat in a shallow lagoon on the west coast of the Sea of Azov. It is capable of autotrophic (CO_2+H_2), methylotrophic (betaine, trimethylamine) or heterotrophic (formate, lactate) acetogenesis under neutral pH and up to saturated NaCl conditions [3.273]. Interestingly, recent genome-wide analyses have pointed out the expansion of the genetic code of *A. arabaticum* toward the 21st amino acid, pyrrolysine (Pyl), which is incorporated in the catalytic site of methyltransferases during trimethylamine metabolization [3.195] [3.211] in a similar mechanism as in several methylotrophic methanogens [3.234].

Autotrophic conversion of CO_2/CO to acetate by using H_2 as electron donors via the reductive acetyl-CoA (Wood-Ljungdahl) pathway appears as being thermodynamically and kinetically outcompeted in habitats where it coexists with autotrophic (i.e., hydrogenotrophic) sulfate reducers and methanogens [3.183]. In spite of their seemingly uncompetitive energetics, acetogens have broad phylogenetic diversity and are widespread in various anoxic subsurface habitats [3.147]. In contrast to the specialized sulfate reducers and methanogens, acetogenic bacteria are more versatile in adopting mixotrophic lifestyle by additionally fermenting a broad range of low molecular weight substrates, thus possibly alleviating ATP limitations under thermodynamically unfavorable conditions [3.147]. Moreover, autotrophic acetogens were shown to employ stepwise ion gradient- (i.e., chemiosmotic) coupled ATP synthesis that, opposed to single-step substrate level phosphorylation requiring at least 31 kJ/ mol ATP (Table 3.2), might be hypothetically initiated by threefold lower energy amounts freed from exergonic steps of the Wood-Ljungdahl pathway [3.167]. Nonetheless, intracellular K^+ accumulation as an energetically cheap haloadaptive strategy employed by acetogens from Halanaerobiales could solidly explain their resilience under anoxic and hypersaline conditions [3.184].

3.5.2.3 Methanogenesis

It has long been postulated that in anoxic, sulfate-rich sediments, sulfate reduction coexists with methanogenesis, as the first is fueled by acetate and/or hydrogen, whereas methanogens have a competitive advantage when methylated substrates (e.g., methanol, methylamines, methionine) become available [3.187]. When sulfate (alongside nitrate or ferric iron) is depleted, methanogenesis takes over as the main terminal step of anaerobic organic matter degradation [3.49]. Salinity, mainly as chlorinity, was found to severely inhibit acetoclastic methanogenesis (<1 M Cl^-) [3.256] and, to a lesser extent, hydrogenotrophic methanogenesis (<2 M Cl^-) [3.177]. On the contrary, methylotrophic methanogenesis by dismutation of C_1 methylated compounds is the dominant terminal step of organic matter degradation in organic-rich saline, marine sediments with limiting SO_4^{2-} concentrations [3.256], alkaline saline lacustrine sediments [3.216] [3.274], and in Cl-based hypersaline lacustrine [3.35] and deep-sea sediments [3.277], where methylated compounds are abundant. In saline aquatic systems, trimethylamines are produced by members of Halanaerobiales that are able to ferment the large pool of glycine betaine, an organic osmolyte that is released from decomposition of betaine-accumulating organisms [3.179]. Typical salt-loving methanogens are the halotolerant hydrogenotrophic *Methanocalculus* spp. (Methanomicrobiales order) and the obligate halophilic methylotrophic *Methanosalsum* spp. (1–12% w/v NaCl), *Methanohalophilus* spp. (2–25% w/v NaCl), and *Methanohalobium* spp. (15–30% w/v NaCl) within Methanosarcinales order (Euryarchaeota phylum) (see review in [3.160]). Whereas circumneutral hypersaline sediments appear as strictly constraining the methanogenesis to the non-competitive methylotrophic pathway, both methylotrophic and hydrogenotrophic methanogenesis are active up to saturated soda salt concentrations in haloalkaline sediments [3.214] [3.216].

Most known halotolerant and halophilic methanogens employ "salt-out" osmoadaptive strategy being capable of *de novo* synthesis and accumulation of organic compatible solutes, mainly glycine betaine, alongside K^+ as intracellular counterion [3.160]. Energy balance (i.e., between energy gain of cell's catabolism and energy requirement for maintenance)

Table 3.2 Metabolic reactions with low Gibbs free energy changes that potentially constrain the anaerobic microbial diversity in saline and hypersaline anoxic sediments.

Metabolic type	Reaction	Standard Gibbs free energy change, $\Delta G^{o'}$ (kJ/ mol)	Environmental conditions
Methylotrophic methanogenesis (with trimethylamine - TMA)	$4TMA + 9H_2O + H^+ \rightarrow 9CH_4 + 3HCO_3^- + 4NH_4^+$	-191.1[b]	pH-neutral and alkaline low-to-hypersaline, sulfate-depleted
Heterotrophic sulfate reduction with incomplete lactate oxidation	$2CH_3CH(OH)COO^- + SO_4^{2-} \rightarrow 2CH_3COO^- + HS^- + 2HCO_3^- + H^+$	-160.1[b]	pH-neutral and alkaline low-to-hypersaline, sulfate-rich
Hydrogenotrophic sulfate reduction	$H_2 + SO_4^{2-} + H^+ \rightarrow HS^- + 4H_2O$	-152.2[a]	pH-neutral and alkaline low-to-moderate saline
Autotrophic acetogenesis	$4H_2 + 2HCO_3^- + H^+ \rightarrow CH_3COO^- + 4H_2O$	-104.6[a]	pH-neutral and alkaline low-to-hypersaline
Heterotrophic sulfate reduction with complete acetate oxidation	$CH_3COO^- + SO_4^{2-} \rightarrow HS^- + 2HCO_3^-$	−47.7[b]	Low-to-moderate saline, sulfate-rich
Hydrogenotrophic methanogenesis	$4H_2 + HCO_3^- + H^+ \rightarrow CH_4 + 3H_2O$	-34.0[b]	Low-saline
Acetoclastic methanogenesis	$CH_3COO^- \rightarrow CH_4 + HCO_3^-$	-31.1[b]	Freshwater, low chlorinity, low temperature
Synthesis of ATP	ADP + Pi→ ATP + H_2O	+31.8[a]	
Minimum biological energy quantum*		-10 to -20[c]	

*Note: Minimum biological energy quantum is defined as the minimum amount of free energy change required to initiate ATP synthesis via chemiosmotic pathway [3.167].

References used: (a) [3.243]; (b) [3.180]; (c) [3.107].

seems essential in shaping distribution and activity of methanogens across saline habitats [3.181]. At same total Na^+ concentrations, the $NaCO_3/NaHCO_3$ (soda)-based solution exerts a significantly (i.e., ca. twofold) lower osmotic pressure than the analogue NaCl solution [3.16]. Consequently, less energy for cell maintenance is spent by soda-adapted organisms (natronophiles and haloalkaliphiles) as compared to true halophiles [3.14] [3.15], thus explaining, at least in part, the broader functional diversity of methanogens (and other microbial metabolic types) in alkaline saline habitats.

Recently, a novel methanogenic euryarchaeal lineage, "Methanonatronarchaeia," capable of using C_1 methylated compounds as electron acceptors and H_2 and formate as electron donors, and having the "salt-in" haloadaptive strategy, was suggested as playing significant roles in completion of organic matter degradation in hypersaline anoxic environments [3.226]. Although methyl-reducing methanogenesis was inferred in members of euryarchaeal Thermoplasmata class [3.137] and "*Ca.* Bathyarchaeota" [3.76], groups frequently associated with anoxic saline sediments [3.5] [3.122] [3.244] [3.251] [A. Baricz personal communication], their active involvement in methane formation under such conditions remains largely uncertain.

In anoxic environments where adverse conditions (e.g., elevated temperature, high ammonia and VFA concentrations) impede conversion of fermentation products (e.g., ethanol, acetate and other VFAs, etc.) by conventional methanogenesis or sulfate reduction, the end step of organic matter degradation is achieved by syntrophic associations. Syntrophic degradation of fermentative intermediates is a cooperative energy-conserving strategy where H_2 or formate released by secondary fermenting bacteria act as interspecies electron carriers being consumed by hydrogenotrophic methanogenic archaea or sulfate-reducing bacteria [3.207]. For example, in syntrophic acetate oxidation, the acetate oxidation by reversed Wood-Ljungdahl pathway is an endergonic process (Table 3.2), but the subsequent use of H_2/formate by hydrogenotrophs shifts the overall Gibbs free change toward negative (exergonic) values. Co-culture enrichments from sediments of soda lakes in Kulunda Steppe (Siberia, Russia) have revealed syntrophic associations capable of oxidizing ethanol, acetate and other VFAs under methanogenic [3.213] and sulfate-reducing conditions [3.215] [3.219] and up to near-saturated soda concentrations. Haloalkaliphilic syntrophic partners were identified as closely related to acetate-utilizing members of the Syntrophomonadaceae family (Clostridiales) co-occurring with sulfate-reducing *Desulfonatronospira* sp., propionate-utilizing Syntrophobacteraceae (Deltaproteobacteria) in association with *Desulfonatronobacter* sp. and methanogenic *Methanocalculus* sp., respectively [3.213] [3.215] [3.220]. Further comparative analyses of metagenomes recovered from soda lake sediments additionally confirmed the presence of Syntrophomonadaceae-related genomes [3.251] and indicated that under extreme haloalkaline conditions, acetate is preferentially consumed by syntrophic acetate oxidation that prevails over acetoclastic methanogenesis or heterotrophic sulfate reduction [3.242].

3.5.2.4 *Biological Sulfur Reduction and Oxidation*

Reportedly, methanogenesis is the major terminal process in carbon flow in freshwater systems whereas sulfate reduction is the dominant end metabolism in marine and lacustrine saline sediments [3.120] [3.262]. The reductive steps of the sulfur cycle leading to sulfidogenesis are active in uppermost layers of anoxic sediments of pH-neutral saline [3.28]

[3.121] and moderately or hypersaline alkaline lakes [3.80] [3.81] [3.135], and are carried out mainly by bacteria of Deltaproteobacteria class within Proteobacteria phylum [3.184] [3.224]. The high rates of sulfidogenesis observed in saline, anoxic organic-rich sediments [3.228] causes their typical "rotten egg" odor [3.267].

During dissimilatory reduction of sulfur compounds, the electrons are transferred from low molecular weight VFAs or H_2, respectively, to sulfate, thiosulfate, sulfite or elemental sulfur with the formation of predominantly volatile H_2S (at low pH) or dissolved HS^- (at high pH). Heterotrophic sulfate reduction by incomplete oxidation of substrates (e.g., lactate) to acetate is associated with a larger free Gibbs energy change than hydrogenotrophic sulfate reduction or complete acetate oxidation, respectively [3.184] (Table 3.2). Similar to salt-loving methanogens, the balance between energy gain and consumption dictates the upper salinity limit for different metabolic types of sulfate-reducing bacteria, microbes that are known to accumulate organic compatible solutes for their osmoregulation [3.180]. For example, the Desulfohalobiaceae and Desulfovibrionaceae families include halotolerant or halophilic sulfate-reducing bacteria that incompletely oxidize low molecular weight substrates (mostly lactate) to acetate, and are active up to near-salt saturation (2–4 M of total Na^+) in both chloride-sulfate- (e.g., *Desulfohalobium retbaense, Desulfovibrio halophilus*) [3.44] [3.178] and soda-based brines (e.g., *Desulfonatronospira* spp., *Desulfonatronovibrio* spp.) [3.224] [3.229]. Noteworthy, under haloalkaline conditions, disproportionation of intermediately oxidized sulfur compounds (e.g., thiosulfate, sulfite) to HS^- and SO_4^{2-} might represent a feasible alternative route for yielding biochemical energy in the absence of other electron donors, a process that has been demonstrated in members of several haloalkaliphilic sulfidogenic bacteria (e.g., *Desulfonatronospira* spp., *Desulfonatrovibrio* spp., *Desulfonatronum* spp.) [3.224] [3.229] [3.230].

Dissimilatory sulfate reduction with complete acetate oxidation or lithotrophic H_2 utilization by halotolerant members of Desulfobacteraceae (e.g., *Desulfobacter halotolerans, Desulfosalsimonas propionicica*) is confined to low or moderate (Cl-based) salinity (i.e., up to 2 M total Na^+) [3.27] [3.128]. Interestingly, mesocosm experiments using surficial samples from saline anoxic sediments followed by SSU rRNA gene sequencing showed that at *in-situ* salinity (ca. 1 M NaCl), Desulfobacteraceae are enriched when acetate is supplied, while Desulfovibrionaceae and Desulfohalobiaceae are successful competitors for lactate [3.35].

In contrast to the relatively moderate salinity limit of complete acetate oxidizers in chloride-based brines, complete oxidation of VFAs (e.g., acetate, propionate or butyrate) coupled with sulfidogenesis was demonstrated under extreme haloalkaline conditions (4 M of total Na^+, pH 9.5–10) in pure cultures of *Desulfonatronobacter acetoxydans* (Desulfobacteraceae) [3.221]. However, an alternative way of complete oxidation of VFAs and other fermentative products is probably attained by syntrophic trades between secondary fermenters and hydrogenotrophic sulfate reducers as discussed above.

Recently, novel metabolic types of obligately anaerobic sulfur-respiring archaea of Halobacteria class were discovered in chloride/sulfate-type hypersaline sediments in the Kulunda Steppe (Russia). *Halanaeroarchaeum sulfurireducens* and *Halodesulfurarchaeum formicicum* are unique examples of haloarchaea growing up to 4 M NaCl (pH 7) by using VFAs as electron donors and oxidized sulfur compounds as terminal electron acceptors [3.223]. Whereas *Halanaeroarchaeum sulfurireducens* uses acetate as carbon and energy sources, and reduces elemental sulfur, *Halodesulfurarchaeum formicicum* grows with formate or, alternatively, with H_2 as sole electron source and reduces thiosulfate, elemental sulfur or dimethylsulfoxide. Both anaerobic sulfur-respiring archaeal isolates were proposed to belong to a separate lineage within the Halobacteria class and to be widely distributed in various anoxic hypersaline ecosystems [3.225].

The inorganic reduced sulfur compounds (e.g., sulfide, thiosulfate) released in the anoxic sediment layers enter the oxidative step occurring in the overlying low redox or photic zones. Chemolithoautotrophic sulfur-oxidizing Gammaproteobacteria pertaining to moderately halotolerant *Thioalkalimicrobium* spp. and *Thioalkalispira* spp., and to phylogentically and metabolically versatile genus *Thioalkalivibrio*, are key components of dark carbon fixation and sulfur oxidation in moderately and hypersaline alkaline sediments, respectively [3.15] [3.216] [3.217]. Transition zones where gradients of light and reduced sulfur compounds overlap promote development of phototrophic sulfur-oxidizing bacteria [3.249]. Anoxygenic purple sulfur bacteria assigned to halotolerant members of Chromatiales and *Ectothiorhodospira* spp., and moderately and extremely halophilic *Halorhodospira* spp. (Gammaproteobacteria class), thrive in sediments of alkaline saline lakes [3.164] and form red microbial mats overlying black anoxic sediments in shallow salterns [3.43] [3.106].

3.5.2.5 Nitrogen Cycling

The biological nitrogen cycling in saline systems appears as severely truncated as key steps such as the nitrogen fixation by diazotrophy, autotrophic ammonium and nitrite oxidation, and dissimilatory nitrate reduction to ammonium are postulated to be precluded by the accompanying unfavorable bioenergetics [3.180]. Under high salt concentrations (>15% w/v salinity), only anaerobic, heterotrophic denitrification (mostly by the facultatively anaerobic halotolerant *Halomonas* spp.) was unequivocally confirmed. Consequently, in stable moderately saline and hypersaline environments (i.e., >10% w/v total salinity), the ammonium ion (NH_4^+) is by far the most prevalent inorganic nitrogen species while the very low NO_3^- and NO_2^- pools are presumably supplied by inputs from streams or surrounding soils [3.115] [3.216]. Moreover, in highly reducing circumneutral brines, the accumulated ammonium may strongly inhibit N_2 fixation [3.109]. In haloalkaline sediments where NH_4^+ is chemically removed as deprotonated NH_3, light-dependent diazotrophy was demonstrated up to moderate salinity (<10% w/v, total salts) while heterotrophic ("dark") nitrogen fixation prevailed at hypersaline conditions [3.168] [3.203].

Fermentative bacteria can perform dissimilatory reduction of nitrate to ammonium (DNRA), a key step in nitrogen cycling in various anoxic, highly sulfidic environments with high C/N ratios, including coastal [3.83] and salt marsh sediments [3.132]. As explained above, the naturally low nitrate/high ammonium concentrations in saline systems might adversely impact the DNRA process. However, high nitrate pollution was suggested to promote DNRA in organic carbon-rich sediments (total organic carbon >4%) from eutrophic sulfated saline Pétrola Lake (~14% w/v total salinity), indicating that this process may be augmented under excess nitrate supply [3.38].

3.5.3 Uncultured Microbial Diversity of Saline Sapropels and Its Putative Ecological Roles

The advent of next-generation sequencing (NGS) technologies targeting microbial diversity (e.g., amplicon-based sequencing of small-subunit rRNA gene) and/or functionality (e.g., shotgun metagenomics and metatranscriptomics, single-cell genomics) facilitated gaining insightful knowledge mainly into unculturable diversity and its putative functional roles in a large variety of aquatic ecosystems [3.88]. In spite of their biases in either unselectively picking extracellular DNA of exogenous sources (i.e., from soil or water column) alongside intracellular DNA of active resident microbes [3.255], or enriching low-abundant active or dormant, inactive cells, both molecular and conventional culturing approaches are crucial for validating the diversity, metabolic roles, and interactions of members of resident microbial communities [3.278].

3.5.3.1 Uncultured Bacteria

Abundant and highly diverse uncultured prokaryotic lineages with yet-to-be explored functional roles were inferred in organic-matter enriched sediments from alkaline saline [3.164] [3.201] or soda lakes [3.251], hypersaline marine-derived crystallizer ponds [3.155], and inland moderately saline [3.65] [3.96] [3.239] or hypersaline lakes [3.5]. For example, sequences classified as "*Ca.* Parcubacteria" (OD1) phylum within the "*Ca.* Patescibacteria" superphylum (former Candidate Phyla Radiation) have been detected in high relative abundances (i.e., >1% of total reads) in pH-neutral hypersaline sapropel of meromictic Ursu Lake by SSU rRNA amplicon sequencing [3.5], as well as in metagenomes recovered from sediments of soda Tanatar and Bitter lakes [3.251]. "*Ca.* Patescibacteria" was shown to have broad ecological distribution and to comprise phylogenetically diverse microbes covering a substantial (>15%) fraction of the diversity in Bacteria domain [3.30] [3.111]. The genome-resolved metagenomics, single-cell genomics and imaging techniques revealed that many Patescibacteria members have small cell (< 0.2 μm) as well as genome (<1 Mpb) sizes and share common structural and metabolic traits such as the absence of an outer membrane, DNA scavenging capacity, incomplete tricarboxylic acid cycle and restricted biosynthetic capacities (e.g., lack of nucleotide and amino acid synthesis) [3.41] [3.171] [3.111] [3.162]. A broad repertoire of lifestyles was suggested for Patescibacteria, including ectosymbiosis or epibiotic parasitism on archaea, bacteria or protists [3.171] [3.162], fermentative metabolism, and anaerobic respiration based on dissimilatory nitrate reduction to ammonium [3.145], and oxidation of hydroxylamine [3.41]. Although relevant involvement in carbon and hydrogen cycles in subsurface environments is predicted for this group, the biogeochemical roles of Patescibacteria are far from being understood.

Additionally, "*Ca.* Aegiribacteria", "*Ca.* Cloacimonetes" (WWE1), "*Ca.* Hydrogenedentes" (NKB19), "*Ca.* Latescibacteria" (WS3) and "*Ca.* Marinimicrobia" (SAR406) of superphylum FCB (Fibrobacteres, Chlorobi and Bacteroidetes), "*Ca.* Omnitrophica" (OP3) within PVC (Planctomycetes, Verrucomicrobia and Chlamydiae) superphylum, and "*Ca.* Acetothermia" (OP1) within Bipolaricaulota candidate phylum were detected in various abundances and compositions in hypersaline microbial mats [3.264], soda sediments [3.251], pH-neutral saline [3.96] or hypersaline [3.5] organic-rich sediments of meromictic lakes. The majority of the above uncultured bacterial groups were recurrently associated with anoxic, organic matter-rich environments where they have been hypothesized to act in anaerobic biopolymer digestion and primary fermentation (e.g., Latescibacteria) [3.77], free-living or host-attached secondary fermentation (e.g., Acetothermia, Omnitrophica) [3.240] [3.269] or syntrophic associations (e.g., Cloacimonetes, Hydrogenedentes) [3.175].

3.5.3.2 *Uncultured Archaea*

Similar to the Patescibacteria, the archaea of the DPANN radiation, presently comprising about 12 phyla, appear to possess reduced, streamlined genomes corroborated by small cell size and low metabolic capabilities suited for an ectosymbiotic and fermentative way of life (for up-to-date comprehensive reviews see [3.40] and [3.64]). Members of "*Ca.* Nanohaloarchaeota" were recognized as typical inhabitants of hypersaline systems [3.84] [3.169]. Recently, metagenome-assembled genomes (MAGs) related to "*Ca.* Diapherotrites," "*Ca.* Aenigmarchaeota," and "*Ca.* Woesearchaeota" were recovered from soda sediments [3.251], whereas SSU rRNA genes related to "*Ca.* Parvarchaeota" (ARMAN) and "*Ca.* Nanohaloarchaeota" co-occurred in significant abundances in hypersaline sapropels of inland meromictic lakes [3.5].

Molecular investigations employing SSU rRNA gene amplicon-based sequencing have revealed that in hypersaline sediments with significant abundance of Euryarchaeota-related sequences, Halobacteria frequently co-occur with Thermoplasmata [3.5] [3.251] [3.259]. Reversely, in sediments within low-salinity range (i.e., freshwater-brackish-marine), the archaea of Thermoplasmata were recurrently associated with "*Ca.* Bathyarchaeota" (former Miscellaneous Crenarchaeotal Group, MCG) [3.153]. Bathyarchaeota is an uncultured archaeal lineage with a complex intragroup phylogeny, cosmopolitan distribution and versatile metabolism [3.78]. Bathyarchaeota-related sequences were detected in numerous ecosystems, including ancient marine sapropels [3.51], coastal salt marshes [3.42] and inland euxinic water layers and sediments [3.24] [3.78] [3.144]. Bathyarchaeota was proposed as a core generalist group in the anoxic, carbon- and sulfate-rich and low-saline sediments from estuarine and marine environments [3.20] [3.134] [3.154]. The genome-based reconstructed metabolic behavior of Bathyarchaeota indicated an anaerobic heterotrophic lifestyle feeding on detrital protein and aromatic compounds from buried organic matter [3.20] [3.154] [3.163]. Co-occurrence network analysis in datasets from numerous sediments revealed the pivotal role of Bathyarchaeota in structuring the archaeal assemblages and possibly a keystone ecological contribution in anoxic freshwater-to-low-saline sediments acting alongside Thermoplasmata as primary degraders of detrital biopolymers and providing labile, low molecular carbon substrates (e.g., methylated amines) to methanogens [3.78].

Metagenomics and cultivation efforts have recently uncovered a novel lineage of archaea within the "*Ca.* Asgardarchaeota" superphylum that is deeply rooted in the tree of life and possibly placed at the very proximity of the eukaryotic branch [3.31] [3.113]. Aquatic sediments with salinities spanning from brackish to hypersaline were sources for numerous MAGs assigned to the "*Ca.* Lokiarchaeia," "*Ca.* Thorarchaeia," and "*Ca.* Heimdallarchaeia" clades [3.31] [3.157] [our unpublished results] (Figure 3.2). Comparative genomics and metagenome-based metabolic reconstruction of "*Ca.* Heimdallarchaeia" recovered from saline organic-rich sediments suggested that they encode eukaryotic-signature proteins and possess versatile mixotrophic metabolism supported by a broad range of energy harvesting capabilities, including synthesis of extracellular polymer-degrading hydrolases, presence of V-type ATPase and of the light-driven ion-pumping rhodopsins [3.31]. Presently, knowledge on Asgard archaea points to their evolutionary relevance in the understanding of the origin of eukaryotes and restructuring the tree of life [3.39], whereas their ecological roles in the organic matter mineralization of various types of aquatic sediments remain elusive.

Despite the considerable advances supported by "omics" technologies that have endowed unravelling of numerous novel prokaryotic lineages, the understanding of their metabolic diversity and ecological roles is far from being complete [3.39]. Targeting the key genetic and molecular components of energy metabolism and salt-adaptive mechanisms seems essential to specifically comprehend the metabolic capabilities and functional roles of the uncultured microbiome present in high salt environments. To successfully address these gaps, innovative approaches such as long reads sequencing of single molecules, improved strategies to verify the genome-based functional predictions, and reverse-genomics isolation in combination with conventional isolation methods are needed [3.52] [3.158].

3.6 Relevance of Saline Sapropels

Besides the economic value of salty sapropels that have long been used in the so-called "mud therapy" [3.86] [3.174], the systematic study of present-day saline organic matter-rich sediments is critical not only to reconstruct past environmental changes [3.75] [3.205] [3.246] but also to understand processes leading to accumulation of kerogen and subsequent formation of certain types of oil deposits [3.1] [3.100]. For example, refractory biomolecules originating from microbial activity or derived by geochemical conversions of buried organic matter, such as triterpanes (e.g., hopanes), *n*-alkanes, isoprenoids, steranes or carotanes, are seen as biologically informative compounds ("biomarkers") of interest while tracking the origins of organic matter [3.45] [3.253], signs of life in geologically old rock or extraterrestrial materials [3.70] [3.245] [3.260]. Additionally, the fine-grained anoxic sedimentary matrix is an ideal trap for preserving the organic, mineralogical or physical biosignatures whose interpretation is useful to diagnose the past life activity equally on Earth and other Earth-like planetary bodies [3.66] [3.101].

Salt-adapted dwellers of modern saline sapropels, such as halophilic and xerophilic fungi or halophilic archaea of the Halobacteria class, have been proposed as model organisms when prospecting life in the extraterrestrial habitats [3.56] [3.238]. Carotenoid pigments (e.g., bacterioruberin) from extreme haloarchaea that are apparently capable of prolonged survival within halite and gypsum evaporite crystals like those observed on Mars [3.233] were proposed as biomarkers for *in-situ* detection of life traces on extraterrestrial surfaces [3.53]. Numerous halophiles are known to possess mechanisms supporting resilience to extreme conditions, including UV-photoprotection [3.117], coping with low-water activity stress [3.141] [3.238], long-term survival ability [3.236], and chemolithotrophy [3.260], that were suggested as essential requirements for microbial colonization of habitats at the thermodynamic and physicochemical frontiers of life.

3.7 Concluding Remarks

Our current knowledge on the origin, geochemistry and biology of modern organic-rich sediments formed under varying saline conditions allow considering them as very complex habitats. Along with high primary productivity, the water stagnancy and rapid organic matter sinking appear to trigger the formation of present-day and fossil organic-rich sediments worldwide, while the local environmental settings (e.g., ionic and mineral composition, pH, water activity, redox potential, etc.) are strongly reflected in their heterogenous geochemical and microbiological composition. In saline sapropels (i.e., > 3% w/v salinity), most records to date indicated a prevalence of Bacteria over Archaea and core assemblages comprised of generalist taxa pertaining to a few higher-level taxonomy lineages. At lower taxonomic level, however, the intricate geochemistry and polydisperse nature of organomineral matrix seemingly promote metabolic specialization and diversification of microbial interactions. At circumneutral low-to-moderate chloride-based (< 2 M NaCl) as well as at haloalkaline (soda-based) conditions up to saturation (~4 M total Na^+), complete biogeochemical cycling of the main elements (C, N, S) could be predicted as the lower overall osmotic pressure seems to allow most microbial metabolic conversions. On the contrary, the organic carbon-rich sediments of the circumneutral brines (~4 M NaCl) might accommodate incomplete sulfur and nitrogen cycling, apparently due to the lack of microbial functional types unable to withstand the imbalance between energy gained during catabolic processes and increased energy demands of maintenance under high external osmotic pressure.

The understanding of the biogeochemistry of modern and fossil saline sapropels is beneficial to life and planetary sciences, including astrobiology, in two aspects: i) discovery of biomarkers and biosignatures alongside innovation and development of their detection tools, and ii) modeling and hypothesis testing of thermodynamic constraints for energy yields and flow within microbial trophic chains for the prospects of the origin, evolution and dispersion of life on Earth and beyond.

Acknowledgments

The completion of this work was supported by grants from the Romanian National Authority for Scientific Research, CNCS–UEFISCDI, project numbers PN-III-P4-ID-PCE-2016-0303 and PN-III-P4-ID-PCCF-2016-0016. The authors are grateful to Prof. Aharon Oren and Dr. Dimitry Y. Sorokin for their constructive suggestions that helped in improving this chapter.

References

3.1. Aizenshtat, Z., Miloslavski, I., Aschengrau, D., Oren, A., Hypersaline depositional environments and their relation to oil generation, in: *Microbiology and Biogeochemistry of Hypersaline Environments*, A. Oren (Ed.), pp. 89–108, CRC Press, Boca Raton, USA, 1999.

3.2. Alcocer, J. and Hammer, U.T., Saline lake ecosystems of Mexico. *Aquat. Ecosyst. Health Manage.*, 1, 3-4, 291–315, 1998.

3.3. Alexe, M., Şerban, G., Baricz, A., Andrei, A.-Ş., Cristea, A., Battes, K.P., Cîmpean, M., Muntean, V., Porav, S., Banciu, H.L., Limnology and plankton diversity of salt lakes from Transylvanian Basin (Romania): A review. *J. Limnol.*, 77, 1, 17–34, 2018.

3.4. Altschul, S.F., Gish, W., Miller, W., Myers, E.W., Lipman, D.J., Basic local alignment search tool. *J. Mol. Biol.*, 215, 3, 403–410, 1990.

3.5. Andrei, A.-Ş., Baricz, A., Robeson, M.S., Păuşan, M.R., Tămaş, T., Chiriac, C., Szekeres, E., Barbu-Tudoran, L., Levei, E.A., Coman, C. *et al.*, Hypersaline sapropels act as hotspots for microbial dark matter. *Sci. Rep.*, 7, 1, #6150, 2017.

3.6. Andrei, A.Ş., Robeson, M.S., II, Baricz, A., Coman, C., Muntean, V., Ionescu, A., Etiope, G., Alexe, M., Sicora, C.I., Podar, M., Banciu, H.L., Contrasting taxonomic stratification of microbial communities in two hypersaline meromictic lakes. *ISME J.*, 9, 12, 2642–2656, 2015.
3.7. Andrei, A.-Ş., Salcher, M.M., Mehrshad, M., Rychtecký, P., Znachor, P., Ghai, R., Niche-directed evolution modulates genome architecture in freshwater Planctomycetes. *ISME J.*, 13, 1056–1071, 2019.
3.8. Antunes, A., Ferrer, M., Yarza, P., Halorhabdus, in: *Bergey's Manual of Systematics of Archaea and Bacteria*, pp. 1–9, John Wiley & Sons, Ltd., Chichester, UK, 2015.
3.9. Arndt, S., Jørgensen, B.B., LaRowe, D.E., Middelburg, J.J., Pancost, R.D., Regnier, P., Quantifying the degradation of organic matter in marine sediments: A review and synthesis. *Earth Sci. Rev.*, 123, 53–86, 2013.
3.10. Arnosti, C., Hinrichs, K.U., Coffinet, S., Wilkes, H., Pantoja, S., The enduring questions: What's for dinner? Where's my knife? ...and can I use my fingers? (Unanswered) questions related to organic matter and microbes in marine sediments. *Front. Mar. Sci.*, 6, #629, 2019.
3.11. Arthur, M.A. and Sageman, B.B., Marine black shales: Depositional mechanisms and environments of ancient deposits. *Annu. Rev. Earth Planet. Sci.*, 22, 499–551, 1994.
3.12. Ballav, S., Kerkar, S., Thomas, S., Augustine, N., Halophilic and halotolerant actinomycetes from a marine saltern of Goa, India producing anti-bacterial metabolites. *J. Biosci. Bioeng.*, 119, 3, 323–330, 2015.
3.13. Baltres, A. and Medeşan, A., High-magnesian calcite in fecal pellets of *Artemia salina* from Techirghiol Lake. *Sediment. Geol.*, 20, 281–290, 1978.
3.14. Banciu, H.L. and Muntyan, M.S., Adaptive strategies in the double-extremophilic prokaryotes inhabiting soda lakes. *Curr. Opin. Microbiol.*, 25, 73–79, 2015.
3.15. Banciu, H.L. and Sorokin, D.Y., Adaptation in haloalkaliphiles and natronophilic bacteria, in: *Polyextremophiles*, J. Seckbach, A. H. Stan-Lotter (Eds.), pp. 121–178, Springer, Dordrecht, The Netherlands, 2013Oren.
3.16. Banciu, H., Sorokin, D.Y., Galinski, E.A., Muyzer, G., Kleerebezem, R., Kuenen, J.G., *Thioalkalivibrio halophilus* sp. nov., a novel obligately chemolithoautotrophic, facultatively alkaliphilic, and extremely salt-tolerant, sulfur-oxidizing bacterium from a hypersaline alkaline lake. *Extremophiles*, 8, 4, 325–334, 2004.
3.17. Barbe, A., Grimalt, J.O., Pueyo, J.J., Albaiges, J., Characterization of model evaporitic environments through the study of lipid components. *Org. Geochem.*, 16, 4-6, 815–828, 1990.
3.18. Barbieri, R. and Stivaletta, N., Halophiles, continental evaporites and the search for biosignatures in environmental analogues for Mars, in: *Life on Earth and other Planetary Bodies. Cellular Origin, Life in Extreme Habitats and Astrobiology*, vol. 24, A. Hanslmeier, S. Kempe, J. Seckbach (Eds.), pp. 13–26, Springer, Dordrecht, The Netherlands, 2012.
3.19. Baxter, B.K. and Zalar, P., The extremophiles of Great Salt Lake: Complex microbiology in a dynamic hypersaline ecosystem, in: *Model Ecosystems in Extreme Environments*, J. Seckbach and P. Rampelotto (Eds.), pp. 57–99, Academic Press, London, UK, 2019.
3.20. Biddle, J.F., Lipp, J.S., Lever, M.A., Lloyd, K.G., Sørensen, K.B., Anderson, R., Fredricks, H.F., Elvert, M., Kelly, T.J., Schrag, D.P., Sogin, M.L., Heterotrophic Archaea dominate sedimentary subsurface ecosystems off Peru. *Proc. Natl. Acad. Sci.*, 103, 10, 3846–3851, 2006.
3.21. Birgel, D., Guido, A., Liu, X., Hinrichs, K.U., Gier, S., Peckmann, J., Hypersaline conditions during deposition of the Calcare di Base revealed from archaeal di- and tetraether inventories. *Org. Geochem.*, 77, 11–21, 2014.
3.22. Boehrer, B., von Rohden, C., Schultze, M., Physical features of meromictic lakes: Stratification and circulation, in: *Ecology of Meromictic Lakes. Ecological Studies (Analysis and Synthesis)*, vol. 228, R. Gulati, E. Zadereev, A. Degermendzhi (Eds.), pp. 15–34, Springer, Cham, Switzerland, 2017.

3.23. Borgendahl, J. and Westman, P., Cyanobacteria as a trigger for increased primary productivity during sapropel formation in the Baltic Sea—a study of the Ancylus/Litorina transition. *J. Paleolimnol.*, 38, 1, 1–12, 2007.
3.24. Borrel, G., Lehours, A.C., Crouzet, O., Jezequel, D., Rockne, K., Kulczak, A., Duffaud, E., Joblin, K., Fonty, G., Stratification of Archaea in the deep sediments of a freshwater meromictic lake: Vertical shift from methanogenic to uncultured archaeal lineages. *PLoS One*, 7, 8, #e43346, 2012.
3.25. Botz, R., Schmidt, M., Wehner, H., Hufnagel, H., Stoffers, P., Organic-rich sediments in brine-filled Shaban-and Kebrit deeps, northern Red Sea. *Chem. Geol.*, 244, 3-4, 520–553, 2007.
3.26. Bovee, R.J. and Pearson, A., Strong influence of the littoral zone on sedimentary lipid biomarkers in a meromictic lake. *Geobiology*, 12, 6, 529–541, 2014.
3.27. Brandt, K.K. and Ingvorsen, K., *Desulfobacter halotolerans* sp. nov., a halotolerant acetate-oxidizing sulfate-reducing bacterium isolated from sediments of Great Salt Lake, Utah. *Syst. Appl. Microbiol.*, 20, 3, 366–373, 1997.
3.28. Brandt, K.K., Vester, F., Jensen, A.N., Ingvorsen, K., Sulfate reduction dynamics and enumeration of sulfate-reducing bacteria in hypersaline sediments of the Great Salt Lake (Utah, USA). *Microb. Ecol.*, 41, 1, 1–11, 2001.
3.29. Brassell, S.C., Biomarkers in recent and ancient sediments: The importance of the diagenetic continuum, in: *Organic Matter: Productivity, Accumulation, and Preservation in Recent and Ancient Sediments*, J.K. Whelan and J.W. Farrington (Eds.), pp. 339–367, Columbia University Press, New York, USA, 1992.
3.30. Brown, C.T., Hug, L.A., Thomas, B.C., Sharon, I., Castelle, C.J., Singh, A., Wilkins, M.J., Wrighton, K.C., Williams, K.H., Banfield, J.F., Unusual biology across a group comprising more than 15% of domain Bacteria. *Nature*, 523, 7559, 208–211, 2015.
3.31. Bulzu, P.-A., Andrei, A.-Ş., Salcher, M.M., Mehrshad, M., Inoue, K., Kandori, H., Beja, O., Ghai, R., Banciu, H.L., Casting light on Asgardarchaeota metabolism in a sunlit microoxic niche. *Nat. Microbiol.*, 4, 7, 1129–1137, 2019.
3.32. Burns, B.P., Anitori, R., Butterworth, P., Henneberger, R., Goh, F., Allen, M.A., Ibañez-Peral, R., Bergquist, P.L., Walter, M.R., Neilan, B.A., Modern analogues and the early history of microbial life. *Precambrian Res.*, 173, 1-4, 10–18, 2009.
3.33. Bushnell, B., *BBMap short read aligner*, University of California, Berkeley, CA, USA, 2016, http://sourceforge.net/projects/bbmap.
3.34. Bushnell, B., Rood, J., Singer, E., BBMerge - Accurate paired shotgun read merging *via* overlap. *PLoS One*, 12, 10, #e0185056, 2017.
3.35. Cadena, S., García-Maldonado, J.Q., López-Lozano, N.E., Cervantes, F.J., Methanogenic and sulfate-reducing activities in a hypersaline microbial mat and associated microbial diversity. *Microb. Ecol.*, 75, 4, 930–940, 2018.
3.36. Calvert, S.E., Vogel, J.S., Southon, J.R., Carbon accumulation rates and the origin of the Holocene sapropel in the Black Sea. *Geology*, 15, 10, 918–921, 1987.
3.37. Canfield, D.E., Organic matter oxidation in marine sediments, in: *Interactions of C, N, P and S Biogeochemical Cycles and Global Change. NATO ASI Series (Series I: Global Environmental Change)*, vol. 4, R. Wollast, F.T. Mackenzie, L. Chou (Eds.), pp. 333–363, Springer, Berlin, Germany, 1993.
3.38. Carrey, R., RodrÃguez-Escales, P., Otero, N., Ayora, C., Soler, A., Gómez-Alday, J.J., Nitrate attenuation potential of hypersaline lake sediments in central Spain: Flow-through and batch experiments. *J. Contam. Hydrol.*, 164, 323–337, 2014.
3.39. Castelle, C.J. and Banfield, J.F., Major new microbial groups expand diversity and alter our understanding of the tree of life. *Cell*, 172, 6, 1181–1197, 2018.

3.40. Castelle, C.J., Brown, C.T., Anantharaman, K., Probst, A.J., Huang, R.H., Banfield, J.F., Biosynthetic capacity, metabolic variety and unusual biology in the CPR and DPANN radiations. *Nat. Rev. Microbiol.*, 16, 10, 629–645, 2018.

3.41. Castelle, C.J., Brown, C.T., Thomas, B.C., Williams, K.H., Banfield, J.F., Unusual respiratory capacity and nitrogen metabolism in a Parcubacterium (OD1) of the Candidate Phyla Radiation. *Sci. Rep.*, 7, #40101, 2017.

3.42. Castro, H., Ogram, A., Reddy, K.R., Phylogenetic characterization of methanogenic assemblages in eutrophic and oligotrophic areas of the Florida Everglades. *Appl. Environ. Microbiol.*, 70, 6559–6568, 2004.

3.43. Caumette, P., Baulaigue, R., Matheron, R., *Thiocapsa halophila* sp. nov., a new halophilic phototrophic purple sulfur bacterium. *Arch. Microbiol.*, 155, 2, 170–176, 1991.

3.44. Caumette, P., Cohen, Y., Matheron, R., Isolation and characterization of *Desulfovibrio halophilus* sp. nov., a halophilic sulfate-reducing bacterium isolated from Solar Lake (Sinai). *Syst. Appl. Microbiol.*, 14, 1, 33–38, 1991.

3.45. Chairi, R., Biomarkers on sediments in a highly saline aquatic ecosystem: Case of the Moknine Continental Sebkha (Eastern Tunisia). *J. Coast. Zone Manag.*, 21, 2, #463, 2018.

3.46. Chairi, R. and Djellouli, Z.E.A., Characterization of organic matter deposited in the oriental zone from Moknine hypersaline environment, oriental Tunisia. *Chin. J. Geochem.*, 24, 4, 345–351, 2005.

3.47. Cheng, Z., Xiao, L., Wang, H., Yang, H., Li, J., Huang, T., Xu, Y., Ma, N., Bacterial and archaeal lipids recovered from subsurface evaporites of Dalangtan Playa on the Tibetan Plateau and their astrobiological implications. *Astrobiology*, 17, 11, 1112–1122, 2017.

3.48. Cloern, J.E., Cole, B.E., Oremland, R.S., Autotrophic processes in meromictic Big Soda Lake, Nevada. *Limnol. Oceanogr.*, 28, 6, 1049–1061, 1983.

3.49. Conrad, R., Importance of hydrogenotrophic, aceticlastic and methylotrophic methanogenesis for methane production in terrestrial, aquatic and other anoxic environments: A mini review. *Pedosphere*, 30, 1, 25–39, 2020.

3.50. Coolen, M.J. and Overmann, J., Analysis of subfossil molecular remains of purple sulfur bacteria in a lake sediment. *Appl. Environ. Microbiol.*, 64, 11, 4513–4521, 1998.

3.51. Coolen, M.J.L., Cypionka, H., Sass, A.M., Sass, H., Overmann, J., Ongoing modification of Mediterranean Pleistocene sapropels mediated by prokaryotes. *Science*, 296, 5577, 2407–2410, 2002.

3.52. Cross, K.L., Campbell, J.H., Balachandran, M., Campbell, A.G., Cooper, S.J., Griffen, A., Heaton, M., Joshi, S., Klingeman, D., Leys, E. *et al.*, Targeted isolation and cultivation of uncultivated bacteria by reverse genomics. *Nat. Biotechnol.*, 37, 11, 1314–1321, 2019.

3.53. Culka, A., Košek, F., Oren, A., Mana, L., Jehlička, J., Detection of carotenoids of halophilic prokaryotes in solid inclusions inside laboratory-grown chloride and sulfate crystals using a portable Raman spectrometer: Applications for Mars exploration. *FEMS Microbiol. Lett.*, 366, 20, #fnz239, 2019.

3.54. Damsté, J.S.S. and Schouten, S., Biological markers for anoxia in the photic zone of the water column, in: *Marine Organic Matter: Biomarkers, Isotopes and DNA. The Handbook of Environmental Chemistry*, vol. 2N, J.K. Volkman (Ed.), pp. 127–163, Springer, Berlin, Germany, 2006.

3.55. Damsté, J.S.S., Kenig, F., Koopmans, M.P., Köster, J., Schouten, S., Hayes, J.M., de Leeuw, J.W., Evidence for gammacerane as an indicator of water column stratification. *Geochim. Cosmochim. Acta*, 59, 9, 1895–1900, 1995.

3.56. DasSarma, S., DasSarma, P., Laye, V.J., Schwieterman, E.W., Extremophilic models for astrobiology: Haloarchaeal survival strategies and pigments for remote sensing. *Extremophiles*, 24, 1, 31–41, 2020.

3.57. Deckker, P.D. and Last, W.M., Modern dolomite deposition in continental, saline lakes, western Victoria, Australia. *Geology*, 16, 1, 29–32, 1988.
3.58. de Leeuw, J.W. and Damsté, J.S.S., Organic sulfur compounds and other biomarkers as indicators of palaeosalinity, in: *Geochemistry of Sulfur in Fossil Fuels*, W.L. Orr and C.M. White (Eds.), ACS Symposium Series Vol. 429, pp. 417–443, American Chemical Society, Washington D.C, USA, 1990.
3.59. de Leeuw, J.W. and Largeau, C., A review of macromolecular organic compounds that comprise living organisms and their role in kerogen, coal, and petroleum formation, in: *Organic Geochemistry. Topics in Geobiology*, vol. 11, M.H. Engel and S.A. Macko (Eds.), pp. 23–72, Springer, Boston, MA, USA, 1993.
3.60. de Leeuw, J.W., Van Bergen, P.F., Van Aarssen, B.G., Gatellier, J.P.L., Damsté, J.S.S., Collinson, M.E., Resistant biomacromolecules as major contributors to kerogen. *Philos. Trans. R. Soc. Lond. Ser. B: Biol. Sci.*, 333, 1268, 329–337, 1991.
3.61. Deng, S., Dong, H., Lv, G., Jiang, H., Yu, B., Bishop, M.E., Microbial dolomite precipitation using sulfate reducing and halophilic bacteria: Results from Qinghai Lake, Tibetan Plateau, NW China. *Chem. Geol.*, 278, 3-4, 151–159, 2010.
3.62. Detkova, E.N. and Pusheva, M.A., Energy metabolism in halophilic and alkaliphilic acetogenic bacteria. *Microbiology (Moscow)*, 75, 1, 1–11, 2006.
3.63. Dimitriu, P.A., Pinkart, H.C., Peyton, B.M., Mormile, M.R., Spatial and temporal patterns in the microbial diversity of a meromictic soda lake in Washington State. *Appl. Environ. Microbiol.*, 74, 15, 4877–4888, 2008.
3.64. Dombrowski, N., Lee, J.H., Williams, T.A., Offre, P., Spang, A., Genomic diversity, lifestyles and evolutionary origins of DPANN archaea. *FEMS Microbiol. Lett.*, 366, 2, #fnz008, 2019.
3.65. Dong, H., Zhang, G., Jiang, H., Yu, B., Chapman, L.R., Lucas, C.R., Fields, M.W., Microbial diversity in sediments of saline Qinghai Lake, China: Linking geochemical controls to microbial ecology. *Microb. Ecol.*, 51, 1, 65–82, 2006.
3.66. Douglas, S., Mineralogical footprints of microbial life. *Am. J. Sci.*, 305, 6-8, 503–525, 2005.
3.67. Durand, B., Sedimentary organic matter and kerogen. Definition and quantitative importance of kerogen, in: *Kerogen: Insoluble Organic Matter from Sedimentary Rocks*, B. Durand (Ed.), pp. 13–34, Éditions Technip, Paris, France, 1980.
3.68. Durbin, A.M. and Teske, A., Archaea in organic-lean and organic-rich marine subsurface sediments: An environmental gradient reflected in distinct phylogenetic lineages. *Front. Microbiol.*, 3, #168, 2012.
3.69. Edgar, R.C., Search and clustering orders of magnitude faster than BLAST. *Bioinformatics*, 26, 19, 2460–2461, 2010.
3.70. Edwards, H.G.M., Herschy, B., Page, K., Munshi, T., Scowen, I.J., Raman spectra of biomarkers of relevance to analytical astrobiological exploration: Hopanoids, sterols and steranes. *Spectrochim. Acta Part A: Mol. Biomol. Spectrosc.*, 78, 1, 191–195, 2011.
3.71. Emeis, K.C., Black Shales and Sapropels, in: *Encyclopedia of Geochemistry: A Comprehensive Reference Source on the Chemistry of the Earth, Encyclopedia of Earth Sciences Series*, W.M. White (Ed.), pp. 195–196, Springer International Publishing, Cham, Switzerland, 2017.
3.72. Emeis, K.C. and Weissert, H., Tethyan–Mediterranean organic carbon-rich sediments from Mesozoic black shales to sapropels. *Sedimentology*, 56, 1, 247–266, 2009.
3.73. Emerson, S. and Hedges, J.I., Processes controlling the organic carbon content of open ocean sediments. *Paleoceanography*, 3, 5, 621–634, 1988.
3.74. Emmerich, M., Bhansali, A., Lösekann-Behrens, T., Schröder, C., Kappler, A., Behrens, S., Abundance, distribution, and activity of Fe (II)-oxidizing and Fe (III)-reducing microorganisms in hypersaline sediments of Lake Kasin, southern Russia. *Appl. Environ. Microbiol.*, 78, 12, 4386–4399, 2012.

3.75. Evans, M.S., Paleolimnological studies of saline lakes. *J. Paleolimnol.*, 8, 2, 97–101, 1993.
3.76. Evans, P.N., Parks, D.H., Chadwick, G.L., Robbins, S.J., Orphan, V.J., Golding, S.D., Tyson, G.W., Methane metabolism in the archaeal phylum Bathyarchaeota revealed by genome-centric metagenomics. *Science*, 350, 6259, 434–438, 2015.
3.77. Farag, I.F., Youssef, N.H., Elshahed, M.S., Global distribution patterns and pangenomic diversity of the candidate phylum "Latescibacteria" (WS3). *Appl. Environ. Microbiol.*, 83, 10, #e00521-17, 2017.
3.78. Fillol, M., Sànchez-Melsió, A., Gich, F., Borrego, C.M., Diversity of Miscellaneous Crenarchaeotic Group archaea in freshwater karstic lakes and their segregation between planktonic and sediment habitats. *FEMS Microbiol. Ecol.*, 91, 4, #fiv020, 2015.
3.79. Foster, I.S., King, P.L., Hyde, B.C., Southam, G., Characterization of halophiles in natural $MgSO_4$ salts and laboratory enrichment samples: Astrobiological implications for Mars. *Planet. Space Sci.*, 58, 4, 599–615, 2010.
3.80. Foti, M., Sorokin, D.Y., Lomans, B., Mussman, M., Zacharova, E.E., Pimenov, N.V., Kuenen, J.G., Muyzer, G., Diversity, activity, and abundance of sulfate-reducing bacteria in saline and hypersaline soda lakes. *Appl. Environ. Microbiol.*, 73, 7, 2093–2100, 2007.
3.81. Foti, M.J., Sorokin, D.Y., Zacharova, E.E., Pimenov, N.V., Kuenen, J.G., Muyzer, G., Bacterial diversity and activity along a salinity gradient in soda lakes of the Kulunda Steppe (Altai, Russia). *Extremophiles*, 12, 1, 133–145, 2008.
3.82. Garcés, B.L.V., Kelts, K., Ito, E., Oxygen and carbon isotope trends and sedimentological evolution of a meromictic and saline lacustrine system: The Holocene Medicine Lake basin, North American Great Plains, USA. *Palaeogeogr. Palaeoclimatol. Palaeoecol.*, 117, 3-4, 253–278, 1995.
3.83. Giblin, A.E., Tobias, C.R., Song, B., Weston, N., Banta, G.T., The importance of dissimilatory nitrate reduction to ammonium (DNRA) in the nitrogen cycle of coastal ecosystems. *Oceanography*, 26, 3, 124–131, 2013.
3.84. Ghai, R., Pašić, L., Fernández, A.B., Martin-Cuadrado, A.B., Mizuno, C.M., McMahon, K.D., Papke, R.T., Stepanauskas, R., Rodriguez-Brito, B., Rohwer, F., Sánchez-Porro, C., New abundant microbial groups in aquatic hypersaline environments. *Sci. Rep.*, I, #135, 2011.
3.85. Gifford, S.M., Sharma, S., Booth, M., Moran, M.A., Expression patterns reveal niche diversification in a marine microbial assemblage. *ISME J.*, 7, 2, 281–298, 2013.
3.86. Gomes, C., Carretero, M.I., Pozo, M., Maraver, F., Cantista, P., Armijo, F., Legido, J.L., Teixeira, F., Rautureau, M., Delgado, R., Peloids and pelotherapy: Historical evolution, classification and glossary. *Appl. Clay Sci.*, 75, 28–38, 2013.
3.87. Grant, W.D., Life at low water activity. *Philos. Trans. R. Soc. Lond. Ser. B: Biol. Sci.*, 359, 1448, 1249–1267, 2004.
3.88. Grossart, H.P., Massana, R., McMahon, K.D., Walsh, D.A., Linking metagenomics to aquatic microbial ecology and biogeochemical cycles. *Limnol. Oceanogr.*, 65, S1, S1–S20, 2020.
3.89. Großkopf, T. and Soyer, O.S., Microbial diversity arising from thermodynamic constraints. *ISME J.*, 10, 11, 2725–2733, 2016.
3.90. Guan, T.W., Lin, Y.J., Ou, M.Y., Chen, K.B., Isolation and diversity of sediment bacteria in the hypersaline Aiding Lake, China. *BioRxiv*, #638304, 2019.
3.91. Gutierrez, M.F., Tavşanoğlu, Ü.N., Vidal, N., Yu, J., Teixeira-de Mello, F., Çakiroglu, A.I., He, H., Liu, Z., Jeppesen, E., Salinity shapes zooplankton communities and functional diversity and has complex effects on size structure in lakes. *Hydrobiologia*, 813, 1, 237–255, 2018.
3.92. Hagens, M., Slomp, C., Meysman, F., Seitaj, D., Harlay, J., Borges, A., Middelburg, J., Biogeochemical processes and buffering capacity concurrently affect acidification in a seasonally hypoxic coastal marine basin. *Biogeosciences*, 12, 1561–1583, 2015.

3.93. Hall, K.J. and Northcote, T.G., Production and decomposition processes in a saline meromictic lake, in: *Saline Lakes. Developments in Hydrobiology*, vol. 59, F.A. Comín and T.G. Northcote (Eds.), pp. 115–128, Springer, Dordrecht, The Netherlands, 1990.
3.94. Hall, K.J. and Northcote, T.G., Long-term (millennial, centurial) changes in the limnology of a saline meromictic lake: Causes and consequences. *Int. Ver. Theor. Angew. Limnol.: Verh.*, 27, 5, 2628–2632, 2000.
3.95. Hallsworth, C.R. and Knox, R.W.O'B., *BGS Rock Classification Scheme.* Volume 3: Classification of Sediments and Sedimentary Rocks. British Geological Survey Research Report RR 99-03, p. 44 pp., Keyworth, Nottingham, UK, 1999.
3.96. Hamilton, T.L., Bovee, R.J., Sattin, S.R., Mohr, W., Gilhooly, W.P., III, Lyons, T.W., Pearson, A., Macalady, J.L., Carbon and sulfur cycling below the chemocline in a meromictic lake and the identification of a novel taxonomic lineage in the FCB superphylum, *Candidatus* Aegiribacteria. *Front. Microbiol.*, 7, #598, 2016.
3.97. Hammer, U.T., Primary production in saline lakes: A review, in: *Salt Lakes: Proceedings of the International Symposium on Athalassic (Inland) Salt Lakes, held at Adelaide, Australia, October 1979*, W.D. Williams (Ed.), *Developments in Hydrobiology*, vol 5, pp. 47–57, Dr. W. Junk Publishers, The Hague, The Netherlands, 1981.
3.98. Hammer, U.T., *Saline Lake Ecosystems of the World (Vol. 59)*, Dr. W. Junk Publishers, Dordrecht, The Netherlands, 1986.
3.99. Hatcher, P.G., Simoneit, B.R., Mackenzie, F.T., Neumann, A.C., Thorstenson, D.C., Gerchakov, S.M., Organic geochemistry and pore water chemistry of sediments from Mangrove Lake, Bermuda. *Org. Geochem.*, 4, 2, 93–112, 1982.
3.100. Hatcher, P.G., Spiker, E.C., Szeverenyi, N.M., Maciel, G.E., Selective preservation and origin of petroleum-forming aquatic kerogen. *Nature*, 305, 5934, 498–501, 1983.
3.101. Hays, L.E., Graham, H.V., Des Marais, D.J., Hausrath, E.M., Horgan, B., McCollom, T.M., Parenteau, M.N., Potter-McIntyre, S.L., Williams, A.J., Lynch, K.L., Biosignature preservation and detection in Mars analog environments. *Astrobiology*, 17, 4, 363–400, 2017.
3.102. Hedges, J.I. and Keil, R.G., Sedimentary organic matter preservation: An assessment and speculative synthesis. *Mar. Chem.*, 49, 2-3, 81–115, 1995.
3.103. Heinzelmann, S.M., Villanueva, L., Lipsewers, Y.A., Sinke-Schoen, D., Damsté, J.S.S., Schouten, S., van der Meer, M.T.J., Assessing the metabolism of sedimentary microbial communities using the hydrogen isotopic composition of fatty acids. *Org. Geochem.*, 124, 123–132, 2018.
3.104. Henneke, E., Luther, G.W., III, De Lange, G.J., Hoefs, J., Sulphur speciation in anoxic hypersaline sediments from the eastern Mediterranean Sea. *Geochim. Cosmochim. Acta*, 61, 2, 307–321, 1997.
3.105. Hierholtzer, A. and Akunna, J.C., Modelling sodium inhibition on the anaerobic digestion process. *Water Sci. Technol.*, 66, 7, 1565–1573, 2012.
3.106. Hirschler-Rea, A., Matheron, R., Riffaud, C., Mouné, S., Eatock, C., Herbert, R.A., Willison, J.C., Caumette, P., Isolation and characterization of spirilloid purple phototrophic bacteria forming red layers in microbial mats of Mediterranean salterns: Description of *Halorhodospira neutriphila* sp. nov. and emendation of the genus *Halorhodospira. Int. J. Syst. Evol. Microbiol.*, 53, 1, 153–163, 2003.
3.107. Hoehler, T.M., Alperin, M.J., Albert, D.B., Martens, C.S., Apparent minimum free energy requirements for methanogenic Archaea and sulfate-reducing bacteria in an anoxic marine sediment. *FEMS Microbiol. Ecol.*, 38, 1, 33–41, 2001.
3.108. Hollander, D.J., McKenzie, J.A., Hsu, K.J., Huc, A.Y., Application of an eutrophic lake model to the origin of ancient organic-carbon-rich sediments. *Global Biogeochem. Cycles*, 7, 1, 157–179, 1993.

3.109. Howarth, R.W., Marino, R., Cole, J.J., Nitrogen fixation in freshwater, estuarine, and marine ecosystems. 2. Biogeochemical controls. *Limnol. Oceanogr.*, 33, 4 part 2, 688–670, 1988.
3.110. Hubert, H.L., Rankey, E.C., Omelon, C., Organic matter, textures, and pore attributes of hypersaline lacustrine microbial deposits (Holocene, Bahamas). *J. Sediment. Res.*, 88, 7, 827–849, 2018.
3.111. Hug, L.A., Baker, B.J., Anantharaman, K., Brown, C.T., Probst, A.J., Castelle, C.J., Butterfield, C.N., Hernsdorf, A.W., Amano, Y., Ise, K., Suzuki, Y., A new view of the tree of life. *Nat. Microbiol.*, 1, 5, #16048, 2016.
3.112. Humayoun, S.B., Bano, N., Hollibaugh, J.T., Depth distribution of microbial diversity in Mono Lake, a meromictic soda lake in California. *Appl. Environ. Microbiol.*, 69, 2, 1030–1042, 2003.
3.113. Imachi, H., Nobu, M.K., Nakahara, N., Morono, Y., Ogawara, M., Takaki, Y., Takano, Y., Uematsu, K., Ikuta, T., Ito, M., Matsui, Y., Isolation of an archaeon at the prokaryote–eukaryote interface. *Nature*, 577, 519–525, 2020.
3.114. Jellison, R. and Melack, J.M., Algal photosynthetic activity and its response to meromixis in hypersaline Mono Lake, California. *Limnol. Oceanogr.*, 38, 4, 818–837, 1993.
3.115. Jellison, R., Miller, L.G., Melack, J.M., Dana, G.L., Meromixis in hypersaline Mono Lake, California. 2. Nitrogen fluxes. *Limnol. Oceanogr.*, 38, 5, 1020–1039, 1993.
3.116. Jellison, R., Williams, W.D., Timms, B., Aladin, N.V., Salt lakes: Values, threats and future, in: *Aquatic Ecosystems: Trends and Global Prospects*, N.V.C. Polunin (Ed.), pp. 94–110, Cambridge University Press, Cambridge, UK, 2008.
3.117. Jones, D.L. and Baxter, B.K., DNA repair and photoprotection: Mechanisms of overcoming environmental ultraviolet radiation exposure in halophilic archaea. *Front. Microbiol.*, 8, #1882, 2017.
3.118. Jones, B.F. and Deocampo, D.M., Geochemistry of saline lakes, in: *Treatise on Geochemistry, Volume 5: Surface and Groundwater, Weathering, and Soils*, 1st Ed., J.I. Drever (Ed.), pp. 393–424, Elsevier, Amsterdam, The Netherlands, 2005.
3.119. Jones, B.E., Grant, W.D., Duckworth, A.W., Owenson, G.G., Microbial diversity of soda lakes. *Extremophiles*, 2, 3, 191–200, 1998.
3.120. Jørgensen, B.B., Mineralization of organic matter in the sea bed—the role of sulphate reduction. *Nature*, 296, 643–645, 1982.
3.121. Jørgensen, B.B. and Cohen, Y., Solar Lake (Sinai). 5. The sulfur cycle of the benthic cyanobacterial mats 1. *Limnol. Oceanogr.*, 22, 4, 657–666, 1977.
3.122. Ibarra-Sánchez, C.L., Prince, L., Aguirre-Noyola, J.L., Sánchez-Cerda, K.E., Navaro-Noya, Y.E., Luna-Guido, M., Conde-Barajas, E., Dendooven, L., Gomez-Acata, E.S., The microbial community in an alkaline saline sediment of a former maar lake bed. *J. Soils Sediments*, 20, 542–555, 2020.
3.123. Katz, B.J., Controlling factors on source rock development—a review of productivity, preservation and sedimentation rate, in: *The Deposition of Organic-Carbon-Rich Sediments: Models, Mechanisms, and Consequences*, SEPM Special Publication 82, N.B. Harris (Ed.), pp. 7–16, 2005.
3.124. Kelderman, P., Nieuwenhuize, J., Meerman-van de Repe, A.M., Van Liere, J.M., Changes of sediment distribution patterns in Lake Grevelingen, an enclosed estuary in the SW Netherlands. *Neth. J. Sea Res.*, 18, 3-4, 273–285, 1984.
3.125. Kelts, K., Environments of deposition of lacustrine petroleum source rocks: An introduction. *Geol. Soc. London Spec. Publ.*, 40, 1, 3–26, 1988.
3.126. Kevbrin, V., Boltyanskaya, Y., Zhilina, T., Kolganova, T., Lavrentjeva, E., Kuznetsov, B., *Proteinivorax tanatarense* gen. nov., sp. nov., an anaerobic, haloalkaliphilic, proteolytic bacterium isolated from a decaying algal bloom, and proposal of *Proteinivoraceae* fam. nov. *Extremophiles*, 17, 5, 747–756, 2013.

3.127. Kidd, R.B., *Stratigraphy of eastern Mediterranean sapropel sequences recovered during Leg 42A and their paleoenvironmental significance. [Initial Report of Deep Sea Drilling Project 42*, pp. 421–443, 1978.
3.128. Kjeldsen, K.U., Jakobsen, T.F., Glastrup, J., Ingvorsen, K., *Desulfosalsimonas propionicica* gen. nov., sp. nov., a halophilic, sulfate-reducing member of the family *Desulfobacteraceae* isolated from a salt-lake sediment. *Int. J. Syst. Evol. Microbiol.*, 60, 5, 1060–1065, 2010.
3.129. Klepac-Ceraj, V., Hayes, C.A., Gilhooly, W.P., Lyons, T.W., Kolter, R., Pearson, A., Microbial diversity under extreme euxinia: Mahoney Lake, Canada. *Geobiology*, 10, 3, 223–235, 2012.
3.130. Knicker, H. and Hatcher, P.G., Sequestration of organic nitrogen in the sapropel from Mangrove Lake, Bermuda. *Org. Geochem.*, 32, 5, 733–744, 2001.
3.131. Kolpakova, M.N., Gaskova, O.L., Naymushina, O.S., Karpov, A.V., Vladimirdnraov, A.G., Krivonogov, S.K., Saline lakes of Northern Kazakhstan: Geochemical correlations of elements and controls on their accumulation in water and bottom sediments. *Appl. Geochem.*, 107, 8–18, 2019.
3.132. Koop-Jakobsen, K. and Giblin, A.E., The effect of increased nitrate loading on nitrate reduction *via* denitrification and DNRA in salt marsh sediments. *Limnol. Oceanogr.*, 55, 2, 789–802, 2010.
3.133. Kristensen, E., Ahmed, S.I., Devol, A.H., Aerobic and anaerobic decomposition of organic matter in marine sediment: Which is fastest? *Limnol. Oceanogr.*, 40, 8, 1430–1437, 1995.
3.134. Kubo, K., Lloyd, K.G., Biddle, J.F., Amann, R., Teske, A., Knittel, K., Archaea of the Miscellaneous Crenarchaeotal Group are abundant, diverse and widespread in marine sediments. *ISME J.*, 6, 10, 1949–1965, 2012.
3.135. Kulp, T.R., Hoeft, S.E., Miller, L.G., Saltikov, C., Murphy, J.N., Han, S., Lanoil, B., Oremland, R.S., Dissimilatory arsenate and sulfate reduction in sediments of two hypersaline, arsenic-rich soda lakes: Mono and Searles Lakes, California. *Appl. Environ. Microbiol.*, 72, 10, 6514–6526, 2006.
3.136. Laanbroek, H.J. and Veldkamp, H., Microbial interactions in sediment communities. *Philos. Trans. R. Soc. Lond. B, Biol. Sci.*, 297, 1088, 533–550, 1982.
3.137. Lang, K., Schuldes, J., Klingl, A., Poehlein, A., Daniel, R., Brune, A., New mode of energy metabolism in the seventh order of methanogens as revealed by comparative genome analysis of "Candidatus Methanoplasma termitum". *Appl. Environ. Microbiol.*, 81, 4, 1338–1352, 2015.
3.138. Larsen, H., Ecology of hypersaline environments, in: *Hypersaline Brines and Evaporitic Environments. Developments in Sedimentology*, vol. 28, A. Nissenbaum (Ed.), pp. 23–39, Elsevier, Amsterdam, The Netherlands, 1980.
3.139. Last, W.M., Sedimentology of a saline playa in the northern Great Plains, Canada. *Sedimentology*, 36, 1, 109–123, 1989.
3.140. Laybourn-Parry, J. and Bell, E.M., Ace Lake: Three decades of research on a meromictic, Antarctic lake. *Polar Biol.*, 37, 11, 1685–1699, 2014.
3.141. Lebre, P.H., De Maayer, P., Cowan, D.A., Xerotolerant bacteria: Surviving through a dry spell. *Nat. Rev. Microbiol.*, 15, 5, 285–296, 2017.
3.142. Lee, C., Love, G.D., Jahnke, L.L., Kubo, M.D., Des Marais, D.J., Early diagenetic sequestration of microbial mat lipid biomarkers through covalent binding into insoluble macromolecular organic matter (IMOM) as revealed by sequential chemolysis and catalytic hydropyrolysis. *Org. Geochem.*, 132, 11–22, 2019.
3.143. Lee, C.J., McMullan, P.E., O'Kane, C.J., Stevenson, A., Santos, I.C., Roy, C., Ghosh, W., Mancinelli, R.L., Mormile, M.R., McMullan, *et al.*, NaCl-saturated brines are thermodynamically moderate, rather than extreme, microbial habitats. *FEMS Microbiol. Rev.*, 42, 5, 672–693, 2018.

3.144. Lehours, A.C., Evans, P., Bardot, C., Joblin, K., Gérard, F., Phylogenetic diversity of archaea and bacteria in the anoxic zone of a meromictic lake (Lake Pavin, France). *Appl. Environ. Microbiol.*, 73, 6, 2016–2019, 2007.
3.145. León-Zayas, R., Peoples, L., Biddle, J.F., Podell, S., Novotny, M., Cameron, J., Lasken, R.S., Bartlett, D.H., The metabolic potential of the single cell genomes obtained from the Challenger Deep, Mariana Trench within the candidate superphylum Parcubacteria (OD 1). *Environ. Microbiol.*, 19, 7, 2769–2784, 2017.
3.146. Leonova, G.A., Maltsev, A.E., Melenevsky, V.N., Krivonogov, S.K., Kondratyeva, L.M., Bobrov, V.A., Suslova, M.Y., Diagenetic transformation of organic matter in sapropel sediments of small lakes (southern West Siberia and eastern Transbaikalia). *Quat. Int.*, 524, 40–47, 2019.
3.147. Lever, M.A., Acetogenesis in the energy-starved deep biosphere–a paradox? *Front. Microbiol.*, 2, #284, 2012.
3.148. Li, J., Pancost, R.D., Naafs, B.D.A., Yang, H., Zhao, C., Xie, S., Distribution of glycerol dialkyl glycerol tetraether (GDGT) lipids in a hypersaline lake system. *Org. Geochem.*, 99, 113–124, 2016.
3.149. Liaw, H.J. and Mah, R.A., Isolation and characterization of *Haloanaerobacter chitinovorans* gen. nov., sp. nov., a halophilic, anaerobic, chitinolytic bacterium from a solar saltern. *Appl. Environ. Microbiol.*, 58, 1, 260–266, 1992.
3.150. Lipsewers, Y.A., Hopmans, E.C., Meysman, F.J., Damsté, J.S.S., Villanueva, L., Abundance and diversity of denitrifying and anammox bacteria in seasonally hypoxic and sulfidic sediments of the saline lake Grevelingen. *Front. Microbiol.*, 7, #1661, 2016.
3.151. Littke, R. and Zieger, L., Formation of organic-rich sediments and sedimentary rocks, in: *Hydrocarbons, Oils and Lipids: Diversity, Origin, Chemistry and Fate*, H. Wilkes (Ed.), pp. 1–18, Springer International Publishing, Cham, Switzerland, 2019.
3.152. Liu, Y. and Boone, D.R., Effects of salinity on methanogenic decomposition. *Bioresour. Technol.*, 35, 3, 271–273, 1991.
3.153. Liu, Y., Priscu, J.C., Xiong, J., Conrad, R., Vick-Majors, T., Chu, H., Hou, J., Salinity drives archaeal distribution patterns in high altitude lake sediments on the Tibetan Plateau. *FEMS Microbiol. Ecol.*, 92, 3, #fiw033, 2016.
3.154. Lloyd, K.G., Schreiber, L., Petersen, D.G., Kjeldsen, K.U., Lever, M.A., Steen, A.D., Stepanauskas, R., Richter, M., Kleindienst, S., Lenk, S., Schramm, A., Predominant archaea in marine sediments degrade detrital proteins. *Nature*, 496, 7444, 215–218, 2013.
3.155. López-López, A., Yarza, P., Richter, M., Suárez-Suárez, A., Antón, J., Niemann, H., Rosselló-Móra, R., Extremely halophilic microbial communities in anaerobic sediments from a solar saltern. *Environ. Microbiol. Rep.*, 2, 2, 258–271, 2010.
3.156. Lowe, D.J., Green, J.D., Northcote, T.G., Hall, K.J., Holocene fluctuations of a meromictic lake in southern British Columbia. *Quat. Res.*, 48, 1, 100–113, 1997.
3.157. MacLeod, F., Kindler, G.S., Wong, H.L., Chen, R., Burns, B.P., Asgard archaea: Diversity, function, and evolutionary implications in a range of microbiomes. *AIMS Microbiol.*, 5, 1, 48–61, 2019.
3.158. Mardanov, A.V., Kadnikov, V.V., Ravin, N.V., Metagenomics: A paradigm shift in microbiology, in: *Metagenomics: Perspectives, Methods, and Applications*, M. Nagajaran (Ed.), pp. 1–13, Academic Press, London, UK, 2018.
3.159. Mather, C.C., Skrzypek, G., Dogramaci, S., Grierson, P.F., Paleoenvironmental and paleohydrochemical conditions of dolomite formation within a saline wetland in arid northwest Australia. *Quat. Sci. Rev.*, 185, 172–188, 2018.
3.160. McGenity, T.J. and Sorokin, D.Y., Methanogens and methanogenesis in hypersaline environments, in: *Biogenesis of Hydrocarbons*, A.J.M. Stams and D.Z. Sousa (Eds.), pp. 283–309, Springer, Cham, Switzerland, 2019.

3.161. McGlue, M.M., Ellis, G.S., Cohen, A.S., Modern muds of Laguna Mar Chiquita (Argentina): Particle size and organic matter geochemical trends from a large saline lake in the thick-skinned Andean foreland. *Geol. Soc. Am. Spec. Pap.*, 515, 1–18, 2015.
3.162. Mello, M.R., Koutsoukos, E.A., Neto, E.V.S., Telles, A.C.S., Jr., Geochemical and micro-paleontological characterization of lacustrine and marine hypersaline environments from Brazilian sedimentary basins, in: *Source Rocks in a Sequence Stratigraphic Framework. AAPG Studies in Geology*, vol. 37, B.J. Katz and L.M. Pratt (Eds.), pp. 17–34, American Association of Petroleum Geologists, Tulsa, OK, USA, 1993.
3.163. MÃheust, R., Burstein, D., Castelle, C.J., Banfield, J.F., The distinction of CPR bacteria from other bacteria based on protein family content. *Nat. Commun.*, 10, 1, 1–12, 2019.
3.164. Meng, J., Xu, J., Qin, D., He, Y., Xiao, X., Wang, F., Genetic and functional properties of uncultivated MCG archaea assessed by metagenome and gene expression analyses. *ISME J.*, 8, 3, 650–659, 2014.
3.165. Mesbah, N.M., Abou-El-Ela, S.H., Wiegel, J., Novel and unexpected prokaryotic diversity in water and sediments of the alkaline, hypersaline lakes of the Wadi An Natrun, Egypt. *Microb. Ecol.*, 54, 4, 598–617, 2007.
3.166. Messager, M.L., Lehner, B., Grill, G., Nedeva, I., Schmitt, O., Estimating the volume and age of water stored in global lakes using a geo-statistical approach. *Nat. Commun.*, 7, #13603, 2016.
3.167. Müller, V. and Hess, V., The minimum biological energy quantum. *Front. Microbiol.*, 8, #2019, 2017.
3.168. Namsaraev, Z., Samylina, O., Sukhacheva, M., Borisenko, G., Sorokin, D.Y., Tourova, T., Effect of salinity on diazotrophic activity and microbial composition of phototrophic communities from Bitter-1 soda lake (Kulunda Steppe, Russia). *Extremophiles*, 22, 4, 651–663, 2018.
3.169. Narasingarao, P., Podell, S., Ugalde, J.A., Brochier-Armanet, C., Emerson, J.B., Brocks, J.J., Heidelberg, K.B., Banfield, J.F., Allen, E.E., *De novo* metagenomic assembly reveals abundant novel major lineage of *Archaea* in hypersaline microbial communities. *ISME J.*, 6, 1, 81–93, 2012.
3.170. Nawrocki, E.P., *Structural RNA homology search and alignment using covariance models*, Washington University School of Medicine, St Louis, MO, USA, 2009, http://eddylab.org/publications/Nawrocki09b/Nawrocki09b-phdthesis.pdf.
3.171. Nelson, W.C. and Stegen, J.C., The reduced genomes of Parcubacteria (OD1) contain signatures of a symbiotic lifestyle. *Front. Microbiol.*, 6, ##713, 2015.
3.172. Nienhuis, P.H. and De Bree, B.H.H., Carbon fixation and chlorophyll in bottom sediments of brackish Lake Grevelingen, The Netherlands. *Neth. J. Sea Res.*, 18, 3-4, 337–359, 1984.
3.173. Nijenhuis, I.A., *Geochemistry of Eastern Mediterranean Sedimentary Cycles. On the Origin of Miocene to Pleistocene Sapropels, Laminites and Diatomites*, vol. 167, Geologica Ultraiectina, Utrecht, The Netherlands, 1999.
3.174. Nissenbaum, A., Rullkötter, J., Yechieli, Y., Are the curative properties of 'black mud'from the Dead Sea due to the presence of bitumen (asphalt) or other types of organic matter? *Environ. Geochem. Health*, 24, 4, 327–335, 2002.
3.175. Nobu, M.K., Narihiro, T., Rinke, C., Kamagata, Y., Tringe, S.G., Woyke, T., Liu, W.T., Microbial dark matter ecogenomics reveals complex synergistic networks in a methanogenic bioreactor. *ISME J.*, 9, 8, 1710–1722, 2015.
3.176. Noike, T., Endo, G., Chang, J.E., Yaguchi, J.I., Matsumoto, J.I., Characteristics of carbohydrate degradation and the rate-limiting step in anaerobic digestion. *Biotechnol. Bioeng.*, 27, 10, 1482–1489, 1985.
3.177. Ollivier, B., Fardeau, M.L., Cayol, J.L., Magot, M., Patel, B.K., Prensier, G., Garcia, J.L., *Methanocalculus halotolerans* gen. nov., sp. nov., isolated from an oil-producing well. *Int. J. Syst. Evol. Microbiol.*, 48, 3, 821–828, 1998.

3.178. Ollivier, B., Hatchikian, C.E., Prensier, G., Guezennec, J., Garcia, J.L., *Desulfohalobium retbaense* gen. nov., sp. nov., a halophilic sulfate-reducing bacterium from sediments of a hypersaline lake in Senegal. *Int. J. Syst. Evol. Microbiol.*, 41, 1, 74–81, 1991.

3.179. Oren, A., Formation and breakdown of glycine betaine and trimethylamine in hypersaline environments. *Antonie van Leeuwenhoek*, 58, 4, 291–298, 1990.

3.180. Oren, A., Bioenergetic aspects of halophilism. *Microbiol. Mol. Biol. Rev.*, 63, 2, 334–348, 1999.

3.181. Oren, A., The bioenergetic basis for the decrease in metabolic diversity at increasing salt concentrations: Implications for the functioning of Salt Lake ecosystems, in: *Saline Lakes: Publications from the 7th International Conference on Salt Lakes, held in Death Valley National Park, California, U.S.A., September 1999. Developments in Hydrobiology*, vol. 162, J.M. Melack, R. Jellison, D.B. Herbst (Eds.), pp. 61–72, Springer, Dordrecht, The Netherlands, 2001.

3.182. Oren, A., Microbial life at high salt concentrations: Phylogenetic and metabolic diversity. *Saline Syst.*, 4, 1, #2, 2008.

3.183. Oren, A., There must be an acetogen somewhere. *Front. Microbiol.*, 3, #22, 2012.

3.184. Oren, A., Life at high salt concentrations, in: *The Prokaryotes: Prokaryotic Communities and Ecophysiology*, E. Rosenberg, E.F. DeLong, S. Lory, E. Stackebrandt, F. Thompson (Eds.), pp. 421–440, Springer, Berlin, Germany, 2013.

3.185. Oren, A., The order Halanaerobiales, and the families Halanaerobiaceae and Halobacteroidaceae, in: *The Prokaryotes: Firmicutes and Tenericutes*, E. Rosenberg, E.F. DeLong, S. Lory, E. Stackebrandt, F. Thompson (Eds.), pp. 153–177, Springer, Berlin, Germany, 2014.

3.186. Oren, A., The family Natranaerobiaceae, in: *The Prokaryotes: Firmicutes and Tenericutes*, E. Rosenberg, E.F. DeLong, S. Lory, E. Stackebrandt, F. Thompson (Eds.), pp. 261–266, Springer, Berlin, Germany, 2014.

3.187. Oremland, R.S. and Polcin, S., Methanogenesis and sulfate reduction: Competitive and non-competitive substrates in estuarine sediments. *Appl. Environ. Microbiol.*, 44, 6, 1270–1276, 1982.

3.188. Ourisson, G. and Albrecht, P., Hopanoids. 1. Geohopanoids: The most abundant natural products on earth? *Acc. Chem. Res.*, 25, 9, 398–402, 1992.

3.189. Overmann, J., Sandmann, G., Hall, K.J., Northcote, T.G., Fossil carotenoids and paleolimnology of meromictic Mahoney Lake, British Columbia, Canada. *Aquat. Sci.*, 55, 1, 31–39, 1993.

3.190. Paul, V.G. and Mormile, M.R., A case for the protection of saline and hypersaline environments: A microbiological perspective. *FEMS Microbiol. Ecol.*, 93, 8, #fix091, 2017.

3.191. Peters, K.E. and Moldowan, J.M., Effects of source, thermal maturity, and biodegradation on the distribution and isomerization of homohopanes in petroleum. *Org. Geochem.*, 17, 1, 47–61, 1991.

3.192. Phinney, H.K., A peculiar lake sediment of algal origin. *Am. Midl. Nat.*, 35, 2, 453–459, 1946.

3.193. Pontefract, A., Zhu, T.F., Walker, V.K., Hepburn, H., Lui, C., Zuber, M.T., Ruvkun, G., Carr, C.E., Microbial diversity in a hypersaline sulfate lake: A terrestrial analog of ancient Mars. *Front. Microbiol.*, 8, #1819, 2017.

3.194. Potonié, H., Über Faulschlamm-(Sapropel)-Gesteine. *Sber. Ges. Naturf. Freunde Berl.*, 10, 243–245, 1904.

3.195. Prat, L., Heinemann, I.U., Aerni, H.R., Rinehart, J., O'Donoghue, P., Söll, D., Carbon source-dependent expansion of the genetic code in bacteria. *Proc. Natl. Acad. Sci. USA.*, 109, 51, 21070–21075, 2012.

3.196. Pratt, L.M. and Davis, C.L., Intertwined fates of metals, sulfur, and organic carbon in black shales, in: *Geochemistry of Organic Matter in Sediments and Sedimentary Rocks*, L.M. Pratt, J.B. Comer, S.C. Brassell (Eds.), SEPM Short Course No. 27, pp. 1–27, SEPM (Society for Sedimentary Geology), Tulsa, OK, USA, 1992.

3.197. Preston, L.J. and Dartnell, L.R., Planetary habitability: Lessons learned from terrestrial analogues. *Int. J. Astrobiol.*, 13, 1, 81–98, 2014.
3.198. Pruesse, E., Quast, C., Knittel, K., Fuchs, B.M., Ludwig, W., Peplies, J., Glockner, F.O., SILVA: A comprehensive online resource for quality checked and aligned ribosomal RNA sequence data compatible with ARB. *Nucleic Acids Res.*, 35, 21, 7188–7196, 2007.
3.199. Ryzhmanova, Y., Oshurkova, V., Troshina, O., Abashina, T., Ariskina, E., Avtukh, A., Shcherbakova, V., *Anoxynatronum buryatiense* sp. nov., an anaerobic alkaliphilic bacterium from a low mineralization soda lake in Buryatia, Russia. *Int. J. Syst. Evol. Microbiol.*, 67, 11, 4704–4709, 2017.
3.200. Roberts, J.A., Kenward, P.A., Fowle, D.A., Goldstein, R.H., González, L.A., Moore, D.S., Surface chemistry allows for abiotic precipitation of dolomite at low temperature. *Proc. Natl. Acad. Sci.*, 110, 36, 14540–14545, 2013.
3.201. Rojas, P., Rodríguez, N., de la Fuente, V., Sánchez-Mata, D., Amils, R., Sanz, J.L., Microbial diversity associated with the anaerobic sediments of a soda lake (Mono Lake, California, USA). *Can. J. Microbiol.*, 64, 6, 385–392, 2018.
3.202. Rossignol-Strick, M., Nesteroff, W., Olive, P., Vergnaud-Grazzini, C., After the deluge: Mediterranean stagnation and sapropel formation. *Nature*, 295, 5845, 105–110, 1982.
3.203. Samylina, O.S., Namsaraev, Z.B., Grouzdev, D.S., Slobodova, N.V., Zelenev, V.V., Borisenko, G.V., Sorokin, D.Y., The patterns of nitrogen fixation in haloalkaliphilic phototrophic communities of Kulunda Steppe soda lakes (Altai, Russia). *FEMS Microbiol. Ecol.*, 95, 11, #fiz174, 2019.
3.204. Sancetta, C., Mediterranean sapropels: Seasonal stratification yields high production and carbon flux. *Paleoceanography*, 9, 2, 195–196, 1994.
3.205. Saros, J. and Fritz, S., Nutrients as a link between ionic concentration/composition and diatom distributions in saline lakes. *J. Paleolimnol.*, 23, 4, 449–453, 2003.
3.206. Schindler, D.W., Recent advances in the understanding and management of eutrophication. *Limnol. Oceanogr.*, 51, 1 part 2, 356–363, 2006.
3.207. Schink, B. and Stams, A.J., Syntrophism among prokaryotes, in: *The Prokaryotes. Volume 2: Ecophysiology and Biochemistry*, E. Rosenberg, E.F. DeLong, S. Lory, E. Stackebrandt, F. Thompson (Eds.), pp. 471–493, Springer, Berlin, Germany, 2013.
3.208. Schnurrenberger, D., Russell, J., Kelts, K., Classification of lacustrine sediments based on sedimentary components. *J. Paleolimnol.*, 29, 2, 141–154, 2003.
3.209. Shadrin, N.V., The alternative saline lake ecosystem states and adaptive environmental management. *J. Oceanol. Limnol.*, 36, 6, 2010–2017, 2018.
3.210. Shkolyar, S. and Farmer, J.D., Biosignature preservation potential in playa evaporites: Impacts of diagenesis and implications for Mars exploration. *Astrobiology*, 18, 11, 1460–1478, 2018.
3.211. Sikorski, J., Lapidus, A., Chertkov, O., Lucas, S., Copeland, A., Del Rio, T.G., Nolan, M., Tice, H., Cheng, J.F., Han, C. *et al.*, Complete genome sequence of *Acetohalobium arabaticum* type strain (Z-7288). *Stand. Genomic Sci.*, 3, 1, 57–65, 2010.
3.212. Sohlenius, G., Emeis, K.C., Andrén, E., Andrén, T., Kohly, A., Development of anoxia during the Holocene fresh-brackish water transition in the Baltic Sea. *Mar. Geol.*, 177, 3-4, 221–242, 2001.
3.213. Sorokin, D.Y., Abbas, B., Geleijnse, M., Kolganova, T.V., Kleerebezem, R., van Loosdrecht, M.C., Syntrophic associations from hypersaline soda lakes converting organic acids and alcohols to methane at extremely haloalkaline conditions. *Environ. Microbiol.*, 18, 9, 3189–3.202, 2016.
3.214. Sorokin, D.Y., Abbas, B., Geleijnse, M., Pimenov, N.V., Sukhacheva, M.V., van Loosdrecht, M., Methanogenesis at extremely haloalkaline conditions in the soda lakes of Kulunda Steppe (Altai, Russia). *FEMS Microbiol. Ecol.*, 91, 4, # fiv016, 2015.

3.215. Sorokin, D.Y., Abbas, B., Tourova, T.P., Bumazhkin, B.K., Kolganova, T.V., Muyzer, G., Sulfate-dependent acetate oxidation under extremely natron-alkaline conditions by syntrophic associations from hypersaline soda lakes. *Microbiology*, 160, 4, 723–73.2, 2014.
3.216. Sorokin, D.Y., Banciu, H.L., Muyzer, G., Functional microbiology of soda lakes. *Curr. Opin. Microbiol.*, 25, 88–96, 2015.
3.217. Sorokin, D.Y., Banciu, H., Robertson, L.A., Kuenen, J.G., Muntyan, M.S., Muyzer, G., Halophilic and haloalkaliphilic sulfur-oxidizing bacteria, in: *The Prokaryotes: Prokaryotic Physiology and Biochemistry*, E. Rosenberg, E.F. DeLong, S. Lory, E. Stackebrandt, F. Thompson (Eds.), pp. 529–554, Springer, Berlin, Germany, 2013.
3.218. Sorokin, D.Y., Berben, T., Melton, E.D., Overmars, L., Vavourakis, C., Muyzer, G., Microbial diversity and biogeochemical cycling in soda lakes. *Extremophiles*, 18, 5, 791–809, 2014.
3.129. Sorokin, D.Y. and Chernyh, N.A., '*Candidatus* Desulfonatronobulbus propionicus': A first haloalkaliphilic member of the order *Syntrophobacterales* from soda lakes. *Extremophiles*, 20, 6, 895–901, 2016.
3.220. Sorokin, D.Y. and Chernyh, N.A., *Desulfonatronospira sulfatiphila* sp. nov., and *Desulfitispora elongata* sp. nov., two novel haloalkaliphilic sulfidogenic bacteria from soda lakes. *Int. J. Syst. Evol. Microbiol.*, 67, 2, 396–401, 2017.
3.221. Sorokin, D.Y., Chernyh, N.A., Poroshina, M.N., *Desulfonatronobacter acetoxydans* sp. nov.: A first acetate-oxidizing, extremely salt-tolerant alkaliphilic SRB from a hypersaline soda lake. *Extremophiles*, 19, 5, 899–907, 2015.
3.222. Sorokin, D.Y., Foti, M., Pinkart, H.C., Muyzer, G., Sulfur-oxidizing bacteria in Soap Lake (Washington State), a meromictic, haloalkaline lake with an unprecedented high sulfide content. *Appl. Environ. Microbiol.*, 73, 2, 451–455, 2007.
3.223. Sorokin, D.Y., Kublanov, I.V., Yakimov, M.M., Rijpstra, W.I.C., Damsté, J.S.S., *Halanaeroarchaeum sulfurireducens* gen. nov., sp. nov., the first obligately anaerobic sulfur-respiring haloarchaeon, isolated from a hypersaline lake. *Int. J. Syst. Evol. Microbiol.*, 66, 6, 2377–2381, 2016.
3.224. Sorokin, D.Y., Kuenen, J.G., Muyzer, G., The microbial sulfur cycle at extremely haloalkaline conditions of soda lakes. *Front. Microbiol.*, 2, #44, 2011.
3.225. Sorokin, D.Y., Messina, E., La Cono, V., Ferrer, M., Ciordia, S., Mena, M.C., Toshchakov, S.V., Golyshin, P.N., Yakimov, M.M., Sulfur respiration in a group of facultatively anaerobic natronoarchaea ubiquitous in hypersaline soda lakes. *Front. Microbiol.*, 9, #2359, 2018.
3.226. Sorokin, D.Y., Makarova, K.S., Abbas, B., Ferrer, M., Golyshin, P.N., Galinski, E.A., Ciordia, S., Mena, M.C., Merkel, A.Y., Wolf, Y.I. *et al.*, Discovery of extremely halophilic, methyl-reducing euryarchaea provides insights into the evolutionary origin of methanogenesis. *Nat. Microbiol.*, 2, #17081, 2017.
3.227. Sorokin, D.Y., Panteleeva, A.N., Tourova, T.P., Kaparullina, E.N., Muyzer, G., *Natronoflexus pectinivorans* gen. nov. sp. nov., an obligately anaerobic and alkaliphilic fermentative member of Bacteroidetes from soda lakes. *Extremophiles*, 15, 6, 691–696, 2011.
3.228. Sorokin, D.Y., Rusanov, I.I., Pimenov, N.V., Tourova, T.P., Abbas, B., Muyzer, G., Sulfidogenesis under extremely haloalkaline conditions in soda lakes of Kulunda Steppe (Altai, Russia). *FEMS Microbiol. Ecol.*, 73, 2, 278–290, 2010.
3.229. Sorokin, D.Y., Tourova, T.P., Henstra, A.M., Stams, A.J., Galinski, E.A., Muyzer, G., Sulfidogenesis under extremely haloalkaline conditions by *Desulfonatronospira thiodismutans* gen. nov., sp. nov., and *Desulfonatronospira delicata* sp. nov.–a novel lineage of Deltaproteobacteria from hypersaline soda lakes. *Microbiology*, 154, 5, 1444–1453, 2008.
3.230. Sorokin, D.Y., Tourova, T.P., Kolganova, T.V., Detkova, E.N., Galinski, E.A., Muyzer, G., Culturable diversity of lithotrophic haloalkaliphilic sulfate-reducing bacteria in soda lakes and the description of *Desulfonatronum thioautotrophicum* sp. nov.,

Desulfonatronum thiosulfatophilum sp. nov., *Desulfonatronovibrio thiodismutans* sp. nov., and *Desulfonatronovibrio magnus* sp. nov. *Extremophiles*, 15, 3, 391–401, 2011.

3.231. Sorokin, D.Y., Tourova, T.P., Panteleeva, A.N., Kaparullina, E.N., Muyzer, G., Anaerobic utilization of pectinous substrates at extremely haloalkaline conditions by *Natranaerovirga pectinivora* gen. nov., sp. nov., and *Natranaerovirga hydrolytica* sp. nov., isolated from hypersaline soda lakes. *Extremophiles*, 16, 2, 307–315, 2012.

3.232. Sorokin, D.Y., Tourova, T.P., Sukhacheva, M.V., Mardanov, A.V., Ravin, N.V., Bacterial chitin utilisation at extremely haloalkaline conditions. *Extremophiles*, 16, 6, 883–894, 2012.

3.233. Squyres, S.W., Grotzinger, J.P., Arvidson, R.E., Bell, J.F., Calvin, W., Christensen, P.R., Clark, B.C., Crisp, J.A., Farrand, W.H., Herkenhoff, K.E., Johnson, J.R., *In situ* evidence for an ancient aqueous environment at Meridiani Planum, Mars. *Science*, 306, 5702, 1709–1714, 2004.

3.234. Srinivasan, G., James, C.M., Krzycki, J.A., Pyrrolysine encoded by UAG in Archaea: Charging of a UAG-decoding specialized tRNA. *Science*, 296, 5572, 1459–1462, 2002.

3.235. Süß, J., Engelen, B., Cypionka, H., Sass, H., Quantitative analysis of bacterial communities from Mediterranean sapropels based on cultivation-dependent methods. *FEMS Microbiol. Ecol.*, 51, 1, 109–121, 2004.

3.236. Stan-Lotter, H., Survival of subsurface microbial communities over geological times and the implications for astrobiology, in: *Model Ecosystems in Extreme Environments. Astrobiology Exploring Life on Earth and Beyond*, vol. 2, J. Seckbach and P. Rampelotto (Eds.), pp. 169–187, Academic Press, London, UK, 2019.

3.237. Steinsberger, T., Schmid, M., Wüest, A., Schwefel, R., Wehrli, B., Müller, B., Organic carbon mass accumulation rate regulates the flux of reduced substances from the sediments of deep lakes. *Biogeosciences*, 14, 13, 3275–3285, 2017.

3.238. Stevenson, A., Cray, J.A., Williams, J.P., Santos, R., Sahay, R., Neuenkirchen, N., McClure, C.D., Grant, I.R., Houghton, J.D., Quinn, J.P. *et al.*, Is there a common water-activity limit for the three domains of life? *ISME J.*, 9, 6, 1333–1351, 2015.

3.239. Swan, B.K., Ehrhardt, C.J., Reifel, K.M., Moreno, L.I., Valentine, D.L., Archaeal and bacterial communities respond differently to environmental gradients in anoxic sediments of a California hypersaline lake, the Salton Sea. *Appl. Environ. Microbiol.*, 76, 3, 757–768, 2010.

3.240. Suominen, S., Dombrowski, N., Damsté, J.S.S., Villanueva, L., A diverse uncultivated microbial community is responsible for organic matter degradation in the Black Sea sulfidic zone. *Environ. Microbiol.*, 2019. https://doi.org/10.1111/1462-2920.14902.

3.241. ten Haven, H.L., de Leeuw, J.W., Rullkötter, J., Damsté, J.S.S., Restricted utility of the pristane/phytane ratio as a palaeoenvironmental indicator. *Nature*, 330, 6149, 641–643, 1987.

3.242. Timmers, P.H., Vavourakis, C.D., Kleerebezem, R., Damsté, J.S.S., Muyzer, G., Stams, A.J., Sorokin, D.Y., Plugge, C.M., Metabolism and occurrence of methanogenic and sulfate-reducing syntrophic acetate oxidizing communities in haloalkaline environments. *Front. Microbiol.*, 9, #3039, 2018.

3.243. Thauer, R.K., Jungermann, K., Decker, K., Energy conservation in chemotrophic anaerobic bacteria. *Bacteriol. Rev.*, 41, 1, 100–180, 1977.

3.244. Thomas, C., Ionescu, D., Ariztegui, D., DSDDP Scientific Team, Archaeal populations in two distinct sedimentary facies of the subsurface of the Dead Sea. *Mar. Genomics*, 17, 53–62, 2014.

3.245. Toporski, J. and Steele, A., The relevance of bacterial biomarkers in astrobiological research, in: *Proceedings of the First European Workshop on Exo-Astrobiology, 16-19 September 2002, Graz, Austria*, ESA SP-518, pp. 239–242, Huguette Lacoste, Noordwijk, Netherlands, 2002.

3.246. Torgersen, T., De Deckker, P., Chivas, A.R., Bowler, J.M., Salt lakes: A discussion of processes influencing palaeoenvironmental interpretation and recommendations for future study. *Palaeogeogr. Palaeoclimatol. Palaeoecol.*, 54, 1-4, 7–19, 1986.

3.247. Tourova, T.P., Grechnikova, M.A., Kuznetsov, B.B., Sorokin, D.Y., Phylogenetic diversity of bacteria in soda lake stratified sediments. *Microbiology (Moscow)*, 83, 6, 869–879, 2014.
3.248. Twenhofel, V.E. and McKelvey, W.H., Sediments of fresh-water lakes. *AAPG Bull.*, 25, 5, 826–849, 1941.
3.249. van Gemerden, H. and Mas, J., Ecology of phototrophic sulfur bacteria, in: *Anoxygenic Photosynthetic Bacteria. Advances in Photosynthesis and Respiration*, vol. 2, R.E. Blankenship, M.T. Madigan, C.E. Bauer (Eds.), pp. 49–85, Springer, Dordrecht, The Netherlands, 1995.
3.250. Vandenbroucke, M. and Largeau, C., Kerogen origin, evolution and structure. *Org. Geochem.*, 38, 5, 719–833, 2007.
3.251. Vavourakis, C.D., Andrei, A.-S., Mehrshad, M., Ghai, R., Sorokin, D.Y., Muyzer, G., A metagenomics roadmap to the uncultured genome diversity in hypersaline soda lake sediments. *Microbiome*, 6, 1, #168, 2018.
3.252. Vavourakis, C.D., Ghai, R., Rodriguez-Valera, F., Sorokin, D.Y., Tringe, S.G., Hugenholtz, P., Muyzer, G., Metagenomic insights into the uncultured diversity and physiology of microbes in four hypersaline soda lake brines. *Front. Microbiol.*, 7, #211, 2016.
3.253. Volkman, J.K., Holdsworth, D.G., Neill, G.P., Bavor, H.J., Jr., Identification of natural, anthropogenic and petroleum hydrocarbons in aquatic sediments. *Sci. Total Environ.*, 112, 2-3, 203–219, 1992.
3.254. Visscher, P.T., Dupraz, C., Braissant, O., Gallagher, K.L., Glunk, C., Casillas, L., Reed, R.E., Biogeochemistry of carbon cycling in hypersaline mats: Linking the present to the past through biosignatures, in: *Microbial Mats - Modern and Ancient Microorganisms in Stratified Systems*, J. Seckbach and A. Oren (Eds.), pp. 443–468, Springer, Dordrecht, The Netherlands, 2010.
3.255. Vuillemin, A., Horn, F., Alawi, M., Henny, C., Wagner, D., Crowe, S.A., Kallmeyer, J., Preservation and significance of extracellular DNA in ferruginous sediments from Lake Towuti, Indonesia. *Front. Microbiol.*, 8, #1440, 2017.
3.256. Waldron, P.J., Petsch, S.T., Martini, A.M., Nüsslein, K., Salinity constraints on subsurface archaeal diversity and methanogenesis in sedimentary rock rich in organic matter. *Appl. Environ. Microbiol.*, 73, 13, 4171–4179, 2007.
3.257. Wasmund, E., Bitumen, Sapropel und Gyttja. *Geol. Fören. Förh.*, 52, 3, 315–350, 1930.
3.258. Watts, J.M., Swan, B.K., Tiffany, M.A., Hurlbert, S.H., Thermal, mixing, and oxygen regimes of the Salton Sea, California, 1997–1999, in: *Saline Lakes: Publications from the 7th International Conference on Salt Lakes, held in Death Valley National Park, California, U.S.A., September 1999*, J.M. Melack, R. Jellison, D.B. Herbst (Eds.), *Developments in Hydrobiology, vol 162*, pp. 159–176, Springer, Dordrecht, The Netherlands, 2001.
3.259. Weigold, P., Ruecker, A., Loesekann-Behrens, T., Kappler, A., Behrens, S., Ribosomal tag pyrosequencing of DNA and RNA reveals "rare" taxa with high protein synthesis potential in the sediment of a hypersaline lake in Western Australia. *Geomicrobiol. J.*, 33, 5, 426–440, 2016.
3.260. Westall, F., Foucher, F., Bost, N., Bertrand, M., Loizeau, D., Vago, J.L., Kminek, G., Gaboyer, F., Campbell, K.A., Bréhéret, J.G., Gautret, P., Biosignatures on Mars: What, where, and how? Implications for the search for martian life. *Astrobiology*, 15, 11, 998–1029, 2015.
3.261. Weston, N.B. and Joye, S.B., Temperature-driven decoupling of key phases of organic matter degradation in marine sediments. *Proc. Natl. Acad. Sci.*, 102, 47, 17036–17040, 2005.
3.262. Weston, N.B., Vile, M.A., Neubauer, S.C., Velinsky, D.J., Accelerated microbial organic matter mineralization following salt-water intrusion into tidal freshwater marsh soils. *Biogeochemistry*, 102, 1-3, 135–151, 2011.
3.263. Williams, W.D., Environmental threats to salt lakes and the likely status of inland saline ecosystems in 2025. *Environ. Conserv.*, 29, 2, 154–167, 2002.
3.264. Wong, H.L., Smith, D.L., Visscher, P.T., Burns, B.P., Niche differentiation of bacterial communities at a millimeter scale in Shark Bay microbial mats. *Sci. Rep.*, 5, #15607, 2015.

3.265. Wright, D.T. and Wacey, D., Precipitation of dolomite using sulphate-reducing bacteria from the Coorong Region, South Australia: Significance and implications. *Sedimentology*, 52, 5, 987–1008, 2005.
3.266. Wu, J., Guan, T., Jiang, H., Zhi, X., Tang, S., Dong, H., Zhang, L., Li, W., Diversity of actinobacterial community in saline sediments from Yunnan and Xinjiang, China. *Extremophiles*, 13, 4, 623–632, 2009.
3.267. Wurtsbaugh, W.A. and Marcarelli, A., *Hydrogen Sulfide in Farmington Bay and The Great Salt Lake: A Potential Odor-Causing Agent. [Report to the Central Davis Sewer Improvement District]*, pp. 1–30, 2004.
3.268. Wurtsbaugh, W.A., Miller, C., Null, S.E., DeRose, R.J., Wilcock, P., Hahnenberger, M., Howe, F., Moore, J., Decline of the world's saline lakes. *Nat. Geosci.*, 10, 11, 816–821, 2017.
3.269. Youssef, N.H., Farag, I.F., Rudy, S., Mulliner, A., Walker, K., Caldwell, F., Miller, M., Hoff, W., Elshahed, M., The Wood–Ljungdahl pathway as a key component of metabolic versatility in candidate phylum Bipolaricaulota (Acetothermia, OP1). *Environ. Microbiol. Rep.*, 11, 4, 538–547, 2019.
3.270. Zaitseva, S.V., Abidueva, E.Y., Radnagurueva, A.A., Bazarov, S.M., Buryukhaev, S.P., Structure of microbial communities of the sediments of alkaline Transbaikalia lakes with different salinity. *Microbiology (Moscow)*, 87, 4, 559–568, 2018.
3.271. Zajc, J., Zalar, P., Gunde-Cimerman, N., Yeasts in hypersaline habitats, in: *Yeasts in Natural Ecosystems: Diversity*, P. Buzzini, M.-A. Lachance, A. Yurkov (Eds.), pp. 293–329, Springer, Cham, Switzerland, 2017.
3.272. Zhilina, T.N., Kevbrin, V.V., Tourova, T.P., Lysenko, A.M., Kostrikina, N.A., Zavarzin, G.A., *Clostridium alkalicellum* sp. nov., an obligately alkaliphilic cellulolytic bacterium from a soda lake in the Baikal region. *Microbiology*, 74, 5, 557–566, 2005.
3.273. Zhilina, T.N. and Zavarzin, G.A., A new extremely halophilic homoacetogen bacteria *Acetohalobium arabaticum*, gen. nov. sp. nov. *Dokl. Akad. Nauk SSSR*, 311, 745–747, 1990.
3.274. Zhilina, T.N. and Zavarzin, G.A., Extremely halophilic, methylotrophic, anaerobic bacteria. *FEMS Microbiol. Rev.*, 7, 3-4, 315–321, 1990.
3.275. Zhilina, T.N., Zavarzin, G.A., Detkova, E.N., Rainey, F.A., *Natroniella acetigena* gen. nov. sp. nov., an extremely haloalkaliphilic, homoacetic bacterium: A new member of *Haloanaerobiales*. *Curr. Microbiol.*, 32, 6, 320–326, 1996.
3.276. Zhilina, T.N., Zavarzina, D.G., Panteleeva, A.N., Osipov, G.A., Kostrikina, N.A., Tourova, T.P., Zavarzin, G.A., *Fuchsiella alkaliacetigena* gen. nov., sp. nov., an alkaliphilic, lithoautotrophic homoacetogen from a soda lake. *Int. J. Syst. Evol. Microbiol.*, 62, 7, 1666–1673, 2012.
3.277. Zhuang, G.C., Elling, F.J., Nigro, L.M., Samarkin, V., Joye, S.B., Teske, A., Hinrichs, K.U., Multiple evidence for methylotrophic methanogenesis as the dominant methanogenic pathway in hypersaline sediments from the Orca Basin, Gulf of Mexico. *Geochim. Cosmochim. Acta*, 187, 1–20, 2016.
3.278. Zinger, L., Gobet, A., Pommier, T., Two decades of describing the unseen majority of aquatic microbial diversity. *Mol. Ecol.*, 21, 8, 1878–1896, 2012.

4

The Haloarchaea of Great Salt Lake as Models for Potential Extant Life on Mars

Madelyn Bayles†, Bradley C. Belasco†, Alexander J. Breda†, Calli B. Cahill†, Adrik Z. Da Silva†, Michael J. Regan Jr.†, Nicklaus K. Schlamp†, Mariah P. Slagle† and Bonnie K. Baxter*†

Division of Sciences and Great Salt Lake Institute, Westminster College, Salt Lake City, Utah, USA

Abstract

Expansive evaporite mineral deposits and other geological features on Mars are evidence of ancient lacustrine systems before the planet experienced global climatic change (~3.5 Gya). On Mars, as the surface water dried up, hypersaline lakes would have filled the ancient lake basins. On Earth, the Bonneville Basin, in the western United States, tells a similar story in a more recent timeframe. Today, the bottom of this basin is the modern Great Salt Lake (GSL) and the Bonneville Salt Flats. The formation of GSL, in the Pleistocene to Holocene transition, followed climate change affecting the large inland sea, Lake Bonneville. Evaporation of this freshwater lake left large evaporitic mineral deposits that continually supply salt to modern GSL. Parts of the lake are at salt saturation due to shrinking shorelines and human intervention, and it is here that haloarchaea thrive, including inside the mineral deposits. The elevation of GSL is on a downward trajectory, and salinity is rising, leaving behind evaporite minerals as the water recedes. Halite and gypsum may contain fluid inclusions where microorganisms may be entombed over geologic time, managing dormancy in the low water activity of saturated brine. Haloarchaea are also resistant to other extremes such as high radiation doses, and they have lifestyle and metabolic flexibility. All of these things taken together make them excellent analogues for life that could have been in hypersaline lakes on Mars and may remain preserved in the evaporitic minerals there. The current Martian ultraviolet flux, magnetosphere, lack of tectonic activity, and desiccation suggest that continued life would be challenging, except microorganisms such as GSL haloarchaea may resist these extreme conditions, especially if entombed in

**Corresponding author*: bbaxter@westminstercollege.edu
†These authors contributed equally
Madelyn Bayles: mmb0907@westminstercollege.edu
Bradley C. Belasco: bcb0801@westminstercollege.edu
Alexander J. Breda: ajb1209@westminstercollege.edu
Calli B. Cahill: cbc1027@westminstercollege.edu
Adrik Da Silva: azd0423@westminstercollege.edu
Michael J. Regan Jr: mjr0303@westminstercollege.edu
Nicklaus K. Schlamp: nks0704@westminstercollege.edu
Mariah Slagle: mps0712@westminstercollege.edu
Bonnie K. Baxter: greatsaltlakeinstitute.org, https://cs.westminstercollege.edu/~bonbax/

Joseph Seckbach and Helga Stan-Lotter (eds.) Extremophiles as Astrobiological Models, (83–124)

minerals. Exploration with orbiters and rovers has located evaporites on Mars, and future missions should focus on these sites for detection of potential extant life or signs of extinct life.

Keywords: Haloarchaea, halophilic archaea, halophiles, astrobiology, evaporites, Martian analogue, Great Salt Lake

4.1 The Great Salt Lake System in the Bonneville Basin

4.1.1 A Significant Terminal Lake in North America

Great Salt Lake (GSL) is one of the largest terminal lakes in the world at the bottom of an enormous watershed, the lake measuring approximately 122 km in length and 50 km in width [4.117] [4.246]. Situated in the state of Utah in the western United States, this closed basin system experiences cycles of drought and flooding [4.13] [4.179]. Therefore, the shorelines and dry lakebed vary in size as the lake elevation changes and meander lines move. This can be seen from space an image that clearly shows the contrast between the lake water and the reflective nature of the surrounding dry salt playa (Figure 4.1).

Freshwater input enters GSL via precipitation and three major rivers in the watershed: the Bear, Weber, and Jordan Rivers (Figure 4.2) [4.112], but water leaves this terminal lake only through evaporation, which leaves behind a geochemical record in mineral deposits. The water temperature fluctuates, affected by both the high elevation and arid conditions of the desert biome, ranging between 0.5 °C in January to 26.7 °C in July [4.46], even up to 45 °C in the shallow shoreline waters [4.206].

A critical habitat for migratory birds, GSL provides invertebrates that feed shorebirds in the wetlands and playa [4.238] and pelagic birds in the open water [4.43]. The brine shrimp (*Artemia franciscana*) and brine flies (*Ephydra* spp.) depend on the microbial foundation of phytoplankton and microbial mats [4.11]. There are fungi, protists, and eukaryotic autotrophs as well, and this whole system rests on a foundation of diverse prokaryotes, bacteria and archaea [4.17]. In the north arm of the lake, where the water is more saline, the biodiversity is more restrictive to the animal life and further enriched with haloarchaea, which is the focus of this chapter.

We are using the less formal term "haloarchaea" here to mean the family of Halobacteriaceae (in the Halobacteriaceae phylum of the archaea) as the term "halophilic archaea" may be less precise since it may include archaea in methanogenic families as well [4.159]. Most require high salinity for survival or growth (a required range of 2 M to upwards of 5 M NaCl at saturation), are facultative anaerobes, and lyse in water that is lower in ionic strength [4.184].

4.1.2 Great Salt Lake as an Extreme Environment

This robust population of haloarchaea rises to meet the osmotic challenges presented by saline water nearly 10-fold that of the ocean, in addition to dramatic shifts in temperature and elevated ultraviolet (UV) irradiation. We will highlight their presence in the north arm of GSL, although it should be noted that they also inhabit the solar-evaporation ponds of the mineral extraction industry (Figure 4.1b).

Figure 4.1 Great Salt Lake is visible from space in its location in western North America (2018). (a) The north arm haloarchaea-rich pink water is separated by a railroad causeway from the south arm, which is tinted green from diverse consortia of algae. The surrounding lakebed salt playa (white) shows where the meander line was at the historic high of 1986. To the west, on the left side of the lake in the image, are the Bonneville Salt Flats, an evaporite deposit formed when this Pleistocene inland sea evaporated as temperatures rose following the last ice age. (b) A close-up image of the lake's railroad causeway and the separation of the north and south arm microbiota. (Image credit: European Space Agency astronaut, Alexander Gerst, International Space Station, public domain).

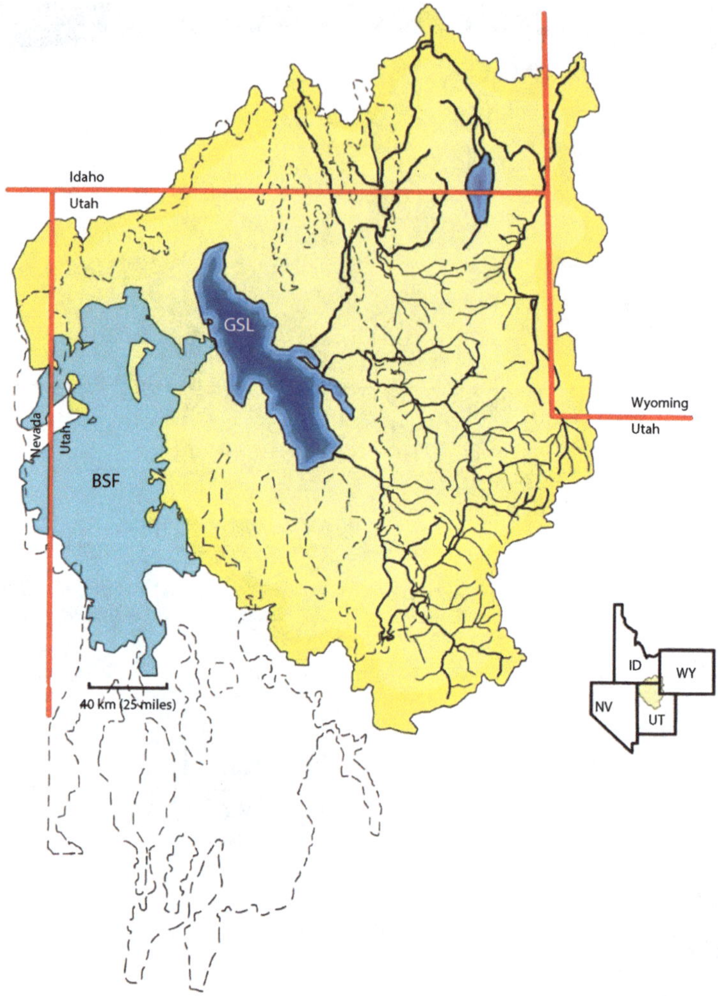

Figure 4.2 Great Salt Lake (GSL) and Bonneville Salt Flats (BSF) in the context of the modern watershed in yellow. The outline of ancient Lake Bonneville is shown with a dashed line. (Image credit: Johanna Bossart).

The North Arm is a Zone of Salt-Saturation

The 1960 construction of rock-filled railroad causeway compartmentalized the north arm and separated it from the less saline south arm [4.15] [4.30] [4.144]. Only 10% of the freshwater input to GSL enters the system in the north arm in the form of precipitation [4.172]. Within just a few years after this structure was built, the north arm became salt-saturated [4.1], and salt began to precipitate to form a thick crust which continues today [4.222]. The brine of the north arm water column has remained close to saturation for most of the decades since, with the exception of flooding in the 1980s; as solubility is dependent on temperature, current measurements fluctuate between 24–34% sodium chloride in seasonal measurements [4.4] [4.118]. The brine in this part of the lake is colored pink by the carotenoid pigments of the resident microorganisms (Figure 4.1) [4.114], which is typical of hypersaline aquatic systems.

Ultraviolet Light Flux at Great Salt Lake

Cellular molecules can be damaged by UV light, which is a particularly important consideration for life at elevation. On Earth, the UV dose increases 9–24% per one thousand meters [4.24]. GSL is at a high altitude; the water level is currently at 1278.4 m above sea level [4.261]. Other factors may also affect the dose of UV radiation, such as time of year, latitude, and even dust conditions. Sodium chloride-rich lakes like GSL may mobilize atmospheric chlorine, which has been known to decrease the ozone concentration, which in turn increases UV exposure [4.251]. During the summer months, longer days will amplify UV exposure. However, since GSL is located at mid-latitude, the UV flux varies throughout the year.

4.1.3 The Declining Elevation of Great Salt Lake

The current trajectory of the elevation of GSL is downward, due to water diversions for consumptive uses coupled with climate change [4.13] [4.179]. However, the natural history of the lake's formation tells us the larger story about long-term evaporation events and climatic shifts.

Ancient Lake Bonneville Becomes Great Salt Lake

Lake Bonneville was a Pleistocene body of water that occupied the hydrographically closed basin where we now find GSL [4.81]. It stretched over much of what is now northern Utah, southern Idaho, and eastern Nevada (Figure 4.2), one of several large lakes that filled this basin over the last 800,000 years [4.196]. At its peak, this pluvial sea covered about 20,000 square miles and was over 1,000 feet deep. During the thaw of the last ice age, Lake Bonneville was subject to a shift in the water budget that diminished the lake level about 100 m per thousand years, where it eventually reached an elevation similar to the modern GSL and remained there [4.195]. As a result, shorelines of ancient lake levels are still visible in the mountain ranges surrounding the valleys of northern Utah, where GSL sits as a saline lake at the bottom of the Bonneville Basin (Figure 4.2) [4.7].

The Modern Terminal Lake Shrinks and Swells

Terminal lakes are characteristically susceptible to changes in local wet and dry seasons, as typified in the American West [4.267]. Moreover, GSL's location in a shallow playa portends that small changes in elevation result in large ones in its area. While these changes are part of the natural fluctuations of GSL, 2016 marked some of the lake's lowest levels in recorded history [4.279] [4.13] [4.179]. Since accurate recordings began during the westward expansion in 1847, the lake level has decreased to about 50% of its pre-industrialized size. Figure 4.1 demonstrates the exposed salty shoreline in white that has historically filled with water in high precipitation years.

4.1.4 Pertinent Mineralogy of Great Salt Lake and the Bonneville Basin

Minerals found at GSL may preserve biology over time (see Section 4.3.4 below) and are thus relevant to the search for extant life or extinct on Mars. In addition, salts or salt-saturated brine may extend the temperature range for the existence of water in the liquid state since it lowers the freezing point, potentially allowing life to exist at much colder temperatures such as the Martian surface [4.256], making GSL salt deposits attractive as an analogue site.

Common GSL evaporite minerals include the salts, halite (NaCl), gypsum ($CaSO_4{\cdot}2H_2O$) and anhydrite ($CaSO_4$, not hydrated), which occur on the shores and in the clay of the lake, and precipitate as water is evaporated (Figure 4.3a-c) [4.61] [4.237]. Both halite [4.118] and gypsum [4.191] have been found to contain communities of halophilic microorganisms.

A precipitating mineral, mirabilite (Na2SO4·10H2O), may develop in mounds when fed by mineral-rich groundwater, a common phenomenon in saline lakes (Figure 4.3d) (e.g., [4.137]). Recently, crystalline mirabilite mounds were documented at GSL [4.262]. These emerged when the seeping mineral-rich ground water met the cold winter air. As the mirabilite crystalized, it formed intricate terraces and mounds (Figure 4.3d). We are currently investigating these crystals for the ability to enrich halophilic life.

The Bonneville Salt Flats (BSF) span Western Utah and extend past the Nevada border (Figure 4.2) [4.260]. Formed from the desiccation of Lake Bonneville [4.141], the flats act as

Figure 4.3 Mineral formation at Great Salt Lake. (a) The beach is expansive as water levels have decreased, with calcium carbonate oolitic sand encased in halite; humans for scale. (b) Gypsum forms in the sediment in clay-rich areas, with the edges of the blades pointing upward; scale bar is ~5 cm. (c) The pink coloration of the halite in some areas is indicative of carotenoid-containing haloarchaea entombed in the minerals; 10 cm tall figure for scale. (d) Mirabilite crystals are sampled from a mound filled with mineral rich fluid. (Image credits: (a) Foad Yousef; (b) Heath Brodie; (c) Bonnie K. Baxter; (d) Calli Cahill).

Figure 4.4 Evaporites on Earth and Mars. (a) Bonneville Salt Flats in the Great Salt Lake desert, featuring evaporite mineral deposits from the lake bottom of Pleistocene Lake Bonneville. (b) Sedimentary rocks in the foreground of Mount Sharp, near Gale Crater, Mars, as imaged by the Curiosity rover (Sol 548, 2014). Gale crater was likely the site of an ancient salty lake about 3.7 billion years ago (Gya), and evaporite minerals such as gypsum and anhydrite have been identified there [4.214]. (Image credits: (a) Ken Krahulec, Utah Geologic Survey; (b) Marco Di Lorenzo and Ken Kremer, NASA/JPL, public domain).

an austere memorial to the inland sea that once occupied the area (Figure 4.4a). The playa can be divided into three geologic zones: the carbonate zone made of clay-sized minerals, the sulfate zone made up of gypsum, and the chloride zone made up of halite. The characteristics of the GSL and BSF are not unique in our solar system and may serve as analogues of sites on our neighbor planet, Mars.

4.2 The Transformation of an Ancient Wet Mars to a Modern Hostile Environment

Haloarchaea have evolved in salty habitats on Earth, such as GSL and its surrounding evaporites. Are there possible habitats on Mars which would support this type of life? As we build this comparison, we must think about the geologic history of Mars, recorded in the rock record of the Red Planet itself. For some time, salts have been recognized as a component of the regolith on Mars, including sulfates, chlorides, carbonates, and other hydrated minerals [4.39]. Widespread presence of salt deposits underscores a high likelihood of the presence of ancient lakes on Mars surface at one time [4.149]. Extensive evaporite geology, along with other clues about the history of water on Mars, suggests that the ancient lakes of Mars likely were habitable hypersaline environments before their demise [4.79] [4.149].

4.2.1 Global Climatic Change on Mars

The lack of significant surface water on modern Mars permits skepticism regarding the presence of life, but four billion years ago (Gya), Mars and Earth were more similar. However, around 3.5 Gya, Mars began a transition, losing its thick atmosphere [4.263] and intrinsic magnetic field, allowing solar and cosmic radiation to strike the surface unimpeded today.

Mars' past global intrinsic magnetic field was analogous to that observed on modern day Earth. Since that time, the Martian core is thought to have cooled more rapidly due to the quicker thermal evolution of smaller bodies [4.264]. This cooling and solidification are hypothesized to be the main driving force behind the Martian intrinsic magnetic field loss

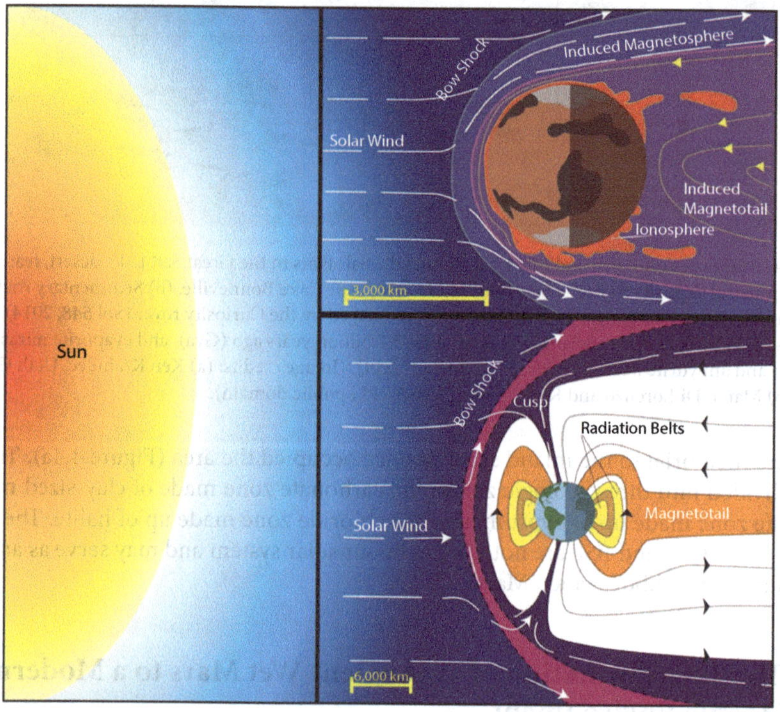

Figure 4.5 Magnetic fields of Mars and Earth. (a) Schematic of the solar wind interaction with Mars [4.28]. The solar wind carries the interplanetary magnetic field as it streams toward the bow shock on the "sunny side" of Mars. Unlike Earth, the solar winds interact directly with the induced magnetosphere and the thin atmosphere of Mars, which contributes to loss of atmosphere and radiation exposure. The Mars Global Surveyor measurements show that Mars does have sources of remnant crustal magnetization [4.21], which is due to the regolith containing high amounts of (about 18%) Fe2+, mostly in the form of oxides, and iron being ferromagnetic [4.107]. Intense crustal magnetic fields (orange) impose structure throughout localized regions of the upper atmosphere and ionosphere [4.42]. (b) Earth's magnetic field interacting with solar wind is depicted based on a current model [4.62]. Note, the sunny side of the magnetosphere is squished close to Earth from the solar winds, while the shady side is an elongated tale. As solar winds, comprised of charged electrons and protons, approach the magnetosphere, they lose velocity interacting with the bow shock region. The bow shock prevents many of these charge particles from stripping away Earth's atmosphere. (Image credit: Johanna Bossart).

[4.21]. Differential interactions between solar wind with Earth [4,62] or Mars [4.28] [4.42] explain the observations regarding modern magnetic field properties (Figure 4.5). Earth's magnetic field is best explained by the dynamo theory. This process describes how the rotation and convection of the iron-rich (electrically conducting) fluid in the outer core creates a long-lasting magnetic field. The magnetic field, without the influence of solar wind, would create a simple dipole much like a bar magnet. However, the solar wind greatly influences the shape of Earth's magnetic field and our atmosphere [4.62]. Figure 4.5 shows that Earth's magnetosphere is shaped by solar winds but remains protective. However, solar winds interact directly with the induced magnetosphere and the thin atmosphere of Mars (and even the surface of the planet), forming a magnetosphere from magnetic field tubes, which does not allow for retention of atmosphere or protection from solar radiation [4.253] [4.263] [4.145].

These interactions can be volatile, causing photochemical loss, resulting in escape of elemental oxygen, nitrogen and carbon [4.108]. Another form of atmospheric stripping is due to ion loss from several processes, such as atmospheric sputtering, which can transfer energy from the solar wind directly to planetary ions and lead to heating, acceleration, and escape. Essentially, solar winds gradually stripped Mars of the necessary elements needed for water and other elements that help create a dense atmosphere with greenhouse conditions [4.145] [4.253].

Loss of water in Martian history has implications for life on the planet's surface and also for sustaining life through carbon cycling as witnessed on Earth. Tectonic activity is necessary to cycle carbon from the lithosphere, to the atmosphere, and back to the biosphere [4.84]. Water acting as a lubricant for the subduction process of plate tectonics potentially helped in sustaining a magnetic dynamo on Mars over geological time scales, which is evidence for the presence of large amounts of liquid water during the time when Mars had an operating dynamo core [4.264]. Water makes the lithosphere malleable enough for one crustal plate to subduct another, while reducing the phase temperature of solid mantle rock, thereby enhancing the cooling of the interior. The habitability of one-plate planets is considered theoretically possible but less likely than one with active tectonics to evolve complex life [4.129]. The geologic history of Mars tells us about its past tectonic activity, including volcanism [4.97] [4.180], but this is not evident today.

4.2.2 The Ancient Lacustrine Environments of Mars

The history of water on Mars is most heavily evidenced by the valleys, craters, and emptied riverbeds on its surface. These extensive environments are widely dispersed all across Mars' surface and many are laden with evaporite salts [4.64] [4.170] [4.193] [4.214] [4.241] [4.258]. Rainfall on ancient Mars has also been studied [4.45]. Hydrogeological modeling showed how widespread late Noachian/early Hesperian waters might have been in Meridiani Planum and the evaporation of these volumes of water would have yielded significant, buried mineralogy on modern Mars [4.5] [4.6] [4.38] [4.259]. Both polar caps reveal abundant internal layers of ice and dust within the spiral-shaped troughs that cut through their volume [4.163]. This layering contains a record of past climates on Mars, in the way that Earth's ice sheets have recorded the climate on our planet [4.75].

The Mars Exploration Rover, *Opportunity*, found evidence of upwelled and dissolved pore spaces that contained former magnesium, iron oxide, and sulfate minerals in some of the largest craters on Mars, such as Eagle, Endurance, Erebus, and potentially Victoria [4.90] [4.161] [4.162]. These groundwater recharge cycles observed in craters show how evaporite minerals can not only exist on the surface via groundwater breaching the top subsurface crustal layers but can also be precipitated as buried evaporite mineral deposits. Based on the scale of these deposits of minerals, it is estimated that these places had thousands of miles of water; scaling up to roughly 130,000 square miles with nearly 2000 feet of strata [4.269].

Evaporite mineralogy and the closed basin nature of many of the regions on Mars have been observed through digital elevation models from the High Resolution Imaging Science Experiment (HiRISE) and the Compact Reconnaissance Imaging Spectrometer for Mars (CRISM) on board the Mars Reconnaissance Orbiter (MRO) [4.64] [4.265] [4.272] [4.280]. The global mineralogical data collected by the CRISM instrument on board MRO demonstrated extensive evaporite deposits, sulfates, sedimentary clay deposits, and phyllosilicates on Mars [4.22] [4.23] [4.33] [4.64] [4.65] [4.66] [4.83] [4.164] [4.170] [4.265], prevalent both

Figure 4.6 Analogue mineralogy: (a) Gypsum "box-work" at Gale Crater on Mars as imaged in 2015 by the Curiosity rover [4.127]; compare with the gypsum blades in the clay-rich sediment at the Great Salt Lake north arm in Figure 4.3b. (b) Landing site of NASA's Mars 2020 Perseverance rover mission, Jezero Crater, as imaged by the Mars Reconnaissance Orbiter. The green coloration shows hydrated sulfate minerals (e.g., gypsum) at the site of an ancient river delta, suggesting evaporite formation [4.83]. (Image credit: NASA-JPL, public domain).

in the southern highlands and the northern lowlands. Gypsum, which can only precipitate from liquid water, is a proliferate mineral on the surface of Mars and is hypothesized to have formed from the percolation through Martian dunes [4.76] [4.213]. In fact, the landing site for the upcoming 2020 NASA Mars Perseverance rover mission, Jezero Crater, appears to be a large lacustrine system with sulfated minerals, as determined by CRISM (Figure 4.6b) [4.83].

On the Martian surface, *in situ* work by rovers, using visual and spectrometer data, and chemoanalyses, have also demonstrated minerals consistent with ancient salt lakes [4.205]. An excellent example of an ancient lake system is Gale Crater, which has the geology and morphology of an ancient saline lake consonant with that of GSL (Figure 4.4b) [4.105] [4.213] [4.214]. Clay-bearing sediments and evaporite deposits, such as sulfated minerals and chloride salts, mark the existence of a past fluvial lake at the lowest portion of the Murray formation [4.214]. The Mars Science Laboratory rover, Curiosity, discovered sulfates in veins in polygonal and box-work type structures with a significant height from the surface from lower Mt. Sharp in the Garden City outcrop (Figure 4.6a) [4.127]. These veins have been interpreted to the late Noachian/early Hesperian boundary in the Martian timeline, approximately 3.5 Gya, and connote an evaporative event [4.64]. A series of lake desiccations are evident in the patterns of carbonate, chloride, and sulfate salt assemblages, which provide geochemical clues to the past chemistry of these lake basins since different mineral results have distinct dissolution properties [4.214]. This is a story related to that of Lake Bonneville and the modern desiccating GSL.

Perchlorate salts were located on Mars by the Phoenix lander, suggesting that some Martian soils may be highly oxidizing [4.100], which is a challenge for microbial life [4.154] [4.190]. However, perchlorates are deliquescent, and this hydroscopic property can enrich liquid water in an area, allowing life to flourish [4.47]. Although these salts are not enriched in the GSL system, we will discuss the potential for haloarchaea resistance in Section 4.5.2.

While there is no physical evidence of crystal mirabilite mounds on Mars, topographical and HiRISE image data have shown white mounds that some experts hypothesize are related to saline groundwater [4.180] [4.236]. Furthermore, diagenetic mineral precipitation was observed in Gale Crater [4.115], akin to mirabilite dehydration to thenardite (Na2SO4) [4.101].

4.2.3 Brine on the Martian Surface

Stable surface water on Mars is not likely, but there may be some small-scale exceptions [4.32] [4.93] [4.152]. Deliquescence occurs when a hygroscopic salt absorbs atmospheric moisture, as a function of pressure differential, and dissolves in the water it has attracted to the local site. Chloride salts, such as halite and also perchlorates, are hygroscopic, and in the Mars atmosphere, this geochemistry could lead to water films or droplets, localized to a region [4.52] [4.215]. For example, water droplets have been observed forming on perchlorate salts at low temperatures (225 K) [4.286]. Pockets of brine formation, and perhaps stable brines, are a consideration for the Martian surface as well as subsurface.

Perhaps the most interesting recent development, with respect to water on Mars, is the discovery of recurring slope lineae, which are narrow, low albedo fluid flows that are observed on Mars; they are temporary in nature but annual in reoccurrence, often reaching the largest size near the end of Martian summer [4.36] [4.157] [4.248]. They are approximately 0.5–5 m in width and are observed on steep (roughly 35° or greater), warm slopes (between 250 K and 273 K) often facing the equator [4.36]. The presence of salts within this water can lower the freezing point of the flowing water by up to 80 K, which could allow the brine to exist as a liquid on the Martian surface for a short period of time [4.181]. The source of fluid for these brine flows may be a subsurface layer of frozen water similar to permafrost [4.248]. Greenhouse-like effects from Martian dust and seasonal high temperatures during the Martian summer could result in a thermal heat exchange significant enough to melt some of this permafrost layer. With respect to the southern hemisphere of Mars, buried ice is only considered to be stable for latitudes of more than 40° S [4.181].

4.2.4 Mars is a Hostile Environment for Life

Despite potential surface or subsurface water [4.32] [4.93] and salt deposits where life may dwell, Mars is cold, dry, and under constant radiation bombardment. It lacks significant tectonic activity and an atmosphere to maintain homeostatic conditions. There are many caveats to the possibility of finding extant life on Mars.

Low and Unstable Temperature

The temperature gradient of Earth, from the poles to the equator and from sea-level to the upper atmosphere, coupled with the Coriolis effect, drive our convective weather patterns. Whereas Earth's convection cycles are driven by the latent heat of moisture, Mars' convective climate pattern is driven by the absorption of solar energy into dust [4.240]. This direct absorption causes the dust to rise vertically and stratify, with the denser layers near the Martian surface and less dense layers in the troposphere 30–50 km above the planet. Because of this non-latent heating process of dust and the thin atmosphere, the temperature gradient is much more expansive on Mars between the sunny side and shady side regions, from a high of around 20° C, but dropping as low as –140 °C. This low temperature at an atmospheric pressure on what would be approximately sea level on Mars (roughly 10 × 104 times lower than pressure at sea level on Earth) would make it unlikely for pure liquid water to exist on the surface of Mars [4.99] [4.165].

Ultraviolet Radiation Flux

A planet's atmosphere is a crucial factor in protecting the planet from UV radiation; carbon dioxide and ozone in Earth's atmosphere absorb and Rayleigh-scatter light in these spectra [4.34]. Modern-day Mars experiences a total integrated UV flux that is comparable to Earth's current UV exposure [4.263]. However, the Martian surface is exposed to intense doses of the lower wavelengths, UVC and UVB light, which are more penetrating and associated with cellular damage [4.40]; the UVC and some of the UV B is filtered out by Earth's atmosphere. Longer days, a lower total atmospheric pressure, and a thin atmosphere lacking significant carbon dioxide and ozone, allow for more of these harmful UV rays to reach the Martian surface [4.263].

Some studies suggest that dust on Mars may mitigate the UV exposure, and thus the dose could be reduced in some areas [4.34] [4.94]. The radiation flux at a point on the surface of Mars also depends on factors such as cloud cover, season, local time, and latitude. Considering potential locations of extant life on Mars, zones under the crustal magnetosphere in the southern hemisphere at higher latitudes where the ozone concentrations are higher in winter months coupled in regions of dense dust (during storms) would be the most ideal conditions to protect from harmful UVB and UVC radiation. Entombment in minerals would provide even more protection [4.72].

4.3 Life in Evaporitic Minerals on Earth

As water evaporates in modern GSL, salts are evident in a thick crust on the floor and the shore of the lake (Section 4.1.4 and Figure 4.3) [4.222]. Additionally, salt domes that are associated with the actions of microorganisms may form [4.207]. For geomicrobiology work directed at preservation of life in the modern dynamic environment, individual crystals can be collected when the conditions at the lake favor their production.

4.3.1 Formation of Halite and Gypsum Primary Crystals in Aquatic Environments

Halite hopper crystals (Figure 4.7a) and primary gypsum crystals (Figure 4.8a) are common features of evaporative environments [4.132], described at evaporative lakes such as the Dead Sea [4.82] and GSL [4.61] [4.118]. Primary crystals form at high ionic concentrations when there is a rapid temperature shift. For example, in the autumn at GSL when the water cools and solubility is lowered, the mineral lattice is compacted as the salts precipitate [4.205] [4.222]. Halite hopper crystals begin to form in the center, salt precipitates around the outside, and the middle then collapses, forming the distinctive pyramid shape (Figure 4.7a) [4.82]. Even small, nascent crystals have this shape (Figure 4.7c). Gypsum crystals are formed under the sediment and halite crust [4.61] [4.205] Figure 4.3b shows gypsum crystals protruding from the sediment at the GSL.

During the formation of halite and gypsum crystals, small pockets of mineral-saturated fluid are trapped within the structure (Figure 4.7b). These structures act as capsules that can seal fluid over geologic time [4.143] [4.242] [4.245]. These fluid inclusions have been shown

Figure 4.7 Halite crystals from Great Salt Lake. (a) Halite hopper crystals, collected from the north arm in October 2019 from the shallow shoreline waters, typically measuring 1-2^2 cm at the base of the pyramid shape. (b) Fluid inclusions are visible inside another halite crystal, marked by a bubble in the sequestered brine; scale bar is 0.5 cm. (c) Nascent halite hopper crystals forming in a sample of mirabilite rich fluid. Secondary growth mirabilite crystals are seen as "horns" emanating from the halite; scale bar is 5 μm. (Image credit: Adrik DaSilva and Calli Cahill).

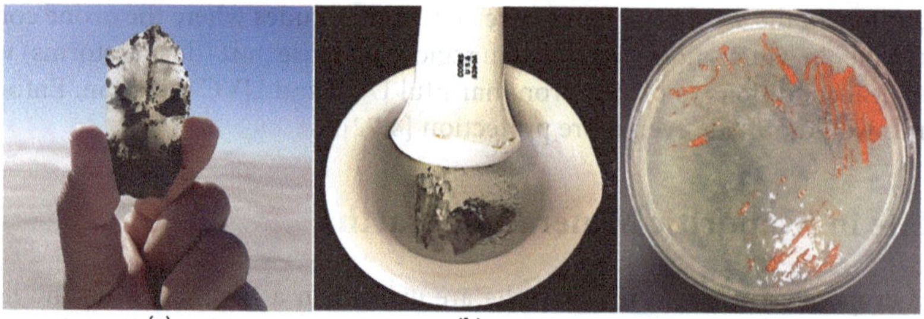

Figure 4.8 Gypsum crystals contain haloarchaea. (a) Blade-like gypsum crystals were obtained from the clay-rich sediments on the Great Salt Lake halite-encrusted north arm beach. (b) Gypsum crystal can be seen in crystal and powder form. Mortar and pestle were autoclaved before crushing a sterilized gypsum crystal, and crystal dust was transferred to 23% salt liquid growth media. (c) Red colonies are indicative of a haloarchaea species, which was isolated from the gypsum cultures and subjected to genetic identification. (Image credit: the authors).

to be adequate habitats for halophilic microorganisms and release significant amounts of brine when crushed [4.177] [4.204]. The fluid can be collected aseptically by drilling into the inclusion, using a dissecting microscope [4.89]. Halite is especially soluble, when primary crystals are exposed to freshwater (rain, snow, or groundwater) or when temperature increases cause the minerals to re-dissolve, the primary crystal structure is compromised. Although fluid may still be trapped inside (Figure 4.7b), for long-term temporal studies, one would want to ensure the fluid is from the primary crystal to make conclusions about life sequestered over time [4.266]. Inclusions inside halite hopper crystals can be very abundant, as many as 10^{10} cm^{-3} fluid pockets are estimated [4.220].

4.3.2 Haloarchaea May Be Entombed in Halite

Within the fluid inclusions, haloarchaea may be easily trapped since they are present in the sequestered brine. It is possible that motility and chemotaxis aid in the entombment of haloarchaea as they may move to follow the water [4.255]. Fluid inclusions can entrap significant concentrations of cells; a halite cube measuring 1 cm per side can contain around six million trapped microbes [4.143]. Past studies have shown that halite fluid inclusions retain fluids for long periods of time with haloarchaea shown to be motile for several weeks and viable in inclusions for at least 6 months [4.177] [4.204]. Furthermore, prokaryotes can be observed *in situ* in (even ancient) fluid inclusions [4.228] and cultivated from them [4.143]. At GSL, mineral entrapment of haloarchaea has been observed [4.14] [4.118], including on halite crystals formed on the wings of birds [4.118]. Although cultivation methods need to be coupled with methods that limit contamination [4.197] [4.198], many researchers have been isolating microorganisms, especially haloarchaea, from halite fluid inclusions over much longer geologic time periods (see Section 4.3.4).

4.3.3 Haloarchaea May Be Entombed in Gypsum

Although halite entrapment of archaea has been extensively examined, gypsum entrapment of archaea has been less studied. Most of the work done with gypsum has analyzed cyanobacteria and microorganisms in gypsum crust, not inside fluid inclusions [4.191], with few exceptions [4.199]. Some have used spectroscopy to identify the biomolecules of microorganisms in gypsum [4.63]. Current studies in our lab have isolated a haloarchaeal species from surface sterilized gypsum primary crystals (Figure 4.8), but not knowingly from an inclusion. Gypsum crystals, collected from the GSL north arm (Figure 4.8a), were surface sterilized [4.225] and then crushed in a sterile crucible (Figure 4.8b), or placed directly into media. We cultivated a single haloarchaea species (Figure 4.8c), which matched *Halobacterium noricense* strain DAB-31, according to NCBI BLAST results of a region of the 16S rRNA gene [4.118], which was identical to that of this species in Black Lake, China [4.8]. When we do similar experiments with GSL halite crystals, we isolate 20–30 species of cultivatable haloarchaea species [4.118], thus our results suggest modern gypsum may not have the preservation potential of diverse communities as does modern halite. This may or may not hold true over geologic time.

4.3.4 Haloarchaea Can Survive over Geologic Time in Fluid Inclusions

Numerous studies have reported evidence of haloarchaea survival inside halite crystals up to hundreds of millions of years old [4.55] [4.87] [4.92] [4.123] [4.143] [4.158] [4.167] [4.173] [4.177] [4.200] [4.225] [4.228] [4.229] [4.242] [4.243] [4.244] [4.266]. Such long-term survival of life, beyond the Permian time period, suggests some sort of metabolic dormancy, as well as flexibility, regarding available nutrient sources and exceedingly robust DNA protection and repair (these haloarchaeal superpowers that relate to persistence in the geologic record will be discussed in Section 4.4). In addition to cellular protective strategies, haloarchaea are further protected by the minerals in which they dwell from environmental assaults, especially UV radiation [4.72]. Recent experiments, which monitored haloarchaeal communities over months in laboratory-produced halite crystals, raised the issue that the highly diverse species assemblages in brine may be restricted as storage time increases, and some species may survive longer than others [4.104]. We do not yet understand the timeline of succession of diversity changes nor the identity of genera that are candidates for long-term survival.

4.4 Great Salt Lake Haloarchaea

In the mineral-rich geochemistry of GSL, haloarchaea thrive even in the most saturated fluids or, as described above, inside minerals. The visible pink coloration of the GSL north arm brine (Figure 4.1) and halite deposits (Figure 4.3c) suggests the presence of an abundance of carotenoid-containing microorganisms [4.12]. Recent investigations on microbial diversity studies at GSL have demonstrated that haloarchaea are the dominant prokaryotes in this salt-saturated system [4.2] [4.4] [4.15] [4.201] [4.202] [4.203] [4.252] [4.271]. A search of "Great Salt Lake" in the genbank database for deposited DNA information provides hits for 1,318 archaeal sequences [4.175]. Table 4.1 shows a list of species that have been detected through direct sequencing, cloning, or cultivation. However, early microbiology studies of lake water detected these pigmented microorganisms even before the segmenting of the hypersaline north arm by a railroad causeway, which were very likely haloarchaea, though

Table 4.1 Haloarchaea species diversity discovered in the Great Salt Lake north arm through cultivation, cloning, or direct amplification from brine [4.4] [4.118] [4.252]. These were typed by closest (16S rRNA gene) matched strains. Note: *Haloarcula* and *Halorubrum* are typically overrepresented in cultivation experiments [4.18] [4.19], which explains the numerous species identified for these genera, due to better data available in database analyses.

***Haloarcula* species**	***Halorubrum* species**	**Other haloarchaea species**
Haloarcula amylolytica	*Halorubrum alkaliphilum*	*Halobacterium salinarum*
Haloarcula argentinensis	*Halorubrum arcis*	*Halobiforma lacisalsi*
Haloarcula hispanica	*Halorubrum californiense*	*Halococcus morrhuae*
Haloarcula japonica	*Halorubrum chaoviator*	*Haloferax lucentense*
Haloarcula marismortui	*Halorubrum ejinorense*	*Halogeometricum borinquense*
Haloarcula quadrata	*Halorubrum ezzemoulense*	*Halomicrobium mukohataei*
Haloarcula salaria	*Halorubrum lacusprofundi*	*Haloplanus natans*
Haloarcula sinaiiensis	*Halorubrum litoreum*	*Haloquadratum walsbyi*
Haloarcula tradensis	*Halorubrum orientale*	*Halorhabdus utahensis*
Haloarcula vallismortis	*Halorubrum tebenquichense*	*Halorhabdus carlsbadense*
	Halorubrum terrestre	*Halosarcina* sp.
	Halorubrum tibetense	*Halosimplex* sp.
	Halorubrum trapanicum	*Haloterrigena thermotolerans*
	Halorubrum xinjiangense	*Natrialba aegyptia*
		Natronobacterium gregory
		Natronococcus occultus
		Natronococcus xinjiangense
		Natronomonas pharaonis

this taxon was not known at the time [4.49] [4.78] [4.119] [4.206]. Recent work on BSF indicates a rich community of haloarchaea as well [4.160].

The haloarchaea consortia in north arm brine, evaluated by 16S rRNA gene cloning and sequencing, are relatively stable over seasonal environmental changes, responding to vacillating salinity and temperature [4.4]. Cultivation of microorganisms in this area of GSL demonstrated predominantly archaea, representing the genera *Haloarcula*, *Halobacterium*, *Halococcus*, and *Halorubrum* [4.2] [4.118]. The stable archaeal assemblages in the north arm appear to possess a lower phylogenetic diversity relative to communities in the south arm [4.201] [4.202] [4.203]. Certainly, high salinity (upwards of 18%) is the most important environmental variable for the enrichment of GSL haloarchaea, and gene regulation allows these microorganisms to respond to a fluctuating environment yet maintain a stable presence [4.3].

Studies on GSL haloarchaea suggest they are not only in the water column. Their presence was demonstrated inside the brine shrimp and their cysts [4.14] [4.217] and sequestered inside halite crystals [4.14] [4.205]. Evidence of haloarchaea were noted in the microbialite structures that actively precipitate calcium carbonate on the benthic region of GSL [4.140]. Interestingly, they may be transported to other sites on the feathers of birds [4.118], which can explain the biogeography of similar species genetically related to GSL haloarchaea.

The GSL haloarchaea species live in hypersaline brine mostly devoid of invertebrates, and their populations are likely controlled by haloviruses [4.16] [4.208] [4.232], which may impact the presence of a particular organism due to specific virus-host interactions. The ratio of virus particles to cells in GSL hypersaline brine is 100:1, suggesting a significant role for these predators [4.16], which may include gene exchange [4.168].

We do not know much regarding the metabolic activities of GSL haloarchaea, but proxy measurements indicate slower reaction rates as the salinity increases. Hydrocarbon degradation experiments in hypersaline evaporation ponds at GSL showed lower amounts of turnover as salinity increased [4.268]. Also, glutamate as a substrate is degraded more slowly at higher salinities [4.209] [4.250] [4.268]. A study that looked at the salinity gradient between the (archaea-rich) north arm and (bacteria-rich) south arm demonstrated a slower rate of conversion of glucose, glycerol, and acetate to CO_2 in the higher salinity brine [4.69].

4.5 Haloarchaea Have Superpowers for Extreme Lifestyles

Highly specialized adaptations allow halophilic organisms to thrive in GSL. These strategies allow haloarchaea to overcome challenges such as high salt, low water availability (or water activity, often abbreviated a_w), limited nutrients, low dissolved oxygen, and high radiation. Survival in minerals over geologic time (Section 4.3.4) is perhaps a result of all of these superpowers combined (Table 4.2), which reinforce the notion that haloarchaea are excellent candidates for extant life on Mars.

4.5.1 Haloarchaea May Show Versatility with Respect to Temperature and pH

The lifestyles of haloarchaea may involve extremes of temperature (especially cold) and extremes of pH (especially alkaline). Haloarchaea in GSL can certainly survive the high desert temperatures of the Utah summers and the cold temperatures of the snowy winters [4.3] [4.46], but what is their complete range of tolerance? This is an important question, particularly concerning the low temperature threshold, to assess the possibility of extant life on chilly Mars. A haloarchaea species and psychrophile, *Halobacterium lacusprofundi,* was recovered from Deep Lake in Antarctica, a hypersaline waterbody that does not freeze due to the salinity and reaches the low temperature of −20 °C [4.77]. This ability of cold temperature survival may be the result of the unusual adaptation allowing the synthesis of unsaturated diether lipids [4.80]. On the other end of the spectrum, hyperthermophilic extreme haloarchaea have not been identified, perhaps because of the permeability to ions created when cell membranes are heated, making the required osmoregulation difficult [4.27]. Although several species of haloarchaea have been noted to have optima temperatures above 50 °C.

The pH optima of haloarchaea is typically between 5.5 and 8.5 [4.88], but these separate into two main clusters: neutrophilic with an optimal pH around 7.0, or strongly alkaliphilic

Table 4.2 Haloarchaea superpowers; survival strategies for life at high salinity. These strategies are listed by categories in order to consider each challenge separately, though some cellular mechanisms may overlap. A representative reference or review is provided.

Category	Strategy	Reference
Lifestyle flexibility	Use of light energy	[4.249]
	Anaerobic growth	[4.96]
	Cold temperatures	[4.77]
	Warm temperatures	[4.27]
	Alkaline pH	[4.27]
	Perchlorate reduction	[4.190]
Osmophily	Growth in low a_w	[4.247]
	Cell morphology	[4.228]
	Cell wall structure	[4.273]
	Protein aa composition	[4.67]
	Potassium accumulation	[4.231]
	Compatible solutes	[4.48]
Radiation resistance	UV resistance	[4.10]
	IR Resistance	[4.126]
	Carotenoid pigments	[4.114]
	Bipyrimidine limitation	[4.113]
	Polyploidy	[4.239]
DNA Repair	Nucleotide Excision Repair	[4.155]
	Photoreactivation	[4.103]
	Base Excision Repair	[4.31]
	Homologous Repair	[4.156]
Long-term survival	Mineral entombment	[4.245]
	Chemotaxis and motility	[4.255]
	DNA as a nutrient	[4.187]
	Necromass as a nutrient	[4.254]
	Algal glycerol as a nutrient	[4.228]
Space conditions	Exposure to space	[4.147]
	Freeze-thaw cycles	[4.146]

with pH at or exceeding 8.5 [4.27]. Bowers and Wiegel identified four taxa of haloarchaea with both high pH optima and optimal temperatures above 50 °C: *Haloarcula quadrata, Haloferax elongans, Haloferax mediterranei* and *Natronolimnobius aegyptiacus.*

4.5.2 Haloarchaea are Metabolically Versatile

The GSL haloarchaea may exhibit a broad array of energy generation modes. Although none of this taxon have exhibited photosynthesis, light energy in place of chemical energy is utilized by some genera. Utilizing the transmembrane protein-carotenoid complex, bacteriorhodopsin, *Halobacteria* spp. (and other haloarchaea) can capture light energy and pump protons across the membrane, generating ATP [4.249]. Also, many haloarchaea have adapted to grow anaerobically, despite being aerobic heterotrophs [4.96] [4.186], and methanogenesis has been detected in the GSL north arm [4.15]. This is critical in systems like GSL and not only inside minerals since hypersalinity results in low levels of dissolved oxygen [4.233].

The reduction of perchlorate by haloarchaea is certainly relevant to Mars (see Section 4.2.2). *Haloferax* species, which are represented in GSL (Table 4.1), have demonstrated this activity [4.153] [4.182]. In fact, many haloarchaea, including *Haloferax mediterranei, Haloferax denitrificans, Haloferax gibbonsii, Haloarcula marismortui,* and *Haloarcula vallismortis,* can grow in the presence of high concentrations of perchlorate and can use the substrate as an electron acceptor for anaerobic respiration [4.190].

4.5.3 Haloarchaea Are Adapted to High Salinity and Desiccating Conditions

The availability of water in the local environs is crucial to microorganisms, and thus the low water activity in salt-saturated water restricts the growth of most prokaryotes, which cannot grow below an a_w value of 0.900 [4.29] [4.148] [4.169] [4.275]. Though a previous study suggested that the water activity in brines on Mars would be too low even for haloarchaea [4.257], this is contrary to more recent calculations. In fact, several species of haloarchaea have demonstrated division at much lower values; researchers observed several species with a_w measurements between 0.687–0.728, and extrapolation of growth curves suggests values as low as 0.611 [4.247]. How do haloarchaea achieve osmophily and this ability to thrive in low water availability?

Some adaptations may be external and structural; even the haloarchaeal cell morphology may change to become smaller and more spherical in low water activity [4.71] [4.228] [4.229]. Shape may be the secret to the success of one of the most common haloarchaea in hypersaline sites around the world, *Haloquadratum walsbyi,* a relative of which inhabits the GSL north arm (Table 4.1) [4.4]. This species is notable for its square, flat shape, which likely serves to increase its surface area to absorb nutrients in its depleted environment [4.25] [4.59]. The outer armor of haloarchaea may be important; the cell wall structures of all haloarchaea are distinct from bacteria, composed of glycoprotein instead of a network of peptidoglycan and outer membrane. The glycoprotein layer has an excess of acidic (negatively charged) amino acids and sometimes negatively charged sulfate, suggesting a hydration shell could surround the cells and allow for nutrient import [4.142] [4.273]. An external protective feature of *Haloquadratum* is the production of halomucin, a secretion that protects the outside of these cells, which may work as a shield against low a_w [4.283].

Even molecular structure inside the cell is important as haloarchaeal proteins contain high concentrations of acidic amino acids, glutamic acid and aspartic acid [4.131]. The majority of these acidic amino acids are located on the external surface of the protein and/or at the N-terminus and may be found around the N-termini, which may stabilize helices through salt bridges [4.60] [4.67] and may help form a hydration shell due to the intracellular accumulation of K+ ions [4.56], reducing the electrostatic free energy of the proteins [4.131].

The ability to live in an osmotically challenging solvent such as the waters of GSL or a fluid inclusion inside minerals (Section 4.4.3) requires a balance of solute on both sides of the membrane [4.86] [4.185]. In all cells, Na+ ions from the environment will passively diffuse across the membrane into the cell, but haloarchaea will pump out the Na+ and exchange with K+ in response to the electrochemical gradient [4.142]. This is one way that haloarchaea can adjust the intracellular fluids to be compatible with the external saltwater. Some species accumulate high (molar) concentrations of K+ and Cl− ions within the cell [4.231]. A more prevalent strategy is the accumulation of the electroneutral organic "compatible solutes," such as sugars or lipids, that provide osmotic balance [4.48] [4.150] [4.218].

Some of the aforementioned adaptations can certainly aid in surviving desiccation. Many saline water bodies, including GSL, have levels that fluctuate with the seasons, and haloarchaea have adapted the ability to survive the dry spells under very low water activity in regions of evaporation. Desiccation typically causes cellular damage, but haloarchaea have developed specialized mechanisms to survive and repair this (see below), which is typically fatal [4.114] [4.126]. We have discussed the ability of haloarchaea to remain in minerals over geologic time (Section 4.3.4), and there are likely strategies that enable this, including chemotaxis and metabolic dormancy, though they are poorly understood and difficult to assess. Chemotaxis may enable entombment in minerals as lake water dries, by allowing cells to swim towards the fluid as the mineral lattice is forming. Researchers observed haloarchaeal cells moving toward the fluid inclusions as they were embedded in halite in the lab [4.73]. Recent work on a GSL *Haloarcula* species using computer simulations and holographic microscopy shows that the cells have a one hundred-fold higher efficiency in chemotaxis than the average computed for bacterial model systems [4.255]. When fluids and nutrients are limited, haloarchaea can modify the path to fulfill its needs, which translates into a survival strategy for desiccating conditions. Once sequestered, haloarchaea likely transition to a dormant state. We hypothesize that this involves gene regulation changes that allow the cells to survive but do not expend energy currency. This state of reduced metabolic activity during environmental stress may involve setting in place "machinery" for transitioning, including that which may respond to signals that conditions are favorable to be resuscitated and actively grow and reproduce [4.133]. Consuming the necromass of other organisms within mineral fluid inclusions may allow haloarchaea to meet their greatly diminished metabolic needs during this dormancy [4.254]; there will be more about this in Section 4.5.7.

4.5.4 Haloarchaea Are Resistant to Ultraviolet Radiation

During the summer months when the GSL levels are at their lowest, and sunlight levels are at their highest, UV resistance is an essential survival mechanism in this hostile environment. Though some haloarchaea species use the abundant visible light to produce ATP with the rudimentary light-collecting protein pump, bacterioruberin [4.231], the detrimental effects of UV irradiation must be dealt with. Haloarchaea are highly resistant to UV radiation, an observation first made in 1963 [4.58], and later measured in a GSL *Halorubrum* species [4.14]. Haloarchaea can endure high doses of UV with little impact on viability [4.10] [4.14] [4.151]. This is due to not only a robust suite of DNA repair, but also specialized photoprotective mechanisms. If entombed in salt, they may receive some shielding from the environment [4.72].

Photoprotection in haloarchaea is remarkable. These extremophiles display a combination of tricks, including pigmentation, radical oxygen species (ROS) scavenging, genome signatures, and polyploidy, which attenuate potential photodamage and reduce the burden on the DNA repair systems. Carotenoid pigments in the membranes of haloarchaea, the most visible characteristic of these cells in their environments (Figure 4.1), prevent damage to DNA and other cellular components from ROS created by photons of light from the UV spectrum [4.114] [4.224] [4.282]. They may also function as secondary pigments and light collectors such as they do in plants [4.281] to dissipate any excess excitation energy, which could protect light-gathering proteins such as bacteriorhodopsin [4.114]. Haloarchaea have evolved genome signatures that feature bipyrimidine limitations, which can ameliorate the cyclization of most susceptible adjacent pyrimidine nucleotides (5' to 3', TC then TT) when exposed to high UV doses [4.113]. Interestingly, haloarchaea are typically polyploid, up to 25 copies during their fastest growth phase [4.54], and this could also be a strategy for protecting genes simply by duplication [4.114] [4.126] [4.239].

All haloarchaea genomes examined show evidence of repair systems to address UV damage to DNA, but then so do all other organisms. The haloarchaeal suite is unique because it includes multiple excision repair and homologous recombination pathways that share lineages with both eukaryotic and bacterial cells [4.51] [4.155] [4.114] [4.121]. The genomes show multiple types of each of the enzymes necessary for excision repair pathways (e.g., *uvrABCD, mutY, alkA, nth*) [4.31] [4.121] [4.285]. A recent review [4.114] details each of these repair processes, which we will briefly outline here. Photoreactivation can repair UV-induced photoproducts in DNA directly, while nucleotide excision repair (NER) removes and replaces a variety of DNA lesions caused directly by sunlight. UV light can also cause indirect DNA damage through the creation of ROS, which is also repaired by NER or base excision repair that can fix oxidative base lesions and single-strand breaks. The homologous recombination pathway repairs double-strand breaks (DSB) in DNA simultaneously through exonuclease resection and subsequent ligation of broken ends. Genetics experiments in model organisms demonstrate that genes in excision repair and homologous recombination pathways are required for survival under conditions that cause ROS and DSBs [4.26] [4.54] [4.120] [4.156] [4.277].

4.5.5 Haloarchaea Are Resistant to Ionizing Radiation

The photoprotective strategies of haloarchaea and their DNA repair machinery discussed above may have broader impacts on survivability of other extremes, and they underscore in part the observation of ionizing radiation (IR) resistance of haloarchaea [4.219]. IR causes DSBs, which can be repaired by homologous recombinational repair, for example. There is certainly overlap with fixing ROS damage from both UV and IR (which can produce ROS by the radiolysis of water) [4.126], therefore the protection by carotenoids extends to IR. Also, the accumulation of intracellular salts (e.g., K+ and Cl- discussed in Section 4.5.3) in some haloarachaea species may mitigate the effects of ROS as well [4.122]. In *Halobacterium salinarum*, high intracellular concentration of Mn+ ions, probably associated with manganese–antioxidant complexes, provided IR protection [4.270]. Haloarchaeal polyploidy would also be applicable to IR resistance.

4.5.6 Haloarchaea May Have Novel Nutrient Attainment Strategies for Dormancy Periods

The mechanism for survival over geologic time (see Section 4.3.4) has not been explained from a metabolism standpoint. We do know that haloarchaea shrink in size under starvation [4.71] [4.228] [4.229], and that low-level metabolism occurs in haloarchaea entrapped in halite under laboratory conditions [4.245]. Perhaps metabolism is slow inside minerals, versus completely dormant, based on analogous work on bacteria in permafrost [4.111]. Not only are water and oxygen limited inside minerals, so are nutrients. Dormancy in nutrient-depleted fluid inclusions requires strategies that could involve shutting down metabolic functions, processing extracellular nutrients, consuming intracellular nutrients, or all of these. Polyploidy could not only provide protection from DNA damage, it may also provide nutrients, acting as a phosphorus storage polymer [4.37] [4.187]. DNA degradation by haloarchaea suggests cells may obtain external nutrients from their briny environs this way [4.37] [4.116]. However, starved haloarchaea may also degrade the extra DNA inside the cell for nutrient obtainment [4.284]. Since the brine in mineral fluid inclusions would contain all material from the solution, entombed haloarchaea would have access to other organisms simultaneously encapsulated, such as the glycerol-rich *Dunaliella* algae [4.143] [4.189] [4.228] or even bacteria such as *Salinibacter* species [4.85]. They would also have access to the necromass nutrients of microbes which have died in the fluid and any larger detritus lodged in the salt crust [4.158] [4.254].

4.5.7 Haloarchaea Are Resistant to Space Conditions

Many of the superpowers that we have described above would allow haloarchaea to live under some of the extremes associated with Mars. More pertinent data may be obtained in experiments on survival and growth of haloarchaea following exposure to space or simulation of space conditions since these may also include microgravity and atmospheric effects, for example [4.102]. Dried cultures of *Halorubrum chaoviator* spent 15 days in the European Space Agency's BIOPAN capsule that enabled exposure of samples to conditions of outer space [4.146] [4.147]. Though this species grew poorly and accumulated DNA damage, it did recover and resumed vegetative growth based on a small percentage of cells that survived in the culture. Perhaps isolated species is not the best way to test survivability, as it is possible that co-cultures mimic nature and different species may aid in protection. *Halococcus morrhuae* was co-cultured with biofilm-forming bacterium *Halomonas muralis* and exposed to space conditions in the EXPOSE-R2 mission on the International Space Station for more than a year [4.135]. This worked to the advantage of the haloarchaeon, which survived even better in the co-culture, but not the bacterium which died. Other haloarchaea have been exposed to space conditions [4.134] [4.188], but the results are not available.

Few experiments have been directed specifically at cold tolerance of haloarchaea (see Section 4.5.1), and this is a consideration for Mars. Brine habitats such as mineral inclusions are especially enticing given that the presence of salt would depress the freezing point of water [4.256]. Two haloarchaea species, *Halobacterium salinarum* and *Halorubrum chaoviator*, survived being dried and subjected to freezing temperatures as well as repeated freeze-thaw cycles, outperforming the control bacterial strains which died after the first round [4.146]. Liquid (salt-media) cultures of *H. salinarum* also survived exposure to Earth's cold stratosphere on helium balloons, but its viability was not as impressive as the psychrophile, *Halorubrum lacusprofundi*, which demonstrated almost complete resistance to these cold temperatures [4.50]. This haloarchaea species adjusted gene regulation, which is likely the secret of surviving environmental stress. We have observed this feature in a GSL *Haloarcula* strain when temperatures were shifted away from optimal conditions [4.3].

4.6 Extant or Extinct Haloarchaea on Mars?

Investigations suggested for exploration by NASA for this decade, 2013–2022, include the search for extant life, especially on Mars [4.211]. A finding of extant, or existing, life on Mars would imply that the microorganisms are flourishing in the conditions of the habitat. Whereas, we have found ancient life on Earth in halite (Section 4.3.4) that might be better described as dormant. Most would agree that if we found haloarchaea in Martian salt, even if in metabolic dormancy, it would still be considered extant life. Another possibility is that we locate molecular fossils of biosignatures, indicating life was there, but the timeline or changing conditions caused it to expire. A finding that suggests extinct life would point to life that thrived on Mars at one time but is now gone. Either discovery would be significant as the first example of life beyond Earth.

4.6.1 Haloarchaea as Extant Life

It has been recognized for some time that the disappearance of surface water, billions of years ago, on Mars during global climatic change would have resulted in a phase of hypersaline liquid water, and that if life were there, it would be halophilic [4.130] [4.221]. Our observations of biological evolution suggest that if aquatic life was present on Mars during that time of flowing rivers, oceans, and lakes, as it was on Earth, life would have evolved mechanisms to survive as the water became salty and then even saltier. We have a sample size of n=1, as Earth's lifeforms are all related to the Last Universal Common Ancestor, [4.125]. But what we know of life on our planet is that it exists where water exists in liquid form, and that salt has not been an obstacle. Our best guess for potential extant life on Mars would be microorganisms similar to extremophiles on Earth that can handle low water activity with aplomb, the haloarchaea. It is also a reasonable speculation that these organisms may have lost the hypersaline lakes they once swam in during better conditions for life on Mars, but they may still be alive in the evaporites left behind.

Hypersaline water as a starting place may not be conducive to the origin of life, but if less salty water is concentrated over time, countless microbial evolution studies on Earth life would support that life would evolve when confronted with high salt concentrations. As our astrobiology studies "follow the water," Knauth notes that Cl- ions literally follow the water since Cl- does not incorporate into silicate mineral structures, and it would outgas alongside water [4.124]. This geochemistry points to the likelihood that early oceans in our Solar System may have been quite salty and restricted to microorganisms. On Mars, sites where evaporite minerals could have been periodically subjected to dissolution and re-evaporation [4.192] may feature evaporite-entombed haloarchaea communities that are more diverse than those found in ancient fluid inclusions left undisturbed over the entire geologic time period [4.104].

Earth's analogue sites to Martian lacustrine systems can help us envision this transition. The modern GSL formed as a result of evaporation events in stages where the water became increasingly hypersaline when compared to its predecessor, Pleistocene Lake Bonneville [4.7]. The fauna of the lake changed from fish and snails to brine shrimp. Certainly, the microbiota evolved more quickly, eventually enriching the haloarchaea that we observe today [4.12]. Mars lost its thick atmosphere, surface water, and tectonics (Section 4.2) before the development of complex life forms occurred on its neighboring planet. This timeline of climatic change tells us that if Mars ever had life, it probably never saw the evolution to more complex life forms [4.124] [4.129].

4.6.2 Haloarchaea as Extinct Life

Perhaps extant life is no longer possible on Mars. If life once existed there but is no longer present, this extinct life may have left behind evidence of its existence. Biosignatures encompass all objects, patterns, or substances that could have been created only through biological means [4.35] [4.57]. They are useful tools for locating life elsewhere in the universe, as they can help us to extend timescales. There is a wide variety of potential biosignatures that could be found in evaporites of Mars. Haloarchaea-specific signatures should be considered.

Nucleic Acids as Biosignatures

Nucleic acids hold encoded information that not only dictates protein sequence, but also may give relatedness information for comparative genomics; therefore, this group of biomolecules is the most sought after due to information richness. DNA is more likely than RNA to have been preserved due to its greater stability. On Earth, including at modern GSL and ancient sites, DNA has been found preserved in both halite and gypsum [205]. Predictions of rates of DNA damage and hydrolysis suggest that DNA should be fragmented over the time scale of tens of thousands of years [4.138] [4.139] [4.198] [4.227] [4.274] and surviving molecules would feature intramolecular chemical crosslinking [4.95] [4.197]. However, a number of studies have reported the presence of ancient DNA from minerals. All of these calculations are determined from laboratory conditions and modeling in water (not brine), and the results are likely to be different in the presence of salts. Also, the numbers are calculated *in vitro,* not for DNA inside cells, and thus they do not account for cellular DNA repair mechanisms, which have been noted in ancient entombed microorganisms [4.111]. To date, there are numerous examples of DNA preservation in evaporite minerals, even over hundreds of millions of years [4.74] [4.212].

Proteins as Biosignatures

Proteins also may supply taxa-specific information, which makes them attractive as indicators of past lifeforms on Mars. On Earth, collagen was isolated from *Tyrannosaurus rex* bones, and sequenced to reveal relationships with the modern protein [4.230]. Protein subunits, amino acids, were detected inside halite fluid inclusions [4.194]. It is important to distinguish between amino acids that have been produced biotically or abiotically, however. Biotic processes produce only L-form amino acids, while abiotic processes produce a 50:50

ratio of D- and L-form amino acids, making it easy to distinguish between the two [4.9] [4.106] [4.216] [4.235].

Carbohydrates as Biosignatures

The polysaccharide, cellulose, a particularly promising carbohydrate biosignature, was discovered in Permian halite fluid inclusions [4.89]. This biomolecule is probably the most abundant in the modern Earth environment; nearly 100 gigatons are created by organisms on Earth each year, and some of it by simple cyanobacteria [4.44]. In addition, it is highly stable and relatively resistant to degradation by reagents and harsh conditions [4.128] [4.166] [4.183]. Although gamma radiation can degrade cellulose fibers, this requires very high doses [4.68]. This molecule is not associated with haloarchaea, but co-entombed cyanobacteria, which have been found in ancient minerals, may possess it [4.199]. Also, osmolytes, such as sugars, may be detected by Raman spectroscopy [4.110].

Lipids as Biosignatures

Lipids may have high preservation potential and could be found in evaporites of Mars. Although they do not provide taxa-specific information, they may point to categories of microorganisms. Carotenoids are found in many life forms, from plants to microorganisms, but they are especially appealing when considering Mars evaporites, as they are associated with halophilic life in Earth's hypersaline systems (Figure 4.1) [4.114]. Haloarchaeal carotenoids have Raman spectroscopy signatures [4.72] [4.109] [4.149] and may be detectable by Mars rover instrumentation. This technique was also used to identify ancient preserved carotenoid lipids in halite fluid inclusions [4.276]. Geologists have noted carotenoid preservation over time in GSL sediment cores as well [4.41]. Stability over geologic time is anticipated given the stability of cyclic and branched hydrocarbons [4.234]. It is important to note that sodium chloride does not have Raman signals, so its structure would not interfere with *in situ* studies on Mars.

Contamination and Biosignatures

Unfortunately, contamination has the potential to cloud useful data relating to biosignatures, and on Mars, the burden of proof in life-detection experiments will be even more strenuous. Halite studies on Earth have been criticized for not sufficiently controlling the introduction of modern biological material [4.98] [4.198]. Analyses of environmental samples through amplification via the polymerase chain reaction (PCR) and/or microbial cultivation techniques are of particular concern [4.74] [4.266]. Also, some regions of an evaporite formation may be in contact with the modern environment, and thus the age of the recovered microorganisms is in question and the geological characterization of sampling site locations become critical. Extensive surface sterilization is one possible solution to contamination, but this is difficult to attain on another planet while sampling remotely. One such method of sterilization involves purposeful contamination of the sample, before surface sterilization, to serve as an internal marker and to normalize background signal [4.225]. This may work in the laboratory but not *in situ* on Mars.

Contamination has already had serious implications in the search for life on Mars. In 1976, the Viking landers arrived on Mars, where they carried out several experiments. In one of these experiments, organic compounds, chloromethane and dichloromethane, were discovered when Martian soil samples were heated [4.176] [4.210] [4.136]. Concern was raised that the compounds could have been introduced to the soil through contamination by cleaning products from Earth. This has remained a contentious issue for decades, especially because recent research into perchlorate has altered perceptions of the 1976 experiment. When perchlorate is heated, it acts as an oxidant, and chloromethane and dichloromethane may result. This raises the possibility that the Viking 1 and 2 did find organic compounds after all [4.136]. More rigorous decontamination methods, such as those detailed above, may have avoided this controversy entirely.

4.7 Conclusions and Insights

The search for life begs the question, what is life? Characteristics for this definition include entities that are self-sustained, enclosed, and possessing the ability to undergo evolution [4.20]. Biology as we know it uses small subunits and defined chemistry to encode information and produce proteins to do cellular work. Carl Sagan was very philosophical regarding definitions of life, suggesting a biochemist would be concerned with the concept of metabolic rates through the use of chemical reactions, a physicist would focus on thermodynamics and a continuous increase in organization, and a physiologist would define life regarding response to signaling and outside stimuli [4.223]. We have only thus far observed "CHON life" (based on Carbon, Hydrogen, Oxygen and Nitrogen) on Earth, but other elemental substitutions for life as we do not know it have been considered [4.269]. Life detection approaches should not presuppose a particular biochemistry are critical for astrobiology; however, most of our studies assume the Martian biosphere would be carbon based and would follow nutrient requirements similar to those of terrestrial microorganisms. When we look to Mars, we necessarily use our abundant knowledge of Earth's life, and we inquire regarding which life forms from Earth might best survive on the Red Planet, and where might we find them?

Why are salts good exploration targets in thinking about habitable locations? The mineral–water–organic interface has a lot to offer for the evolution of life [4.173] and the maintenance of bioenergetics and cellular processes [4.226]. The preservation potential of organics in minerals is high, including shielding effects from radiation damage [4.53] [4.72]. As related to Mars, the sheer abundance of chloride and sulfate salts on the surface and subsurface provide an enormous resource, and they are extensively mapped (Section 4.2). Evaporites are distributed at sites with varied geological contexts: higher terrains occurring from surface runoff, groundwater upwellings, and sites associated with hydrothermal activity [4.193]. Liquid water can only exist in Martian surface temperatures when it contains salts, and the freezing temperature is depressed [4.256]. Ancient lake basins are targets for future Mars life-detection rover missions [4.91] [4.171], and we should be practicing our interdisciplinary techniques in similar environs on Earth, such as GSL and the surrounding evaporites. A possible site for a crewed mission to Mars is Columbus Crater in Terra Sirenum, which contains evaporite salts [4.278].

Why should we consider haloarchaea? It has been suggested for some time that haloarchaea are likely candidates for extant life [4.130] [4.188]. They are certainly not the only extremophile taxon, but they are abundant, versatile, and have many characteristics that suggest they could sustain life in Martian conditions (Table 4.2). For example, consider all of the strategies regarding protection from and repair of UV and IR cellular damage. The resistance of haloarchaeal cells to radiation is exceedingly important when considering the timeline of possible entombment of extant life in minerals on Mars over billions of years. And the detection of haloarchaea or their biosignatures (e.g., carotenoids) might be possible *in situ* within the mineral record.

In our search for extant life beyond our planet, Mars is certainly an attractive location for all the reasons outlined in this chapter, but most important is the shared history in the formation of our solar system and co-existence in our habitable zone. Life exists on Earth, and perhaps it does on Mars, or it once did. The lack of tectonic processes on Mars means that the rock record is less disturbed over time than that of Earth, which bodes well for preservation, but the timescale of billions of years since the loss of standing surface water does not. Martian evaporite minerals on the surface and near-subsurface are accessible to rovers for *in situ* analyses, making these sites a priority for Mars missions using existing instrumentation. Studies of Earth hypersaline lakes, such as GSL, indicate they are all declining in elevation [4.279], and thus it is an opportune time to understand what happened on Mars, especially concerning potential aquatic life, as water became more saline and eventually dried into mineral deposits. Analogues such as GSL and its haloarchaeal life provide a rich resource for astrobiology studies and the search for extant life on Mars.

Acknowledgments

Funding for the primary research reported within was from a NASA Space Grant Prime Award (NNX15A124H/10037896WEST), the W.M. Keck Foundation, and the Lawrence T. and Janet T. Dee Foundation.

References

4.1. Adams, T.C., Salt migration to the northwest body of Great Salt Lake, Utah. *Science*, 143, 3610, 1027–1029, 1964.

4.2. Almeida-Dalmet, S. and Baxter, B.K., Unexpected complexity at salinity saturation: Microbial diversity of the north arm of Great Salt Lake, Utah, in: *Great Salt Lake Biology: A Terminal Lake in a Time of Change*, B.K. Baxter and J.K. Butler (Eds.), Springer, Cham, 2020.

4.3. Almeida-Dalmet, S., Litchfield, C.D., Gillevet, P., Baxter, B.K., Differential gene expression in response to salinity and temperature in a *Haloarcula* strain from Great Salt Lake, Utah. *Genes*, 9, 1, 52, 2018.

4.4. Almeida-Dalmet, S., Sikaroodi, M., Gillevet, P.M., Litchfield, C.D., Baxter, B.K., Temporal study of the microbial diversity of the North Arm of Great Salt Lake, Utah, US. *Microorganisms*, 3, 3, 310–26, 2015.

4.5. Andrews-Hanna, J.C. and Lewis, K.W., Early Mars hydrology: 2. Hydrological evolution in the Noachian and Hesperian epochs. *J. Geophys. Res.: Planets*, 116, E2, 2011.

4.6. Andrews-Hanna, J.C., Phillips, R.J., Zuber, M.T., Meridiani Planum and the global hydrology of Mars. *Nature*, 446, 7132, 163–166, 2007.
4.7. Atwood, G., Wambeam, T.J., Anderson, N.J., The present as a key to the past: Paleoshoreline correlation insights from Great Salt Lake, in: *Lake Bonneville a Scientific Update*, C.G. Oviatt and J.F. Shroder (Eds.), Elsevier, Netherlands, 2016.
4.8. Ayan, B. and Dilbar, T., *Separation and phylogenetic analysis of halophilic acidophilic bacteria in Black Lake in the Korla region of Xinjiang*, 2017, Accessed on 2/18/20. Retrieved from https://www.ncbi.nlm.nih.gov/nucleotide/MG559533.1?report=genbankandlog$=nuclalignandblast_rank=6andRID=3M13EKMK016.
4.9. Bada, J.L., State-of-the-art instruments for detecting extraterrestrial life. *Proc. Natl. Acad. Sci.*, 98, 3, 797–800, 2001.
4.10. Baliga, N.S., Bjork, S.J., Bonneau, R., Pan, M., Iloanusi, C., Kottemann, M.C., Hood, L., DiRuggiero, J., Systems level insights into the stress response to UV radiation in the halophilic asrchaeon *Halobacterium* NRC-1. *Genome Res.*, 14, 1025–1035, 2004.
4.11. Barrett, K.L. and Belovsky, G.E., Invertebrates and phytoplankton of Great Salt Lake: Is salinity the driving factor?, in: *Great Salt Lake Biology: A Terminal Lake in a Time of Change*, B.K. Baxter and J.K. Butler (Eds.), Springer, Cham, 2020.
4.12. Baxter, B.K., Great Salt Lake microbiology: A historical perspective. *Int. Microbiol.*, 1–17, 2018.
4.13. Baxter, B.K. and Butler, J.K., Climate change and Great Salt Lake, in: *Great Salt Lake Biology: A Terminal Lake in a Time of Change*, B.K. Baxter and J.K. Butler (Eds.), Springer, Cham, 2020.
4.14. Baxter, B.K., Eddington, B., Riddle, M.R., Webster, T.N., Avery, B.J., Great Salt Lake halophilic microorganisms as models for astrobiology: Evidence for desiccation tolerance and ultraviolet radiation resistance, in: *Instruments, Methods, and Missions for Astrobiology X*, R.B. Hoover, G.V. Levin, A.Y. Rozanov, P.C.W. Davies (Eds.), p. 6694, SPIE, Bellingham, WA, 2007, 669415.
4.15. Baxter, B.K., Litchfield, C.D., Sowers, K., Griffith, J.D., Dassarma, P.A., Dassarma, S., Microbial diversity of Great Salt Lake, in: *Adaptation to Life at High Salt Concentrations in Archaea, Bacteria, and Eukarya. Cellular Origin, Life in Extreme Habitats and Astrobiology*, vol. 9, N. Gunde-Cimerman, A. Oren, A. Plemenitaš (Eds.), pp. 9–25, Springer, Dordrecht, Netherlands, 2005.
4.16. Baxter, B.K., Mangalea, M.R., Willcox, S., Sabet, S., Nagoulat, M.N., Griffith, J.G., Haloviruses of Great Salt Lake: A model for understanding viral diversity, in: *Halophiles and Hypersaline Environments: Current Research and Future Trends*, A. Ventosa, A. Oren, Y. Ma (Eds.), Springer, Dordrecht, Netherlands, 2011.
4.17. Baxter, B.K. and Zalar, P., The extremophiles of Great Salt Lake: Complex microbiology in a dynamic hypersaline ecosystem, in: *Model Ecosystems in Extreme Environments*, J. Seckbach and P.H. Rampelotto (Eds.), Springer, Netherlands, 2019.
4.18. Benlloch, S., Acinas, S.G., Antón, J., Lopez-Lopez, A., Luz, S.P., Rodríguez-Valera, F., Archaeal biodiversity in crystallizer ponds from a solar saltern: Culture versus PCR. *Microb. Ecol.*, 41, 12–19, 2001.
4.19. Benlloch, S., Martínez-Murcia, A.J., Rodríguez-Valera, F., Sequencing of bacterial and archaeal 16S rRNA genes directly amplified from a hypersaline environment. *Syst. Appl. Microbiol.*, 18, 4, 574–581, 1995.
4.20. Benner, S.A., Defining life. *Astrobiology*, 10, 10, 1021–1030, 2010.
4.21. Bertucci, C., Duru, F., Edberg, N., Fraenz, M., Martinecz, C., Szego, K., Vaisberg, O., The induced magnetospheres of Mars, Venus, and Titan. *Space Sci. Rev.*, 162, 1-4, 113–171, 2011.
4.22. Bibring, J.P., Langevin, Y., Gendrin, A., Gondet, B., Poulet, F., Berthé, M., Soufflot, A., Arvidson, R., Mangold, N., Mustard, J., Drossart, P., Mars surface diversity as revealed by the OMEGA/Mars Express observations. *Science*, 307, 5715, 1576–1581, 2005.

4.23. Bishop, J.L., Loizeau, D., McKeown, N.K., Saper, L., Dyar, M.D., Des Marais, D.J., Parente, M., Murchie, S.L., What the ancient phyllosilicates at Mawrth Vallis can tell us about possible habitability on early Mars. *Planet. Space Sci.*, 86, 130–149, 2013.
4.24. Blumthaler, M., Ambach, W., Ellinger, R., Increase in solar UV radiation with altitude. *J. Photochem. Photobiol. B: Biol.*, 39, 130–134, 1997.
4.25. Bolhuis, H., Palm, P., Wende, A., Falb, M., Rampp, M., Rodriguez-Valera, F., Pfeiffer, F., Oesterhelt, D., The genome of the square archaeon Haloquadratum walsbyi: Life at the limits of water activity. *BMC Genomics*, 7, 1, 169, 2006.
4.26. Boubriak, I., Ng, W.L., DasSarma, P., DasSarma, S., Crowley, D.J., McCready, S.J., Transcriptional responses to biologically relevant doses of UV-B radiation in the model archaeon, *Halobacterium* sp. NRC-1. *Saline Syst.*, 4, 13, 2008.
4.27. Bowers, K.J. and Wiegel, J., Temperature and pH optima of extremely halophilic archaea: A mini-review. *Extremophiles*, 15, 2, 119–128, 2011.
4.28. Brain, D.A., McFadden, J.P., Halekas, J.S., Connerney, J.E.P., Bougher, S.W., Curry, S., Dong, C.F., Dong, Y., Eparvier, F., Fang, X., Fortier, K., The spatial distribution of planetary ion fluxes near Mars observed by MAVEN. *Geophys. Res. Lett.*, 42, 21, 9142–9148, 2015.
4.29. Brown, A.D., Microbial water stress. *Bacteriol. Rev.*, 40, 803–846, 1976.
4.30. Cannon, J.S. and Cannon, M.A., The Southern Pacific Railroad trestle – past and present, in: *Great Salt Lake: An Overview of Change*, J.W. Gwynn (Ed.), pp. 283–294, Special Publication of the Department of Natural Resources, Salt Lake City, 2002.
4.31. Capes, M.D., Coker, J.A., Gessler, R., Grinblat-Huse, V., DasSarma, S.L., Jacob, C.G., Kim, J.M., DasSarma, P., DasSarma, S., The information transfer system of halophilic archaea. *Plasmid*, 65, 77–101, 2011.
4.32. Carr, M.H. and Wänke, H., Earth and Mars: Water inventories as clues to accretional histories. *Icarus*, 98, 1, 61–71, 1992.
4.33. Carter, J., Poulet, F., Bibring, J.P., Mangold, N., Murchie, S., Hydrous minerals on Mars as seen by the CRISM and OMEGA imaging spectrometers: Updated global view. *J. Geophys. Res.: Planets*, 118, 4, 831–858, 2013.
4.34. Catling, D.C., Cockell, C.S., McKay, C.P., *Ultraviolet Radiation on the Surface of Mars [NASA Technical Report]*, 1999, https://ntrs.nasa.gov/search.jsp?R=20000110306.
4.35. Chan, M.A., Hinman, N.W., Potter-McIntyre, S.L., Schubert, K.E., Gillams, R.J., Awramik, S.M., Boston, P.J., Bower, D.M., Des Marais, D.J., Farmer, J.D., Jia, T.Z., Deciphering biosignatures in planetary contexts. *Astrobiology*, 19, 9, 1075–1102, 2019.
4.36. Chevrier, V.F. and Rivera-Valentin, E.G., Formation of recurring slope lineae by liquid brines on present-day Mars. *Geophys. Res. Lett.*, 39, 21, 2012. https://doi.org/10.1029/2012GL054119.
4.37. Chimileski, S., Dolas, K., Naor, A., Gophna, U., Papke, R.T., Extracellular DNA metabolism in *Haloferax volcanii*. *Front. Microbiol.*, 5, 57, 2014.
4.38. Clark, B.C., Morris, R.V., McLennan, S.M., Gellert, R., Jolliff, B., Knoll, A.H., Squyres, S.W., Lowenstein, T.K., Ming, D.W., Tosca, N.J., Yen, A., Chemistry and mineralogy of outcrops at Meridiani Planum. *Earth Planet. Sci. Lett.*, 240, 1, 73–94, 2005.
4.39. Clark, B.C. and Van Hart, D.C., The salts of Mars. *Icarus*, 45, 2, 370–378, 1981.
4.40. Cockell, C.S., The ultraviolet history of the terrestrial planets—Implications for biological evolution. *Planet. Space Sci.*, 48, 2, 203–214, 2000. https://doi.org/10.1016/S0032-0633(99)00087-2.
4.41. Collister, J.W. and Schamel, S., Lipid composition of recent sediments from the Great Salt Lake, in: *Great Salt Lake: An Overview of Change (DNR special publication)*, J.W. Gwynn (Ed.), pp. 128–142, Utah Geological Survey Publication, 2002.
4.42. Connerney, J.E.P., Espley, J., Lawton, P., Murphy, S., Odom, J., Oliversen, R., Sheppard, D., The MAVEN magnetic field investigation. *Space Sci. Rev.*, 195, 1-4, 257–291, 2015.

4.43. Conover, M.R. and Bell, M.E., Importance of Great Salt Lake to pelagic birds: Eared grebes, phalaropes, gulls, ducks, and white pelicans, in: *Great Salt Lake Biology: A Terminal Lake in a Time of Change*, B.K. Baxter and J.K. Butler (Eds.), Springer, Cham, 2020.
4.44. Cox, P.M., Betts, R.A., Jones, C.D., Spall, S.A., Totterdell, I.J., Acceleration of global warming due to carbon-cycle feedbacks in a coupled climate model. *Nature*, 408, 184–187, 2000.
4.45. Craddock, R.A. and Lorenz, R.D., The changing nature of rainfall during the early history of Mars. *Icarus*, 293, 172–179, 2017.
4.46. Crosman, E.T. and Horel, J.D., Modis-derived surface temperature of the Great Salt Lake. *Remote Sens. Environ.*, 113, 73–81, 2009.
4.47. Cull, S.C., Arvidson, R.E., Catalano, J.G., Ming, D.W., Morris, R.V., Mellon, M.T., Lemmon, M., Concentrated perchlorate at the Mars Phoenix landing site: Evidence for thin film liquid water on Mars. *Geophys. Res. Lett.*, 37, 22, 2010.
4.48. Da Costa, M.S., Santos, H., Galinski, E.A., An overview of the role and diversity of compatible solutes in Bacteria and Archaea, in: *Biotechnology of Extremophiles*, pp. 117–153, Springer, Berlin, Heidelberg, 1998.
4.49. Daniels [*sic* Daines], L.L., On the flora of Great Salt Lake. *Am. Nat.*, 51, 499–506, 1917.
4.50. DasSarma, P., Laye, V.J., Harvey, J., Reid, C., Shultz, J., Yarborough, A., Lamb, A., Koske-Phillips, A., Herbst, A., Molina, F., Grah, O., Survival of halophilic Archaea in Earth's cold stratosphere. *Int. J. Astrobiol.*, 16, 4, 321–327, 2017.
4.51. DasSarma, S., Kennedy, S.P., Berquist, B., Ng, W.V., Baliga, N.S., Spudich, J.L., Krebs, M.P., Eisen, J.A., Johnson, C.H., Hood, L., Genomic perspective on the photobiology of *Halobacterium* species NRC-1, a phototrophic, phototactic, and UV-tolerant haloarchaeon. *Photosynth. Res.*, 70, 1, 3–17, 2001.
4.52. Davila, A.F., Duport, L.G., Melchiorri, R., Jaenchen, J., Valea, S., de Los Rios, A., Fairen, A.G., Moehlmann, D., McKay, C.P., Ascaso, C., Wierzchos, J., Hygroscopic salts and the potential for life on Mars. *Astrobiology*, 10, 6, 617–628, 2010.
4.53. Davila, A.F., Skidmore, M., Fairén, A.G., Cockell, C., Schulze-Makuch, D., New priorities in the robotic exploration of Mars: The case for *in situ* search for extant life. *Astrobiology*, 10, 7, 705–710, 2010.
4.54. Delmas, S., Shunburne, L., Ngo, H.P., Allers, T., Mre11-Rad50 promotes rapid repair of DNA damage in the polyploid archaeon *Haloferax volcanii* by restraining homologous recombination. *PLoS Genet.*, 5, #e1000552, 2009.
4.55. Denner, E.B., McGenity, T.J., Busse, H.J., Grant, W.D., Wanner, G., Stan-Lotter, H., *Halococcus salifodinae* sp. nov., an archaeal isolate from an Austrian salt mine. *Int. J. Syst. Evol. Microbiol.*, 44, 4, 774–780, 1994.
4.56. Dennis, P.P. and Shimmin, L.C., Evolutionary divergence and salinity-mediated selection in halophilic archaea. *Microbiol. Mol. Biol. Rev.*, 61, 1, 90–104, 1997.
4.57. Des Marais, D.J., Allamandola, L.J., Benner, S.A., Boss, A.P., Deamer, D., Falkowski, P.G., Farmer, J.D., Hedges, S.B., Jakosky, B.M., Knoll, A.H., Liskowsky, D.R., The NASA astrobiology roadmap. *Astrobiology*, 3, 2, 219–235, 2003.
4.58. Dundas, I.D. and Larsen, H., A study on the killing by light of photosensitized cells of *Halobacterium salinarium*. *Arch. Microbiol.*, 46, 19–28, 1963.
4.59. Dyall-Smith, M.L., Pfeiffer, F., Klee, K., Palm, P., Gross, K., Schuster, S.C., Rampp, M., Oesterhelt, D., *Haloquadratum walsbyi*: Limited diversity in a global pond. *PLoS One*, 6, 6, 2011.
4.60. Dym, O., Mevarech, M., Sussman, J.L., Structural features that stabilize halophilic malate dehydrogenase from an archaebacterium. *Science*, 267, 5202, 1344–1346, 1995.
4.61. Eardley, A.J. and Stringham, B., Selenite crystals in the clays of Great Salt Lake. *J. Sediment. Petrol.*, 22, 4, 234–238, 1952.

4.62. Eastwood, J.P., Hietala, H., Toth, G., Phan, T.D., Fujimoto, M., What controls the structure and dynamics of Earth's magnetosphere? *Space Sci. Rev.*, 188, 1-4, 251–286, 2015.

4.63. Edwards, H.G., Villar, S.E.J., Parnell, J., Cockell, C.S., Lee, P., Raman spectroscopic analysis of cyanobacterial gypsum halotrophs and relevance for sulfate deposits on Mars. *Analyst*, 130, 6, 917–923, 2005.

4.64. Ehlmann, B.L. and Edwards, C.S., Mineralogy of the Martian surface. *Annu. Rev. Earth Planet. Sci.*, 42, 291–315, 2014.

4.65. Ehlmann, B.L., Mustard, J.F., Swayze, G.A., Clark, R.N., Bishop, J.L., Poulet, F., Des Marais, D.J., Roach, L.H., Milliken, R.E., Wray, J.J., Barnouin-Jha, O., Identification of hydrated silicate minerals on Mars using MRO-CRISM: Geologic context near Nili Fossae and implications for aqueous alteration. *J. Geophys. Res.: Planets*, 114, E2, 2009.

4.66. Ehlmann, B.L., Mustard, J.F., Murchie, S.L., Bibring, J.P., Meunier, A., Fraeman, A.A., Langevin, Y., Subsurface water and clay mineral formation during the early history of Mars. *Nature*, 479, 7371, 53–60, 2011.

4.67. Eisenberg, H., Mevarech, M., Zaccai, G., Biochemical, structural, and molecular genetic aspects of halophilism, in: *Advances in Protein Chemistry*, vol. 43, pp. 1–62, Academic Press, 1992.

4.68. Ershov, B.G., Radiation-chemical degradation of cellulose and other polysaccharides. *Russ. Chem. Rev.*, 67, 4, 315–334, 1998.

4.69. Fendrich, C. and Schink, B., Degradation of glucose, glycerol, and acetate by aerobic bacteria in surface water of Great Salt Lake, Utah, U.S.A. *Syst. Appl. Microbiol.*, 11, 94–96, 1988.

4.70. Fendrihan, S., Bérces, A., Lammer, H., Musso, M., Rontó, G., Polacsek, T.K., Holzinger, A., Kolb, C., Stan-Lotter, H., Investigating the effects of simulated Martian ultraviolet radiation on *Halococcus dombrowskii* and other extremely halophilic archaebacteria. *Astrobiology*, 9, 1, 104–112, 2009. https://doi.org/10.1089/ast.2007.0234.

4.71. Fendrihan, S., Dornmayr-Pfaffenhuemer, M., Gerbl, F.W., Holzinger, A., Grösbacher, M., Briza, P., Erler, A., Gruber, C., Plätzer, K., Stan-Lotter, H., Spherical particles of halophilic archaea correlate with exposure to low water activity–implications for microbial survival in fluid inclusions of ancient halite. *Geobiology*, 10, 5, 424–433, 2012.

4.72. Fendrihan, S., Musso, M., Stan-Lotter, H., Raman spectroscopy as a potential method for the detection of extremely halophilic archaea embedded in halite in terrestrial and possibly extraterrestrial samples. *J. Raman Spectrosc.*, 40, 12, 1996–2003, 2009.

4.73. Fendrihan, S. and Stan-Lotter, H., Survival of halobacteria in fluid inclusions as a model of possible biotic survival in Martian halite, in: *Mars and Planetary Science and Technology*, H.N. Teodorescu and H.S. Griebel (Eds.), pp. 9–18, Performantica Press, Iasi, Romania, 2004.

4.74. Fish, S.A., Shepherd, T.J., McGenity, T.J., Grant, W.D., Recovery of 16S ribosomal RNA gene fragments from ancient halite. *Nature*, 420, 432–436, 2002.

4.75. Fishbaugh, K.E., Byrne, S., Herkenhoff, K.E., Kirk, R.L., Fortezzo, C., Russell, P.S., McEwen, A., Evaluating the meaning of "layer" in the Martian north polar layered deposits and the impact on the climate connection. *Icarus*, 205, 1, 269–282, 2010.

4.76. Fishbaugh, K.E., Poulet, F., Chevrier, V., Langevin, Y., Bibring, J.P., On the origin of gypsum in the Mars north polar region. *J. Geophys. Res.: Planets*, 112, E7, 2007.

4.77. Franzmann, P.D., Stackebrandt, E., Sanderson, K., Volkman, J.K., Cameron, D.E., Stevenson, P.L., McMeekin, T.A., Burton, H.R., *Halobacterium lacusprofundi* sp. nov., a halophilic bacterium isolated from Deep Lake, Antarctica. *Syst. Appl. Microbiol.*, 11, 1, 20–27, 1988.

4.78. Frederick, E., *On the bacterial flora of Great Salt Lake and the viability of other microorganisms in Great Salt Lake water.* (Master's thesis), 65 pp., University of Utah, 1924.

4.79. Fukushi, K., Sekine, Y., Sakuma, H., Morida, K., Wordsworth, R., Semiarid climate and hyposaline lake on early Mars inferred from reconstructed water chemistry at Gale. *Nat. Commun.*, 10, 1, 1–11, 2019.

4.80. Gibson, J.A., Miller, M.R., Davies, N.W., Neill, G.P., Nichols, D.S., Volkman, J.K., Unsaturated diether lipids in the psychrotrophic archaeon *Halorubrum lacusprofundi*. *Syst. Appl. Microbiol.*, 28, 1, 19–26, 2005.
4.81. Gilbert, G.K., *Lake Bonneville. U.S. Geological Survey Monograph 1*, 438 pp., 1890.
4.82. Gornitz, V.M. and Schreiber, B.C., Displacive halite hoppers from the Dead Sea; some implications for ancient evaporite deposits. *J. Sediment. Res.*, 51, 3, 787–794, 1981.
4.83. Goudge, T.A., Mustard, J.F., Head, J.W., Fassett, C.I., Wiseman, S.M., Assessing the mineralogy of the watershed and fan deposits of the Jezero crater paleolake system, Mars. *J. Geophys. Res.*, 120, 4, 775–805, 2015.
4.84. Goudie, A.S. and Viles, H.A., Weathering and the global carbon cycle: Geomorphological perspectives. *Earth-Sci. Rev.*, 113, 1-2, 59–71, 2012.
4.85. Gramain, A., Díaz, G.C., Demergasso, C., Lowenstein, T.K., McGenity, T.J., Archaeal diversity along a subterranean salt core from the Salar Grande (Chile). *Environ. Microbiol.*, 13, 2105–2121, 2011.
4.86. Grant, W.D., Life at low water activity. *Philos. Trans. Royal Soc. Lond. Ser. B: Biol. Sci.*, 359, 1448, 1249–1267, 2004.
4.87. Grant, W.D., Gemmell, R.T., McGenity, T.J., Halobacteria: The evidence for longevity. *Extremophiles*, 2, 3, 279–287, 1998.
4.88. Grant, W.D. and Larsen, H., Group Ill. Extremely halophilic archaeobacteria. Order Halobacteriales ord. nov, in: *Bergey's Manual of Systematic Bacteriology*, J.T. Staley, M.P. Bryant, N. Pfennig, J.G. Holt (Eds.), pp. 2216–2233, Williams and Wilkins, Baltimore, Maryland, USA, 1989.
4.89. Griffith, J.D., Willcox, S., Powers, D.W., Nelson, R., Baxter, B.K., Discovery of abundant cellulose microfibers encased in 250 Ma Permian halite: A macromolecular target in the search for life on other planets. *Astrobiology*, 8, 2, 215–228, 2008.
4.90. Grotzinger, J.P., Gupta, S., Malin, M.C., Rubin, D.M., Schieber, J., Siebach, K., Sumner, D.Y., Stack, K.M., Vasavada, A.R., Arvidson, R.E., Calef, F., Deposition, exhumation, and paleoclimate of an ancient lake deposit, Gale crater, Mars. *Science*, 350, 6257, aac7575, 2015.
4.91. Grotzinger, J.P., Sumner, D.Y., Kah, L.C., Stack, K., Gupta, S., Edgar, L., Rubin, D., Lewis, K., Schieber, J., Mangold, N., Milliken, R., A habitable fluvio-lacustrine environment at Yellowknife Bay, Gale Crater, Mars. *Science*, 343, 6169, 1242777, 2014.
4.92. Gruber, C., Legat, A., Pfaffenhuemer, M., Radax, C., Weidler, G., Busse, H.J., Stan-Lotter, H., *Halobacterium noricense* sp. nov., an archaeal isolate from a bore core of an alpine Permian salt deposit, classification of *Halobacterium* sp. NRC-1 as a strain of *H. salinarum* and emended description of *H. salinarum*. *Extremophiles*, 8, 6, 431–439, 2004.
4.93. Haberle, R.M., McKay, C.P., Schaeffer, J., Cabrol, N.A., Grin, E.A., Zent, A.P., Quinn, R., On the possibility of liquid water on present-day Mars. *J. Geophys. Res.*, 106, E10, 23317–23326, 2001.
4.94. Hagen, C.A., Hawrylewicz, E.J., Anderson, B.T., Cephus, M.L., Effect of ultraviolet on the survival of bacteria airborne in simulated Martian dust clouds. *Life Sci. Space Res.*, 8, 53–58, 1970.
4.95. Hansen, A.J., Mitchell, D.L., Wiuf, C., Paniker, L., Brand, T.B., Binladen, J., Gilichinsky, D.A., Rønn, R., Willerslev, E., Crosslinks rather than strand breaks determine access to ancient DNA sequences from frozen sediments. *Genetics*, 173, 2, 1175–1179, 2006.
4.96. Hartmann, R., Sickinger, H.D., Oesterhelt, D., Anaerobic growth of halobacteria. *Proc. Natl. Acad. Sci.*, 77, 7, 3821–3825, 1980.
4.97. Hauber, E., Brož, P., Jagert, F., Jodłowski, P., Platz, T., Very recent and wide-spread basaltic volcanism on Mars. *Geophys. Res. Lett.*, 38, 10, 2011. https://doi.org/10.1029/2011GL047310.

4.98. Hebsgaard, M.B., Phillips, M., Willerslev, E., Geologically ancient DNA: Fact or artefact? *TRENDS Microbiol.*, 13, 212–220, 2005.
4.99. Hecht, M.H., Metastability of liquid water on Mars. *Icarus*, 156, 2, 373–386, 2002.
4.100. Hecht, M.H., Kounaves, S.P., Quinn, R.C., West, S.J., Young, S.M.M., Ming, D.W., Catling, D.C., Clark, B.C., Boynton, W.V., Hoffman, J., DeFlores, L.P., Detection of perchlorate and the soluble chemistry of Martian soil at the Phoenix lander site. *Science*, 325, 5936, 64–67, 2009.
4.101. Herrero-Fernández, M.J., Escavy Fernández, J.I., Schreiber, B.C., Thenardite after mirabilite deposits as a cool climate indicator in the geological record: lower Miocene of central Spain. *Clim. Past*, 11, 1, 1–13, 2015.
4.102. Horneck, G., Klaus, D.M., Mancinelli, R.L., Space microbiology. *Microbiol. Mol. Biol. Rev.*, 74, 1, 121–156, 2010.
4.103. Hescox, M.A. and Carlberg, D.M., Photoreactivation in *Halobacterium cutirubrum*. *Can. J. Microbiol.*, 18, 981–985, 1972.
4.104. Huby, T.J., Clark, D.R., McKew, B.A., McGenity, T.J., Extremely halophilic archaeal communities are resilient to short-term entombment in halite. *Environ. Microbiol.* 00(00), 00-00, 2020.
4.105. Hurowitz, J.A., Grotzinger, J.P., Fischer, W.W., McLennan, S.M., Milliken, R.E., Stein, N., Vasavada, A.R., Blake, D.F., Dehouck, E., Eigenbrode, J.L., Fairen, A.G., Redox stratification of an ancient lake in Gale crater, Mars. *Science*, 356, 6341, eaah6849, 2017.
4.106. Hutt, L.D., Glavin, D.P., Bada, J.L., Mathies, R.A., Microfabricated capillary electrophoresis amino acid chirality analyzer for extraterrestrial exploration. *Anal. Chem.*, 71, 18, 4000–4006, 1999.
4.107. Hviid, S.F., Madsen, M.B., Gunnlaugsson, H.P., Goetz, W., Knudsen, J.M., Hargraves, R.B., Smith, P., Britt, D., Dinesen, A.R., Mogensen, C.T., Olsen, M., Magnetic properties experiments on the Mars Pathfinder lander: Preliminary results. *Science*, 278, 5344, 1768–1770, 1997.
4.108. Jakosky, B.M. and Phillips, R.J., Mars' volatile and climate history. *Nature*, 412, 6843, 237, 2001.
4.109. Jehlička, J. and Oren, A., Raman spectroscopy in halophile research. *Front. Microbiol.*, 4, 380, 2013.
4.110. Jehlička, J., Oren, A., Edwards, H.G., Raman spectra of osmotic solutes of halophiles. *J. Raman Spectrosc.*, 43, 8, 1134–1140, 2012.
4.111. Johnson, S.S., Hebsgaard, M.B., Christensen, T.R., Mastepanov, M., Nielsen, R., Munch, K., Rønn, R., Ancient bacteria show evidence of DNA repair. *Proc. Natl. Acad. Sci.*, 104, 36, 14401–14405, 2007.
4.112. Jones, B.F., Naftz, D.L., Spencer, R.J., Oviatt, C.G., Geochemical evolution of Great Salt Lake, Utah, USA. *Aquat. Geochem.*, 15, 1-2, 95–121, 2009.
4.113. Jones, D.L. and Baxter, B.K., Bipyrimidine signatures as a photoprotective genome strategy in G + C-rich halophilic archaea. *Life*, 6, 3, 37, 2016.
4.114. Jones, D.L. and Baxter, B.K., DNA repair and photoprotection: Mechanisms of overcoming environmental ultraviolet radiation exposure in halophilic archaea. *Front. Microbiol.*, 8, 1882, 2017.
4.115. Kah, L.C., Stack, K.M., Eigenbrode, J.L., Yingst, R.A., Edgett, K.S., Syndepositional precipitation of calcium sulfate in Gale Crater, Mars. *Terra Nova*, 30, 6, 431–439, 2018.
4.116. Kakhki, A.M., Amoozegar, M.A., Khaledi, E.M., Diversity of hydrolytic enzymes in haloarchaeal strains isolated from salt lake. *Int. J. Environ. Sci. Technol.*, 8, 4, 705–714, 2011.
4.117. Keck, W. and Hassibe, W., *The Great Salt Lake*, p. 25, U.S. Geological Survey, 1979.
4.118. Kemp, B., Tabish, E., Wolford, A., Jones, D., Butler, J., Baxter, B., The biogeography of Great Salt Lake halophilic archaea: Testing the hypothesis of avian mechanical carriers. *Diversity*, 10, 124, 2018.

4.119. Kirkpatrick, R., *The life of Great Salt Lake, with special reference to the algae.* (Master's thesis), 30pp., University of Utah, 1934.
4.120. Kish, A. and DiRuggiero, J., Rad50 is not essential for the Mre11-dependent repair of DNA double-strand breaks in *Halobacterium* sp. strain NRC-1. *J. Bacteriol.*, 190, 5210–5216, 2008.
4.121. Kish, A. and DiRuggiero, J., DNA replication and repair in halophiles, in: *Advances in Understanding the Biology of Halophilic Microorganisms*, R.H. Vreeland (Ed.), pp. 163–198, Springer, Dordrecht, Netherlands, 2012.
4.122. Kish, A., Kirkali, G., Robinson, C., Rosenblatt, R., Jaruga, P., Dizdaroglu, M., DiRuggiero, J., Salt shield: Intracellular salts provide cellular protection against ionizing radiation in the halophilic archaeon, *Halobacterium salinarum* NRC-1. *Environ. Microbiol.*, 5, 1066–1078, 2009.
4.123. Kminek, G., Bada, J.L., Pogliano, K., Ward, J.F., Radiation-dependent limit for the viability of bacterial spores in halite fluid inclusions and on Mars. *Radiat. Res.*, 159, 722–729, 2003.
4.124. Knauth, L.P., Temperature and salinity history of the Precambrian ocean: Implications for the course of microbial evolution, in: *Geobiology: Objectives, Concepts, Perspectives*, pp. 53–69, Elsevier, 2005.
4.125. Koonin, E.V., Comparative genomics, minimal gene-sets and the last universal common ancestor. *Nat. Rev. Microbiol.*, 1, 2, 127, 2003.
4.126. Kottemann, M., Kish, A., Iloanusi, C., Bjork, S., DiRuggiero, J., Physiological responses of the halophilic archaeon Halobacterium sp. strain NRC1 to desiccation and gamma irradiation. *Extremophiles*, 9, 3, 219–227, 2005.
4.127. Kronyak, R.E., Kah, L.C., Edgett, K.S., VanBommel, S.J., Thompson, L.M., Wiens, R.C., Sun, V.Z., Nachon, M., Mineral-filled fractures as indicators of multigenerational fluid flow in the Pahrump Hills member of the Murray formation, Gale crater, Mars. *Earth Space Sci.*, 6, 2, 238–265, 2019.
4.128. Kvien, I., Tanem, B.S., Oksman, K., Characterization of cellulose whiskers and their nanocomposites by atomic force and electron microscopy. *Biomacromolecules*, 6, 3160–3165, 2005.
4.129. Lammer, H., Bredehöft, J.H., Coustenis, A., Khodachenko, M.L., Kaltenegger, L., Grasset, O., Prieur, D., Raulin, F., Ehrenfreund, P., Yamauchi, M., Wahlund, J.E., What makes a planet habitable? *Astron. Astrophys. Rev.*, 17, 2, 181–249, 2009.
4.130. Landis, G.A., Martian water: Are there extant halobacteria on Mars? *Astrobiology*, 1, 2, 161–164, 2001.
4.131. Lanyi, J.K., Salt-dependent properties of proteins from extremely halophilic bacteria. *Bacteriol. Rev.*, 38, 3, 272, 1974.
4.132. Leitner, C., Neubauer, F., Marschallinger, R., Genser, J., Bernroider, M., Origin of deformed halite hopper crystals, pseudomorphic anhydrite cubes and polyhalite in Alpine evaporites (Austria, Germany). *Int. J. Earth Sci.*, 102, 3, 813–829, 2013.
4.133. Lennon, J.T. and Jones, S.E., Microbial seed banks: The ecological and evolutionary implications of dormancy. *Nat. Rev. Microbiol.*, 9, 2, 119–130, 2011.
4.134. Leuko, S., Rettberg, P., Pontifex, A.L., Burns, B.P., On the response of halophilic archaea to space conditions. *Life*, 4, 1, 66–76, 2014.
4.135. Leuko, S., Stan-Lotter, H., Lamers, G., Sjöström, S., Rabbow, E., Parpart, A., Rettberg, P., On the resistance of the archaeon *Halococcus morrhuae* and the biofilm-forming bacterium *Halomonas muralis* to exposure to low-Earth orbit for 534 days, in: *Extremophiles as Astrobiological Models [EXAM]*, J. Seckbach and H. Stan-Lotter (Eds.), *[EXAM, Volume in the series, Astrobiology Perspectives on Life of the Universe, Series Editors: Richard Gordon and Joseph Seckbach, In preparation]*, Wiley-Scrivener, Beverly, Massachusetts, USA, 2020.

4.136. Levin, G.V. and Straat, P.A., The case for extant life on Mars and its possible detection by the Viking labeled release experiment. *Astrobiology*, 16, 10, 798–810, 2016.

4.137. Li, Z., Li, Y., Cheng, H., Huang, R., Younger Dryas event recorded by the mirabilite deposition in Huahai lake, Hexi Corridor, NW China. *Quat. Int.*, 250, 93–99, 2012.

4.138. Lindahl, T., Instability and decay of the primary structure of DNA. *Nature*, 362, 6422, 709–715, 1993.

4.139. Lindahl, T. and Nyberg, B., Rate of depurination of native deoxyribonucleic acid. *Biochemistry*, 11, 19, 3610–3618, 1972.

4.140. Lindsay, M.R., Anderson, C., Fox, N., Scofield, G., Allen, J., Anderson, E., Bueter, L., Poudel, S., Sutherland, K., Munson-McGee, J.H. *et al.*, Microbialite response to an anthropogenic salinity gradient in Great Salt Lake, Utah. *Geobiology*, 15, 1, 131–145, 2017.

4.141. Lines, G.C., *Hydrology and surface morphology of the Bonneville Salt Flats and Pilot Valley playa, Utah, Vol. 2057*, Dept. of the Interior, Geological Survey, 1979.

4.142. Litchfield, C.D., Survival strategies for microorganisms in hypersaline environments and their relevance to life on early Mars. *Meteorit. Planet. Sci.*, 33, 4, 813–819, 1998.

4.143. Lowenstein, T.K., Schubert, B.A., Timofeeff, M.N., Microbial communities in fluid inclusions and long-term survival in halite. *Geol. Soc. Am. Today*, 21, 1, 4–9, 2011.

4.144. Madison, R.J., Effects of a causeway on the chemistry of the brine in Great Salt Lake, Utah, in: *Utah Geological and Mineralogical Survey Water-Resources Bulletin*, p. 14, 1970.

4.145. Maltagliati, L., Montmessin, F., Fedorova, A., Korablev, O., Forget, F., Bertaux, J.L., Evidence of water vapor in excess of saturation in the atmosphere of Mars. *Science*, 333, 6051, 1868–1871, 2011.

4.146. Mancinelli, R.L., Landheim, R., Sanchez-Porro, C., Dornmayr-Pfaffenhuemer, M., Gruber, C., Legat, A., Ventosa, A., Radax, C., Ihara, K., White, M.R., Stan-Lotter, H., *Halorubrum chaoviator* sp. nov., a haloarchaeon isolated from sea salt in Baja California, Mexico, Western Australia and Naxos, Greece. *Int. J. Syst. Evol. Microbiol.*, 59, Pt 8, 1908, 2009.

4.147. Mancinelli, R.L., White, M.R., Rothschild, L.J., Biopan-survival. I. Exposure of the osmophilic *Synechococcus* sp. (Nageli) and *Haloarcula* sp. to the space environment. *Adv. Space Res.*, 22, 327–334, 1998.

4.148. Manzoni, S., Schimel, J.P., Porporato, A., Responses of soil microbial communities to water-stress: Results from a meta-analysis. *Ecology*, 93, 930–938, 2012.

4.149. Marshall, C.P., Leuko, S., Coyle, C.M., Walter, M.R., Burns, B.P., Neilan, B.A., Carotenoid analysis of halophilic archaea by resonance Raman spectroscopy. *Astrobiology*, 7, 4, 631–643, 2007.

4.150. Martin, D.D., Ciulla, R.A., Roberts, M.F., Osmoadaptation in archaea. *Appl. Environ. Microbiol.*, 65, 5, 1815–1825, 1999.

4.151. Martin, E.L., Reinhardt, R.L., Baum, L.L., Becker, M.R., Shaffer, J.J., Kokjohn, T.A., The effects of ultraviolet radiation on the moderate halophile *Halomonas elongata* and the extreme halophile *Halobacterium salinarum*. *Can. J. Microbiol.*, 46, 180–187, 2000.

4.152. Martínez, G.M. and Renno, N.O., Water and brines on Mars: Current evidence and implications for MSL. *Space Sci. Rev.*, 175, 1-4, 29–51, 2013.

4.153. Martinez-Espinosa, R.M. and Bonete, M.J., Bioremediation of chlorate and perchlorate salted water using *Haloferax mediterranei*. *J. Biotechnol.*, 2, 131, S227, 2007.

4.154. Matsubara, T., Fujishima, K., Saltikov, C.W., Nakamura, S., Rothschild, L.J., Earth analogues for past and future life on Mars: Isolation of perchlorate resistant halophiles from Big Soda Lake. *Int. J. Astrobiol.*, 16, 3, 218–228, 2017.

4.155. McCready, S. and Marcello, L., Repair of UV damage in *Halobacterium salinarum*. *Biochem. Soc. Trans.*, 31, 3, 694–698, 2003.

4.156. McCready, S., Müller, J.A., Boubriak, I., Berquist, B.R., Ng, W.L., DasSarma, S., UV irradiation induces homologous recombination genes in the model archaeon, *Halobacterium* sp. NRC-1. *Saline Syst.*, 1, 3, 2005.

4.157. McEwen, A.S., Ojha, L., Dundas, C.M., Mattson, S.S., Byrne, S., Wray, J.J., Cull, S.C., Murchie, S.L., Thomas, N., Gulick, V.C., Seasonal flows on warm Martian slopes. *Science*, 333, 6043, 740–743, 2011.

4.158. McGenity, T.J., Gemmell, R.T., Grant, W.D., Stan-Lotter, H., Origins of halophilic microorganisms in ancient salt deposits. *Environ. Microbiol.*, 2, 243–250, 2000.

4.159. McGenity, T.J. and Oren, A., Hypersaline environments, in: *Hypersaline Environments in Life at Extremes: Environments, Organisms, and Strategies for Survival*, E.M. Bell (Ed.), pp. 402–437, CAB International, Wallingford, UK, 2012.

4.160. McGonigle, J.M., Bernau, J.A., Bowen, B.B., Brazelton, W.J., Robust archaeal and bacterial communities inhabit shallow subsurface sediments of the Bonneville Salt Flats. *mSphere*, 4, 4, e00378–19, 2019.

4.161. McLennan, S.M., Bell Iii, J.F., Calvin, W.M., Christensen, P.R., Clark, B.D., De Souza, P.A., Farmer, J., Farrand, W.H., Fike, D.A., Gellert, R., Ghosh, A., Provenance and diagenesis of the evaporite-bearing Burns formation, Meridiani Planum, Mars. *Earth Planet. Sci. Lett.*, 240, 1, 95–121, 2005.

4.162. McLennan, S.M. and Grotzinger, J.P., The sedimentary rock cycle of Mars, in: *The Martian Surface: Composition, Mineralogy, and Physical Properties*, Bell, J. (Ed.), pp. 541–577, 2008.

4.163. Milkovich, S.M. and Head, J.W., North polar cap of Mars: Polar layered deposit characterization and identification of a fundamental climate signal. *J. Geophys. Res.: Planets*, 110, E1, 2005, https://doi.org/10.1029/2004JE002349.

4.164. Millken, R.E., Grotzinger, J.P., Thomson, B.J., Paleoclimate of Mars as captured by the stratigraphic record in Gale Crater. *Geophys. Res. Lett.*, 37, 4, 2010.

4.165. Mischna, M.A. and Piqueux, S., The role of atmospheric pressure on Mars surface properties and early Mars climate modeling. *Icarus*, 113496, 2019.

4.166. Morán, J.I., Alvarez, V.A., Cyras, V.P., Vázquez, A., Extraction of cellulose and preparation of nanocellulose from sisal fibers. *Cellulose*, 15, 1, 149–159, 2008.

4.167. Mormile, M.R., Biesen, M.A., Gutierrez, M.C., Ventosa, A., Pavlovich, J.B., Onstott, T.C., Fredrickson, J.K., Isolation of *Halobacterium salinarum* retrieved directly from halite brine inclusions. *Environ. Microbiol.*, 5, 11, 1094–1102, 2003.

4.168. Motlagh, A.M., Bhattacharjee, A.S., Coutinho, F.H., Dutilh, B.E., Casjens, S.R., Goel, R.K., Insights of phage-host interaction in hypersaline ecosystem through metagenomics analyses. *Front. Microbiol.*, 8, 352, 2017.

4.169. Moyano, F.E., Manzoni, S., Chenu, C., Responses of soil heterotrophic respiration to moisture availability: An exploration of processes and models. *Soil Biol. Biochem.*, 59, 72–85, 2013.

4.170. Murchie, S.L., Mustard, J.F., Ehlmann, B.L., Milliken, R.E., Bishop, J.L., McKeown, N.K., Noe Dobrea, E.Z., Seelos, F.P., Buczkowski, D.L., Wiseman, S.M., Arvidson, R.E., A synthesis of Martian aqueous mineralogy after 1 Mars year of observations from the Mars Reconnaissance Orbiter. *J. Geophys. Res.: Planets*, 114, E2, 2009.

4.171. Mustard, J.F., Adler, M., Allwood, A., Bass, D.S., Beaty, D.W., Bell, J.F., III, Brinckerhoff, W.B., Carr, M., Des Marais, D.J., Drake, B. *et al.*, [Report of the Mars 2020 Science Definition Team], 154 pp., posted July, 2013, by the Mars Exploration Program Analysis Group (MEPAG), 2013, http://mepag.jpl.nasa.gov/reports/MEP/Mars_2020_SDT_Report_Final.pdf.

4.172. Naftz, D.L., Millero, F.J., Jones, B.F., Green, W.R., An equation of state for hypersaline water in Great Salt Lake, Utah, USA. *Aquat. Geochem.*, 17, 809–820, 2011.

4.173. Nakashima, S., Kebukawa, Y., Kitadai, N., Igisu, M., Matsuoka, N., Geochemistry and the origin of life: From extraterrestrial processes, chemical evolution on Earth, fossilized life's records, to natures of the extant life. *Life*, 8, 4, 39, 2018.
4.174. NASA, (n.d.). SHERLOC. Retrieved February 18, 2020, from https://mars.nasa.gov/mars2020/mission/instruments/sherloc/.
4.175. National Library of Medicine, 2020. National Center for Biotechnology Information <https://www.ncbi.nlm.nih.gov/nuccore/?term=%22great+salt+lake%22/> Accessed 02.14.2020.
4.176. Navarro-González, R., Vargas, E., Rosa, J.D.L., Raga, A.C., McKay, C.P., Reanalysis of the Viking results suggests perchlorate and organics at midlatitudes on Mars. *J. Geophys. Res.*, 115, E12, 2010.
4.177. Norton, C.F. and Grant, W.D., Survival of halobacteria within fluid inclusions in salt crystals. *J. Gen. Microbiol.*, 134, 1365–1373, 1988.
4.178. Norton, C.F., McGenity, T.J., Grant, W.D., Archaeal halophiles (halobacteria) from two British salt mines. *Microbiology*, 139, 5, 1077–1081, 1993.
4.179. Null, S.E. and Wurtsbaugh, W.A., Water development, consumptive uses, and the Great Salt Lake, in: *Great Salt Lake Biology: A Terminal Lake in a Time of Change*, B.K. Baxter and J.K. Butler (Eds.), Springer, Cham, 2020.
4.180. Oehler, D.Z. and Allen, C.C., Evidence for pervasive mud volcanism in Acidalia Planitia, Mars. *Icarus*, 208, 2, 636–657, 2010.
4.181. Ojha, L., Wilhelm, M.B., Murchie, S.L., McEwen, A.S., Wray, J.J., Hanley, J., Massé, M., Chojnacki, M., Spectral evidence for hydrated salts in recurring slope lineae on Mars. *Nat. Geosci.*, 8, 11, 829–832, 2015, https://doi.org/10.1038/ngeo2546.
4.182. Okeke, B.C., Giblin, T., Frankenberger, W.T., Jr., Reduction of perchlorate and nitrate by salt tolerant bacteria. *Environ. Pollut.*, 118, 3, 357–363, 2002.
4.183. Okuda, K., Kudlicka, K., Kuga, S., Brown, R.M., Jr., [beta]-Glucan synthesis in the cotton fiber (I. Identification of [beta]-1,4- and [beta]-1,3-Glucans synthesized *in vitro*). *Plant Physiol.*, 101, 1131–1142, 1993.
4.184. Oren, A., The ecology of the extremely halophilic archaea. *FEMS Microbiol. Rev.*, 13, 4, 415–439, 1994.
4.185. Oren, A., Diversity of halophiles, in: *Extremophiles Handbook*, K. Horikoshi (Ed.), pp. 309–325, Springer, Tokyo, Japan, 2011.
4.186. Oren, A., Life at high salt and low oxygen: How do the Halobacteriaceae cope with low oxygen concentrations in their environment?, in: *Polyextremophiles*, pp. 531–548, Springer, Dordrecht, Netherlands, 2013.
4.187. Oren, A., DNA as genetic material and as a nutrient in halophilic Archaea. *Front. Microbiol.*, 5, 539, 2014.
4.188. Oren, A., Halophilic archaea on Earth and in space: Growth and survival under extreme conditions. *Philos. Trans. Royal Soc. A: Math. Phys. Eng. Sci.*, 372, 2030, 20140194, 2014.
4.189. Oren, A., Glycerol metabolism in hypersaline environments. *Environ. Microbiol.*, 19, 851–863, 2017.
4.190. Oren, A., Bardavid, R.E., Mana, L., Perchlorate and halophilic prokaryotes: Implications for possible halophilic life on Mars. *Extremophiles*, 18, 1, 75–80, 2014.
4.191. Oren, A., Sørensen, K.B., Canfield, D.E., Teske, A.P., Ionescu, D., Lipski, A., Altendorf, K., Microbial communities and processes within a hypersaline gypsum crust in a saltern evaporation pond (Eilat, Israel). *Hydrobiologia*, 626, 1, 15–26, 2009.
4.192. Orosei, R., Lauro, S.E., Pettinelli, E., Cicchetti, A., Coradini, M., Cosciotti, B., Di Paolo, F., Flamini, E., Mattei, E., Pajola, M., Soldovieri, F., Radar evidence of subglacial liquid water on Mars. *Science*, 361, 490–493, 2018.

4.193. Osterloo, M.M., Anderson, F.S., Hamilton, V.E., Hynek, B.M., Geologic context of proposed chloride-bearing materials on Mars. *J. Geophys. Res.*, 115, E10012, 2010, https://agupubs.onlinelibrary.wiley.com/doi/full/10.1029/2010JE003613.
4.194. Osterrothová, K. and Jehlička, J., Investigation of biomolecules trapped in fluid inclusions inside halite crystals by Raman spectroscopy. *Spectrochim. Acta Part A: Mol. Biomol. Spectrosc.*, 83, 1, 288–296, 2011.
4.195. Oviatt, C.G. and Shroder, J.F. (Eds.), *Lake Bonneville: A Scientific Update Vol. 20*, Elsevier, Netherlands, 2016.
4.196. Oviatt, C.G., Thompson, R.S., Kaufman, D.S., Bright, J., Forester, R.M., Reinterpretation of the Burmester core, Bonneville basin, Utah. *Quat. Res.*, 52, 2, 180–184, 1999.
4.197. Pääbo, S., Ancient DNA: Extraction, characterization, molecular cloning, and enzymatic amplification. *Proc. Natl. Acad. Sci.*, 86, 6, 1939–1943, 1989.
4.198. Pääbo, S., Poinar, H., Serre, D., Jaenicke-Després, V., Hebler, J., Rohland, N., Kuch, M., Krause, J., Vigilant, L., Hofreiter, M., Genetic analyses from ancient DNA. *Annu. Rev. Genet.*, 38, 645–679, 2004.
4.199. Panieri, G., Lugli, S., Manzi, V., Roveri, M., Schreiber, B.C., Palinska, K.A., Ribosomal RNA gene fragments from fossilized cyanobacteria identified in primary gypsum from the late Miocene, Italy. *Geobiology*, 8, 2, 101–111, 2010.
4.200. Park, J.S., Vreeland, R.H., Cho, B.C., Lowenstein, T.K., Timofeeff, M.N., Rosenzweig, W.D., Haloarchaeal diversity in 23, 121 and 419 MYA salts. *Geobiology*, 7, 5, 515–523, 2009.
4.201. Parnell, J.J., Crowl, T.A., Weimer, B.C., Pfrender, M.E., Bio-diversity in microbial communities: System scale patterns and mechanisms. *Mol. Ecol.*, 18, 1455–1462, 2009.
4.202. Parnell, J.J., Rompato, G., Crowl, T.A., Weimer, B.C., Pfrender, M.E., Phylogenetic distance in Great Salt Lake microbial communities. *Aquat. Microb. Ecol.*, 64, 267–273, 2011.
4.203. Parnell, J.J., Rompato, G., Latta, L.C., Pfrender, M.E., Van Nostrand, J.D., He, Z., Zhou, J., Andersen, G., Champine, P., Ganesan, B., Weimer, B.C., Functional biogeography as evidence of gene transfer in hypersaline microbial communities. *PLoS ONE*, 5, 9, e12919, 2010.
4.204. Pasteris, J.D., Freeman, J.J., Wopenka, B., Qi, K., Ma, Q., Wooley, K.L., With a grain of salt: What halite has to offer to discussions on the origin of life. *Astrobiology*, 6, 4, 625–643, 2006.
4.205. Perl, S. and Baxter, B.K., Great Salt Lake as an astrobiology analogue for ancient Martian hypersaline aqueous systems, in: *Great Salt Lake Biology: A Terminal Lake in a Time of Change*, B.K. Baxter and J.K. Butler (Eds.), Springer, Cham, 2020.
4.206. Post, F.J., The microbial ecology of the Great Salt Lake. *Microb. Ecol.*, 3, 143–165, 1977.
4.207. Post, F.J., Oxygen-rich gas domes of microbial origin in the salt crust of the Great Salt Lake, Utah. *Geomicrobiol. J.*, 2, 127–139, 1980.
4.208. Post, F.J., Microbiology of the Great Salt Lake North Arm. *Hydrobiologia*, 81, 59–69, 1981.
4.209. Post, F.J. and Stube, J.C., A microcosm study of nitrogen utilization in the Great Salt Lake, Utah. *Hydrobiolgia*, 158, 89–100, 1988.
4.210. Quinn, R.C., Martucci, H.F., Miller, S.R., Bryson, C.E., Grunthaner, F.J., Grunthaner, P.J., Perchlorate radiolysis on mars and the origin of Martian soil reactivity. *Astrobiology*, 13, 6, 515–520, 2013.
4.211. Quinn, R.C., Ricco, A.J., Davila, A., Koehne, J.E., McKay, C.P., Dateo, C.E., Fonda, M.L., *Scientific and Technological Approaches to Searching for Extant Life in the Solar System [NASA technical report]*, 2017, https://ntrs.nasa.gov/archive/nasa/casi.ntrs.nasa.gov/20180008711.pdf.
4.212. Radax, C., Gruber, C., Stan-Lotter, H., Novel haloarchaeal 16S rRNA gene sequences from Alpine Permo-Triassic rock salt. *Extremophiles*, 5, 221–228, 2001.
4.213. Rampe, E.B., Ming, D.W., Blake, D.F., Bristow, T.F., Chipera, S.J., Grotzinger, J.P., Morris, R.V., Morrison, S.M., Vaniman, D.T., Yen, A.S., Achilles, C.N., Mineralogy of an ancient

lacustrine mudstone succession from the Murray formation, Gale crater, Mars. *Earth Planet. Sci. Lett.*, 471, 172–185, 2017.

4.214. Rapin, W., Ehlmann, B.L., Dromart, G., Schieber, J., Thomas, N.H., Fischer, W.W., Fox, V.K., Stein, N.T., Nachon, M., Clark, B.C., Kah, L.C., An interval of high salinity in ancient Gale crater lake on Mars. *Nat. Geosci.*, 12, 11, 889–895, 2019.

4.215. Rennó, N.O., Bos, B.J., Catling, D., Clark, B.C., Drube, L., Fisher, D., Goetz, W., Hviid, S.F., Keller, H.U., Kok, J.F., Kounaves, S.P., Possible physical and thermodynamical evidence for liquid water at the Phoenix landing site. *J. Geophys. Res.: Planets*, 114, E1, 2009.

4.216. Richardson, C.D., Hinman, N.W., Scott, J.R., Effect of thenardite on the direct detection of aromatic amino acids: Implications for the search for life in the solar system. *Int. J. Astrobiol.*, 8, 4, 291–300, 2009.

4.217. Riddle, M.R., Baxter, B.K., Avery, B.J., Molecular identification of microorganisms associated with the brine shrimp, *Artemia franciscana. Aquat. Biosyst.*, 9, 1–11, 2013.

4.218. Roberts, M.F., Osmoadaptation and osmoregulation in archaea. *Front. Biosci.*, 5, 362, 796–812, 2000.

4.219. Robinson, K., *Resistance to ionizing radiation and oxidative stress in Halobacterium Salinarum NRC-1.* Thesis, University of Maryland, 2009.

4.220. Roedder, E., The fluids in salt. *Am. Mineral.*, 69, 413–439, 1984.

4.221. Rothschild, L.J., Earth analogs for Martian life. Microbes in evaporites, a new model system for life on Mars. *Icarus*, 88, 1, 246–260, 1990.

4.222. Rupke, A. and Boden, T., *Great Salt Lake north arm salt crust monitoring, spring 2017 update.* Utah Geological Survey Open-File Report 714, 13 p., 2020, https://doi.org/10.34191/OFR-714.

4.223. Sagan, C., Definitions of life, in: *The Nature of Life: Classical and Contemporary Perspectives from Philosophy and Science*, M.A. Bedau and C.E. Cleland (Eds.), Cambridge University Press, U.K, 2010.

4.224. Saito, T., Miyabe, Y., Ide, H., Yamamoto, O., Hydroxyl radical scavenging ability of bacterioruberin. *Radiat. Phys. Chem.*, 50, 3, 267–269, 1997.

4.225. Sankaranarayanan, K., Timofeeff, M.N., Spathis, R., Lowenstein, T.K., Lum, J.K., Ancient microbes from halite fluid inclusions: Optimized surface sterilization and DNA extraction. *PLoS One*, 6, 6, e20683, 2011.

4.226. Schoepp-Cothenet, B., Van Lis, R., Atteia, A., Baymann, F., Capowiez, L., Ducluzeau, A.L., Duval, S., Ten Brink, F., Russell, M.J., Nitschke, W., On the universal core of bioenergetics. *Biochimica et Biophysica Acta (BBA)-Bioenergetics*, 1827, 2, 79–93, 2013.

4.227. Schroeder, G.K., Lad, C., Wyman, P., Williams, N.H., Wolfenden, R., The time required for water attack at the phosphorus atom of simple phosphodiesters and of DNA. *Proc. Natl. Acad. Sci. U.S.A.*, 103, 11, 4052–4055, 2006.

4.228. Schubert, B.A., Lowenstein, T.K., Timofeeff, M.N., Microscopic identification of prokaryotes in modern and ancient halite, Saline Valley and Death Valley, California. *Astrobiology*, 9, 5, 467–482, 2009.

4.229. Schubert, B.A., Lowenstein, T.K., Timofeeff, M.N., Parker, M.A., Halophilic archaea cultured from ancient halite, Death Valley, California. *Environ. Microbiol.*, 12, 2, 440–454, 2010.

4.230. Schweitzer, M.H., Suo, Z., Avci, R., Asara, J.M., Allen, M.A., Arce, F.T., Horner, J.R., Analyses of soft tissue from Tyrannosaurus rex suggest the presence of protein. *Science*, 316, 5822, 277–280, 2007.

4.231. Shahmohammadi, H.R., Asgarani, E., Terato, H., Saito, T., Ohyama, Y., Gekko, K., Yamamoto, O., Ide, H., Protective roles of bacterioruberin and intracellular KCl in the resistance of Halobacterium salinarium against DNA-damaging agents. *J. Radiat. Res.*, 39, 4, 251–262, 1998.

4.232. Shen, P.S., Domek, M.J., Sanz-García, E., Makaju, A., Taylor, R.M., Hoggan, R., Culumber, M.D., Oberg, C.J., Breakwell, D.P., Prince, J.T., Belnap, D.M., Sequence and structural characterization of Great Salt Lake Bacteriophage CW02, a member of the T7-like supergroup. *J. Virol.*, 86, 15, 7907–7917, 2012.
4.233. Sherwood, J.E., Stagnitti, F., Kokkinn, M.J., Williams, W.D., Dissolved oxygen concentrations in hypersaline waters. *Limnol. Oceanogr.*, 36, 2, 235–250, 1991.
4.234. Simoneit, B.R., Biomarkers (molecular fossils) as geochemical indicators of life. *Adv. Space Res.*, 33, 8, 1255–1261, 2004.
4.235. Skelley, A.M. and Mathies, R.A., Chiral separation of fluorescamine-labeled amino acids using microfabricated capillary electrophoresis devices for extraterrestrial exploration. *J. Chromatogr. A*, 1021, 1, 191–199, 2003.
4.236. Socki, R.A., Sun, T., Niles, P.B., Harvey, R.P., Bish, D.L., Tonui, E., *Antarctic mirabilite mounds as Mars analogs: The Lewis Cliffs Ice Tongue revisited*, 2012, https://ntrs.nasa.gov/archive/nasa/casi.ntrs.nasa.gov/20120001959.pdf.
4.237. Sonnenfeld, P., *Brines and Evaporites*, University of California, Academic Press, Oakland, CA, 1984.
4.238. Sorenson, E.D., Hoven, H.M., Neil, J., Great Salt Lake shorebirds, their habitats and food base, in: *Great Salt Lake Biology: A Terminal Lake in a Time of Change*, B.K. Baxter and J.K. Butler (Eds.), Springer, Cham, 2020.
4.239. Soppa, J., Ploidy and gene conversion in Archaea. *Biochem. Soc. Trans.*, 39, 1, 150–154, 2011.
4.240. Spiga, A., González-Galindo, F., López-Valverde, M.Á., Forget, F., Gravity waves, cold pockets and CO2 clouds in the Martian mesosphere. *Geophys. Res. Lett.*, 39, 2, 2012.
4.241. Squyres, S.W. and Knoll, A.H., Sedimentary rocks at Meridiani Planum: Origin, diagenesis, and implications for life on Mars. *Earth Planet. Sci. Lett.*, 240, 1, 1–10, 2005.
4.242. Stan-Lotter, H. and Fendrihan, S., Halophilic archaea: Life with desiccation, radiation and oligotrophy over geological times. *Life*, 5, 3, 1487–1496, 2015.
4.243. Stan-Lotter, H., McGenity, T.J., Legat, A., Denner, E.B., Glaser, K., Stetter, K.O., Wanner, G., Very similar strains of *Halococcus salifodinae* are found in geographically separated Permo-Triassic salt deposits. *Microbiology*, 145, 12, 3565–3574, 1999.
4.244. Stan-Lotter, H., Pfaffenhuemer, M., Legat, A., Busse, H.-J., Radax, C., Gruber, C., *Halococcus dombrowskii* sp. nov., an archaeal isolate from a Permo-Triassic alpine salt deposit. *Int. J. Syst. Bacteriol.*, 52, 1807–1814, 2002.
4.245. Stan-Lotter, H., Radax, C., McGenity, T.J., Legat, A., Pfaffenhuemer, M., Wieland, H., Gruber, C., Denner, E.B.M., From intraterrestrials to extraterrestrials: Viable haloarchaea in ancient salt deposits, in: *Halophilic Microorganisms*, A. Ventosa (Ed.), pp. 89–102, Springer, Berlin, Germany, 2004.
4.246. Stephens, D.W., Changes in lake levels, salinity and the biological community of Great Salt Lake (Utah, USA), 1847–1987, in: *Developments in Hydrobiology*, vol., 59, pp. 139–146, 1990.
4.247. Stevenson, A., Cray, J.A., Williams, J.P., Santos, R., Sahay, R., Neuenkirchen, N., McClure, C.D., Grant, I.R., Houghton, J.D., Quinn, J.P. *et al.*, Is there a common water-activity limit for the three domains of life? *ISME J.*, 9, 6, 1333–1351, 2015.
4.248. Stillman, D.E., Michaels, T.I., Grimm, R.E., Harrison, K.P., New observations of martian southern mid-latitude recurring slope lineae (RSL) imply formation by freshwater subsurface flows. *Icarus*, 233, 328–341, 2014, https://doi.org/10.1016/j.icarus.2014.01.017.
4.249. Stoeckenius, W. and Bogomolni, R.A., Bacteriorhodopsin and related pigments of halobacteria. *Annu. Rev. Biochem.*, 51, 1, 587–616, 1982.
4.250. Stube, J.C., Post, F.J., Procella, D.B., *Nitrogen cycling in microcosms and application to the biology of the north arm of Great Salt Lake.* (Publication No. PRJSBA-016-1), Utah Water Research Laboratory, Utah State University, Logan, UT, 1976.

4.251. Stutz, J., Ackermann, R., Fast, J.D., Barrie, L., Atmospheric reactive chlorine and bromine at the Great Salt Lake, Utah. *Geophys. Res. Lett.*, 29, 10, 18–1, 2002.

4.252. Tazi, L., Breakwell, D.P., Harker, A.R., Crandall, K.A., Life in extreme environments: Microbial diversity in Great Salt Lake, Utah. *Extremophiles*, 18, 525–535, 2014.

4.253. Terada, N., Kulikov, Y.N., Lammer, H., Lichtenegger, H.I., Tanaka, T., Shinagawa, H., Zhang, T., Atmosphere and water loss from early Mars under extreme solar wind and extreme ultraviolet conditions. *Astrobiology*, 9, 1, 55–70, 2009.

4.254. Thomas, C., Grossi, V., Antheaume, I., Ariztegui, D., Recycling of archaeal biomass as a new strategy for extreme life in Dead Sea deep sediments. *Geology*, 47, 5, 479–482, 2019.

4.255. Thornton K,L., Butler J.K., Davis SJ, Baxter B.K., Wilson L.G. Haloarchaea Swim Slowly for Optimal Chemotactic Efficiency in Low Nutrient Environments. *Nature Communications* 11:4453. 2020 https://doi.org/10.1038/s41467-020-18253-7

4.256. Toner, J.D., Catling, D.C., Light, B., The formation of supercooled brines, viscous liquids, and low-temperature perchlorate glasses in aqueous solutions relevant to Mars. *Icarus*, 233, 36–47, 2014.

4.257. Tosca, N.J., Knoll, A.H., McLennan, S.M., Water activity and the challenge for life on early Mars. *Science*, 320, 5880, 1204–1207, 2008.

4.258. Tosca, N.J. and McLennan, S.M., Chemical divides and evaporite assemblages on Mars. *Earth Planet. Sci. Lett.*, 241, 1-2, 21–31, 2006.

4.259. Tosca, N.J., McLennan, S.M., Clark, B.C., Grotzinger, J.P., Hurowitz, J.A., Knoll, A.H., Schröder, C., Squyres, S.W., Geochemical modeling of evaporation processes on Mars: Insight from the sedimentary record at Meridiani Planum. *Earth Planet. Sci. Lett.*, 240, 1, 122–148, 2005.

4.260. Turk, L.J., Davis, S.N., Bingham, C.P., Hydrogeology of lacustrine sediments, Bonneville Salt Flats, Utah. *Econ. Geol.*, 68, 1, 65–78, 1973.

4.261. United States Geological Survey, 2020. https://waterdata.usgs.gov/ < Accessed February 19, 2020>.

4.262. Utah Geological Survey, Interesting facts about the mirabilite spring mounds near Great Salt Lake marina, 2020. https://geology.utah.gov/interesting-facts-about-the-mirabilite-spring-mounds-near-great-salt-lake-marina/ < Accessed January 14, 2020>.

4.263. Valeille, A., Bougher, S.W., Tenishev, V., Combi, M.R., Nagy, A.F., Water loss and evolution of the upper atmosphere and exosphere over Martian history. *Icarus*, 206, 1, 28–39, 2010.

4.264. Vandaele, A.C., Korablev, O., Daerden, F., Aoki, S., Thomas, I.R., Altieri, F., López-Valverde, M., Villanueva, G., Liuzzi, G., Smith, M.D., Erwin, J.T., Martian dust storm impact on atmospheric H_2O and D/H observed by ExoMars Trace Gas Orbiter. *Nature*, 568, 7753, 521–525, 2019.

4.265. Viviano-Beck, C.E., Seelos, F.P., Murchie, S.L., Kahn, E.G., Seelos, K.D., Taylor, H.W., Taylor, K., Ehlmann, B.L., Wiseman, S.M., Mustard, J.F., Morgan, M.F., Revised CRISM spectral parameters and summary products based on the currently detected mineral diversity on Mars. *J. Geophys. Res.: Planets*, 119, 6, 1403–1431, 2014.

4.266. Vreeland, R.H., Rosenzweig, W.D., Powers, D.W., Isolation of a 250 million-year-old halotolerant bacterium from a primary salt crystal. *Nature*, 407, 897–900, 2000.

4.267. Wang, S.Y., Gillies, R.R., Reichler, T., Multidecadal drought cycles in the Great Basin recorded by the Great Salt Lake: Modulation from a transition-phase teleconnection. *J. Clim.*, 25, 5, 1711–1721, 2012.

4.268. Ward, D.M. and Brock, T.D., Hydrocarbon biodegradation in hypersaline environments. *Appl. Environ. Microbiol.*, 35, 353–359, 1978.

4.269. Ward, P., *Life as We Do Not Know It: The NASA Search for (and Synthesis of) Alien Life*, Penguin Group, New York, 2007.

4.270. Webb, K.M., Yu, J., Robinson, C.K., Noboru, T., Lee, Y.C., DiRuggiero, J., Effects of intracellular Mn on the radiation resistance of the halophilic archaeon *Halobacterium salinarum*. *Extremophiles*, 17, 485–497, 2013.

4.271. Weimer, B.C., Rompato, G., Parnell, J., Gann, R., Ganesan, B., Navas, C., Gonzalez, M., Clavel, M., Albee-Scott, S., Microbial biodiversity of Great Salt Lake, Utah. *Nat. Resour. Environ. Issues*, 15, 15–22, 2009.

4.272. Weitz, C.M. and Bishop, J.L., Formation of clays, ferrihydrite, and possible salts in Hydrae Chasma, Mars. *Icarus*, 319, 392–406, 2019.

4.273. Wieland, F., The cell surfaces of halobacteria, in: *Halophilic Bacteria Vol. II*, F. Rodriguez-Valera (Ed.), pp. 55–65, CRC Press, Boca Raton, Florida, USA, 1988.

4.274. Willerslev, E., Hansen, A.J., Rønn, R., Brand, T.B., Barnes, I., Wiuf, C., Gilichinsky, D., Mitchell, D., Cooper, A., Long-term persistence of bacterial DNA. *Curr. Biol.*, 14, 1, R9–R10, 2004.

4.275. Williams, J.P. and Hallsworth, J.E., Limits of life in hostile environments; no limits to biosphere function? *Environ. Microbiol.*, 11, 3292–3308, 2009.

4.276. Winters, Y.D., Lowenstein, T.K., Timofeeff, M.N., Identification of carotenoids in ancient salt from Death Valley, Saline Valley, and Searles Lake, California, using laser Raman spectroscopy. *Astrobiology*, 13, 11, 1065–1080, 2013.

4.277. Woods, W.G. and Dyall-Smith, M.L., Construction and analysis of a recombination-deficient (radA) mutant of Haloferax volcanii. *Mol. Microbiol.*, 23, 4, 791–797, 1997.

4.278. Wray, J.J., Milliken, R.E., Dundas, C.M., Swayze, G.A., Andrews-Hanna, J.C., Baldridge, A.M., Chojnacki, M., Bishop, J.L., Ehlmann, B.L., Murchie, S.L., Clark, R.N., Columbus crater and other possible groundwater-fed paleolakes of Terra Sirenum, Mars. *J. Geophys. Res.: Planets*, 116, E1, 2011.

4.279. Wurtsbaugh, W.A., Miller, C., Null, S.E., DeRose, R.J., Wilcock, P., Hahnenberger, M., Howe, F., Moore, J., Decline of the world's saline lakes. *Nat. Geosci.*, 10, 11, 816, 2017.

4.280. Ye, C. and Glotch, T.D., Spectral properties of chloride salt-bearing assemblages: Implications for detection limits of minor phases in chloride-bearing deposits on Mars. *J. Geophys. Res.: Planets*, 124, 2, 209–222, 2019.

4.281. Young, A.J., The photoprotective role of carotenoids in higher plants. *Physiol. Plant.*, 83, 702–708, 1991.

4.282. Young, A.J. and Lowe, G.M., Antioxidant and prooxidant properties of carotenoids. *Arch. Biochem. Biophys.*, 385, 1, 20–27, 2001.

4.283. Zenke, R., von Gronau, S., Bolhuis, H., Gruska, M., Pfeiffer, F., Oesterhelt, D., Fluorescence microscopy visualization of halomucin, a secreted 927 kDa protein surrounding Haloquadratum walsbyi cells. *Front. Microbiol.*, 6, 249, 2015.

4.284. Zerulla, K., Chimileski, S., Näther, D., Gophna, U., Papke, R.T., Soppa, J., DNA as a phosphate storage polymer and the alternative advantages of polyploidy for growth or survival. *PLoS One*, 9, e94819, 2014.

4.285. Zhao, A., Gray, F.C., MacNeill, S.A., ATP- and NAD+-dependent DNA ligases share an essential function in the halophilic archaeon *Haloferax volcanii*. *Mol. Microbiol.*, 59, 743–752, 2006.

4.286. Zorzano, M.P., Mateo-Martí, E., Prieto-Ballesteros, O., Osuna, S., Renno, N., Stability of liquid saline water on present day Mars. *Geophys. Res. Lett.*, 36, 20, 2009.

5

Arsenic-and Light Hydrocarbon-Rich Hypersaline Soda Lakes and Their Resident Microbes as Possible Models for Extraterrestrial Biomes

Ronald S. Oremland

Department of Microbiology and Environmental Toxicology, University of California at Santa Cruz, Santa Cruz, California, USA

Abstract

Asking a microbiologist to nominate a favorite microorganism as an example of life they think likely to be encountered beyond Earth, is by definition an exercise in personal bias. Yet it is also perhaps a necessary thing to do when we contemplate what type of terrestrial microorganism could prove adaptable to the environmental rigors encountered on an alien orb. There are practical reasons as well. Planetary protection is a consideration in the context of how clean a lander need be to avoid the mistake of inserting our own microbes onto the planet's surface. Pitfalls, however remote, would include generating false positives, or even seeding the planet with one of our own adaptive microbes, thereby initiating an unwanted runaway exercise in planetary eco-transformation. Yet sterilizing a spacecraft is very expensive. In this context, I recall being part of a small panel, chaired by the late Chuck Klein (NASA), to tackle the question of just how clean a spacecraft need be to land on Mars without generating such problems. After a couple of days of running around the central unanswered question of whether or not our terrestrial microbes could survive on Mars, he threw up his hands and asked *"What would happen if I placed a cubic meter of fresh horse manure onto the Martian surface?"* The panel members kept silent, until I finally raised my hand and answered: *"I can't say anything about the survival of the critters in the manure, but I will state positively that if you can convince Congress to fund such a mission, I would be the first astronaut to proudly step into that pile."* The question remains unanswered to this day.

Keywords: Arsenophiles, Mars, acetylenophiles, Enceladus, alien acetylenophiles, Titan

5.1 Introduction

Soda lakes are alkaline (e.g., pH $\geq$ 9.5), often hypersaline (e.g., salinity >> 35 g/L), mineral-rich water bodies that undergo bouts of lake-wide anoxia or if deep enough, retain perennially anoxic, highly sulfidic bottom waters. Hence, they have aroused the curiosity of

Email: roremland@comcast.net
Ronald S. Oremland: https://www.researchgate.net/profile/Ronald_Oremland2/stats

Joseph Seckbach and Helga Stan-Lotter (eds.) Extremophiles as Astrobiological Models, (125–136)

microbiologists, because despite the harsh chemical conditions, their biota sustain high rates of primary production and host a microbial flora actively involved in biogeochemical mineral/nutrient cycling. They are by definition "extreme" environments, and when located in volcanic regions (as they often are), contain an abundance of toxic elements in solution, including arsenic and boron. These elements enter the lake ecosystems via volcanic sources associated with the hydrothermal leaching of minerals in the surrounding watersheds. When co-located in arid regions (as they often are), evapo-concentration results in soda lake salinities reaching saturation, with brines that have relatively low water activities (a_w). This condition presents further challenges to life and makes the resident microbial denizens that thrive in these harsh milieus poly-extremophiles [5.2].

Nonetheless, the chemistry of soda lakes likely serves as an example of what was once the Earth's first global ocean [5.9]. Water flowing across early volcanic rock surfaces would have caused release of sodium and potassium ions preferentially to the more tightly mineral-bound halides. When coupled with the selective absorption of protons into the rock matrix, the stage would be set for an increase in pH and a dearth of anions in solution. This dearth would be compensated by the solvation of atmospheric CO_2 along with any local volcanic CO_2 outgassing, resulting in a bicarbonate/carbonate-rich matrix. Over the eons of time, as the continents weathered and the Earth's crust chemically exchanged with its mantle, more chloride dissolved into seawater, the seawater pH lowered, and calcium and magnesium came into the mix. In the context of extraterrestrial worlds, especially the ones we will discuss, such planetary mixing of their hydro- and lithospheres never had the time or scale to occur as they did on Earth. Hence, their waters were more likely to be akin to chemistry of soda lakes than to the Earth's current global ocean.

Yet, are these microbes likely to be the first types of "alien" (i.e., non-terrestrial) life we will encounter when humans (or our robotic probes) step off the Earth's surface and onto (or beneath) the surface of other "habitable" planetary bodies? This is a matter of personal conjecture, often advocated by the nature of the broad research scope and the imaginations of the contributors to this book. Hence, there is a collegial degree of "*my little guys work better in this context than your little guys*" competition built into this book's conception. Nonetheless, there is a plausible case for terrestrial soda lakes and their resident microbes as the likely alien microbiota that we could encounter when searching for biosignatures of extant or extinct life in our Solar System. In this context we will consider the planet(oid)s Mars, Enceladus, and Titan as theoretical targets to scour for life. In addition, we will briefly consider the metabolism of light hydrocarbons because these gases also are abundant in such examples as Big Soda Lake [5.17] and Mono Lake [5.16]. These celestial orbs are a prime subset of the broader aspects of prokaryotic adaptations to tolerate extremes to physical and chemical conditions likely encountered in the Solar System's most propitious sites for encountering alien microbes [5.13]. As we transit from Mars to the icy orbs of the outer Solar System, I will make a brief case that acetylene could serve as a plausible metabolic font for life therein [5.19].

First, however, a personal digression

Currently I find myself at the tail end of my career. My position is referred to as "emeritus," which is just a nice way of saying that I am retired and no longer have an active research team or close affiliation with the USGS where I labored diligently for 42 years. With the indulgence of the editors Joseph Seckbach and Helga Stan-Lotter (who are slightly desperate for my contribution) I would like to reminisce about how, as a child, I became interested in space exploration.

I grew up in Coney Island, the southern-most portion of Brooklyn. The region's geography is composed of the amusement area on the west, Brighton Beach in the middle, and the residential area of Manhattan Beach (my home) on its eastern extension. Coney Island lies surrounded by water: the New York Bight lays to the south, Sheepshead Bay/Gravesend Bay to the north, with Jamaica Bay and the Narrows, respectively, to the east and west. Hence, from a tender age I have always had water, specifically seawater, in my thoughts and often splashing at my toes.

My natal day roughly coincided with the commercialization of television, so that by 1949 a Philco, equipped with its 9-inch black and white screen (and numerous vacuum tubes), occupied our living room for well into the 1960s. Early programming geared for children (aside from cartoons) included cowboys (Hopalong Cassidy; Roy Rogers; Gene Autry), and science fiction adventures (Captain Video; Tom Corbett and the Space Cadets). The latter category also aired serialized film episodes from the 1930s (Flash Gordon; Buck Rogers) featuring Buster Crabbe taking on the wicked minions of either Ming the Merciless or Killer Kane, and with every episode Flash/Buck fought their diabolical plans to bend the entire Solar System to their will.

These episodes curried my young mind with escapist play with my many neighborhood friends...if we weren't cowboys, fighting GIs, or superman and his pals (FYI: surplus army rubberized ponchos served us as our *de rigueur* superhero capes) we could be "spacemen" (aka: astronauts) alit for the first time on one of the Solar System's unexplored planets. We never quite figured out what we would do once we got there other than to pick up rocks and chase after the neighborhood feral cats who served us well as putative dangerous alien life forms. While the evil Ming had long-since been vanquished on TV by the likes of Flash, Dale, and Dr. Zarkov, I did enjoy girding myself up for the rigors of prolonged space flight (Figure 5.1).

With time, the television programming of this genre became a tad less fanciful and bit more predictive/speculative, especially when Disney first entered the TV arena with his Sunday evening shows. My favorites were the rare ones from "Tomorrowland" which sometimes featured space flight as portrayed by the likes of Wernher von Braun, and of possible encounters with curious alien life as conjectured by Carl Sagan. These episodes fired up my imagination with regard to the often-posed question by a visiting aunt:

"So, what do you want to be when you grow up, young man?"

I sensed that a cowboy occupation would not go over well, while a response of "surgeon or dentist" was likely to elicit a smile and a playful request for a future appointment, albeit at a discounted rate. After all, we had several proud representatives of these occupations in my family. Yet when I replied:

"I want to be a spaceman!"

The response was a groan with the admonition to my parents that I had been watching too much television.

Movies were also an influence, perhaps none more so than Disney's *Fantasia*. Released in 1940, it became profitable only when it began postwar cycling in neighborhood theaters. To

Figure 5.1 The author (second from left) attired as a spaceman and budding astrobiologist in a Camp Dellwood (Honesdale, PA) costume event, summer 1953. Note his detailed preparation for facing the rigors of spaceflight, including a plastic communicator helmet, terry cloth bath shirt, pajama bottoms, and high-top Keds sneakers. Weapons include a Dick Tracy cartoon pistol in his belt as well as a ray gun (aka flashlight) capable of shooting diffuse beams of incoherent, visible light for several feet ahead of his trek on the wooded path back to his bunk. (Image credit: the author).

me, Stravinsky's discordant and pounding "The Rite of Spring" sequence held my imagination. It portrayed the Earth's accretionary phase, crashing meteor-bolide impacts, bubbling molten rock fumaroles, and the violent water/lava interactions that ultimately gave rise to the origin of (microbial) life. Follow that passage over the eons with the rise of metazoans, vertebrates, and ultimately the age of the dinosaurs as portrayed by the *Tyrannosaurus rex* vs. stegosaurus battle.

Spoiler alert: I rooted for the doomed stegosaurus.

In a blend of deep fascination combined with mortal fear, I certainly could not get these creatures out of my mind. When my parents would take me into Manhattan to the American Museum of Natural History (and the Hayden Planetarium), the Hall of the Dinosaurs was something I approached with absolute dread, but I could not turn back. It didn't help that my father told me that after the museum closed at night, the skeletons would walk out into Central Park, eat up all the trees, and become whole once again. This stratagem seemed to have a positive influence on my consumption of any proffered broccoli spears my mother served up at dinnertime.

Yet it did cause me some considerable nocturnal worries. At the tender age of five I would go to bed each night with my toy pistols lying beside me just in case a *T. rex* wandered into my Coney Island neighborhood, as it once happened in the movies (re: *The Beast from 20,000 Fathoms*). My older brother Mel pointed out to me that my arsenal items were useless as weapons, being only toys. I countered that while both he and I both knew that fact, the *T. rex* certainly did not and might thereby be dissuaded from gobbling me up.

As I grew older, my summers were consumed by reading science fiction, an especially enjoyable genre of fantasy, while I worked as a City lifeguard at Coney Island. The works of Asimov, Clarke, Pohl, Bradbury and others helped pass the time and fire my imagination, but I sensed that any manned missions to Mars and beyond were unlikely in my lifetime, so I turned my nascent scientific interests towards oceanography. The Apollo Program moon landings coincided with the release of the movie *2001: A Space Odyssey*, which more than anything fired up my brain with regard to the possibilities of discoveries to be made beyond low Earth orbit. An excellent book account of the movie's making was written by Benson [5.3].

I was awarded my PhD in marine sciences by the University of Miami in 1976 and drove out to California for a postdoc at the NASA Ames Research Center in Moffett Field, California [5.21]. This coincided with the successful landing of the two Viking missions that searched for experimental evidence for microbial life on the Martian surface. While enroute through the deserts of Utah and Neveda along Interstate 80, I looked out at the sands and mused that I could certainly do a career shift to exobiology from oceanography if Viking turned up evidence of extant life from the tested Martian regolith "soils."

Alas, when I arrived at NASA, the scientists all wore long faces, owing to the lack of any positive signs of life in the tested soils of the Viking experiments. The critical result of the gas exchange experiment ran by Vance Oyama was to detect production of metabolic gases (e.g., N_2, N_2O, H_2S, H_2, CH_4, CO_2) as soils were wetted, given nutrients, and dark-incubated. Yet in a quick reponse, once wetted the soils released a burst of O_2, a result that implied the existance of superoxide molecules in the soil, and strongly shifted interpretation of the results of all four experiments as being caused by the chemistry of this strong oxidant, rather than that of any Martian microbial activity. With the lack of evidence for microbial activity, the possibility of a future for myself as a NASA scientist dimmed at this point. So when a job opened up for me a year later at the US Geological Survey in Menlo Park, California, I was perfectly OK with going back to my original career path of aquatic/marine microbiology. And back down at NASA, this is where things stood for nearly 20 years until new technologies emerged and the Red Planet once again drew the interest of the scientific community.

Over my career I was able to conduct research on a variety of microbial processes that involve metabolism of gases as well as toxic elements like arsenic and selenium. While the main thrust of the work was relevant to the overall missions of the USGS (e.g., water quality; ecosystem and human health), I was also able to dabble in extreme environments and isolate extremophiles. These research efforts were supported in part by NASA's Exobiology Research Program, which allowed me to remain on the periphery of the larger effort of looking for signs of life beyond Earth in our Solar System.

So with a throwback to the enthusiastic 6-year-old "spaceman" version of myself shown in Figure 5.1, let us briefly visit the planet(oid)s Mars, Enceladus, and Titan in search of "alien" microbes. Or perhaps just some neighborhood feral cats will suffice?

5.2 Mars

When the accretionary discs of the Solar System were "handed out" Mars got the proverbial short end of the stick. With a diameter of 6,779 km, Mars is about half that of Earth's

(12,742 km), and while that sounds large, a simple trigometric volume calculation puts it at about 11 % of Earth's volume, or about 12% of the Earth's mass owing to the abundance of dense basalt rocks. This mass constraint really calls into question whether microbial life ever existed on the Red Planet. At present, owing to its low temperatures and low atmospheric pressure (less than 1% of Earth's and mostly composed of CO_2), liquid water does not exist on the surface, occurring only as a solid or a vapor. At its warmest, on a blistering summer day along its equator Mars can hit about 16°C, but this would only be at the rough location of one's shins. Owing to the aforementioned thin atmosphere and negligible latent heat, the thermometer would plunge well below zero by the time we reach the altitude at one's head. Mars lacks a protective magnetic field, so it is subject to the full force of the solar wind along with deadly UV radiation, owing to the lack of a protective ozone layer. Not a likely place to find thriving microbial communities, let alone any feral cats!

And yet, there are enough tantalizing bits of information to suggest that Mars once was far more clement, with a denser atmosphere and liquid water on it's surface for some prolonged amount of time. Way back in its Noachian epoch (4.1–3.7 Ga), this appears to have been the case. Close observations from orbit have revealed the presence of clay minerals, a geological prerequisite for life, and perhaps a once rudimentary protective magnetic field that dissipated as time progressed to the later Hesperian and Amazonian epochs. More recent revelations from the Curiosity rover include the detection of complex organic material in the sediments of Gale Crater [5.5], as well as seasonal signs for the presence of methane in the crater's atmosphere [5.30], both of which could be interpreted as signs of past microbial life, although there are also many non-biogenic possiblities as well.

For the sake of this book chapter, however, let us accept the possibility that microbial life on Mars got its start in the wet Noachian era, and as things became inclement, extremophilic life managed to hang on. Given the much smaller mass of Mars, it cooled more quickly than Earth but soon faced the loss of its atmosphere owing to the erosive effects of the solar wind and the lack of an internal dynamo that would have generated a protective magnetic field. Giant shield volcanoes, like Olympus Mons, came on the scene somewhat later (3.6–2.5 Ga), perhaps during the Hesperian Epoch [5.8], and if like what occurs on Earth, hydrothermal fluids may have mobilized arsenic oxyanions within the aquifers. As these aquifers seeped onto the Martian surface, evapo-concentration would render them highly saline, alkaline, and arsenic-rich, making the last surface biome not unlike Searles Lake (Figure 5.2) in California's Mohave Desert [5.18].

So I nominate a bacterial denizen isolated from Searles Lake as my candidate for a credible extremophile that could survive in one of Mars' last surface biomes: *Halarsenatibacter silvermanii* strain SLAS-1 [5.18] [5.27]. This haloalkaliphilic arsenotroph grows only at high salinities (250–350 g/L), pH (8.7–9.8), and is an obligate respiratory anaerobe, capable of using As(V), Fe(III), or S(0) as electron acceptors. Electron donors include simple organic acids and sugars (e.g., lactate, pyruvate, galactose, malate, glucose), but strain SLAS-1 can also grow as a chemoautotroph, using sulfide as its electron donor for the reduction of As(V) to As(III), and obtaining its carbon by the incorporation of ^{14}C-bicarbonate into cellular material, although not by Rubisco activity (i.e., Calvin-Benson-Bassham pathway). Another reason I favor this microbe is that it has an unusual mode of motility, able to make sinuous motions akin to that of an eel (or perhaps a hula dancer) to weave its way forward through the salt-saturated brine. The accumulated As(III) in the brine pool could be recycled to As(V) by serving as an electron donor for anoxygenic photosynthetic bacteria,

Figure 5.2 Three photographs of the playa of Searles Lake, California, with its pigmentation most evident. These photos were taken by Professor Chad Saltikov during a field trip/sample collection excursion with his students from the University of California at Santa Cruz. (Image credit: Chad Saltikov).

thereby perpetuating a closed arsenic cycle operating under a salt crust, protecting the brine to a certain degree from complete desiccation [5.6] [5.7]. The red colors evident in the playa brines of Searles Lake (Figure 5.2) were a consequence of the multitude of extremely halophilic phototrophs residing therein with their associated carotenoids evident.

One more extremophile we isolated from the Searles Lake milieu is *Desulfohalophilus alkaliarsenatis* strain SLSR-1 [5.28]. As its name implies, SLSR-1 can grow via sulfate-reduction, although it "prefers" arsenate respiration under the *in-situ* conditions of Searles Lake's saturated brine coupled with its high boron content. Those conditions also include growth in artificial medium with a measured low water activity (a_w) of 0.75, as well as demonstration of *in-situ* arsenotrophic activity in brine sediments ($a_w = 0.71$), values that are considerably lower than calculated for ancient Martian surface runoff (0.78–0.86).

Before I gather my friends in Figure 5.1 together to blast off for our next celestial destination, I would like to make one final remark. Mars has recently been touted (on national media and TED talks) as ripe for radical terraforming, essentially classed as a "fixer-upper" planet, no less than by the likes of such eminent people as Elon Musk (Founder, SpaceX) and Gwynne Shotwell (COO, SpaceX). Touted are such weird ideas as hitting the Martian polar caps with nukes to melt the ice caps and liberate CO_2 and installing giant mirrors in orbit to heat its surface. Yet nearly 45 years ago, two NASA scientists and colleagues of mine, Mel Averner and Bob McElroy (now long deceased) published one of the first rational intellectual studies on the possibility of terraforming Mars to suit human habitation [5.2]. These public servants did their work as a matter of course, and to virtually no public fanfare. Elon Musk was all of 4 years old at the time the study came out.

5.3 Enceladus

Even with a diameter of only 505 km, Enceladus is the 6th largest of Saturn's some 82 moons. The Cassini mission to the Saturnian system has revealed that Enceladus is likely to be the celestial body in our Solar System with the greatest potential for encountering alien microbial life [5.25], perhaps even a "second genesis." Under its smooth ice cover lies a planetary ocean composed of liquid water, roughly equivalent in pH and salinity to terrestrial seawater. How long that ocean has existed is not clear, but perhaps for a few million years. Whether this is sufficient time to allow for the possibility of primitive life arising is anyone's guess. The orbit that Enceladus weaves around Saturn, as well as gravitational forces exerted by other Saturnian moons on the planetoid, squeezes and distorts its silicate core, causing frictional heating and dissolution of solid minerals into the subsurface liquid ocean. It also caused long fissures to occur near the satellite's south pole (the four "tiger stripes") which are breached by the spectacular, eruptive release of water jets carrying materials from the ocean interior into the void of space. The contents of these jets provide scientific excitement because they contain evidence that beneath its icy shell, Enceladus is a "habitable" world, if not "inhabited" for life-as-we-know-it (i.e., water based). These contents, as identified by instrumentation on Cassini, detected CH_4 and other light hydrocarbons, H_2, NH_3, CO_2, sulfur and phosphorus compounds [5.29], as well as higher molecular weight complex organics [5.23]. In contrast to Mars, which has an abundance of strong oxidants on its surface (e.g., perchlorates, iron III) but lacks organic reductants (i.e., potential electron donors), Enceladus is the reverse, heavy on putative electron donors but lacking in electron acceptors to oxidize them. Hence, of the many putative electron donors spewed out by the jets, it is acetylene and its biopotential that intrigues me [5.10].

So in this case my candidate "extremophile" would be an anaerobic bacterium like *Pelobacter acetylenicus* or *Pelobacter* strain SFB93 that can grow by fermenting acetylene [5.1] [5.4] [5.24]. It is a two-stage process that first hydrates the acetylene with water (enzyme: acetylene hydratase) to form acetaldehyde, and then the second enzyme (acetaldehyde dismutase) splits acetaldehyde into equimolar ethanol and acetate (plus some H_2). While these microbes cannot be considered as extremophiles in the traditional sense of resistance to high salinity, temperature, pH, metals, radiation, etc., they are well-adapted for life in the Enceladus ocean and are capable of making a metabolic living therein. Furthermore, the products of acetaldehyde dismutase can feed into methanogenesis, as well as any anaerobes that can link to respiratory electron acceptors, like nitrate or sulfate, should they exist in the brine. Thus, acetylene could serve as the basis of an Enceladus foodchain, and a life-detection concept has been proposed that would exploit the ability of these microbes to achieve a stable carbon isotopic fractionation of acetylene and its metabolic products, perhaps leaving behind a detectable biosignature [5.15]. The fact that Enceladus spews forth its ocean brine contents into space may ease the collection of these materials from orbit, or by collection as snow falls on the moon's surface, for subsequent return to Earth or orbit for analysis. This would be far easier than drilling through its ice crust which even at the south pole is still some 5 km deep, especially if the odd feral cat lands in the collection net.

5.4 Titan

With a diameter of 5,150 km, Titan is the largest moon in the Saturnian system, and as planetary moons go for the Solar System, its size is only exceeded by Jupiter's Ganymede (5,262 km). Titan is an odd, yet fascinating, alien world which nonetheless shares some curious similarities with Earth. It hosts a dense atmosphere (about 1.5x that of Earth), the main component of which, like Earth, is N_2 (~94%) with CH_4 (~5.5%) as its secondary component in lieu of Earth's O_2. Although it receives only 1% of Earth's insolation, the methane in Titan's atmosphere undergoes photolytic reactions caused by bombardment of high energy photon's emanating from the Sun. This forms complex organic materials that fall onto the orb's surface and impart an orange haze to its atmosphere. The temperature on Titan's surface is about 98 K (−179 °C), which means that any water there has the consistency of granite. *You can almost hear the mothers of the kids in Figure 5.1 screaming for us to dress extra warmly before we ventured out from our rocket ship to explore Titan's surface.* The Cassini satellite and Huygens probe have revealed a wealth of new information, one of which is that Titan has the equivalent of a hydrological cycle, but in lieu of water, it is centered upon liquid methane. It rains methane on Titan, which also has river systems and large lakes composed primarily of liquified methane and ethane. Titan also tilts on its axis as it orbits Saturn and therefore has seasons. During summer, the liquid methane evaporates into the atmosphere, leaving behind diminished lakes composed mostly of ethane. The uncovered lake "playas" have complex associations of benzene molecules that function as the equivalent of an evaporative salt crust in the terrestrial soda lake summer analog. Altogether a thoroughly alien but oddly familiar world. If microbial life exists there, it would be life as we do not know it, with liquid methane/ethane rather than water serving as it universal solvent. Far fetched?

Perhaps not. Conclusive observations made in Chile using the Atacama Large Millimeter/submillimeter Array (ALMA) reported the presence of vinyl cyanide (C_2H_3CN), formed photochemically in Titan's atmosphere, and thence falling and dissolving into its surface lakes/seas, like Ligeia Mare [5.22]. Since vinyl cyanide can form polymers, it can act in a similar fashion to the way the lipid membranes of terrestrial organisms separate the internal workings of cells from their aqueous milieu. All well and good, but what could serve as a food source to sustain such alien life? Here we can turn once again to acetylene, the third most abundant hydrocarbon in Titan's atmosphere.

McKay and Smith [5.11] proposed an alien, one-step methanogenic pathway for life on Titan that would achieve the reduction and cleavage of acetylene to form methane, using the abundant hydrogen present in the satellite's atmosphere:

$$C_2H_2 + 3\ H_2 \text{----------}\rightarrow 2\ CH_4 \qquad \Delta G = -\ 334\ \text{kJoules/mole.}$$

Curiously, the reaction is thermodynamically highly favorable at the ridiculously low temperature of 90 K. It is analogous to what can occur in terrestrial aqueous systems between a fermentative acetylenotroph (e.g., *P. acetylenicus*) and a methanogen [5.14]. The McKay/Smith paper may have remained as a simple backwater intellectual exercise, had not the Huygens probe found evidence for removal of H_2 and C_2H_2 from the atmosphere during its descent from orbit to the moon's surface [5.26]. One interpretation of these observations was that there was something alive on Titan that was consuming (aka "eating") these gases in a manner consistent with that predicted by McKay and Smith [5.11]. It did generate some excitement for a while, although it did not present conclusive proof for an alien biological sink for these compounds. The matter has been reviewed extensively by McKay [5.12] especially with regard to the possibilities of life as we do not know it occurring upon Titan's surface and in its lakes, rivers, and seas. Clearly there are exciting things to be discovered in future missions to Titan, including the possibility of an alien, acetylene-based life resident thereon.

Then again, considering Titan's organically-hued surface as revealed by the Huygens lander, perhaps we should be looking more closely for feral orange tabbies?

References

5.1. Akob, D.M., Sutton, J.M., Fierst, J.L., Haase, K.B., Baesman, B., Luther, G.W., III, Miller, L.G., Oremland, R.S., Acetylenotrophy: A hidden but ubiquitous microbial metabolism? *FEMS Microbiol. Ecol.*, 94, 2018.

5.2. Averner, M.M. and McElroy, R.L., *On the Habitability of Mars: An Approach to Planetary Ecosynthesis [NASA SP-414]*, 1976.

5.3. Benson, M., *Space Odyssey*, S. Kubrick, A.C. Clarke, the Making of a Masterpiece (Eds.), 444 pp., Simon and Schuster, New York, NY, 2018.

5.4. Culbertson, C.W., Strohmaier, F.E., Oremland, R.S., Acetylene as a substrate in the development of primordial bacterial communities. *Orig. Life Evol. Biosph.*, 18, 397–407, 1988.

5.5. Eigenbrode, J.L., Summons, R.E., Steele, A., Freissinet, C., Millan, M., Navarro-González, R., Sutter, B., McAdam, A.C., Franz, H.B., Glavin, D.P. *et al.*, Organic matter preserved in 3-billion-year-old mudstones at Gale Crater, Mars. *Science*, 360, 1096–1101, 2018.

5.6. Hoeft, S.E., Kulp, T.R., Han, S., Lanoil, B., Oremland, R.S., Coupled arsenotrophy in a photosynthetic hot spring biofilm from Mono Lake, California. *Appl. Environ. Microbiol.*, 76, 4633–4639, 2010.

5.7. Hoeft McCann, S., Conrad, A., Hernandez-Maldonado, J., Stoneburner, B., Saltikov, C.W., Stolz, J.F., Oremland, R.S., Arsenite as an electron donor for anoxygenic photosynthesis: Description of three strains of *Ectothiorhodospira* from Mono Lake, California and Big Soda Lake, Nevada. *Life*, 7, 1, 2017.

5.8. Isherwood, R.J., Jozwiak, L.M., Jansen, J.C., Andrews-Hanna, J.C., The volcanic history of Olympus Mons from a paleo-topography and flexural modeling. *Earth Planet. Sci. Lett.*, 363, 88–96, 2012.

5.9. Kempe, S. and Degens, E.T., An early soda ocean? *Chem. Geol.*, 53, 95–108, 1985.

5.10. Matson, D.L., Castillo, J.C., Lunine, J., Johnson, T.V., Enceladus plume: Compositional evidence for a hot interior. *Icarus*, 187, 569–573, 2006.

5.11. McKay, C.P. and Smith, H.D., Possibilities for methanogenic life in liquid methane on the surface of Titan. *Icarus*, 178, 274–276, 2005.

5.12. McKay, C.P., Titan as the abode of life. *Life*, 6, 2016.

5.13. Merino, N., Aronson, H.S., Bojanova, D.P., Feyhi-Buska, J., Wong, M.J., Zhang, S., Giovannelli, D., Living at extremes: Extremophiles and the limits of life in a planetary context. *Front. Microbiol.*, 2019.

5.14. Miller, L.G., Baesman, S., Kirshstein, J., Voytek, M.A., Oremland, R.S., A biogeochemical and genetic survey of acetylene fermentation by environmental samples and bacterial isolates. *Geomicrobiol. J.*, 30, 501–516, 2013.

5.15. Miller, L.G., Baesman, S.M., Oremland, R.S., Stable carbon isotope fractionation during bacterial acetylene fermentation: Potential for life detection in hydrocarbon-rich volatiles of icy planet(oid)s. *Astrobiology*, 15, 977–986, 2015.

5.16. Oremland, R.S., Miller, L., Whiticar, M., Sources and flux of natural gases from Mono Lake, California. *Geochim. Cosmochim. Acta*, 51, 29152929, 1987.

5.17. Oremland, R.S. and Des Marais, D.J., Distribution, abundance, and carbon isotopic composition of gaseous hydrocarbons in Big Soda Lake, Nevada: An alkaline, meromictic lake. *Geochim. Cosmochim. Acta*, 47, 2107–2114, 1983.
5.18. Oremland, R.S., Kulp, T.R., Switzer Blum, J., Hoeft, S.E., Baesman, S., Miller, L.G., Stolz, J.F., A microbial arsenic cycle in a salt-saturated, extreme environment. *Science*, 308, 1305–1308, 2005.
5.19. Oremland, R.S. and Voytek, M.A., Acetylene as fast-food: Implications for development of life on the anoxic primordial Earth and planet(oid)s of the outer solar system. *Astrobiology*, 8, 1–14, 2008.
5.20. Oremland, R.S., A random biogeochemical walk into three soda lakes of the western USA: with an introduction to a few of their microbial denizens, in: *Polyextremophiles – Life under multiple forms of stress*, J. Seckbach, A. Oren, H. Stan-Lotter (Eds.), pp. 181–199, Springer, Dordrecht, Germany, 2013.
5.21. Oremland, R.S., Why I never worked on anaerobic oxidation of methane (AOM) beyond the unsuccessful attempts of my NRC postdoc at NASA Ames Research Center (Sept. 1976 – Sept. 1977). *FEMS Microbiol. Lett.*, 366, 2019.
5.22. Palmer, M.Y., Cordiner, M.A., Nixon, C.A., Charnley, S.B., Teanby, N.A., Kisiel, Z., Irin, P.G.J., Mumma, M.J., ALMA detection and astrobiological potential of vinyl cyanide on Titan. *Sci. Adv.*, 3, e1700022, 2017.
5.23. Postberg, F., Khawaja, N., Abel, B., Choblet, G., Glein, C.R., Gudipati, M.S., Henderson, B.L., Hsu, H.W., Kempf, S., Klenner, F. *et al.*, Macromolecular organic compounds from the depths of Enceladus. *Nature*, 558, 564–568, 2018.
5.24. Schink, B., Fermentation of acetylene by an obligate anaerobe, *Pelobacter acetylenicus* sp. nov. *Arch. Microbiol.*, 142, 295–301, 1985.
5.25. Spilker, L., Cassini-Huygens' exploration of the Saturn system: 13 years of discovery. *Science*, 364, 1046–1051, 2019.
5.26. Strobel, D.F., Molecular hydrogen in Titan's atmosphere: implications of the measured tropospheric and thermospheric mole fractions. *Icarus*, 208, 878–886, 2010.
5.27. Switzer Blum, J., Han, S., Lanoil, B., Saltikov, C., Witte, B., Tabita, F.R., Langley, S., Beveridge, T.J., Stolz, J.F., Jahnke, L., Oremland, R.S., Ecophysiology of "*Halarsenatibacter silvermanii*" strain SLAS-1^T, gen. nov., sp. nov., a facultative chemo-autotrophic arsenate-respirer from salt-saturated Searles Lake, California. *Appl. Environ. Microbiol.*, 75, 1950–1960, 2009.
5.28. Switzer Blum, J., Kulp, T.R., Han, S., Lanoil, B., Saltikov, C.W., Stolz, J.F., Miller, L.G., Oremland, R.S., *Desulfohalophilus alkaliarsenatis* gen., nov., sp. nov., an extremely halophilic sulfate- and arsenate-respiring bacterium from Searles Lake, California. *Extremophiles*, 16, 727–742, 2012.
5.29. Waite, H.J., Jr., Lewis, W.S., Magee, B.A., Lunine, J.I., McKinnon, W.B., Glein, C.R., Mousis, O., Young, D.T., Brockwell, T., Westlake, J. *et al.*, Liquid water on Enceladus from observations of ammonia and ^{40}Ar in the plume. *Nature*, 460, 487–490, 2009.
5.30. Webster, C.R., Mahaffy, P.R., Atreya, S.K., Moores, J.E., Flesch, G.J., Malespin, C., McKay, C.P., Martinez, G., Smith, C.L., Martin-Torres, J. *et al.*, Background levels of methane in Mars; atmosphere showing strong seasonal variations. *Science*, 360, 1093–1096, 2018.

6

Antarctic Bacteria as Astrobiological Models

Carmel Abbott and David A. Pearce*

Faculty of Health and Life Sciences, Northumbria University at Newcastle, Northumberland Road, Newcastle-upon-Tyne, U.K.

Abstract

Antarctica contains many different types of habitat that would traditionally be considered harsh from a human perspective; it can be extremely cold, have low levels of liquid water, low humidity, low nutrient availability, high levels of salinity and high levels of non-ionizing radiation. Yet a wide variety of bacteria have been found living there, despite these harsh conditions; some of them are believed to be unique to the continent, others more cosmopolitan in distribution. When we compare aspects of these Antarctic habitats to conditions known to occur on Mars, or to what is known of the icy moons of Jupiter and Saturn, we find notable similarities even though, clearly, significant differences remain. It is therefore unsurprising that scientists have used bacteria isolated from the Antarctic as astrobiological models. The extent to which this has been done to date, however, is perhaps surprisingly limited despite the enormous potential in this approach. In this chapter, we examine the differences and similarities between specific habitats in Antarctica and those which they might mimic on Mars, Europa and Enceladus. It considers the nature of the microbiological adaptions found in these Antarctic habitats and the experiments carried out to date on bacteria isolated from them. The chapter concludes by discussing the future potential of Antarctic bacterial species as well as the lessons learnt in understanding the limits of life here on Earth and the possibility of finding evidence of microbial life elsewhere in the solar system.

Keywords: Antarctica, bacteria, extremophile, endemic, astrobiology, models, Mars, icy moon

Abbreviations

A_w	water activity
BOSS	Biofilm Organisms Surfing Space
CHNOPS	combined elements of carbon, hydrogen, nitrogen, oxygen, phosphorus and sulfur
ESA	European Space Agency
EuTEF	European Technology Exposure Facility

**Corresponding author*: david.pearce@northumbria.ac.uk
Carmel Abbott: https://www.researchgate.net/profile/Carmel_Abbott
David Pearce: https://www.northumbria.ac.uk/about-us/our-staff/p/david-pearce

Joseph Seckbach and Helga Stan-Lotter (eds.) Extremophiles as Astrobiological Models, (137–160)

ISS	International Space Station
MDVs	McMurdo Dry Valleys
MVT	Miers Valley transect
NASA	National Aeronautics and Space Administration
ClO_4^-	perchlorate
T_{opt}	temperature at which growth is fastest (assumed to be optimal temperature for growth)
UV	ultraviolet
WMO	World Meteorological Organization

6.1 Introduction

Astrobiology is concerned with the search and potential for life beyond the Earth's biosphere. Fundamental questions in the field of astrobiology include the possibility of life existing on other planets, the origin and evolution of life on Earth, and planetary protection—preventing the transfer of terrestrial microorganisms to other planetary bodies during space missions (forward contamination) and the potential for cross-contamination of Earth's ecosystem from extra-terrestrial materials returned (backward contamination). In the first instance, it is important to understand and to clearly define the envelope or boundaries for life on Earth [6.43] so we can develop the required techniques to look for life elsewhere and also to investigate the adaptations of microorganisms exposed to extremes that could pre-adapt them to similar environments elsewhere.

Research from astronomical observations and planetary missions has determined the approximate environmental conditions on the Moon, Mars, and other celestial bodies, such as the icy moons of Europa (smallest of the four Galilean moons orbiting Jupiter) and Enceladus (the sixth largest moon of Saturn), and indeed the space environment itself, all of which are extremely hostile to life (Table 6.1). As humans we have evolved in the Earth's atmosphere, and so are suited to ~101 kPa pressure, gaseous concentrations of 21% oxygen and 78% nitrogen, an average global temperature of about 15 °C, and limited exposure to ultraviolet (UV) and ionizing radiation [6.73].

Table 6.1 Comparison of main environmental data of Earth to space and celestial bodies of astrobiological interest.

	Earth	Mars	Moon	Europa	Enceladus	Space
Mean Temperature (°C) (range)	15 [6.73] (-89.2 to 56.7) [6.110]	-63 (-89 to -31) [6.74]	-20 (-178 to 117) [6.75]	-183 (-227 to -177) [6.4]	-203 [6.82]	-270 in deep space [6.33] (-157 to 121 at LEO) [6.76]
Pressure (Pa) (range)	1.01×10^5 (8.7×10^4 to 1.09×10^5) [6.73]	636 (400 to 870) [6.74]	3×10^{16} [6.75]	no data	no data	1.460×10^{16} [6.39]
Atmospheric composition (%) by volume	N_2 78.08 O_2 20.95 Ar 0.93 CO_2 0.04 [6.73]	CO_2 95.32 N_2 2.70 Ar 1.6 O_2 0.13 [6.74]	^4He 25.8 ^{20}Ne 25.8 H_2 22.6 ^{40}Ar 1.3 [6.75]	no data	no data	no data
Cosmic radiation	3 mGy/y [6.11]	76 mGy/yr [6.45]	no data	no data	no data	500mGy/h [6.15]

The Earth does, however, contain diverse environments and habitats, and these include extreme environments. These are areas with one or more extreme conditions such as temperature, pH or pressure [6.60]. Examples include the sulfuric hot springs of Yellowstone National Park in America, dry and hot deserts like Atacama in the high northern Andean plains of Chile, dry and cold deserts like the Antarctic dry valleys, the cold and high altitude of the Tibetan plateau, hypersaline environments such as Mono Lake in California, and the cold and dark deep sea environment of the Mariana Trench which is characterized by high pressure as well as hydrothermal vents which release hot and acidic water, causing surrounding temperatures of up to 300 °C. Earth's atmosphere itself can also be considered an extreme environment with higher levels of ultraviolet (UV) radiation and low nutrient availability, as can human environments where, for example, acid mine drainage and heavy metal pollution have also been generated. However, what is an extreme and uninhabitable environment for human life is not necessarily extreme for other forms of life, especially microorganisms. In particular, the extremophiles, which are organisms that live in extreme environments and include primarily prokaryotic organisms (bacteria and predominately the archaea) with few eukaryotic examples. Indeed, many extremophiles are also called polyextremophiles due to their ability to tolerate conditions which include more than one environmental extreme.

In the study of astrobiology, extreme habitats on Earth can be used as analogues for space environments and provide a wealth of information about possible extinct and/or extant life. The environmental extremes in Antarctica, across all the habitable niches, share similar characteristics to aspects of the environments on Mars and the icy moons Enceladus and Europa (Table 6.1).

While many parts of the Antarctic are extreme, with low temperatures year-round and relatively high levels of ultraviolet and cosmic ray radiation, the intensity of scientific investigation varies considerably, which can be due to logistical considerations such as ease of access or historical due to national interests and features of particular significance. Vostok and the McMurdo Dry Valleys (MDVs) are regions that have been compared to some aspects of both Mars and the icy moons and have therefore received particular attention as astrobiological models. Indeed, extremophiles adapted to those environments meet some of the criteria for survival and even growth on Mars.

6.2 Antarctica as an Analogous Environment for Astrobiology

Antarctica forms part of the cryosphere on Earth. The cryosphere is an all-encompassing term for those portions of Earth's surface where water is in solid form and includes areas of snow, ice, permafrost, seasonally frozen ground, glaciers and ice shelves. Estimates of the extent of the cryosphere vary and this can partially be down to how it is defined, and there is also seasonal variation as the greater land area of the Northern Hemisphere leads to greater increases in seasonal frozen ground and snow cover during the winter than the growth of the Antarctic sea ice during the Southern Hemisphere winter. Using recent estimates of the extent of glaciers, Antarctic ice, snow and ice in the Northern Hemisphere suggest that the cryosphere varies seasonally between 10.5 and 30% of the surface of the globe [6.73] [6.16] [6.3].

One of the focus areas, Vostok in East Antarctica, is the place with the coldest recorded temperature on Earth at–89.2 °C (with a mean temperature range of–10 °C at the coast to

-60 °C at the highest altitude in the continental interior). The region is totally covered by ice up to 4.8 km thick and spans an area of 14 million km^2 with lakes between 500 m and 800 m in depth beneath it [6.18]. Despite containing 90% of the world's fresh water, it is mostly in solid form, also making Antarctica a hyperarid environment [6.61]. In terms of nutrients, total dissolved solids are estimated to be 1 to 40 mM and the oldest ice is ~420,000 years old [6.18]. While this is a very different environment to the McMurdo Dry Valleys, it might be a better analogue for the less well explored environments of the icy moons.

Antarctica also has a wide variety of different regional habitats, which can expand the range of adaptations that can be found in Antarctic bacteria. Some of these are found elsewhere, such as active volcanoes and microbial mats, others are more unique to the region such as deep subglacial lakes and the hyperarid soils of the McMurdo Dry Valleys. Margesin and Collins [6.63] provide a good review of the environmental niches to be found in Antarctica.

6.2.1 Temperature

Soil temperatures in the McMurdo Dry Valleys range from -59.2 °C to 25.7 °C and average about -25 °C. The air temperatures recorded are between -65.7 °C and +10.0 °C [6.28]. These temperature ranges are similar to those reported by various mission locations across Mars (Table 6.2). Psychrophilic bacteria might therefore be considered as candidates for this aspect of astrobiological research.

6.2.2 Soil as a Life-Supporting Medium

The elements carbon, hydrogen, nitrogen, oxygen, phosphorus and sulfur (collectively termed CHNOPS) form the minimum essential elements for maintenance of life. This was illustrated in the European Space Agency's (ESA) MELiSSA closed loop ecological life support system experiment, which produced an ecosystem of chemical and biochemical degradation and synthetic reactions that were both mass and stoichiometrically balanced for these six elements to ensure its long-term stability [6.86]. CHNOPS are available in different forms, such as macromolecules from geochemical cycles due to physical processes (plate tectonics and atmospheric chemical processes) and from redox reactions of microbes in elemental cycles, including methanogenesis, chemoautotrophy, iron and sulfate reduction [6.31]. On Mars, micrometeorites present on the surface provide some of the same elements as the geochemical processes on Earth [6.104]. For a comprehensive review of elements on Earth, Mars, Enceladus and Europa and the mechanisms by which they can be derived see Cockell *et al.* [6.20].

As with soil elsewhere, the soil physiochemical properties among the McMurdo Dry Valleys of Antarctica have a high degree of heterogeneity. Soil salinity, pH and organic matter content influence microbial distribution and community composition, with pH, salinity, phosphorus and nitrogen content common limiting factors [6.111]. The soil at Miers Valley is more alkaline than the other valleys and has a high carbon to nitrogen ratio, high carbon content, high copper concentration and low gravimetric water content [6.58]. Miers Valley soil also stands out in terms of salinity (Table 6.2), although analysis of the physiochemical properties of soil shows that even for samples taken from the same site there is a high level of variation [6.94].

Table 6.2 Comparison of some of the environmental extremes found in the Antarctic Dry Valleys as an analogue environment to specific aspects of the conditions across Mars. Martian data from the Chryse Planitia (19 °N), Vastitas Borealis (68 °N) and Gale Crater (-5 °N) regions are from the missions Mars Pathfinder, Phoenix Lander and the combined Mars Science Laboratory (MSL) and Curiosity Rover mission respectively.

Observations	Antarctica (Dry Valleys)	Reference	Mars	Mars location	Reference
Air temp range (°C)	-65.7 °C to +10.0 °C	[6.28]	-78 °C to -13 °C	Chryse Planitia	[6.64]
Surface temp range (°C)	-59.2 °C to +25.7 °C	[6.28]	-97.7 °C to -19.6 °C	Vastitas Borealis	[6.23] [6.41]
			-83 °C to 13 °C	Gale Crater	[6.41]
Soil pH	7.1 to 9.1	[6.58] [6.94]	Neutral	Vastitas Borealis and Gale Crater	[6.46] [6.37]
Presence of CHNOPS	Yes	[6.58] [6.94]	Yes	Gale Crater	[6.37] [6.20] [6.104]
Organic Carbon	0.06 to 0.46 %	[6.58] [6.94]	50 nmol	Gale Crater	[6.29]
Salts and minerals in soil	Salinity varies widely (109-6130 μS, conductivity used as indirect measure)	[6.58]	10 mM dissolved salts comprised of Na^+, K^+ Mg^{2+}, Ca^{2+}, Cl^- and ClO_4^- ions	Vastitas Borealis	[6.46]
			Clay minerals with low salinity	Gale Crater	[6.37]
Presence of perchlorates	No		0.4 to 0.6 wt %	Vastitas Borealis (thought to be ubiquitous across Mars)	[6.46] [6.105] [6.87]
Surface radiation			76 mGy/yr	Gale Crater	[6.45]

Despite the very low levels of organic carbon and moisture found in the soil, Hart *et al.* [6.44] found a wide variety of organic biomarkers such as lipids and aliphatic compounds derived from bacteria in the paleosols (formed as part of a soil from a past geological age) of the McMurdo Dry Valleys. They considered how these ancient soils compare to Martian soils from the known lander sites, commenting that currently Antarctic soils show much less history of glacial retreat than those of Mars. Mars is also estimated to have experienced approximately 1000 times more weathering of its surface, probably resulting in almost all of the iron being oxidized to Fe^{3+} rather than the bioactive Fe^{2+} cation, whereas in Antarctica only 20% of the iron is oxidized. Nevertheless, the presence of biomarkers in the soil is an indication that microbial life may have been present, even if it is currently impossible to isolate and identify it.

6.2.3 Soil Salinity

There are a wide variety of lake types in Antarctica, some of which are considered briny with chloride concentrations as high as 4.8 M (compared to 0.50 M in seawater [6.66]). Yet polyextremophilic bacteria and archaea have been isolated from the sediments under such lakes [6.103] and even within the water column itself [6.19]. There are also reports of possible microbial growth and metabolism of deep-sea halophiles in magnesium chloride saturated brine and in the calcium chloride-rich Don Juan Pond (water activity (a_w) <0.450), as reviewed in Stevenson *et al.* [6.101], and other brines of astrobiological interest [6.40].

Halophilic and xerophilic bacteria are of particular interest in life detection and survivability studies on Mars due to the presence of perchlorates $\left(ClO_4^-\right)$ in Martian Arctic soil samples, most probably in the form of calcium perchlorate ($Ca(ClO_4)_2$) and magnesium perchlorate ($Mg(ClO_4)_2$), at levels of 0.4 to 0.6 wt% [6.46] [6.105]. The a_w of $Mg(ClO_4)_2$ and $Ca(ClO_4)_2$ brines at 25 °C are 0.39 and 0.29, which increases to 0.54 and 0.51 respectively when the temperatures decrease towards temperatures found on Mars [6.106].

The ubiquity of perchlorate on Mars may have implications for the presence of liquid water in those areas of the planet where the continued presence of water has been found [6.53] [6.85]. Perchlorate brines enter a supercooled liquid phase before vitrifying at extremely low temperatures (such as -135 °C), typically lower than those observed on Mars, even at night in the Martian winter. This suggests that briny, liquid water might be available in some habitats year-round [6.106]. Life has shown it can persist and remain viable even in such circumstances by entering stasis during vitrification [6.102].

Perchlorate exposed to ultraviolet radiation on the surface of Mars is reactive and has been shown to be highly deleterious to cells [6.108], as have other components of the Martian surface such as silicate [6.5]. However, while the surface might be more hostile to life, it is irrelevant for shielded water and the potential for life even short distances below the surface due to shielding by the Martian regolith (i.e., the layer of loose, heterogeneous superficial deposits covering solid rock). Antarctic extremophiles could potentially survive such conditions.

6.3 Astrobiological Environments of Interest

Atmospheric composition, energy source and ultraviolet radiation are also important considerations for bacteria that have adapted to live at the physiochemical boundaries of life. Clearly, any comparison between the conditions in Antarctica and those in space draw significant differences. Space is much colder (-270 °C), lacks nutrients and has lower pressure (1.46 aPa) [6.39]. However, dust, rocks, comets and other such material can provide a substrate for bacterial adhesion and survival [6.48]. Spacecraft also provide a potential substrate and polyextremophiles from Antarctica are robust and able to adapt. For this reason, there has been interest in determining whether they can survive or even grow while exposed to the space environment. Several experiments, such as EXPOSE-E, EXPOSE-EuTEF (European Technology Exposure Facility) and EXPOSE-R2/BOSS (Biofilm Organisms Surfing Space) aboard the International Space Station (ISS) and Lithopanspermia/BIOPAN aboard the Foton satellites, have investigated the survival of a wide range of microbial life, including extremophilic bacteria when exposed to space [6.88] [6.83] [6.26] [6.89] [6.24]. Furthermore, the planned ICE-EXPOSE project will study the adaptation strategies of microorganisms to the extreme conditions of Antarctica and of astrobiologically relevant extremophiles and human commensals to spaceflight conditions. The results from ICE-EXPOSE will also inform planetary protection policies in terms of monitoring survival, outgrowth and potential contamination issues in spaceflight environments which could impact forward contamination.

The Galileo mission revealed that Europa, long considered to be a solid lump of ice, is actually composed of an ice cover over a liquid center. The surface is hostile, with temperatures ranging between -227 °C (poles) and -177 °C (equator), and is constantly bombarded by high energy electrons from Jupiter [6.4]. The liquid interior may be mostly hydrated magnesium sulfate with some hydrated sodium sulfate and the possibility of some hydrogen sulfate (sulfuric acid) [6.67]. The potential for life on Europa was reviewed in terms of access to redox energy sources and CHNOPS elements, concluding that life beneath the crust is at least theoretically possible within the parameters as defined from life on Earth [6.42].

The Cassini mission showed the plumes of ejecta (effectively huge geysers) from Enceladus, an icy moon of Saturn, contain a mix of organic carbon, nitrogen, inorganic salts and energy sources suitable for life in liquid water. Like Europa, Enceladus has an icy shell, but it appears that under that shell is a water reservoir that contains a mix of components that could contain life in a form that we might recognize (as reviewed in more detail in McKay *et al.* [6.68] and Hsu *et al.* [6.52]). Enceladus itself has a surface temperature of -203 °C and the icy shell is estimated to be 92 km thick. Modeling suggests that it is not in thermal equilibrium unless it is tectonically active [6.82].

Comparing the environments found in Antarctica with those on Mars and the icy moons is a useful exercise in appreciating the challenges microbial life may have faced for continued survival. Species of bacteria endemic to Antarctica have been used as astrobiological models focusing primarily on the conditions of Mars. A variety of cosmopolitan bacteria have also been used as astrobiological models and are briefly mentioned below, however, these are not the main focus of this chapter.

6.4 Bacterial Adaptations to Extreme Environments as Analogues for Astrobiology

Space missions, especially those that intend to land on other planets or moons, require long journeys. Studies from the ISS and experiments on both the living quarters themselves and how bacteria can survive in space have been undertaken. The resultant data have changed our understanding of the potential for forward contamination from unmanned missions. Experiments dating back to the early 1960s have attempted to characterize adaptations and the survival of microorganisms to a set of extraterrestrial extremes. For example, the survival of *Bacillus subtilis* in space vacuum and ultraviolet irradiation [6.6] and *Halomonas* spp. to desiccation and radiation on a Mars surface analogue [6.72].

Analogue studies often use extremophiles such as halophiles, barophiles, osmophiles or psychrophiles (Figure 6.1). They differ from other microorganisms by being adapted to extremes (that is, towards the limits of the envelope of life) and thriving rather than simply tolerating or resisting stress. Many extremophiles are polyextremophiles, meaning they are adapted to multiple environmental stressors. Any life found elsewhere in the solar system will have to cope with a variety of extreme environmental conditions simultaneously. Antarctic bacteria show such adaptations; for example, *Psychrobacter arcticus* 273-4 has no genes for glycolysis or phosphotransferase and relies wholly on other carbon sources [6.107]. Meanwhile, other organisms found in these environments simply alter the expression of their genes. For example, *Psychrobacter* sp. PAMC 21119 has a full repertoire of metabolic genes but in Antarctica, where it was isolated, it upregulates expression of genes suited to this environment while downregulating expression of genes involved in primary metabolism [6.107]. This illustrates that not all organisms found in extreme environments are extremophiles, rather they could be well adapted or physiologically flexible mesophiles.

6.4.1 Psychrophiles

Psychrophiles are common extremophiles adapted to low temperatures and can play an important role in biogeochemical cycles, especially efflux of carbon in seasonally frozen ecosystems contributing to greenhouse gases and climate change [6.80]. In the literature,

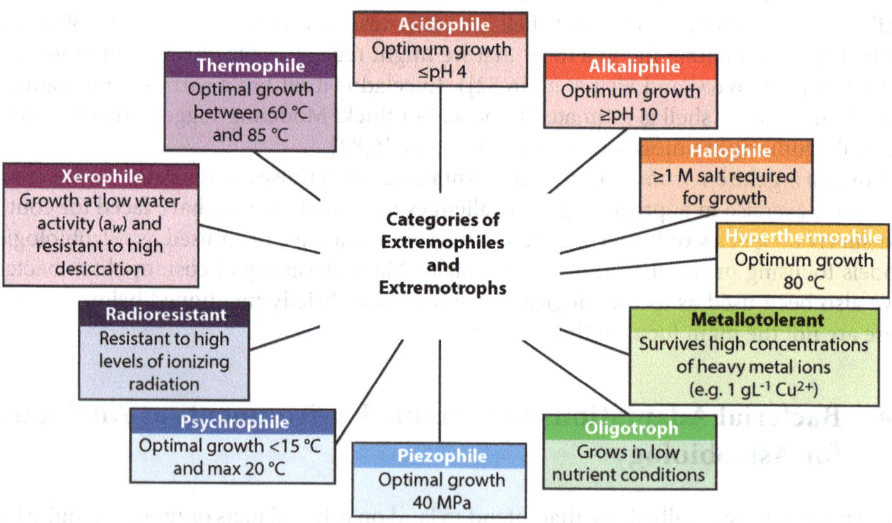

Figure 6.1 Extremophiles and Extremotrophs by category. Characterizing the parameters for life at extremes [6.38] [6.50] [6.90]. (Image credit: the authors).

there are three working definitions of a psychrophile; either an organism that can only grow at temperatures <20 °C, organisms that grow at ≤5 °C, or organisms with a growth optimum at a low temperature [6.13]. Generally, they can proliferate between -20 °C to +20 °C with an optimal temperature for which growth is fastest (T_{opt}) of <15 °C [6.38]. It should be noted that caution should be applied when describing optimal growth conditions where they have been defined *in vitro* but are referring to growth *in situ*. Psychrophiles have a T_{opt} that is regularly higher than that of its natural environment. This can cause some psychrophiles to experience and exhibit physiological stress [6.109]. For example, *Psychrobacter arcticus* 273-4, which has a T_{opt} of 17 °C but can grow below 0 °C, essentially has two completely differently metabolic systems and switches between them, as indicated by a shift in the metabolism of acetate, amino acid biosynthesis, oxidative stress and changes in the expression of over 1,000 genes when grown at 6 °C, 17 °C and 22 °C [6.7]. Defining psychrophiles by T_{opt} also ignores other factors that are found in the microorganism's natural environment such as substrate availability and concentration [6.13]. Psychrophiles are surprisingly uncommon when we consider that in most of the Earth's biosphere, including most of the marine environment, the atmosphere and the cryosphere itself provide such selection pressures.

6.5 Antarctic Bacteria as Analogues for Astrobiology

Molecular analysis of samples from extreme environments, such as the Antarctic permafrost, have revealed an abundance of microorganisms, particularly in habitable niches, with up to 10^3–10^7 viable cells/g dry weight of soil [6.35]. The predominant phylotypes are Acidobacter, Actinobacteria, Proteobacteria, Firmicutes and Bacteriodetes [6.9] [6.1] [6.79]. Within this catalogue of microbial diversity are not only cosmopolitan bacteria but also bacteria thought to be endemic to the Antarctic. Microbes considered endemic to Antarctica have been isolated from a range of environments. For example, *Methylobacterium* sp. V3 from drilling Lake Vostok [6.17], *Arthrobacter flavus* from a microbial mat in Lake Fryxell [6.8], *Bacillus megaterium* from Schirmacher Oasis lake sediment [6.99], *Planococcus antarcticus* in a cyanobacterial mat sample from Canopus Lake [6.91] and *Deinococcus* spp. from Antarctic soil, sandstone, marble and seawater [6.47] [6.27] [6.100].

Early studies with molecular techniques, such as 16S rRNA sequencing, revealed the presence of DNA from at least 14 different bacterial phyla in the McMurdo Dry Valleys [6.12]. They suggest, as might be expected from the wide variety of environmental conditions, that there are ecological niches within the Dry Valleys where different populations of microorganisms can be observed. For example, GeoChip analysis revealed specialized hypolithic, chaemoendolithic and cryptoendolithic microbial communities associated with distinct rocky niches [6.14]. Mojib *et al.* [6.70] isolated bacteria from the lakes of the Schirmacher Oasis, Queen Maud Land, and East Antarctica, and used 16S rRNA typing to categorize them. They identified two species and six different phyla, with some families identified as well. It is estimated that 85% of the many species isolated are of novel lineages and are yet to be cultured, which makes full characterization by methods such as 16S rRNA typing challenging [6.57]. A recent review of the microbial diversity of bacteria found in the surface snow in Antarctica reported nearly 7,000 different taxa, many of which could not be identified beyond the phylum level [6.62]. High-throughput sequencing techniques and metagenome assembly can reveal uncultured prokaryotes, which adds to the sequence data. Based on these techniques, a recent estimate suggests Antarctic ice-free soils contain at least 63 bacteria phyla and over 8000 species [6.57].

A polyphasic approach, mixing both cultivation and meta-omics techniques to fully characterize a wide range of bacteria isolated in all regions of Antarctica which have been identified to the genus level, and identifying those that are viable, has been suggested since the extreme cold and dry conditions are ideal for preserving DNA [6.57]. It is clear that more dynamic studies of biological activity to determine the diversity of living bacteria need to be undertaken. A full culture and meta-omic study of these bacteria and their stress response/survival pathways and strategies across more diverse sites in Antarctica will not only improve our understanding of microbial diversity and extremophiles on Earth, it will help shape our understanding of astrobiology [6.12] [6.70] [6.57].

As well as the endemic bacteria, cosmopolitan extremophiles and mesophiles have also been isolated from the Antarctic. Particularly interesting is the identification of *Hydrogenophilus thermoluteolus,* which was originally isolated from hot springs in Japan [6.36] and later in ice obtained from coring activity at Lake Vostok [6.9] [6.84]. Initially, the presence of a thermophilic, facultative chemolithoautotroph in an accretion ice sample from Lake Vostok may seem surprising and indeed has proved somewhat controversial. However, a chemoautotrophic lifestyle is better suited to the low nutrient and organic carbon content habitat and the presence of *H. thermoluteolus* may be indicative of a geothermal environment underneath the lake [6.9]. In which case, this observation may support the hypothesized ancient evolutionary history of microbes with the Antarctic, before the first Antarctic glaciation 34 Ma, which led to geographic isolation. However, it is also feasible that the requirements involved to live in hot springs may be so restrictive that the 16S rRNA sequence type may have independently evolved in both environments. Some caution should be taken when considering the reported detection of microorganisms from low biomass environmental samples such as ice core samples. Metagenomic techniques are highly sensitive and contamination can be a problem, whether that is contamination of DNA from sources including the environment, researchers or laboratory reagents; or cross-contamination from sample processing and sequencing errors; which can confound microbiome data [6.30].

Despite an increase in molecular methods to detect microbial diversity (i.e., high throughout 16S rRNA amplicon sequencing), most identification is only classified to the phylum level. This has a detrimental impact on identifying the functional adaptations of polyextremophilic microbes. Some of the main physiological adaptations common in extremophiles that enable them to survive in the Antarctic, and examples of bacteria living under these conditions, are presented in Table 6.3.

6.6 Endemic Antarctic Bacteria used in Astrobiology

Bacteria thought to be endemic to Antarctica have been used to model aspects of astrobiology (Table 6.4). To date, these models have mostly used bacteria isolated from the McMurdo Dry Valleys that simulate conditions on Mars.

Parageobacillus thermantarcticus (DSM 9572) strain M1 (originally *Geo/Bacillus thermoantarcticus*) was originally isolated from the geothermal soils of Mount Melbourne in Antarctica [6.78]. It is a thermophile and produces exopolysaccharides which help it

Table 6.3 Antarctic environmental conditions potentially driving bacterial adaptation.

Environmental condition	Description	Reference	Example bacteria	Reference
Cold	Low temperature in winter between -40 °C and -60 °C	[6.12]	*Herminiimonas glaciei* Grows at 1 to 37 °C Survives very long term >100,000 years at ~-10 °C	[6.49]
			Parageobacillus thermantarcticus (synonym *Bacillus thermoantarcticus*) (Optimal growth at 90 °C) Viable after storage at -196 °C	[6.32]
			Carnobacterium pleistocenium Long-term survival in temperatures of ~-10 °C	[6.49]
			Carnobacterium pleistocenium Growth at low temperature 0 °C	[6.77]
			Chryseobacterium greenlandensis Long-term survival in temperatures of ~-10 °C	[6.49]
Extreme temperature fluctuation and freeze thaw cycle	>20 °C temperature range in one day. Many freeze thaws per day in summer months	[6.12]	*Gloeocapsa* spp.; *Hormathonema–Gloeocapsa* spp.; *Chroococcidiopsis* spp.; *Cyanothece* spp.; *Nostoc* spp.; *Acaryochloris* spp. Lithobionts all isolated in Antarctica. Possibly able to survive attached to rocks in hostile environments on other planets too	[6.25]
	Rocks reach as high as 50 °C in the summer sun	[6.25]		
Low water availability	Soil <2% mass water content, precipitation <10 cm yr^{-1} mostly as snow	[6.12]		

(*Continued*)

Table 6.3 Antarctic environmental conditions potentially driving bacterial adaptation. (*Continued*)

Environmental condition	Description	Reference	Example bacteria	Reference
Desiccation	Low relative humidity, e.g., 5%	[6.112]	*Halomonas* sp. Radiation resistance presumed to be secondary to desiccation resistance. No mechanism determined D_{10} ~1 to 2 days	[6.72]
			Parageobacillus thermantarcticus Dried and stored at -20 °C for 3 months or at RT for 1 year. Remained viable after both treatments	[6.32]
			Deinococcus radiodurans D_{10} 26 days	[6.72]
Radiation	>20 kJm^{-2}	[6.56]	*Parageobacillus thermantarcticus* Viable after exposure to UV at 254 nm, doses up to 1,000 J/m^2 D_{10} 0.55 to 2.66 kGy Spores viable after exposure to 1 kGy (D_{10} value 2.50 kGy)	[6.32]
			Bacillus subtilis Radiation inactivation constant (wet pH 7) 1.22 ± 0.04, Dry 1.01 ± 0.03	[6.55]
			Bacillus thuringiensis Radiation inactivation constant Dry 1.00 ± 0.01	[6.55]

(*Continued*)

Table 6.3 Antarctic environmental conditions potentially driving bacterial adaptation. (*Continued*)

Environmental condition	Description	Reference	Example bacteria	Reference
			Salibacillus marismortui Radiation inactivation constant (wet pH 7) 1.23 ± 0.02, Dry 1.14 ± 0.03	[6.55]
			Deinococcus radiodurans D_{10} 50 kGy	[6.72]
			Gloeocapsa spp.; *Hormathonema–Gloeocapsa* spp.; *Chroococcidiopsis* spp.; *Cyanothece* spp.; *Nostoc* spp.; *Acaryochloris* spp.	[6.25]
			Parageobacillus thermantarcticus Viable after exposure to UV at 254 nm, doses up to 1,000 J/m^2	[6.32]
Halotolerant	Grows in <2 0% NaCl or similar	[6.16]	*Salibacillus marismortui* (formerly *Bacillus marismortui*) Requires salt for growth. Optimal growth at 10% salt (w/v) but one grows at 5% to 25% salt (w/v)	[6.3] [6.55]
Atmospheric pressure	~ 1000 mbar (McMurdo Dry Valleys)	[6.61]		

Table 6.4 Bacteria thought to be endemic to the Antarctic and used as astrobiological models.

Phylum	Bacteria	Isolated from	Reference	Used as an astrobiological model
Proteobacteria	*Brevundimonas* sp. MV.7*	Miers Valley, McMurdo Dry Valleys	[6.22]	[6.72]
		La Gorce Mountains soil	[6.1]	
	Pseudomonas sp. MV.27	Miers Valley, McMurdo Dry Valleys	[6.22]	[6.22]
	Halomonas spp. MVT 161; 463; 464; 468; 461	Miers Valley Transect (MVT), McMurdo Dry Valleys	[6.72]	[6.72]
	Psychromonas antarcticus	Pond sediment, McMurdo Ice Shelf	[6.71]	[6.93]
Firmicutes	*Parageobacillus thermantarcticus* (synonyms *Bacillus thermoantarcticus* and *Geobacillus thermantarcticus*)	Mount Melbourne, Antarctica	[6.78]	[6.32]
				[6.95]
				[6.65]

*The 16S rRNA genome sequence of *Brevundimonas* sp. MV.7 matches a partial 16S rRNA genome sequence deposited to GenBank (Accession (DQ675501)) by Liu *et al.* [6.59]. This sequence was isolated from an uncultured sample from glacial meltwaters in Mount Everest. Therefore, *Brevundimonas* sp. MV.7 may not be endemic to the Antarctic but there is no clear evidence yet that it is not. It also shared 99% homology with another uncultured bacterium from a spacecraft assembly cleanroom at Kennedy Space Center.

survive in a wide range of temperatures and saline concentrations as well as improving surface adhesion. *P. thermantarcticus* (DSM 9572) strain M1 was also able to survive extreme cold, desiccation, ultraviolet radiation and a simulated Martian atmosphere [6.65]. Its spore-forming ability enables it to survive extreme environment conditions and is therefore an interesting microorganism to use as a model for astrobiology.

Both *Brevundimonas* sp. MV.7 and *Halomonas* spp. (see Table 6.4) were isolated from the McMurdo Dry Valleys. Kim *et al.* [6.54] state that all halomonads are mesophiles. While that may be true for these strains, Musilova *et al.* [6.72] did not try to characterize their optimal growth and it is possible that they have adapted sufficiently well to their environment to be considered psychrophiles. There is a strong correlation between the resistance to radiation and desiccation of the two strains, suggesting a common repair mechanism [6.72]. Interestingly, Musilova *et al.* [6.72] state that radiation resistance is not an adaptive strategy by extremophiles, but a consequence of natural selection because radiation- and desiccation-induced DNA damage are hypothesized to have the same type of repair mechanism used by cells for survival.

Related *Brevundimonas* spp. have also been isolated in unusual environments such as inside NASA cleanrooms. The genus is regarded as an emerging global pathogen which may have implications in terms of both astronaut health and colonization of materials of space habitat infrastructures.

Pseudomonas sp. MV.27, which was also isolated in the McMurdo Dry Valleys [6.92], were assessed for their ability to withstand high doses of radiation at -79 °C in a Martian surface analogue experiment. *Brevundimonas* sp. MV.7 was used in this experiment too. They both showed better survival than at room temperature. Their ability to survive freeze-thaw was also examined. Despite the literature defining *Pseudomonas* sp. MV.27 as psychrophilic [6.92], they are probably best regarded as psychrotolerant as their reported T_{opt} was >20 °C, considerably higher than they will typically experience in their natural environment. As mentioned previously, the T_{opt} needs to be regarded with healthy skepticism for these organisms; no metabolic profiling was done to ascertain stress response of these *in-vitro* growth rates.

Psychromonas antarcticus has not, to date, been used in any published astrobiological research. It natively grows as a mat in a subglacial lake in the McMurdo Dry Valleys and is recommended for use in modeling growth on icy moons because of this [6.93].

6.7 Cosmopolitan Bacteria Found in Antarctica and used in Astrobiology

Other microbial species that are not endemic to Antarctica, but have been isolated under analogous environmental selection pressures, have been used in astrobiological modeling. *Rhodococcus* spp. which were isolated from the McMurdo Dry Valleys have been used in a two-part experiment examining possible models for panspermia (life originating elsewhere in the Universe and being bought to Earth), including survival after hypervelocity impact into chalk and granite and simulated aerocapture in a planetary atmosphere [6.10]. They were also used in the desiccation and radiation resistance experiments by Musilova *et al.* [6.72]. Unusual strains of *Bacillus thuringiensis* were isolated in Northern Victoria Land [6.34] and a different strain of this species was used to model the effect of ionizing radiation on the long-term survival of spores in a simulated Martian environment [6.55].

6.8 Conclusion

While five organisms remain a surprisingly small pool, all of the bacteria endemic to Antarctica and used in astrobiological experiments to date are motile rods. While being motile has obvious advantages in low-nutrient environments, there could be an advantage to being a rod that has yet to be elucidated. Similarly, although physiological and biochemical analyses were done to characterize the Antarctic isolated bacteria, and exopolysaccharide production was documented, astrobiology-related experiments to date have mainly focused on desiccation and radiation resistance. The production of a biofilm has both offensive and defensive roles that limit the vulnerability of bacteria to external stressors. Polyextremophile biofilm producing bacteria have been investigated for their ability to survive the harsh environmental conditions of space and Mars, including the BOSS (Biofilm Organisms Surfing Space) experiments, part of the EXPOSE-R2 mission mentioned above [6.21]. However, none of these have considered endemic Antarctic strains which may already be adapted to conditions bearing similarities to those on Mars and to a lesser degree, space, a factor which should not be overlooked in future experiments.

Extremophiles are continuously being isolated from around the world. Some are endemic to where they are isolated but most are cosmopolitan with pre-adapted survival strategies. Antarctica is a rich source of extremophiles, not just psychrophiles but polyextremophiles exposed to unusually high level of UV radiation for Earth, hyperarid conditions, hypersaline conditions, extremely low nutrient levels, frequent freeze-thaw cycling, desiccation, and usually several or most of these simultaneously. Some of the most extreme environments in Antarctica could be considered analogues for elements of the conditions found on Mars and there have been a number of experiments conducted that exploit this similarity. There are also recommendations for future work to further extend this list of similar experiments [6.93].

To date, a range of extremophiles have been used in astrobiological models, using facilities such as EXPOSE and experimental methods of which are reviewed in Olsson-Francis and Cockell [6.81]. Polyextremophiles endemic to the Antarctic have been isolated which could be used for much of this work that is being undertaken. For example, *Deinococcus* spp. are widely used in astrobiological models for their resistance to radiation and desiccation, and there are at least five *Deinococcus* spp. endemic to Antarctica. However, none of the Antarctic endemic *Deinococcus* spp. have, to date, been used in astrobiology experiments. *Bacillus subtilis* is, likewise, commonly used as a model sporulating organism and there are many potentially endemic *Bacillus* spp. in Antarctica which would form a model extremophile sporulating species.

Bacteria that have the potential to grow in the challenging environment of space are of potential concern to astronaut health upon the ISS and future human missions; to the structural integrity of space habitats and mission modules, such as the Mars Science Laboratory; to planetary protection when sending missions to investigate target bodies in the solar system and receiving return mission samples; and in implicating any investigations into the search for extraterrestrial life. This is because space environments and habitats which support human life will, by definition, also support microbial life, but under highly selective and artificial conditions.

The knowledge gained about extremophiles is also informing research into the evolution of life on Earth and its possible presence elsewhere with regards to the lithopanspermia hypothesis. Lithopanspermia extends the concept of life being transferred between planets (panspermia, above) by suggesting rocks are a medium of transfer and these are generated during impacts by meteors [6.51].

Future experiments need to bear in mind that focusing on one or two variables will not account for the combinatorial effects that this can have on survivability. For example, an increase in pressure can increase heat tolerance, even above 100 °C [6.69]. There are possible links between low temperature and surviving radiation [6.22] and links between desiccation and radiation resistance [6.72]. There may be a pattern that can be elucidated between all three and an underlying mechanism that can help with predicting patterns of life elsewhere in the solar system. In the next few years the ESA ICE-EXPOSE project will build upon the knowledge of the physiological adaptations of microorganisms isolated from the Antarctic, on their survival during spaceflight and exposure under space- and Mars-like conditions.

High-throughput sequencing and metagenome-assembly techniques are revealing a greater diversity of prokaryotes in the Antarctic than previously estimated. Therefore, there is a large potential to isolate, characterize and build upon the current knowledge of endemic Antarctic bacteria. The environmental conditions of habitats in Antarctica and the stress response/survival pathways of bacteria living there suggest they are an excellent resource for candidates for growth in analogue conditions for Mars and the icy moons.

References

6.1. Aislabie, J.M., Broady, P.A., Saul, D.J., Culturable aerobic heterotrophic bacteria from high altitude, high latitude soil of La Gorce Mountains (86 30'S, 147 W), Antarctica. *Antarct. Sci.*, 18, 3, 313–321, 2006.

6.2. Allison, I., Hock, R., King, M.A., Mackintosh, A.N., Future Earth and the cryosphere, in: *Global Change and Future Earth: The Geoscience Perspective*, T. Beer, J. Li, K. Alverson (Eds.), pp. 91–113, Cambridge University Press, Cambridge, 2018.

6.3. Arahal, D.R., Marquez, M.C., Volcani, B.E., Schleifer, K.H., Ventosa, A., *Bacillus marismortui* sp. nov., a new moderately halophilic species from the Dead Sea. *Int. J. Syst. Evol. Microbiol.*, 49, 2, 521–530, 1999.

6.4. Ashkenazy, Y., The surface temperature of Europa. *Heliyon*, 5, 6, #e01908, 2019.

6.5. Bak, E.N., Larsen, M.G., Moeller, R., Nissen, S.B., Jensen, L.R., Nørnberg, P., Jensen, S.J., Finster, K., Silicates eroded under simulated martian conditions effectively kill bacteria—A challenge for life on Mars. *Front. Microbiol.*, 8, #1709, 2017.

6.6. Bücker, H., Horneck, G., Wollenhaupt, H., Schwager, M., Taylor, G.R., Viability of *Bacillus subtilis* spores exposed to space environment in the M-191 experiment system aboard Apollo 16. *Life Sci. Space Res.*, 12, 209–213, 1974.

6.7. Bergholz, P.W., Bakermans, C., Tiedje, J.M., *Psychrobacter arcticus* 273-4 uses resource efficiency and molecular motion adaptations for subzero temperature growth. *J. Bacteriol.*, 191, 7, 2340–2352, 2009.

6.8. Brambilla, E., Hippe, H., Hagelstein, A., Tindall, B.J., Stackebrandt, E., 16S rDNA diversity of cultured and uncultured prokaryotes of a mat sample from Lake Fryxell, McMurdo Dry Valleys, Antarctica. *Extremophiles*, 5, 1, 23–33, 2001.
6.9. Bulat, S.A., Alekhina, I.A., Blot, M., Petit, J.R., De Angelis, M., Wagenbach, D., Lipenkov, V.Y., Vasilyeva, L.P., Wloch, D.M., Raynaud, D., Lukin, V.V., DNA signature of thermophilic bacteria from the aged accretion ice of Lake Vostok, Antarctica: implications for searching for life in extreme icy environments. *Int. J. Astrobiol.*, 3, 1, 1–12, 2004.
6.10. Burchell, M.J., Shrine, N.R., Mann, J., Bunch, A.W., Brandao, P., Zarnecki, J.C., Galloway, J.A., Laboratory investigations of the survivability of bacteria in hypervelocity impacts. *Adv. Space Res.*, 28, 4, 707–712, 2001.
6.11. Cai, G.G., Geng, K., Wang, Q., The environmental monitoring of the natural radiation background in Antarctica with LiF: Mg, Cu, P TLD and X-gamma radiometry. *Radiat. Prot. Dosim.*, 60, 3, 259–262, 1995.
6.12. Cary, S.C., McDonald, I.R., Barrett, J.E., Cowan, D.A., On the rocks: The microbiology of Antarctic Dry Valley soils. *Nat. Rev. Microbiol.*, 8, 2, 129–38, 2010.
6.13. Cavicchioli, R., On the concept of a psychrophile. *ISME J.*, 10, 4, 793–4, 2016.
6.14. Chan, Y., Van Nostrand, J.D., Zhou, J., Pointing, S.B., Farrell, R.L., Functional ecology of an Antarctic dry valley. *Proc. Natl. Acad. Sci.*, 110, 22, 8990–8995, 2013.
6.15. Chancellor, J.C., Scott, G.B., Sutton, J.P., Space radiation: The number one risk to astronaut health beyond low earth orbit. *Life*, 4, 3, 491–510, 2014.
6.16. Chen, Y.H., Shyu, Y.T., Lin, S.S., Characterization of candidate genes involved in halotolerance using high-throughput omics in the halotolerant bacterium *Virgibacillus chiguensis*. *PLoS One*, 13, 8, #e0201346, 2018.
6.17. Christner, B.C., Mosley-Thompson, E., Thompson, L.G., Reeve, J.N., Isolation of bacteria and 16S rDNAs from Lake Vostok accretion ice. *Environ. Microbiol.*, 3, 9, 570–577, 2001.
6.18. Christner, B.C., Royston-Bishop, G., Foreman, C.M., Arnold, B.R., Tranter, M., Welch, K.A., Lyons, W.B., Tsapin, A.I., Studinger, M., Priscu, J.C., Limnological conditions in subglacial lake Vostok, Antarctica. *Limnol. Oceanogr.*, 51, 6, 2485–2501, 2006.
6.19. Clocksin, K.M., Jung, D.O., Madigan, M.T., Cold-active chemoorganotrophic bacteria from permanently ice-covered Lake Hoare, McMurdo Dry Valleys, Antarctica. *Appl. Environ. Microbiol.*, 73, 9, 3077–3083, 2007.
6.20. Cockell, C.S., Bush, T., Bryce, C., Direito, S., Fox-Powell, M., Harrison, J.P., Lammer, H., Landenmark, H., Martin-Torres, J., Nicholson, N., Noack, L., Habitability: A review. *Astrobiology*, 16, 1, 89–117, 2016.
6.21. Cottin, H. and Rettberg, P., EXPOSE-R2 on the International Space Station (2014–2016): Results from the PSS and BOSS Astrobiology Experiments. *Astrobiology*, 19, 8, 975–978, 2019.
6.22. Dartnell, L.R., Hunter, S.J., Lovell, K.V., Coates, A.J., Ward, J.M., Low-temperature ionizing radiation resistance of *Deinococcus radiodurans* and Antarctic Dry Valley bacteria. *Astrobiology*, 10, 7, 717–732, 2010.
6.23. Davy, R., Davis, J.A., Taylor, P.A., Lange, C.F., Weng, W., Whiteway, J., Gunnlaugson, H.P., Initial analysis of air temperature and related data from the Phoenix MET station and their use in estimating turbulent heat fluxes. *J. Geophys. Res.: Planets*, 115, E3, 2010.
6.24. de La Torre, R., Sancho, L.G., Horneck, G., de los Ríos, A., Wierzchos, J., Olsson-Francis, K., Cockell, C.S., Rettberg, P., Berger, T., de Vera, J.P.P. *et al.*, Survival of lichens and bacteria exposed to outer space conditions–results of the *Lithopanspermia* experiments. *Icarus*, 208, 2, 735–748, 2010.
6.25. De Los Ríos, A., Wierzchos, J., Ascaso, C., The lithic microbial ecosystems of Antarctica's McMurdo Dry Valleys. *Antarct. Sci.*, 26, 5, 459–477, 2014.
6.26. De Vera, J.P., Lichens as survivors in space and on Mars. *Fungal Ecol.*, 5, 4, 472–479, 2012.

6.27. Dong, N., Li, H.R., Yuan, M., Zhang, X.H., Yu, Y., *Deinococcus antarcticus* sp. nov., isolated from soil. *Int. J. Syst. Evol. Microbiol.*, 65, 2, 331–335, 2015.
6.28. Doran, P.T., McKay, C.P., Clow, G.D., Dana, G.L., Fountain, A.G., Nylen, T., Lyons, W.B., Valley floor climate observations from the McMurdo Dry Valleys, Antarctica, 1986–2000. *J. Geophys. Res.: Atmos.*, 107, D24, 2002. #ACL-13.
6.29. Eigenbrode, J.L., Summons, R.E., Steele, A., Freissinet, C., Millan, M., Navarro-González, R., Sutter, B., McAdam, A.C., Franz, H.B., Glavin, D.P. *et al.*, Organic matter preserved in 3-billion-year-old mudstones at Gale crater, Mars. *Science*, 360, 6393, 1096–1101, 2018.
6.30. Eisenhofer, R., Minich, J.J., Marotz, C., Cooper, A., Knight, R., Weyrich, L.S., Contamination in low microbial biomass microbiome studies: Issues and recommendations. *Trends Microbiol.*, 27, 2, 105–117, 2019.
6.31. Falkowski, P.G., Fenchel, T., Delong, E.F., The microbial engines that drive Earth's biogeochemical cycles. *Science*, 320, 5879, 1034–1039, 2008.
6.32. Finore, I., Lama, L., Di Donato, P., Romano, I., Tramice, A., Leone, L., Nicolaus, B., Poli, A., *Parageobacillus thermantarcticus*, an Antarctic cell factory: From crop residue valorization by green chemistry to astrobiology studies. *Diversity*, 11, 8, 128, 2019.
6.33. Fixsen, D.J., The temperature of the cosmic microwave background. *Astrophys. J.*, 707, 2, 916–920, 2009.
6.34. Forsyth, G. and Logan, N.A., Isolation of *Bacillus thuringiensis* from Northern Victoria Land, Antarctica. *Lett. Appl. Microbiol.*, 30, 3, 263–266, 2000.
6.35. Gilichinsky, D.A., Wilson, G.S., Friedmann, E.I., McKay, C.P., Sletten, R.S., Rivkina, E.M., Vishnivetskaya, T.A., Erokhina, L.G., Ivanushkina, N.E., Kochkina, G.A., Shcherbakova, V.A., Microbial populations in Antarctic permafrost: biodiversity, state, age, and implication for astrobiology. *Astrobiology*, 7, 2, 275–311, 2007.
6.36. Goto, E., Kodama, T., Minoda, Y., Isolation and culture conditions of thermophilic hydrogen bacteria. *Agric. Biol. Chem.*, 41, 4, 685–690, 1977.
6.37. Grotzinger, J.P., Sumner, D.Y., Kah, L.C., Stack, K., Gupta, S., Edgar, L., Rubin, D., Lewis, K., Schieber, J., Mangold, N. *et al.*, A habitable fluvio-lacustrine environment at Yellowknife Bay, Gale Crater, Mars. *Science*, 343, 6169, #1242777, 2014.
6.38. Gupta, G.N., Srivastava, S., Khare, S.K., Prakash, V., Extremophiles: An overview of microorganism from extreme environment. *Int. J. Agric. Environ. Biotechnol.*, 7, 2, 371–380, 2014.
6.39. Halliday, D., Resnick, R., Walker, J., *Fundamentals of Physics*, 6th ed., John Wiley & Sons, Inc, NJ, USA, 2004.
6.40. Hallsworth, J.E., Microbial unknowns at the saline limits for life. *Nat. Ecol. Evol.*, 3, 11, 1503–1504, 2019.
6.41. Hamilton, V.E., Vasavada, A.R., Sebastián, E., de la Torre Juarez, M., Ramos, M., Armiens, C., Arvidson, R.E., Carrasco, I., Christensen, P.R., De Pablo, M.A., Goetz, W., Observations and preliminary science results from the first 100 sols of MSL Rover Environmental Monitoring Station ground temperature sensor measurements at Gale Crater. *J. Geophys. Res.: Planets*, 119, 4, 745–770, 2014.
6.42. Hand, K.P., Chyba, C.F., Priscu, J.C., Carlson, R.W., Nealson, K.H., Astrobiology and the potential for life on Europa, in: *Europa*, R.T. Pappalardo, W.B. McKinnon, K.K. Khurana (Eds.), p. 589, University of Arizona Press, Tucson, 2009.
6.43. Harrison, J.P., Gheeraert, N., Tsigelnitskiy, D., Cockell, C.S., The limits for life under multiple extremes. *Trends Microbiol.*, 21, 4, 204–212, 2013.
6.44. Hart, K.M., Szpak, M.T., Mahaney, W.C., Dohm, J.M., Jordan, S.F., Frazer, A.R., Allen, C.C., Kelleher, B.P., A bacterial enrichment study and overview of the extractable lipids from paleosols in the Dry Valleys, Antarctica: implications for future Mars reconnaissance. *Astrobiology*, 11, 4, 303–321, 2011.

6.45. Hassler, D.M., Zeitlin, C., Wimmer-Schweingruber, R.F., Ehresmann, B., Rafkin, S., Eigenbrode, J.L., Brinza, D.E., Weigle, G., Böttcher, S., Böhm, E. *et al.*, Mars' surface radiation environment measured with the Mars Science Laboratory's Curiosity rover. *Science*, 343, 6169, #1244797, 2014.
6.46. Hecht, M.H., Kounaves, S.P., Quinn, R.C., West, S.J., Young, S.M.M., Ming, D.W., Catling, D.C., Clark, B.C., Boynton, W.V., Hoffman, J., DeFlores, L.P., Detection of perchlorate and the soluble chemistry of Martian soil at the Phoenix lander site. *Science*, 325, 5936, 64–67, 2009.
6.47. Hirsch, P., Gallikowski, C.A., Siebert, J., Peissl, K., Kroppenstedt, R., Schumann, P., Stackebrandt, E., Anderson, R., *Deinococcus frigens* sp. nov., *Deinococcus saxicola* sp. nov., and *Deinococcus marmoris* sp. nov., low temperature and draught-tolerating, UV-resistant bacteria from continental Antarctica. *Syst. Appl. Microbiol.*, 27, 6, 636–645, 2004.
6.48. Hoover, R.B. and Gilichinsky, D., Significance to astrobiology of micro-organisms in permafrost and ice, in: *Permafrost Response on Economic Development, Environmental Security and Natural Resources*, R. Paepe, V.P. Melnikov, E. Van Overloop, V.D. Gorokhov (Eds.), pp. 553–579, Springer, Dordrecht, 2001.
6.49. Hoover, R.B. and Pikuta, E.V., Life in ice: Implications to astrobiology, in: *Instruments and Methods for Astrobiology and Planetary Missions XII*, p. 7441, 2009, #74410P.
6.50. Horikoshi, K. and Bull, A., Prologue: Definition, categories, distribution, origin and evolution, pioneering studies, and emerging fields of extremophiles, in: *Extremophiles Handbook*, K. Horikoshi (Ed.), pp. 4–15, Springer, Tokyo, 2011.
6.51. Horneck, G., Stöffler, D., Ott, S., Hornemann, U., Cockell, C.S., Moeller, R., Meyer, C., De Vera, J.P., Fritz, J., Schade, S., Artemieva, N.A., Microbial rock inhabitants survive hypervelocity impacts on Mars-like host planets: first phase of lithopanspermia experimentally tested. *Astrobiology*, 8, 1, 17–44, 2008.
6.52. Hsu, H.W., Postberg, F., Sekine, Y., Shibuya, T., Kempf, S., Horányi, M., Juhász, A., Altobelli, N., Suzuki, K., Masaki, Y., Kuwatani, T., Ongoing hydrothermal activities within Enceladus. *Nature*, 519, 7542, 207–210, 2015.
6.53. Jones, R.M., Goordial, J.M., Orcutt, B.N., Low energy subsurface environments as extraterrestrial analogs. *Front. Microbiol.*, 9, #1605, 2018.
6.54. Kim, K.K., Lee, J.S., Stevens, D.A., Microbiology and epidemiology of *Halomonas* species. *Future Microbiol.*, 8, 12, 1559–1573, 2013.
6.55. Kminek, G., Bada, J.L., Pogliano, K., Ward, J.F., Radiation-dependent limit for the viability of bacterial spores in halite fluid inclusions and on Mars. *Radiat. Res.*, 159, 6, 722–729, 2003.
6.56. Köhler, H., Contreras, R.A., Pizarro, M., Cortés-Antíquera, R., Zúñiga, G.E., Antioxidant responses induced by UVB radiation in *Deschampsia antarctica* Desv. *Front. Plant Sci.*, 8, #921, 2017.
6.57. Lambrechts, S., Willems, A., Tahon, G., Uncovering the uncultivated majority in Antarctic soils: Toward a synergistic approach. *Front. Microbiol.*, 10, #242, 2019.
6.58. Lee, C.K., Barbier, B.A., Bottos, E.M., McDonald, I.R., Cary, S.C., The inter-valley soil comparative survey: The ecology of Dry Valley edaphic microbial communities. *ISME J.*, 6, 5, 1046–1057, 2012.
6.59. Liu, Y., Yao, T., Jiao, N., Kang, S., Zeng, Y., Huang, S., Microbial community structure in moraine lakes and glacial meltwaters, Mount Everest. *FEMS Microbiol. Lett.*, 265, 1, 98–105, 2006.
6.60. Maccario, L., Sanguino, L., Vogel, T.M., Larose, C., Snow and ice ecosystems: Not so extreme. *Res. Microbiol.*, 166, 10, 782–795, 2015.

6.61. MacClune, K.L., Fountain, A.G., Kargel, J.S., MacAyeal, D.R., Glaciers of the McMurdo dry valleys: Terrestrial analog for Martian polar sublimation. *J. Geophys. Res.: Planets*, 108, E4, 2003.
6.62. Malard, L.A., Šabacká, M., Magiopoulos, I., Mowlem, M., Hodson, A., Tranter, M., Siegert, M.J., Pearce, D.A., Spatial variability of Antarctic surface snow bacterial communities. *Front. Microbiol.*, 10, #461, 2019.
6.63. Margesin, R. and Collins, T., Microbial ecology of the cryosphere (glacial and permafrost habitats): Current knowledge. *Appl. Microbiol. Biotechnol.*, 103, 6, 2537–2549, 2019.
6.64. Martínez, G., Valero, F., Vázquez, L., Characterization of the Martian surface layer. *J. Atmos. Sci.*, 66, 1, 187–198, 2009.
6.65. Mastascusa, V., Romano, I., Di Donato, P., Poli, A., Della Corte, V., Rotundi, A., Bussoletti, E., Quarto, M., Pugliese, M., Nicolaus, B., Extremophiles survival to simulated space conditions: An astrobiology model study. *Orig. Life Evol. Biosph.*, 44, 3, 231–237, 2014.
6.66. Matsubaya, O., Sakai, H., Torii, T., Burton, H., Kerry, K., Antarctic saline lakes—Stable isotopic ratios, chemical compositions and evolution. *Geochim. Cosmochim. Acta*, 43, 1, 7–25, 1979.
6.67. McCord, T.B., Teeter, G., Hansen, G.B., Sieger, M.T., Orlando, T.M., Brines exposed to Europa surface conditions. *J. Geophys. Res.: Planets*, 107, E1, 4–1, 2002.
6.68. McKay, C.P., Anbar, A.D., Porco, C., Tsou, P., Follow the plume: The habitability of Enceladus. *Astrobiology*, 14, 4, 352–355, 2014.
6.69. Merino, N., Aronson, H.S., Bojanova, D.P., Feyhl-Buska, J., Wong, M.L., Zhang, S., Giovannelli, D., Living at the extremes: Extremophiles and the limits of life in a planetary context. *Front. Microbiol.*, 10, #780, 2019.
6.70. Mojib, N., Huang, J., Hoover, R.B., Pikuta, E.V., Storrie-Lombardi, M., Sattler, B., Andersen, D., Bej, A.K., Diversity of bacterial communities in the lakes of Schirmacher Oasis, Antarctica, in: *Instruments and Methods for Astrobiology and Planetary Missions XII*, p. 7441, 2009, #74410J.
6.71. Mountfort, D.O., Rainey, F.A., Burghardt, J., Kaspar, H.F., Stackebrandt, E., *Psychromonas antarcticus* gen. nov., sp. nov., a new aerotolerant anaerobic, halophilic psychrophile isolated from pond sediment of the McMurdo Ice Shelf, Antarctica. *Arch. Microbiol.*, 169, 3, 231–238, 1998.
6.72. Musilova, M., Wright, G., Ward, J.M., Dartnell, L.R., Isolation of radiation-resistant bacteria from Mars analog Antarctic Dry Valleys by preselection, and the correlation between radiation and desiccation resistance. *Astrobiology*, 15, 12, 1076–1090, 2015.
6.73. National Aeronautics and Space Administration (NASA), Earth Fact Sheet, 2020. https://nssdc.gsfc.nasa.gov/planetary/factsheet/earthfact.html.
6.74. National Aeronautics and Space Administration (NASA), Mars Fact Sheet, 2020. https://nssdc.gsfc.nasa.gov/planetary/factsheet/marsfact.html.
6.75. National Aeronautics and Space Administration (NASA), Moon Fact Sheet, 2020. https://nssdc.gsfc.nasa.gov/planetary/factsheet/moonfact.html.
6.76. National Aeronautics and Space Administration (NASA), Staying Cool on the ISS, 2020. https://science.nasa.gov/science-news/science-at-nasa/2001/ast21mar_1.
6.77. Nicholson, W.L., Krivushin, K., Gilichinsky, D., Schuerger, A.C., Growth of *Carnobacterium* spp. from permafrost under low pressure, temperature, and anoxic atmosphere has implications for Earth microbes on Mars. *Proc. Natl. Acad. Sci.*, 110, 2, 666–671, 2013.
6.78. Nicolaus, B., Lama, L., Esposito, E., Manca, M.C., Gambacorta, A., di Prisco, G., *Bacillus thermoantarcticus*" sp. nov., from Mount Melbourne, Antarctica: A novel thermophilic species. *Polar Biol.*, 16, 2, 101–104, 1996.

6.79. Niederberger, T.D., McDonald, I.R., Hacker, A.L., Soo, R.M., Barrett, J.E., Wall, D.H., Cary, S.C., Microbial community composition in soils of Northern Victoria Land, Antarctica. *Environ. Microbiol.*, 10, 7, 1713–1724, 2008.
6.80. Nikrad, M.P., Kerkhof, L.J., Häggblom, M.M., The subzero microbiome: Microbial activity in frozen and thawing soils. *FEMS Microbiol. Ecol.*, 92, 6, #fiw081, 2016.
6.81. Olsson-Francis, K. and Cockell, C.S., Experimental methods for studying microbial survival in extraterrestrial environments. *J. Microbiol. Methods*, 80, 1, 1–13, 2010.
6.82. O'Neill, C. and Nimmo, F., The role of episodic overturn in generating the surface geology and heat flow on Enceladus. *Nat. Geosci.*, 3, 2, 88–91, 2010.
6.83. Onofri, S., de la Torre, R., de Vera, J.P., Ott, S., Zucconi, L., Selbmann, L., Scalzi, G., Venkateswaran, K.J., Rabbow, E., Sánchez Iñigo, F.J., Horneck, G., Survival of rock-colonizing organisms after 1.5 years in outer space. *Astrobiology*, 12, 5, 508–516, 2012.
6.84. Petit, J.R., Alekhina, I., Bulat, S., Lake Vostok, Antarctica: Exploring a subglacial lake and searching for life in an extreme environment, in: *Lectures in Astrobiology. Advances in Astrobiology and Biogeophysics*, M. Gargaud, B. Barbier, H. Martin, J. Reisse (Eds.), pp. 227–288, Springer, Berlin, Heidelberg, 2005.
6.85. Piqueux, S., Buz, J., Edwards, C.S., Bandfield, J.L., Kleinböhl, A., Kass, D.M., Hayne, P.O., MCS and THEMIS Teams, Widespread shallow water ice on Mars at high latitudes and mid-latitudes. *Geophys. Res. Lett.*, 46, 24, 14290–14298, 2019.
6.86. Poughon, L., Farges, B., Dussap, C.G., Godia, F., Lasseur, C., Simulation of the MELiSSA closed loop system as a tool to define its integration strategy. *Adv. Space Res.*, 44, 12, 1392–1403, 2009.
6.87. Quinn, R.C., Martucci, H.F.H., Miller, S.R., Bryson, C.E., Grunthaner, F.J., Grunthaner, P.J., Perchlorate radiolysis on Mars and the origin of martian soil reactivity. *Astrobiology*, 13, 6, 515–520, 2013.
6.88. Rabbow, E., Rettberg, P., Barczyk, S., Bohmeier, M., Parpart, A., Panitz, C., Horneck, G., von Heise-Rotenburg, R., Hoppenbrouwers, T., Willnecker, R. *et al.*, EXPOSE-E: An ESA astrobiology mission 1.5 years in space. *Astrobiology*, 12, 5, 374–386, 2012.
6.89. Rabbow, E., Rettberg, P., Parpart, A., Panitz, C., Schulte, W., Molter, F., Jaramillo, E., Demets, R., Weiß, P., Willnecker, R., EXPOSE-R2: The astrobiological ESA mission on board of the International Space Station. *Front. Microbiol.*, 8, #1533, 2017.
6.90. Rasuk, M.C., Ferrer, G.M., Moreno, J., Farias, M., Albarracín, V., The diversity of microbial extremophiles, in: *Molecular Diversity of Environmental Prokaryotes*, T.B. Rodrigues, A. Emiliano, T. Silva (Eds.), pp. 87–126, CRC Press, Taylor & Francis Group, FL, USA, 2016.
6.91. Reddy, G., Prakash, J., Vairamani, M., Prabhakar, S., Matsumoto, G., Shivaji, S., *Planococcus antarcticus* and *Planococcus psychrophilus* spp. nov. isolated from cyanobacterial mat samples collected from ponds in Antarctica. *Extremophiles*, 6, 3, 253–261, 2002.
6.92. Reddy, G.S., Matsumoto, G.I., Schumann, P., Stackebrandt, E., Shivaji, S., Psychrophilic pseudomonads from Antarctica: *Pseudomonas antarctica* sp. nov., *Pseudomonas meridiana* sp. nov. and *Pseudomonas proteolytica* sp. nov. *Int. J. Syst. Evol. Microbiol.*, 54, 3, 713–719, 2004.
6.93. Rettberg, P., Antunes, A., Brucato, J., Cabezas, P., Collins, G., Haddaji, A., Kminek, G., Leuko, S., McKenna-Lawlor, S., Moissl-Eichinger, C. *et al.*, Biological contamination prevention for outer solar system moons of astrobiological interest: What do we need to know? *Astrobiology*, 19, 8, 951–974, 2019.
6.94. Richter, I., Herbold, C.W., Lee, C.K., McDonald, I.R., Barrett, J.E., Cary, S.C., Influence of soil properties on archaeal diversity and distribution in the McMurdo Dry Valleys, Antarctica. *FEMS Microbiol. Ecol.*, 89, 2, 347–359, 2014.

6.95. Romano, I., De Angelis, A., Poli, A., Ragni, P., Lilla, L., Zito, G., Nicolaus, B., De Luca, A.C., Di Donato, P., Resistance and Raman spectroscopy analysis of *Parageobacillus thermantarcticus* spores after γ-ray exposure. *Extremophiles*, 22, 6, 931–941, 2018.

6.96. Romanov, P., Global multisensor automated satellite-based snow and ice mapping system (GMASI) for cryosphere monitoring. *Remote Sens. Environ.*, 196, 42–55, 2017.

6.97. Shepherd, A., Fricker, H.A., Farrell, S.L., Trends and connections across the Antarctic cryosphere. *Nature*, 558, 7709, 223–232, 2018.

6.98. Skansi, M., King, J., Lazzara, M.A., Cerveny, R.S., Stella, J.L., Soloman, S., Jones, P., Bromwich, D., Renwick, J., Burt, C.C. *et al.*, Evaluating highest-temperature extremes in the Antarctic. *Eos*, 98, 2017.

6.99. Shekh, R.M., Singh, P., Singh, S.M., Roy, U., Antifungal activity of Arctic and Antarctic bacteria isolates. *Polar Biol.*, 34, 1, 139–143, 2011.

6.100. Srinivasan, S., Lim, S., Lim, J.H., Jung, H.Y., Kim, M.K., *Deinococcus rubrus* sp. nov., a bacterium isolated from Antarctic coastal sea water. *J. Microbiol. Biotechnol.*, 27, 535–41, 2017.

6.101. Stevenson, A., Burkhardt, J., Cockell, C.S., Cray, J.A., Dijksterhuis, J., Fox-Powell, M., Kee, T.P., Kminek, G., McGenity, T.J., Timmis, K.N., Timson, D.J., Multiplication of microbes below 0.690 water activity: Implications for terrestrial and extraterrestrial life. *Environ. Microbiol.*, 17, 2, 257–277, 2015.

6.102. Storey, J.M. and Storey, K.B., Cold hardiness and freeze tolerance, in: *Functional Metabolism: Regulation and Adaptation*, K.B. Storey (Ed.), pp. 473–503, John Wiley & Sons, Inc, NJ, USA, 2005.

6.103. Tang, C., Madigan, M.T., Lanoil, B., Bacterial and archaeal diversity in sediments of west Lake Bonney, McMurdo Dry Valleys, Antarctica. *Appl. Environ. Microbiol.*, 79, 3, 1034–1038, 2013.

6.104. Tomkins, A.G., Genge, M.J., Tait, A.W., Alkemade, S.L., Langendam, A.D., Perry, P.P., Wilson, S.A., High survivability of micrometeorites on Mars: Sites with enhanced availability of limiting nutrients. *J. Geophys. Res.: Planets*, 124, 7, 1802–1818, 2019.

6.105. Toner, J.D., Catling, D.C., Light, B., Soluble salts at the Phoenix Lander site, Mars: A reanalysis of the Wet Chemistry Laboratory data. *Geochim. Cosmochim. Acta*, 136, 142–168, 2014.

6.106. Toner, J.D., Catling, D.C., Light, B., The formation of supercooled brines, viscous liquids, and low-temperature perchlorate glasses in aqueous solutions relevant to Mars. *Icarus*, 233, 36–47, 2014.

6.107. Tribelli, P.M. and López, N.I., Reporting key features in cold-adapted bacteria. *Life*, 8, 1, 8, 2018.

6.108. Wadsworth, J. and Cockell, C.S., Perchlorates on Mars enhance the bacteriocidal effects of UV light. *Sci. Rep.*, 7, 1, 1–8, 2017.

6.109. Williams, T.J., Lauro, F.M., Ertan, H., Burg, D.W., Poljak, A., Raftery, M.J., Cavicchioli, R., Defining the response of a microorganism to temperatures that span its complete growth temperature range (-2°C to 28°C) using multiplex quantitative proteomics. *Environ. Microbiol.*, 13, 8, 2186–2203, 2011.

6.110. World Meteorological Organization (WMO), (n.d.). World Meteorological Organization Global Weather & Climate Extremes Archive. https://wmo.asu.edu/content/world-meteorological-organization-global-weather-climate-extremes-archive.

6.111. Zeglin, L.H., Sinsabaugh, R.L., Barrett, J.E., Gooseff, M.N., Takacs-Vesbach, C.D., Landscape distribution of microbial activity in the McMurdo Dry Valleys: Linked biotic processes, hydrology, and geochemistry in a cold desert ecosystem. *Ecosystems*, 12, 4, 562–573, 2009.

6.112. Zhang, Q. and Yan, T., Correlation of intracellular trehalose concentration with desiccation resistance of soil *Escherichia coli* populations. *Appl. Environ. Microbiol.*, 78, 20, 407–7413, 2012.

7

Extremophilic Life in Our Oceans as Models for Astrobiology

Robert Y. George[1,2]

[1]*University of North Carolina Wilmington, Wilmington, North Carolina, USA*
[2]*George Institute for Biodiversity and Sustainability, Wake Forest, North Carolina, USA*

Abstract
Earth harbors vast ecosystems that support life (psychrophilic, barophilic and chemosynthetic) adapted to conditions that may occur on other planets in the universe. We define organisms that inhabit severe ecosystems as extremophiles whether this life persists on Earth or other planets. The Southern Ocean harbors marine life which has adapted to near-freezing conditions, coupled with new low oxygen zones that can lead to hypoxic zones as in the upwelling low-latitude areas of the eastern edges of the world's oceans. In my research in the Southern Ocean around Antarctica, I encountered austral summer blooms of diatoms and vast biomass of krill aggregations (swarms) along the western Antarctic Peninsula (WAP); but now krill and its predators (penguins, baleen whales and seals) are threatened by the rapid warming of recent decades and further adversely impacted by formation of low pH and low oxygen zones in the deep water column of the Amundsen Sea, where ice shelves melted (1973–2003), creating the new Elizabeth City State University Bay. Secondly, in the past four decades we have also discovered unique hydrothermal warm deep-sea chemosynthetic communities and cold seepage ecosystems with peculiar marine life adapted to a toxic marine environment. This chapter discusses how life originated, and comments on the evolution of life that perhaps exists on planets other than Earth. A third ecosystem discussed in this chapter is the so-called ultra-abyssal or hadal zone that supports unique fauna in extreme depths (6,000–10,000 meters) and, therefore, under perpetual hyperbaric conditions in the trenches of the world's oceans. This chapter also focuses on the Puerto Rico Trench, the only trench in the North Atlantic Ocean. Climate-change-induced processes may even slow down major ocean currents such as the Antarctic bottom water (AABW) flowing into the lower hadal or ultra-abyssal zone at a depth greater than 8000 meters in the Puerto Rico Trench, with potential creation of hypoxic hadal zones at extreme ocean depths.

Keywords: Marine life, psychrophile, barophile, chemosynthetic, Puerto Rico Trench, Antarctic bottom water (AABW)

Email: georgeryt@cs.com
Robert Y. George: www.GIBSconservation.org

Joseph Seckbach and Helga Stan-Lotter (eds.) Extremophiles as Astrobiological Models, (161–182)

7.1 Introduction

In the early seventies, Menzies, George and Rowe [7.40] made an effort to define deep-sea ecosystems as abyssal and ultra-abyssal environments, with vertical zones that are different in bathymetric range at different latitudes. Now, in 2020, the deep-sea is clearly defined as the largest ecosystem on our planet, covering approximately 60% of the Earth's surface, most of which is unexplored. The high biodiversity in the abyssal mud habitat is astonishing in abyssal plains of all the world's oceans, with one square meter of deep-sea floor with as many as 20 macrofaunal genera (polychaetes, paracarid isopods, amphipods, cumaceans), the same genera with as many as 9 sympatric species that evolved by natural selection pressure as defined by Darwin (1859). Now we know that the deep-sea provides services that are essential to the functioning of our planet. The deep sea also has a wealth of resources, including mineral deposits, fish and invertebrates as food sources, oil and gas, novel and rare minerals that, through technological developments of the deep-sea mining process, are now being exploited with permits from the International Seabed Authority (ISA) of the United Nations [7.23].

The pros and cons of the impact of climate change on life in the seas around us, coupled with ongoing ocean acidification, ocean de-oxygenation, as well as synergistic and antagonistic effects of these environmental parameters, including hydrostatic pressure, are already having significant impacts in the highly vulnerable deep ocean ecosystems that constitute the bulk of our planet both in volume and area. There is thus far no evidence that such vast and deep oceans occur elsewhere in the oceans of other planets. Perhaps we have not yet explored the universe to come to this conclusion. However, I am of the opinion that planet Earth is unique. I have discussed this idea in a chapter I wrote titled "Confluence of Biblical Truth and Scientific Facts Must Emerge for Wise Stewardship of God's Creation" [7.24] in a recent book, *Theology and Science: From Genesis to Astrobiology*, edited by Seckbach and Gordon [7.24].

7.2 Southern Ocean Ecosystem: West Antarctic Peninsula Region

Warming trends over the last four decades around Antarctica clearly show evidence of rapid shrinking of sea ice since the 1980s, and the most vulnerable region affected by surface seawater temperature is the west Antarctic Peninsula (Figure 7.1). George [7.23] described the rate of melting of the ice shelf in the Amundsen Sea from 1972 to 2003, thus creating the Elizabeth City State University Bay. George [7.23] also hypothesized that an increase in pCO_2 at mid-water column depths will threaten krill egg development and by 2100 the collapse of the krill fisheries near the tip of the Antarctic Peninsula (Figure 7.1) will be imminent.

7.3 Sea Ice Decline in WAP and Ice Shelf Collapse in Amundsen Sea

Using Landsat satellite image data from NASA dating back to 1972, LeCompte *et al.* [7.34] and George and Hayden [7.25] established a grounding line; a boundary between disintegrating floating ice shelf and the ground ice that feeds it in the Amundsen Sea (Figure 7.2). This project significantly improved our knowledge of cryospheric processes by comparing

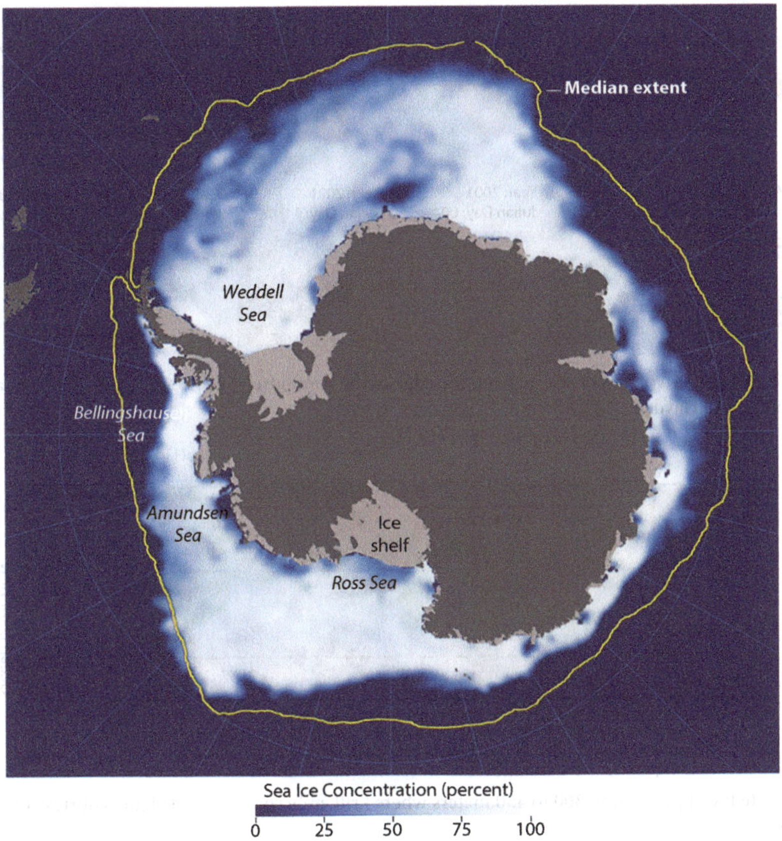

Figure 7.1 Note the disappearance of sea ice at the tip of the Antarctic Peninsula. (Image credit: National Snow and Ice Data Center (NSIDC; [7.11]).

30 years of records from 1972 to 2003. Pine Island, located south of Larsen, equivalent to the size of two-thirds of the UK, is a key study site of Elizabeth City State University's Center of Excellence in Remote Sensing Education and Research (CERSER). This study proved the break-up of Larsen A ice shelf in 1995 and Larsen B ice shelf in 2002. The warming of the waters off western Antarctic Peninsula (WAP) has influenced density and distribution of the Antarctic krill *Euphausia superba*. Fevolden and George [7.12] established the size frequency patterns of *E. superba* in WAP in the austral summer of 1983. In 2009-2010, Bernard *et al.* [7.5] started to provide evidence to support the conclusion that WAP is a region of rapid climate change that altered plankton community structure, suggesting ecosystem evolution in some parts of the Southern Ocean. While *E. superba* is a major grazer in the coastal regions of WAP, *Salpa thompsoni* dominated grazing offshore. *E. superba* will likely become less dominant if the pteropods *Limacina helicina* and *S. thompsoni* expand their ranges in the WAP region. Krill distribution and abundance also depends on primary production in the preceding years. George [7.18] [7.22] reported that krill egg development is influenced by temperature, pressure and pH. The potential collapse of Antarctic

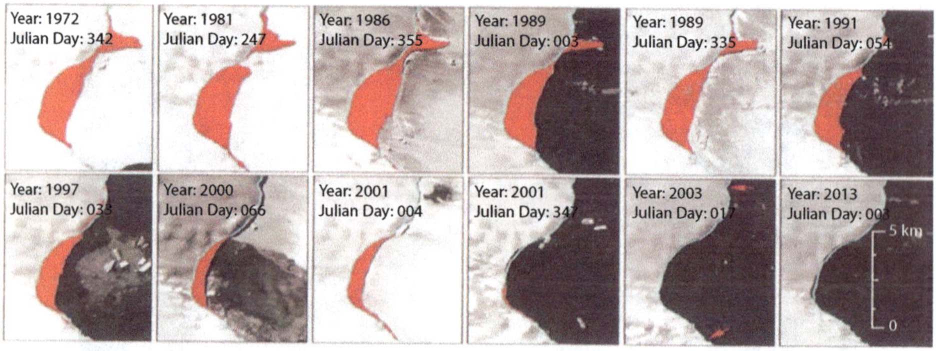

Figure 7.2 Progression of loss of sea ice from the edge of the Amundsen Sea from 1972 to 2003. (Image credit: George and Hayden [7.25]).

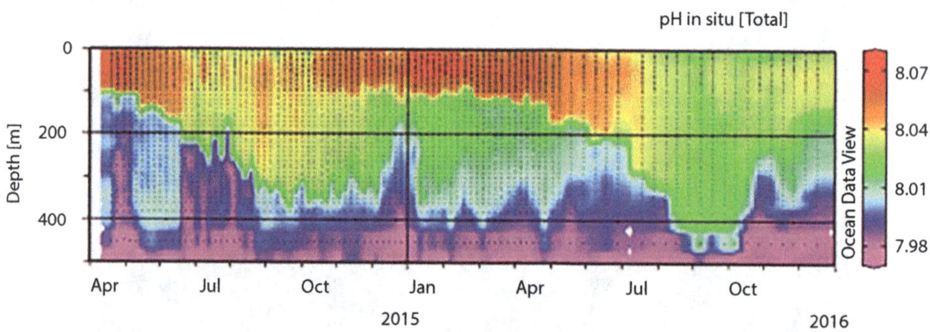

Figure 7.3 Reduced pH around 300 to 450 meters where krill embryos develop. (Figure courtesy of Dr. Chris Langdon).

krill *E. superba* may not occur in 2100 with atmospheric CO_2 at 770 ppm, as hypothesized by Kawaguchi *et al.* [7.33], but eggs of pteropod *Limacina helicina antarctica* may become far more susceptible with atmospheric CO_2 reaching 770 ppm, since 80% of pteropod eggs failed to undergo organogenesis [7.36].

The long-term monitoring of environmental parameters in the NSF LTER site revealed that below a depth of 300 m there was clear evidence of decreasing pH (Figure 7.3) and this decline in pH also had a significant seasonal signature.

7.4 Deoxygenation Leading toward Hypoxic Zone in Amundsen Sea

It is now becoming clear that the western Antarctic Peninsula region may face a severe threat from sea ice decline, warming and increase in $pC0_2$ in mesopelagic ecosystems from 200 to 900 m. In addition, recent research [7.23] revealed that the deoxygenation process is in progress in the deep-water column of the Amundsen Sea, as shown in Figures 7.4 and 7.5 below.

If this trend continues in the coming decades, it is inevitable that the pH now at 7.93 will decrease to 7.67 in 40 years (2060) and the pCO_2 now at 531 micrograms will increase to

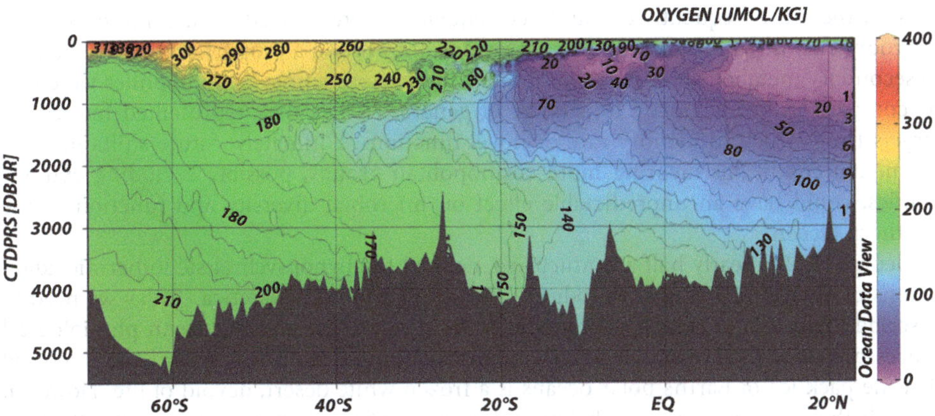

Figure 7.4 Transect from 20 N to 65 N showing low oxygen (hypoxia zone) only in lower latitudes. (Figure courtesy of Dr. Chris Langdon).

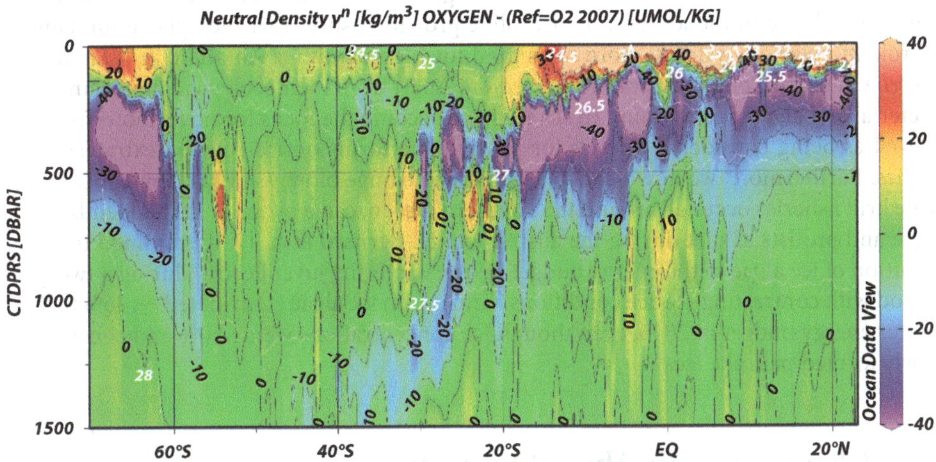

Figure 7.5 Transect from 20 N to 65 N, showing appearance of reduced oxygen zones in the latitudes above 65 N. (Figure courtesy of Dr. Chris Langdon).

1260 micrograms. Such a change can have an adverse impact on organisms at the physiological and molecular level.

7.5 Microbial Extremophiles in Southern Ocean

Dickinson *et al.* [7.9] focused on the extremophile community in the boundary region located between the Polar Front and the Southern Antarctic Circumpolar Current in the Southern Ocean, to explore the potential of metagenomic approaches as a tool for bioprospecting in the search for novel functional activity. They assessed the microbial composition and diversity from a region north of the current limit for winter sea ice, north of the Southern Antarctic Circumpolar Front (SACCF) but south of the Polar Front. Although most of the more frequently encountered sequences were derived from common marine microorganisms, within these dominant groups, they found a proportion of genes related to secondary metabolism of potential interest in bioprospecting. Extremophiles were rare by comparison but belonged to a range of genera. Hence, they represented interesting targets from which to identify rare or novel functions. Ultimately, future shifts in environmental conditions favoring more cosmopolitan groups, possibly excluding extremophiles, could have an unpredictable effect on microbial diversity and function in the Southern Ocean.

George [7.15], nearly half a century ago, defined the hypopsychral stenothermic adaptations of organisms living at the edge of the ice all around Antarctica. Today we need to describe these unique adaptive processes both at microbial and metazoan physiological processes as models at various levels of complexity. Thomas and Dieckmann [7.46] reported that the pack ice of Earth's polar oceans is a frozen white desert, devoid of life. However, beneath the snow lies a unique habitat for a group of psychrophilic organisms. Bacteria and microscopic plants and animals are encased in an ice matrix at low temperatures and light levels, with the only liquid being pockets of concentrated brines. Survival in these conditions requires a complex suite of physiological and metabolic adaptations, but sea-ice organisms thrive in the ice, and their prolific growth ensures that they play a fundamental role in polar ecosystems, more so as food for the ubiquitous Antarctic krill *Euphausia superba* in the prolonged austral winter season. Apart from their ecological importance, the bacterial and algal species found in sea ice have become the focus for novel biotechnology, as well as being considered proxies for possible life forms on ice-covered extraterrestrial bodies. Furthermore, we now need to learn how the huge biomass of krill, on which large-scale krill fisheries occur all around the Southern Ocean, can exist since krill depends on ice algae and protists as diet during the austral winter. Bernard *et al.* [7.5] examined the contribution of ice algae to the energy budget of overwintering juvenile krill during a two-year period with contrasting sea-ice conditions. Grazing on ice algae contributed ~146% to their winter energy budget in 2015, even though ice concentrations were ≤50% and consisted of newly formed pancake ice.

7.6 Chemosynthetic Abyssal Ecosystems

7.6.1 Methane Cold Seep Deep-Sea Communities in the Abyss

First discovered in 1983 on the Florida Escarpment in the Gulf of Mexico, cold seeps are now known to be located in all major oceans along continental margins. Cold seeps are found along the Cascadia fault in the Northeast Pacific, where tectonic activity enhances methane production. Temperatures at cold seeps are approximately the same as the surrounding seawater, ranging from 2–4 degrees Celsius along the Cascadia fault. Methane—the main component of natural gas—and other hydrocarbons form the basis of cold seep ecosystems, acting as a chemical energy source that fuels microbes through a process called chemosynthesis. Microbes within sediment at these sites convert chemicals released at the seep into forms that can be used by other species and animal communities while also creating a filter that absorbs greenhouse gases, which would otherwise be released.

During this initial stage, when methane is relatively abundant, dense mussel beds also form near the cold seep. Mostly composed of species in the genus *Bathymodiolus*, these mussels do not directly consume food. Instead, they are nourished by symbiotic bacteria that also produce energy from methane, similar to their relatives that form mats. Chemosynthetic bivalves are prominent constituents of the fauna of cold seeps and are represented in that setting by five families: Solemyidae, Lucinidae, Vesicomyidae, Thyasiridae and Mytilidae.

7.6.2 Are Hydrothermal Vents Abyssal Gold Mines?

Hot vents on the ocean floor are basically fissures found near areas of high volcanic activities with plumes arising from tectonic plates moving apart. The Earth's crust holds vast quantities of water and in land these are seen as geysers and in deep seas as hot vents or black smokers. Complex biological communities were discovered during the first *Alvin* dives of the East Pacific Rise by Woods Hole oceanographers in deep-sea biology (Holger Jannasch, Robert Ballard, Howard Sanders and Fred Grassle). These benthic communities are fueled by chemicals dissolved in vent fluids. The base of the food chain is indeed chemosynthetic bacteria and archaea that nourish giant tube worms (*Riftia pachyptila*), huge clams (*Calyptogena magnifica)*, limpets, vent fish (*Diplacanthaopora* sp.), mussels (*Bathymodiolus*), polychaetes (*Alvinella pompejana*), yeti crabs and shrimps (*Remicaris exoculata)* from the Lost City vents in the Mid-Atlantic Ridge.

Historically, the very first hot vent ecosystem in the deep sea was discovered in the East Pacific Rise in 1977, at the junction of the spreading tectonic plates of the Pacific and Rivera. Subsequently, several vents were discovered one after the other, as follows: vent off the Galapagos, vent off Mexico and California, vent off Oregon and British Columbia, Explorer Ridge at Juan de Fuca, and also in the Atlantic, the so-called "Lost City vent," at 30 N in 800 m in the Mid-Atlantic Ridge, at hadal depth in Mariana Trench and Cayman Trench, the Indian Ocean vent, and the Antarctic vent of the Antarctic peninsula, north of Greenland in the Arctic (Kipinovitch-Gakkai Ridge). Figure 7.6 shows a typical vent community from the new vents discovered in the Mid-Atlantic Ridge (see [7.43]).

Hydrothermal fluids, seeping out or oozing out, reach the deep-sea floor with a temperature as high as 400 °C (750 F), and this fluid at such high temperature cannot boil because of the extreme high hydrostatic pressure. This hot plume, while encountering the cold abyssal waters, generates a series of chemical reactions. As a consequence, sulfur and other metals precipitate, leading to creation of metallic deposits on the deep-sea floor and the metals such as gold and copper lure companies, like Australian Blue Metal and Nautilus Minerals in Canada, to go after these precious and semiprecious metals for commercial exploitation. There is no fishery potential in vent or seep ecosystems and the few species, such as *Riftia pachyptila* (Figure 7.7) and zoarcids (Figure 7.8), may even be poisonous! Taxonomically these chemosynthetic ecosystems are intriguing, with a discovery of 375 new species of diverse taxa thus far.

Tube worms (Figures 7.7 and 7.8) that grow near the boundary where hot vent fluid mixes with cold seawater on the ocean floor are an example of extremophiles that broaden our perspective on where to look for life. Chemosynthetic bacteria living inside the tube worms derive energy from chemicals emitted in the hot water of hydrothermal vents. One major exciting discovery in the so-called Beebe Vent Field in the Cayman Trough was a new

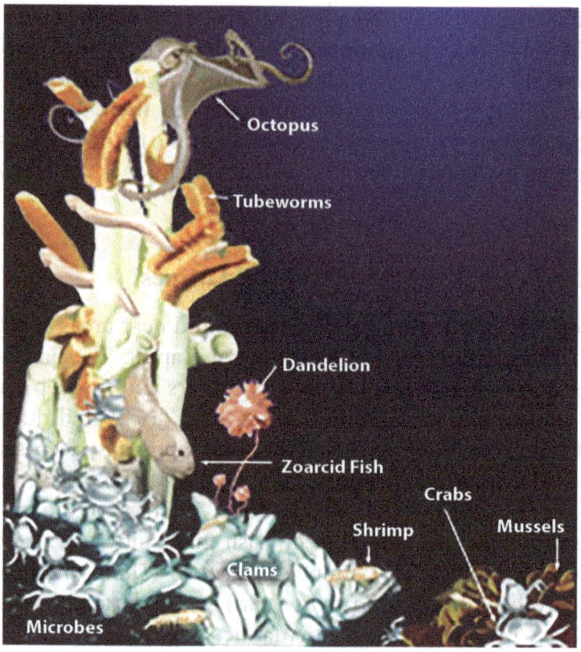

Figure 7.6 Typical vent community. (Figure courtesy of Dr. Fred Grassle).

Figure 7.7 Giant tube worms *Riftia pachyptila* clustering on hydrothermal vent. (Figure courtesy of Dr. Fred Grassle).

Figure 7.8 Vent zoarcid fish *Diplacanthaopora* sp. with tube worm *R. pachyptila* in the background. (Figure courtesy of Dr. Fred Grassle).

species of pale blind shrimp, *Rimicaris hybisae* (Figure 7.9a), which are characterized by dorsal light-sensing organs. Previously, a closely related species, *Rimicaris exoculata* (Figure 7.9b), was discovered 2500 miles north in the Mid-Atlantic Ridge vents.

These vent ecosystems have been characterized thus far as ephemeral and now are under industrial commercial exploitation for mining of polymetallic sulfide deposits. However, the back-arc basin vents off the west coast of Tonga in the South Pacific are now found to be stable and long-lasting with diverse vent-associated species [7.10]. The vent ecosystems in the mid-ocean ridges are also rich in species composition [7.2] and so are the vents near the Galapagos Rift, East Pacific Rise and Juan de Fuca Ridge [7.43]. Recently, the International Seabed Authority (ISA) has given permits to mine in vents near Fiji, Papua New Guinea and the Solomon Islands [7.35]. Obviously, mining activities by Nautilus Minerals Niugini Ltd., as approved by ISA and national government bodies, may cause damage to vent ecosystems and there is still a need to arrive at methods to recover damages from mining impacts [7.47]. The vent edifice sulfide deposits host faunal groups of chemosynthetic bacteria-containing invertebrates such as the mussel *Bathymodiolus septemdierum* and provannid snails *Alviconcha* spp.

According to Peter Aldhous, a deep sea gold rush is now inevitable from mining hydrothermal vents such as the vent edifices which are rich in sulfides.

Fitzpatrick [7.13] postulated a hypothesis that life originated in extreme environment, such as the vent ecosystem, *where life thrives without relying on the sun as an energy source. Hydrogen sulfide in the deep-sea vent* turned out to be the food source for vent bacteria. Chemosynthesis is the biological conversion of carbon molecules and nutrients into organic matter. Chemosynthesis uses inorganic molecules (such as hydrogen sulfide) or methane

Figure 7.9 (a) *Rimicaris hybisae* and (b) *Rimicaris exoculata.* (Figures courtesy of Fred Grassle).

and combines them with an oxygen source (in this case seawater) to create simple sugars. Those sugars are the fuel that bacteria live off, and the bacteria are the fuel that larger organisms, such as mussels in seeps and *R. pachyptila* tube worms in vents, need to survive.

As mining companies prepare to exploit the copper and gold in this seabed, we explore the fate of the unique ecosystems around tectonic boundaries 1600 meters below the waves with a 150-ton tank-like vehicle that wields a robotic arm armed with rotating rock cutters. After it has ground down the volcanic chimneys that rise from the sea floor, other mining robots follow in its tracks. This underwater tractors can obliterate the vent ecosystems while extracting the ores of copper and gold.

7.7 Hydrothermal Activity in Hrad Vallis on Mars

Hamilton *et al.* [7.29] reported that "Hrad Vallis" is an Amazonian-aged outflow channel located in the northwestern part of the Elysium volcanic province of Mars. The formation of Hrad Vallis may have been associated with catastrophic aqueous flooding and volcanism, which makes determining its emplacement history important for constraining the planet's hydrological and thermal evolution during the Amazonian Period. Through geological mapping, geomorphologic analysis, and numerical simulations they assessed whether Hrad Vallis formed in association with mudflows (i.e., lahars), effusive volcanism, or a combination of both processes. Among the three youngest flows in the region, the lower and upper flow units are inferred to be the products of catastrophic aqueous floods, whereas the middle unit is interpreted to be pāhoehoe-like lava flow (a smooth and billowy type of lava) formed through the process of inflation over the course of decades. The latter flow unit is similar to terrestrial lava-rise plateaus, but its emplacement may have involved interactions with surficial ice deposits, which affected the flow path and resulted in a range of lava-water interactions. We conclude that Hrad Vallis had a complex history involving episodes of both aqueous flooding and effusive volcanism and that the interactions between lava and ice in these regions may have generated ephemeral hydrothermal systems with potential astrobiological significance. However, we have no evidence of any prokaryotic or microbial life persisting in this Martian ecosystem, and it obviously calls for more research and collaborative international Martian investigations in the future. A satellite image of the lava rise in Mars is shown in Figure 7.10.

By studying Martian rocks or meteorites or satellite imagery, scientists have pieced together some of the natural history of the Red Planet, dating back millions of years. Perhaps the Martian research may help to identify regions that are possibly hospitable for microbial life. Hamilton *et al.* [7.29] studied a 370-meter-deep and 800-kilometer-long Martian valley called Hrad Vallis that is young in geological time, formed during the Amazonian age within the past 3 billion years. Some of the lava flows may have interacted violently with ice deposits on a surface, creating lakes and sinkholes. This type of landscape, called thermokarst terrain, occurs on Earth (our planet) in Arctic and sub-Arctic regions with permafrost melts. This could lead to a warm, moist environment, providing a habitat capable of sustaining life.

The Smithsonian's geologist, John Grant, was inspired by the landing of NASA's *Viking* spacecraft on Mars. There is evidence that 4 billion years ago Mars had lakes such as Lake Champlain on the northern shores of New York and Vermont where John Grant grew up.

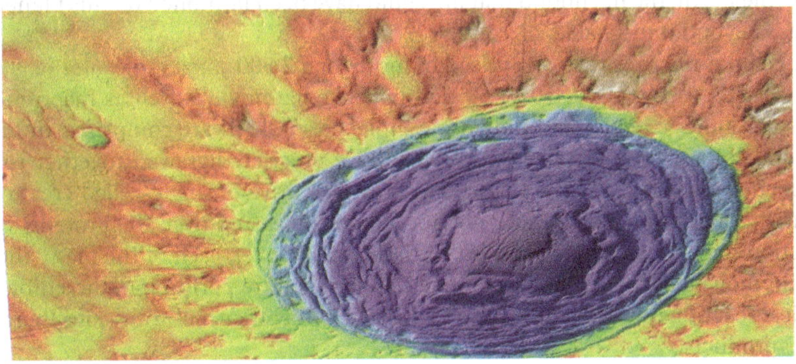

Figure 7.10 Color image of a lava-rise pit formed by gradual thickening of a lava around a high-standing topographic obstacle in Hrad Valley on Mars. Reddish-yellowish colors indicate relatively high elevations and blueish-greenish colors represent lower elevations; see [7.50]. (Image credit: University of Arizona and NASA – Creative Commons).

However, in my opinion, these lakes then were like lakes on Earth with conducive habitat for supporting the origin and evolution of life. But the lakes on Mars became extinct and life originated and evolved on planet Earth, culminating in higher apes such as chimpanzees. I also opine that "we are alone" as the only intelligent life on Earth in the entire universe.

Nearly half a century ago, Carl Sagan (Professor of Astronomy at Cornell University) pointed out, on the basis of 7000 photographs taken by NASA Mariner 9 of the Martian surface, that this "Red Planet" is a dead planet which may reveal fossil life that can provide evidence of early life and its evolution and eventual extinction. Carl Sagan also commented on the need to explore the two moons of Mars, Phobos and Deimos. Just recently, in mid-November, 2018, NASA's *InSight* lander touched down on the surface of Mars, plunged through the thin Martian atmosphere, and probed the interior of Mars using a mechanical mole to tunnel 16 feet deep to measure internal temperature, and used a seismometer to monitor quakes, if any, besides exploring evidence of meteorite strikes. These new initiatives in the exploration of Mars may or may not reveal any evidence of life on Mars that existed in the distant past.

7.8 Why Chemosynthetic Ecosystems Remind Us of Environmental Conditions When Life Originated in the Universe

Is it possible life could not have begun on Earth [7.32]? Life was present on this planet from the very beginning, with evidence of biological activity present dating back 4.28 to 3.8 billion years ago during a period when the surface of the Earth was being repeatedly pulverized and vaporized by a constant bombardment of meteors, comets, and cosmic debris. Given that Earth is approximately 4.6 billion years old, the lack of essential ingredients for creating RNA, DNA and basic proteins, the thermal conditions which would have destroyed all biological molecules, coupled with evidence of life within 300 to 800 million years after earth became Earth, the likelihood that life could have been fashioned on this planet during this time is beyond improbable.

Single cellular microbes are comprised of more than 2,500 small molecules (e.g., including amino acids consisting of 10 to 50 tightly packed atoms), as well as macromolecules (proteins and nucleic acids) and polymeric molecules (which are comprised of hundreds to thousands of small molecules), all of which are precisely jigsawed together to form a single-celled organism [7.7]. The tiniest and most primitive of single-celled creatures contain a variety of micro-, macro-, and polymeric molecules and over 700 proteins which fit and function together as a living mosaic of tissues. Moreover, each of the many thousands of different molecules that make up a single cellular creature perform an incredible variety of chemical reactions—usually in concert with that cell's other molecules and their protein (enzyme) products.

J.B.S. Haldane was one of the first to consider the possibility of the central role of RNA in the origin of life, an idea that became popular as early as the late 1920s [7.27] and during the mid-1960s [7.28]. As a young college student in Madras (India), I heard Haldane speak on his theory on the origin of life. His hypothesis was related to his interest in that the first synthetic organisms may have been something like a tobacco mosaic virus, but including the enzyme or enzymes needed for its own replication [7.28].

Linus Carl Pauling, an American biochemist and peace activist and the recipient of the Nobel Prize in Chemistry and the Nobel Peace Prize, diligently used X-ray crystallography to decipher and describe the alpha helix that eventually led to the discovery of the double-helical structure of DNA by James Watson, Francis Crick and Rosalind Franklin.

Nobel laureate Francis Crick believed that even 10 billion years would not be enough time from Big Bang to the emergence of DNA molecule in the prolonged chemical evolution that preceded organic evolution [7.8]. According to Crick, estimates of the time needed for these chance combinations to have produced life have ranged from 100 billion to over 1 trillion years.

The discovery of hydrothermal vents has revived the discussion on the various theories for the origin of life. Researchers such as Günter Wächtershäuser have proposed that life began in submarine hydrothermal vents [7.49]. Wächtershäuser's theory requires a cycle of chemical reactions to take place, releasing energy in a form that allows the exploitation by other processes. To account for a hydrothermal origin of life, amino acid synthesis has been proposed to occur in the Earth's crust, where they were consequently injected into cooler water with hydrothermal fluids. The presence of clay minerals at these lower temperatures may have promoted the genesis of peptides and protocells from amino acids, and hence the first origins of life.

This hypothesis is especially promising, as the hydrothermal vents emit multitudes of methane (CH_4) and ammonia (NH_3) gases which were lacking in the Earth's primitive atmosphere. Presently, numerous species of extremophiles live nearby or even on the deep-sea vents, possibly providing evidence of the formation and continuation of life at such extreme temperatures.

The origin of life still remains a mystery. Gabel [7.14] proposed a realistic chemical basis for the origin of life. He indicated that polyphosphate complexes serve as template and depolymerizing agent for the origin of proteins. Previously, in 1978, a workshop was held in Hyderabad, India, to look at chemical evolution, origin of life and evolution of life processes [7.6]. In this conference, Dr. Ponnamperuma reported that Antarctica is a vast storehouse of meteorites and three of these meteorites were found to be carbonaceous chondrites [7.42]. The question arises from these findings whether extraterrestrial life exists in other planets in the universe. Scientists also opine that deep-sea hydrothermal vents may hold the clues to decipher the origin of life and also reveal such scenario existing elsewhere beside the deepest depths of the ocean in fracture zones or in mid-ocean ridges.

7.9 Ultra-Abyssal Ecosystem: Puerto Rico Trench

In his classic book entitled *Half Mile Down*, Beebe [7.3] narrated so elegantly the curious quest by Alexander the Great in 333 BC to dive down boldly in a "Glass Bell" (the first bathyscaphe) to see what ocean life exists in the depths firsthand. His bold dive to deep depths came from the stories he heard from Greek sponge divers about strange fishes that they saw in the deepest depths. Alexander strictly told his soldiers before the dive that "he should be drawn up only if the chain, connected to his glass diving bell, vibrated." Almost two millennia later, there was a genuine quest in the fifties to explore the abyss below a depth of 6000 meters and the leadership to achieve this goal came from Russian and Danish oceanographers, the former under the guidance of Dr. George Belyaev in Moscow and the later under the guidance of Dr. Torben Wolff in Copenhagen.

In his book entitled *The Greatest Depths: Probing the Seas to 20,000 Feet and Below*, Soule [7.44] discussed the deep dives of submersible *Archimède* into the deepest trench in the North Atlantic Ocean. This dive into the abyss achieved a record for man going down to 7000 meters but virtually did not really reveal what life exists at such great depths. Belyaev [7.4] coined the word "ultra-abyssal zone" to report on life found at the greatest depths in as many as 23 trenches in the world's oceans but he told me in Moscow that Russians had difficulty sampling the Puerto Rico Trench. Likewise, Dr. Torben Wolff confessed to me in Copenhagen that the Danish explorers lost their trawls in the Puerto Rico Trench since this trench was formed because of tension between two tectonic plates rather than collision. The sea floor at depths below 6000 meters was too rough in the Puerto Rico Trench, with ridges and troughs, and it was hard to study what kind of ultra-abyssal life exists there. Wolff [7.51] coined the term "hadal community" to describe marine life existing in the deepest depths of the oceans below 6000 meters.

Menzies and George [7.38| reported that the most important adaptation of hadal or ultra-abyssal animals is at the physiological, biochemical or molecular level in order to live at pressures as high as 1000 atm. Also, all trenches have a vertical wall on their coastal edge and this topographic configuration allows accumulation of terrestrial debris and this condition serves as an ecological stimulus in a hadal zone with one or two species dominating the niche. George and Higgins [7.26] successfully defined and described the eutrophic hadal community in the Puerto Rico Trench. This study revealed that the standing crop of meiofauna is 91 to 97% nematodes and harpacticoid copepods, between 17,059 and 17,303 individuals per square meter of sea floor. Osmotrophic organisms, such as pogonophorans, dominated the mud floor. Deposit-feeding animals such as holothurians, isopods and amphipods represented the dominant trophic type in Gilliss Deep at 7600 m (see illustration in Figure 7.11). This study in the Puerto Rico Trench also discovered great abundance of lyssianasid amphipods as scavengers feeding on food parcels on the flat floor of the Puerto Rico Trench. This study also revealed the presence of Antarctic bottom water (AABW) at the hadal depths and the presence of a new species of ultra-abyssal asellote munnopsid isopod, *Storthyngura torbeni* n. sp., named and described by this author in honor of the hadal biologist Dr. Torben Wolff [7.20]. Menezes *et al.* [7.37] reported that the AABW that flows from the Southern Ocean far beyond the equator into the Puerto Rico Trench is slowing down because of climate change processes and freshening due to melting glaciers, and this process may shut the flow of AABW into the Puerto Rico Trench, creating hypoxic conditions at hadal depths. This hypothesis is still not supported by any data on current velocity and the deoxygenation can only be proven with new measurement of dissolved oxygen at the hadal depths. If deoxygenation occurs in the floor of the Puerto Rico Trench (Figure 7.11), it will be a threat to the eutrophic hadal community discovered by George and Higgins [7.26].

George [7.21] explained how animals living under high hydrostatic pressure in extreme depths of the ocean are adapted physiologically and biochemically both at molecular and whole organism level. George [7.16] reported on the depth- and pressure-related differences in energy flow, expressed in calories per unit weight and unit time, discovered in deep-sea crab *Parapagurus pilosimanus* (2000–5500 m). Abyssal-ultra-abyssal amphipod *Eurythenes gryllus* exhibited no significant change in metabolic rate over a wide range of

Figure 7.11 Artist's interpretation of ultra-abyssal life in the Puerto Rico Trench. Note the red amphipod, *Eurythenes gryllus*, around 7200 meters. (Original painting by author's wife, Mrs. Chandra George).

hydrostatic pressure from 100 to 325 atm [7.17]. Evidently these deep-sea animals are barophilic and their proteins depolymerize when exposed to shallow-water conditions and even pressure below their habitat depths, for example 200 atm or less. While writing a chapter in a book edited by Vernberg and Vernberg [7.48] on functional adaptations of marine organisms, George [7.17] clearly stated that low temperature and high pressure directly affect the kinetics of biochemical reactions and cellular chemistry of organisms. Under extremely low temperature and high pressure, membrane structure gets tightened and this synergistic effect also leads to slowing of sodium transport. George [7.19] discussed basal and active metabolic rates of deep-sea animals under fed and starved conditions and discovered that pressure exerts significant influence on ammonium excretion rates in deep-sea animals.

7.10 Affiliations of Abyssal Life to Astrobiology: Some Perspectives

George [7.23] discussed the question of "time," beginning with the "Big Bang" about 15 billion years ago, and also pointed out that we have no evidence to conclude how much longer human beings will exist on planet Earth, based on neither biblical nor scientific grounds. However, Hawking and Mlodinow [7.30], in their book entitled *The Grand Design*, which takes us through a whirlwind tour of fundamental physics and cosmology, hypothesized that "if the current understanding of cosmic expansion driven by dark energy is correct, the universe will continue to expand more and more rapidly until even atoms are ripped apart. The time or duration for this to come about can perhaps be as little as 20 billion years in the future." I seriously doubt that such a speculation can be proven by any mathematical genius today. Hawking also presented another provocative theory that in six billion years the solar system will collapse and the sun will die, and the Earth will evaporate. Hawking was notorious for speculating, but philosopher Francis Bacon [7.1] emphatically defined the fundamental need for scientific inquiry which is not speculation but clearly formulated observations to put forth a hypothesis. While speculating on the existence of intelligent life on other planets, Stephen Hawking stated: "What are the chances that we will encounter some alien form of life as we explore the galaxy? If the argument about the time scale for the appearance of life on Earth is correct, there ought to be many other stars, whose planets have life on them. Some of these stellar systems could have formed 5 billion years before the Earth."

What can be hypothesized, in my opinion, can only come from what we know for sure. If viruses and microbes have fallen and continue to fall to Earth, it can be predicted that they have or will transfer genes to various species of Eukaryotes, including humans and those species who have not yet evolved. It can also be predicted, based on the number of silent viral prokaryote genes within the eukaryotic—including the human—genome, that evolution is an ongoing process which will continue billions of years into the future.

"Evolution" metamorphosis does not stop with modern humans. If humans do not self-destruct and become extinct, it can be predicted that this process of metamorphosis will lead to increasingly intelligent and technologically advanced species who in turn will be targeted by viruses whose DNA/RNA genomes are templates fashioned and obtained from life which evolved in worlds much older than our own.

Europa (Figure 7.12) is thought to have subsurface liquid water. Scientists hypothesize that Europa's hidden ocean is salty, tidal, and causes its ice surface to move, resulting in large fractures which are clearly visible in the above image. Although Europa is thought to have the ingredients needed to support life—water, energy, organic compounds—it lies outside of our solar system's habitable zone.

Microorganisms have been detected in great depth in subterranean environments such as granite, sediments, permafrost areas, caves, and rocks in gold and uranium mines [7.45]. Currently, Earth is the only *known* planet (or moon) to have consistent, stable bodies of liquid water on its surface. In our solar system, Earth orbits around the sun in an area called the habitable zone. The temperature, along with an ample amount of atmospheric pressure within this zone, allows water to be liquid for long periods of time.

Figure 7.12 Europa. (Image credit: NASA).

Evidence points to water on other planets in our solar system. In 2015, NASA confirmed that liquid water flows intermittently on present-day Mars. Also, in 2015, scientists used data from NASA's Cassini mission to discover that a global ocean lies beneath the icy crust of Saturn's moon Enceladus. Scientists believe that Jupiter's moon Europa has a subsurface ocean as well.

Finding liquid water beyond our solar system may indicate the existence of extraterrestrial life. Using data collected by NASA's Kepler spacecraft and the W. M. Keck Observatory, scientists have determined that 22 percent of solar-type stars in our galaxy have Earth-sized planets in their habitable zone and could host life. Some of these planets are a mere 12 light years away and can be seen with the naked eye in the night sky. In 2013, NASA's Hubble Space Telescope spotted what may be plumes of water vapor spouting from a possible subsurface ocean on Europa.

Microbiologist Dr. Mark Lever of Aarhus University in Denmark, while working with rock samples from the ocean floor discovered an entire ecosystem living without light or oxygen flourishing beneath the ocean floor. Scientists call it the dark biosphere, and it's potentially one of the biggest ecosystems on the planet. Buried oceanic crust covers 60 percent of Earth's surface. For the first time, researchers have pulled up pieces of the crust and examined the life within. In its rocks, microbial communities thrive, eating altered minerals for food. They're gaining energy from chemical reactions from water with rock. Mark Lever remarked: "Our evidence suggests that this is an ecosystem that is based on chemosynthesis and not on photosynthesis. While bacteria and other microbes have been noted in deep boreholes drilled into the seafloor, the discovery confirms the extent of the life within the oceanic crust, as well as the possibility of life on other planets." Lever asserted that "I think it's quite likely there is similar life on other planets. On Mars, even though we don't have oxygen, we have rocks there that are iron-rich. It's feasible that similar reactions could be occurring on other planets and perhaps in the deep subsurface of these planets." NASA scientists also announced the discovery of the chemical ingredients for life in Mars rocks, including sulfur, nitrogen, hydrogen, oxygen, phosphorus and carbon. On November 2018, NASA's spacecraft landed on Mars after a journey of 300 million miles in six months. The venture was a genuine international (USA, Russian, Germany, France, etc.) collaboration. I suggest we continue this remarkable joint venture involving many nations. We must not waste billions of dollars to develop a military "Space Force."

I am tempted to propose that with modern space technology today we may soon have exciting times if we diligently probe beyond the Earth to distant planets in space for microbial life or even unicellular organisms living in psychrophilic conditions adapted to extremely low temperature. Instead of wasting billions of dollars on a "Space Force" as recently advocated, we may use money prudently to explore other planets in the universe.

7.11 Can We Find Protozoans Such as Xenophyophores on Other Planets?

Many scientists infer that liquid water once covered large areas of Mars and Venus. Water is thought to exist as liquid beneath the surface of some planetary bodies, similar to groundwater on Earth. Are we likely to discover that water bodies in other planets support life with viruses, bacteria and peculiar unicellular organisms as found in the deep seas of planet Earth?

Figure 7.13 Xenophyophore from the Galapagos Rift. The "giant amoeba" (belonging to the Foraminifera) is about 20 cm in diameter. (Image credit: NOAA Ocean Explorer Gallery; public domain).

The intertidal zones and continental shelves of the world's oceans do not support the presence of unique protozoans known as xenophyophores (Figure 7.13), which are unicellular multinucleate organisms found on the ocean floor throughout the world's oceans, over the continental slopes (500 to 1000 m), in abyssal plains (5000 to 5400 m) and in trenches down to 10,600 meters [7.35]. They are a kind of foraminifera that extract minerals from their surroundings and use them to form an exoskeleton known as a test. The unique organisms are pressure-adapted and cannot inhabit shallow depths in the seas around us.

7.12 Barophilic Organisms in the Deep-Sea

My colleague, Dr. Arthur Yayanos (Scripps Institution of Oceanography, La Jolla, California), told me, while giving a tour of his laboratory with cultures from various trenches, including the Mariana Trench, that piezophilic bacteria (piezophiles) are adapted to high pressures and are prominent among the inhabitants of the cold deep ocean below depths of approximately 2000 m [7.52] [7.31]. Recently we witnessed descriptions of a diverse assortment of prokaryotic species that thrive under extreme environments that used to be considered inhospitable. We now know that a number of archaea and bacteria live under extreme environmental conditions that include high pressure, ionizing radiation, both high and low temperature, hypersaline waters, as well as low and high levels of pH.

ZoBell and Johnson [7.53] demonstrated that *Desulfovibrio indonesiensis* strain P23 was able to grow under hydrostatic pressure of up to 40 MPa (*in-situ* pressure ~30 MPa), even after cultivation at atmospheric pressure for more than three years. *D. indonesiensis* strain P23 obviously has not lost its piezophilic properties, which may be a common feature of pressurized deep-sea organisms sampled and isolated under decompressed conditions. Dr. Claude ZoBell coined the word "barophilism" for such pressure-adapted bacteria living at depths of 10,000 m in hadal zones in trenches of the world's oceans. Whereas their maximum growth rate is close to the upper limiting growth temperature, it is also at a pressure close to that of their deep-sea habitat. In contrast, bacteria in the upper 2000 m of the ocean and in terrestrial habitats have their maximum growth rate at atmospheric pressure. Two decades later, in 1969, I met Dr. ZoBell at a conference on "Barobiology" in Wrightsville Beach, North Carolina, where he presented a paper on adaptations of bacteria to high pressure and I presented a paper on adaptations of metazoan organisms [7.39]. Mota *et al.* [7.41] pointed out that microorganisms have the ability to adapt to extreme hydrostatic pressure. For instance, the modulation of microbial metabolic pathways, as a response to different pressure conditions, may even lead to the production of novel compounds with potential biotechnological and industrial applications. This discovery paves the way for our future exploration in space research that may open new avenues of biotechnological research.

Acknowledgments

I wish to thank Prof. Joseph Seckbach of Hebrew University in Israel for inviting me to write a chapter in an earlier book on "Theology and Science" and this collaboration led to the writing of this chapter in the book dealing with "Extremophiles as Astrobiological Models." I also thank Dr. Helga Stan-Lotter (Austria) for improving the book manuscript in its initial phase. The research reported in this chapter about the Southern Ocean around the Antarctic Peninsula was supported by several grants to the author from the Office of Polar Program (OPP) of the US National Science Foundation (NSF). I thank Dr. George Llano, Dr. Frank Williamson, Dr. Richard Williams and Dr. Christian Fritsen, all polar scientists from the Office of Polar Program (OPP) of the US National Science Foundation. I also wish to acknowledge two great hadal biologists who inspired me to explore the Puerto Rico Trench, one of whom is Dr. George Belyaev of Moscow University, whom I visited in 1979 as a visiting professor of the USSR-USA Academy of Sciences at Shirshov Institute of Oceanology of Moscow University. The other hadal biologist that I am immensely thankful to is Dr. Torben Wolff, who led the *Galathea* Expedition around the world to sample the hadal depths. I wish to thank the University of Miami for providing their ship R/V GILLISS for studying hadal life in the Puerto Rico Trench, with two 30-day cruises. I am also deeply grateful to two deep-sea biologists, my former colleague Dr. Robert J. Menzies of Florida State University, for inspiring me to collaborate with him to study life in the deep sea, and Dr. Howard L. Sanders of Woods Hole Oceanographic Institution, for exposing me to the new discoveries on chemosynthetic deep sea. I also wish to thank Prof. Chris Langdon of the University of Miami for providing the data on low oxygen in the Amundsen Sea.

It is with great pleasure I dedicate this chapter to deep-sea oceanographer Dr. Fred Grassle whom I first met at Duke University Marine Laboratory in 1996 where he did his doctoral research while I served there as a research oceanographer. I met Fred at Woods Hole Oceanographic Institution when I visited him in 1978 and traveled with him to Sweden for the Deep-Sea Symposium and also to Edinburgh for the 1970 International Ocean Assembly. Dr. Grassle is one of the founders of Census of Marine Life and also the Ocean Biogeographic Information System (OBIS). Fred was the first to explore the deep-sea hydrothermal vents while diving in the submersible *Alvin* into the Galapagos Rift in the deep sea. Dr. Fred Grassle died at age 78 on July 6, 2018.

References

7.1. Bacon, F., *Novum Organum and other parts of "The Great Instauration*, P. Urbach and J. Gibson (Eds.), Open Court, Chicago, 1627, 1994.

7.2. Baker, E.T. and German, C.R., *On the Global Distribution of Hydrothermal Vent Fields in Mid-Ocean Ridges*, C.P. German and L.M. Parson (Eds.), pp. 245–266, American Geophysical Union, Washington DC, 2004.

7.3. Beebe, W., *Half Mile Down*, Cadmus Books, E.M. Hale and Company, Chicago, 1934.

7.4. Belyaev, G.M., Bottom fauna of the ultra-abyssal depths of the world oceans. (In Russian). *Akad. Nauk. SSSR, Trudy Inst. Okeanol.*, 103, 259–280, 1966.

7.5. Bernard, K.S., Gunther, L.A., Mahaffey, S.H., Qualls, K.M., Sugla, M., Saenz, B.T., Cossio, A.M., Walsh, J., Reiss, C.S., The contribution of ice algae to the winter energy budget of juvenile Antarctic krill in years with contrasting sea ice conditions. *ICES J. Mar. Sci.*, 76, 1, 206–216, 2019, https://doi.org/10.1093/icesjms/fsy145.

7.6. Chada, M.S., *Summary of Proceedings of Workshop on Chemical Evolution, Origin of Life and Evolution of Life Processes, Hyderabad, India*, Jan. 29-31, 1978, Indian Institute of Science, Bangalore, India, 1978.

7.7. Cowan, M.K. and Talaro, K.P., *Microbiology. A Systems Approach*, McGraw-Hill-Education, Columbus, Ohio, USA, 2008.

7.8. Crick, F. and Orgel, L.E., Directed panspermia. *Icarus*, 19, 341–346, 1973.

7.9. Dickinson, I., Goodall-Copestake, W., Thorne, M.A., Schlitt, T., Ávila-Jiménez, M.L., Pearce, D.A., Extremophiles in an Antarctic marine ecosystem. *Microorganisms*, 4, 1, pii: E8, 2016.

7.10. Du Preez, C.D. and Fisher, C.R., Long-term stabiltiy of back-arc basin hydrothermal vents. *Front. Mar. Sci.*, 2018.

7.11. Fetterer, F., Knowles, K., Meier, W.N., Savoie, M., Windnagel, A.K., *updated daily. Sea Ice Index, Version 3*, NSIDC: National Snow and Ice Data Center, Boulder, Colorado USA, 2017, https://doi.org/10.7265/N5K072F8. [Accessed May 31,2020].

7.12. Fevolden, S. and George, R.Y., Size frequency pattern of *Euphausia superba* in the waters west of the Antarctic Peninsula in the austral summer of 1983. *J. Crust. Biol.*, 4, Spec. No 1, 107–122, 1984.

7.13. Fitzpatrick, G., Early life may have originated from deep sea vents. *Astrobiol. Mag.*, 2013. Jan 25, 2013.

7.14. Gabel, N.W., A realistic chemical basis for the origin of life. *Evol. Theory*, 9, 181–209, 1990.

7.15. George, R.Y., Thermal sensitivity of hypopsychral species of Antarctic and high arctic marine Crustacea. *Proceeding of the Joint Oceanographic Assembly, Tokyo 1970*, 474 pp., 1971.

7.16. George, R.Y., Behavioral and metabolic adaptations of polar and deep sea crustaceans: A hypothesis concerning physiological basis for evolution of cold adapted crustaceans. *Bull. Biol. Soc. Wash.*, 3, 283–296, 1979.

7.17. George, R.Y., Functional adaptations of deep-sea organisms, in: *Functional Adaptations of Marine Organisms*, J. Vernberg and W. Vernberg (Eds.), pp. 279–332, Academic Press, 1981.

7.18. George, R.Y., Ontogenetic adaptations in growth and respiration of *Euphausia superba in* relation to temperature and pressure. *J. Crust. Biol.*, 4, Spec.No. 1, 252–262, 1984.

7.19. George, R.Y., Basal and active metabolic rates of deep sea animals in relation to pressure and food ration, in: *Proceedings of the Nineteenth European Marine Biology Symposium, Plymouth, UK 1984*, P.E. Gibbs (Ed.), Cambridge University Press, pp. 173–182, 1985.

7.20. George, R.Y., *Storthyngura torbeni* n. sp., a hadal isopod from the Puerto Rico Trench and hypothesis on its origin. *Proc. Biol. Soc. Wash.*, 100, 4, 681–686, 1987.

7.21. George, R.Y., Pressure-induced metabolic adaptive strategies in deep-sea mega fauna. *Paper presented at the Seventh Deep-Sea Biology Symposium in Crete*, Greece, 1994.

7.22. George, R.Y., Potential impact of carbonate chemistry change (pCO_2) on krill and krill-based food-chain in the Southern Ocean with emphasis on embryogenesis of Antarctic krill. *Biogeosci. Discuss.*, 2017.
7.23. George, R.Y., Theoecology in today's kairos moment to develop connectivity with our creator and creation: Is it tragedy of commons or commodity. *Theoecology J.*, 3, 24–42, 2018.
7.24. George, R.Y., Confluence of biblical truth and scientific facts must merge for wise stewardship of God's creation, in: *Theology and Science: Genesis to Astrobioogy*, J. Seckbach and R. Gordon (Eds.), World Scientific Publishers, Singapore, 2019.
7.25. George, R.Y. and Hayden, L., Krill distribution in response to decadal warming evidenced by ice-shelf disintegration off the Western Antarctic Peninsula, and changing carbonate chemistry in the Southern Ocean threatening krill egg development. *Paper presented at the 3rd International Symposium on Krill*, St Andrews, Scotland, 12-16, June 2017, 2017.
7.26. George, R.Y. and Higgins, R.P., Eutrophic hadal benthic community in the Puerto Rico Trench. *Ambio Special Report* 6, 51–58, 1979.
7.27. Haldane, J.B.S., Origin of life. *Ration. Annu.*, 148, 3–10, 1929.
7.28. Haldane, J.B.S., Data needed for a blueprint of the first organism, in: *The Origins of Prebiological Systems and Their Molecular Matrices*, S.W. Fox (Ed.), p. 11, Academic Press, New York, 1965.
7.29. Hamilton, C.W., Mouginis-Mark, P.J., Sori, M.M., Scheidt, S.P., Bramson, A.M., Episodes of aqueous flooding and effusive volcanism, associated with Hrad Vallis, Mars. *J. Geophys. Res.*, 123, 6, 1484–1510, 2018.
7.30. Hawking, S. and Mlodinow, L., *The Grand Design*, Bantam Book Trade Paper Backs, New York, 2010.
7.31. Islas, S., Velasco, A.M., Becerra, A., Delaye, L., Lazcano, A., Extremophiles and the origin of life, in: *Physiology and Biochemistry of Extremophiles*, C. Gerday and N. Glansdorff (Eds.), pp. 3–10, ASM Press, Washington DC, 2007.
7.32. Joseph, R., *Life on Earth Came from Other Planets*, University Press, California, 1997.
7.33. Kawaguchi, S., Ishida, A., King, R., Raymond, B., Waller, N., Constable, A., Nicol, S., Wakita, M., Ishimatsu, A., Risk maps for Antarctic krill under projected Southern Ocean acidification. *Nat. Clim. Change*, 3, 843–847, 2013.
7.34. LeCompte, M., Bindschadler, R., Bailey Hayden, L., Jefferson, M., Bridgers, Y.S., Lawrence, R., Bevins, J., Brownlow, J., Evans, R., Hawk, K., Koch, G., Reduction and loss of an ice shelf in Elizabeth City State University Bay, Antarctica: 1972–2003. *IEEE J. Sel. Top. Appl. Earth Obs. Remote Sens.*, 6, 3, 1509–1515, 2013.
7.35. Levin, L.A., Mengerink, K., Gjerde, K.M., Rowden, A.A., Van Dover, C.L., Clark, M.R., Ramirez-Llodra, E., Currie, B., Smith, C.R., Sato, K.N. *et al.*, Defining "serious harm" to the marine environment in the context of deep-sea-bed mining. *Mar. Policy*, 74, 245–259, 2016.
7.36. Manno, C., Peck, V.L., Tailing, G.A., Pteropod eggs released to high pCO_2 lack resilience to ocean acidification. *Sci. Rep.*, 6, 25752, 2016.
7.37. Menezes, V.V., Macdonald, A.M., Schatzman, C., Accelerated freshening of Antarctic Bottom Water over the last decade in the Southern Indian Ocean. *Sci. Adv.*, 3, 1, e1601426, 2017.
7.38. Menzies, R.J. and George, R.Y., A re-evaluation of the concept of hadal or ultra-abyssal fauna. *Deep Sea Res.*, 14, 703–723, 1967.
7.39. Menzies, R.J., George, R.Y., Avent, R., Responses of selected aquatic organisms to increased hydrostatic pressure: Preliminary results, in: *Barobiology and the Experimental Biology of the Deep Sea*, R.W. Brauer (Ed.), pp. 37–5, University of North Carolina Press, 1972.
7.40. Menzies, R.J., George, R.Y., Rowe, G.T., *Abyssal Environment and Ecology of the World Oceans*, 540 pp., John Wiley Interscience Publishers, New York, 1973.

7.41. Mota, M., Lopes, R.P., Delgadillo, I., Saraiva, J.A., Microrganisms under high pressure – adaptation, growth and biotechnological potential. *Biotechnol. Adv.*, 31, 8, 1426–34, 2013.
7.42. Ponnamperuma, C., Organic matter in meteorites, in: *Chemical Evolution, Origin of Life and Evolution of Life Processes, Summary of Proceedings of Workshop, Hyderabad*, M.S. Chadha (Ed.), Indian Institute of Science, Bangalore, pp. 19–21, 1978.
7.43. Shank, T.M., Fornari, D.J., von Damm, K.L., Lilley, M.D., Haymon, R.M., Lutz, R.A., Temporal and spatial patterns of biological community development at nascent deep sea hydrothermal vents (9.5' N East Pacific Rise). *Deep-Sea Res. Part II Top. Stud. Oceanogr.*, 45, 465–515, 1998.
7.44. Soule, G., *The Greatest Depths: Probing the Seas to 20,000 Feet and Below*, Macrae Smith Company, Philadelphia, 1970.
7.45. Stan-Lotter, H., Fendrihan, S., Dornmayr-Pfaffenhuemer, M., Gerbl, F., Legat, A., Gruber, C., Weidler, G., Microorganisms in ancient terrestrial environment and outer space?, in: *From Fossils to Astrobiology: Records of Life on Earth and Search for Extraterrestrial Biosignatures*, J. Seckbach and M. Walsh (Eds.), pp. 235–248, Springer, Netherlands, 2009.
7.46. Thomas, D.N. and Dieckmann, G.S., Antarctic sea ice–a habitat for extremophiles. *Science*, 295, 5555, 641–644, 2002.
7.47. Van Dover, C.L., Ardron, J.A., Escobar, E., Gianni, M., Gjerde, K.M., Jaeckel, A., Jones, D.O.B., Levin, L.A., Niner, H.J., Pendleton, L. *et al.*, Biodiversity loss from deep sea mining. *Nat. Geosci.*, 10, 464–465, 2017.
7.48. Vernberg, F. and Vernberg, W.B. (Eds.), *Functional Adaptations of Marine Organisms*, Academic Press, New York, London, 1981.
7.49. Wächtershäuser, G., Evolution of the first metabolic cycles. *Proc. Natl. Acad. Sci. U.S.A.*, 87, 1, 200–204, 1990.
7.50. Witman, S., Tracing the steps of hydrothermal activity in Hrad Vallis, Mars. *Eos*, 99, 2018, https://doi.org/10.1029/2018EO100647.
7.51. Wolff, T., The hadal community, an introduction. *Deep Sea Res.*, 6, 2, 95–107, 1960.
7.52. Yayanos, A., Dietz, A.S., Von Boxtel, R., Isolation of a deep-sea barophilic bacterium and some of its growth characteristics. *Science*, 205, 808–810, 1979.
7.53. ZoBell, C.E. and Johnson, F.H., The influence of hydrostatic pressure on the growth and viability of terrestrial and marine bacteria. *J. Bacteriol.*, 57, 179–189, 1949.

Part II

EXTREMOPHILES IN SPACE (INTERNATIONAL SPACE STATION, OTHERS) AND SIMULATED SPACE ENVIRONMENTS

8

Challenging the Survival Thresholds of a Desert Cyanobacterium under Laboratory Simulated and Space Conditions

Daniela Billi

Department of Biology, Laboratory of Astrobiology and Molecular Biology of Cyanobacteria, University of Rome Tor Vergata, Rome, Italy

Abstract
Knowledge of the limit of life's adaptability to extreme environments is essential for identifying potentially habitable niches in planets and moons in our Solar System or in planetary systems around other stars. Dryness is one of the main life-threatening factors since water removal causes membrane phase transition and production of reactive oxygen species that cause lipid peroxidation, protein oxidation and DNA damage, which are lethal to most organisms. Anhydrobiotic cyanobacteria of the genus *Chroococcidiopsis* possess a remarkable resistance to desiccation and radiation; as such they have extended the limits of life as we know it in several new directions. Investigating the threshold of such resistance can help us in understanding not only the limits of life on Earth but also in assessing the potential habitability of Mars, of icy moons, and of exoplanets characterized by high doses of radiation and transient availability of liquid water. New insights have been provided by experiments exposing cyanobacteria to laboratory simulations that mimic planetary conditions, and also to real space conditions.

Keywords: Anhydrobiosis, radioresistance, astrobiology, *Chroococcidiopsis*, BOSS space experiment, BIOMEX space experiment, EXPOSE-R2 space mission

8.1 Introduction

Back in July 1976, the failure of NASA's Viking mission in finding evidence of life on the surface of Mars severely dampened the interest in searching for extraterrestrial life. A few months later, Imre Friedmann and Roseli Ocampo-Friedmann reported the discovery of endolithic communities in the Dry Valleys of Antarctica [8.29], an environment previously considered too hostile to support life. This finding revived hopes of finding life on Mars. One evening in 1978, Imre Friedmann, glued to the television, cried out to his wife/

Email: billi@uniroma2.it
Daniela Billi: https://en.wikipedia.org/wiki/Daniela_Billi, https://www.researchgate.net/profile/Daniela_Billi, https://scholar.google.com/citations?user=hTMjWykAAAAJ&hl=it

Joseph Seckbach and Helga Stan-Lotter (eds.) Extremophiles as Astrobiological Models, (185–196)

colleague Roseli: "Walter Cronkite is talking about us." The famous anchorman was reporting how a Florida State University microbiologist had accounted for the Viking mission's failure: they were looking in the wrong place, the soil.

The discovery of the cryptoendolithic communities in Dry Valleys in Antarctica was highly dramatic. At that time Imre Friedmann had no funding to go there himself, thus he asked Wolf Vishniac to look for colonized rocks in his stead. Unfortunately, he died while climbing in Antarctica, but some months later, Imre Friedmann received a box full of rocks from his widow. As it happened, they were colonized by cyanobacteria.

In 1989, while attending the International Association for Cyanophyte Research held in Plön, Germany, Imre Friedmann met Maria Grilli Caiola, a professor of botany at the University of Rome Tor Vergata (Figure 8.1). Shortly afterwards began a collaboration to decipher the basis of the desiccation tolerance of desert cyanobacteria of the genus *Chroococcidiopsis* [8.30]. Later on, in 1993, Maria Grilli Caiola was the supervisor of a PhD student, the author of this chapter, who had started working on desert strains of *Chroococcidiopsis*. Finally, in 2006, part of the Culture Collection of Microorganisms from Extreme Environments (CCMEE) set up by Imre Friedmann and Roseli Ocampo-Friedmann was transferred to the University of Rome Tor Vergata (Figure 8.2).

8.2 Endurance of *Chroococcidiopsis* Under Air-Drying and Space Vacuum

Dryness is one of the main factors threatening life since water removal is lethal to the majority of the organisms [8.20]. Upon air-drying, lipid membranes undergo phase transition that results in a decreased fluidity leading to cell leakage upon rehydration; carbohydrates, proteins, and nucleic acids undergo so-called Maillard reactions, i.e., amino-carbonyl reactions that create irreversible crosslinked products [8.20]. These structural changes are paralleled by functional changes, such as inhibited or altered enzyme activity, that cause the production of reactive oxygen species [8.44].

Figure 8.1 E. Imre Friedmann and Maria Grilli Caiola while attending the International Association for Cyanophyte Research in Plön, Germany, 1989. (Image credit: the author).

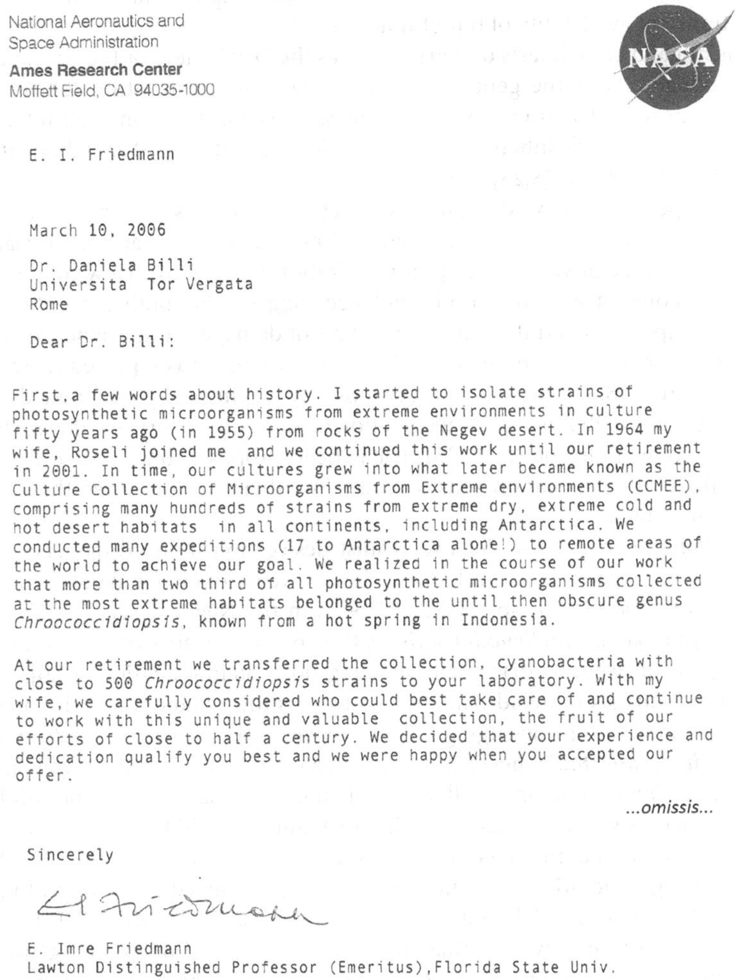
National Aeronautics and
Space Administration
Ames Research Center
Moffett Field, CA 94035-1000

E. I. Friedmann

March 10, 2006

Dr. Daniela Billi
Universita Tor Vergata
Rome

Dear Dr. Billi:

First, a few words about history. I started to isolate strains of photosynthetic microorganisms from extreme environments in culture fifty years ago (in 1955) from rocks of the Negev desert. In 1964 my wife, Roseli joined me and we continued this work until our retirement in 2001. In time, our cultures grew into what later became known as the Culture Collection of Microorganisms from Extreme environments (CCMEE), comprising many hundreds of strains from extreme dry, extreme cold and hot desert habitats in all continents, including Antarctica. We conducted many expeditions (17 to Antarctica alone!) to remote areas of the world to achieve our goal. We realized in the course of our work that more than two third of all photosynthetic microorganisms collected at the most extreme habitats belonged to the until then obscure genus *Chroococcidiopsis*, known from a hot spring in Indonesia.

At our retirement we transferred the collection, cyanobacteria with close to 500 *Chroococcidiopsis* strains to your laboratory. With my wife, we carefully considered who could best take care of and continue to work with this unique and valuable collection, the fruit of our efforts of close to half a century. We decided that your experience and dedication qualify you best and we were happy when you accepted our offer.

...omissis...

Sincerely

E. Imre Friedmann
Lawton Distinguished Professor (Emeritus), Florida State Univ.

Figure 8.2 E. Imre Friedmann's letter on the history of desert strains of *Chroococcidiopsis*. (Image credit: the author).

Only the so-called anhydrobiotes can withstand complete water removal [8.20]. To do so they accumulate trehalose, and/or sucrose, which according to the water replacement hypothesis, avoid the phase transition of lipid membranes and stabilize subcellular components [8.19]. In addition, the high intracellular trehalose and sucrose concentration is considered instrumental in vitrification, a process that allows anhydrobiotes to dry without dying [8.19] [8.49].

Despite the interest in anhydrobiotes, it is not known how long they can persist in the air-dried state. The appreciation of the desiccation threshold is relevant not only to understanding the limits of life on Earth, but also to assessing the habitability of planetary bodies with transient availability of liquid water [8.53].

In extreme hot and cold deserts on Earth, such as the Dry Valleys or the Atacama Desert in Chile, cyanobacteria of the genus *Chroococcidiopsis* dominate lithic communities in microscopic fissures or structural cavities of rocks, as well as the stone-soil interface (see photos in [8.6]) These rock-inhabiting communities are wetted and metabolically active only for a few hours per year [8.28] [8.51].

Laboratory experiments have demonstrated that desert strains of *Chroococcidiopsis* can be stored in a dried, dormant state for prolonged periods of time (at least 4 years); after that they return to vegetative growth upon rehydration [8.4] [8.26]. How they manage to do so is not yet completely understood. Evidence suggests that both the stabilization of dried cellular components and the limitation/repair of damage act in a synergic way [8.4]. In response to water stress, *Chroococcidiopsis* sp. CCMEE 029 was reported to accumulate trehalose and sucrose [8.32], these two sugars possibly playing a key role in its desiccation tolerance. In fact, although during salt acclimation most cyanobacteria synthesize trehalose and sucrose as compatible solutes [8.35], it has been proposed that, unlike desiccation-tolerant organisms, the desiccation-sensitive ones lose their viability when the glassy state has not yet been formed [8.12].

Chroococcidiopsis's first line of defense against desiccation-induced damage was identified as an efficient antioxidant mechanism that prevents oxidative damage to proteins [8.26]. When *Chroococcidiopsis* sp. CCMEE 029 from the Negev desert and the desiccation-sensitive cyanobacterium *Synechocystis* sp. PCC 6803 were air-dried for one year, only *Chroococcidiopsis* survived; its dried cells showing no protein oxidation [8.26]. How this cyanobacterium manages to avoid oxidative damage is unexplained, and the presence of a Mn-dependent system for scavenging reactive oxygen species, such as that reported for *Deinococcus radiodurans* [8.22], needs to be investigated. Insights into the mechanisms that stabilize air-dried *Chroococcidiopsis* will help explain their enhanced resistance under additional stresses such as space radiation and high vacuum. In addition, the presence of an efficient DNA repair system might contribute to restoring, upon rehydration, DNA lesions accumulated during desiccation or accumulated during prolonged dry storage. Indeed, the subcellular localization of DNA lesions in *Chroococcidiopsis* sp. CCMEE 029 stored in the desiccated state for 4 years revealed that survivors avoided and/or limited genome fragmentation as well as genome covalent modifications [8.4]. The absence of genome degradation in dried *Chroococcidiopsis* cells is significant when compared to the DNA degradation occurring in the bacterium *Deinococcus radiodurans* after 6 weeks of drying [8.38].

Desiccation tolerance is a prerequisite for coping with vacuum-induced dehydration. Due to its extreme dehydrating effect, space vacuum is considered one of the factors limiting survival under space conditions [8.34]. Therefore, to cope with this stress, upon rehydration anhydrobiotes must repair DNA lesions accumulated during long-term exposure [8.42].

During the ground-based simulations performed in the context of the EXPOSE-R2 space mission, dried cells of *Chroococcidiopsis* sp. CCMEE 029 were demonstrated to withstand 7 days of space vacuum [8.1]. Such a threshold was stretched further by their survival after 672 days in space vacuum outside the International Space Station [8.9] [8.10] [8.11]. This survival is relevant when compared with the lack of recovery in *Deinococcus radiodurans* after 9-month exposure to space vacuum during the EURECA space mission [8.25]. The endurance of *Chroococcidiopsis* under space vacuum proved that dried cells are efficiently stabilized; it was reported that if bacterial spores were not protected by internal or external substances, when exposed to high vacuum, they experienced dramatic changes in lipids, carbohydrates, proteins, and nucleic acids [8.42]. Indeed, the survival under space vacuum of bacterial spores, which do not accumulate trehalose and sucrose, increased when embedded in chemical protectants such as sugars [8.33].

The exposure of *Chroococcidiopsis* to space vaccum duirng EXPOSE-R2 space mission, caused no genome degradation as suggested by PCR-based assays, whose genomic-PCR profile showed bands ranging from 400 bp to 1000 bp, thus comparable to that of ground control [8.9] [8.10] [8.11]. Indeed, DNA lesions result in decreased/absent PCR amplicons because they block the DNA polymerase progression [8.37]. Nevertheless, DNA lesions were induced by space vacuum, as revealed by amplifying a 4000-bp DNA fragment [8.11]. The reason is that the amplification of a long target is impaired compared to that of a short target because the probability of encountering a blocking DNA lesion increases with the DNA template length [8.39]. The survivability of vacuum-exposed dried cells of *Chroococcidiopsis* upon rehydration revealed that the accumulated DNA damage did not exceed their repair potential.

8.3 Endurance of *Chroococcidiopsis* Under Laboratory Simulated and Space Radiation

Anhydrobiotes are known to cope with high doses of UV and ionizing radiation that do not occur in nature [8.21]. Indeed, radioresistance is considered a by-product of desiccation-tolerance mechanisms, avoiding oxidative damage common to desiccation and radiation [8.21]. Indeed, the difference between *Deinococcus radiodurans* and a radiation-sensitive bacterium such as *Escherichia coli* is based on the presence of an efficient antioxidant system [8.22].

When the radiation-tolerant *Chroococcidiopsis* sp. CCMEE 029 and the radiation-sensitive *Synechocystis* sp. 6803 were exposed to increasing doses of γ-rays, only *Chroococcidiopsis* survived, showing no significant increase in protein oxidation after 25 kGy [8.26]. This suggests that proteins, including those involved in DNA repair as reported for *D. radiodurans* [8.48], were not oxidized. Desert strains of *Chroococcidiopsis* were also reported to cope with doses of x-rays as high as 15 kGy [8.7]. *Chroococcidiopsis* sp. CCMEE 029 can withstand up to 13 kJ/m^2 of UVC radiation, showing a 10% survival rate after exposure to 300 J/m^2, a dose close to that of *Bacillus subtilis* spores (303 J/m^2; [8.41]) and lower than that of *D. radiodurans* (660 J/m^2; [8.3]).

The radioresisatnce treshold of *Chroococcidiopsis* was further extended when dried cells were exposed to radiation. During ground-based simulations of a Martian UV-flux, dried monolayers of *Chroococcidiopsis* sp. CCMEE 029 survived 15 kJ/m^2 [8.16]. This resistance was further extended when dried biofilms survived 1.5×10^3 kJ/m^2 of a Mars-like UV flux simulation performed during the EXPOSE-R2 space mission [8.2]. This result suggests a few hours of survival of radiation-tolerant life at Mars's equator [8.15]. In fact, solar UV radiation is immediately lethal due to its high energy being efficiently absorbed by biological macromolecules, albeit relatively easily shielded against [8.43]. Hence the scored survival of *Chroococcidiopsis* biofilms was ascribed to the shielding provided to the bottom-layer cells by the top-layer cells that were killed by the UV radiation [8.2].

When exposed to γ-radiation in the dried state, *Chroococcidiopsis* sp. CCMEE 029 survived a dose of 24 kGy but not of 46.88 kGy [8.50]. The observed 10% survival after 12 kGy of γ-rays was comparable to that of 12.5 kGy reported for dried cells of *D. radiodurans* [8.3]; their increased radioresistance was explained as due to structural advantages combined with reduced generation of reactive oxygen species and lower rates of potentially harmful chemical reactions [8.3]. When exposed to heavy ions, dried cells of *Chroococcidiopsis* sp. CCMEE 029 survived 2 kGy of Fe ions, the highest dose tested [8.50].

Since NASA's Mars Science Laboratory rover reported a dose rate of 76 mGy/year at the Gale Crater [8.31], it can be speculated that dried *Chroococcidiopsis* could cope with a radiation environment equivalent to 20,0000 years on the Martian surface. In addition, considering that galactic cosmic rays and solar particle events are one of the limiting factors for microbial survival in space [8.27], a prolonged *Chroococcidiopsis* survival in space can be expected.

Since ground-based simulations do not mimic the whole combination of space factors simultaneously, the exposure of terrestrial life forms under real space conditions is the only way to address key questions as to their endurance [8.18].

In the BOSS experiment (Biofilm Organisms Surfing Space) performed during the EXPOSE-R2 space mission, dried *Chroococcidiopsis* biofilms obtained from desert strains of *Chroococcidiopsis*, and including CCMEE 029, survived 18-month exposure to space conditions in low Earth orbit. During the space mission they received a total dose of 0.5 Gy of cosmic ionizing radiation combined with space vacuum (varying between 1.33×10^{-3} to 1.33×10^{-4} Pa) and solar radiation of about 4×10^2 kJ/m^2 [8.45]. During the EXPOSE-R2 space mission, dried *Chroococcidiopsis* biofilms were also exposed to Mars-like conditions, e.g., a Mars-like UV flux of about 4×10^2 kJ/ m^2 combined with a Mars-like atmosphere at 980 Pa [8.9] [8.45]. The occurrence of survivors within the biofilms was due to the shielding provided by top-cell layers to the bottom-cell layers and enhanced by the presence of abundant extracellular polymeric substances and additional features acquired upon drying [8.9].

In the BIOMEX experiment (BIOlogy and Mars EXperiment) performed during the EXPOSE-R2 space mission, dried cells of *Chroococcidiopsis* sp. CCMEE 029 were exposed to Mars-like conditions; survivors were scored only when samples were mixed with Martian soil simulants P-MRS (phyllosilicatic Martian regolith simulant) and S-MRS (sulfatic Martian regolith simulant) that mimic the supposed wetter (early) and dryer (late) Mars soil, respectively [8.10], or when mixed with lunar regolith analogue [8.11]. The scored genome integrity suggests that DNA might still be detectable after prolonged periods in microbes shielded by minerals. These results are relevant to future lunar astrobiology activities. Indeed, even if the Earth's natural satellite is too hostile for life, it provides a unique test bed for addressing key questions regarding the limits and adaptation of life, the testing of life-detection technologies and the development of life-support systems [8.14].

Overall, the results obtained for *Chroococcidiopsis* during the EXPOSE-R2 space mission also have implications for supporting the hypothesis that, if life ever appeared on Mars, during the planet's climatic history, a desiccation-, radiation-tolerant life form could have survived the transport from habitable niches subsequently turned uninhabitable, to new habitable ones [8.52]. Results demonstrate that when cells were shielded against UV radiation by minerals or by the biofilm organization, 18-month exposure to space vacuum and space radiation did not impair their survival. These data have further implications in the lithopanspermia scenario, e.g., the transport of microorganisms from one celestial body to another aboard planetary ejecta [8.40]. This phenomenon is constrained by three events: (i) the launch of microbe-bearing rocks from a donor planet into space; (ii) the transit through space to a recipient planet, and (iii) the entry into a recipient planet [8.34]. Indeed, since a large amount of terrestrial ejecta have reached the lunar surface 3.8–3.9 Gyr ago [8.36], a possible Earth-to-Moon transfer of microbial life inside rocks might have occurred. In the context of the Earth-to-Moon lithopanspermia, relevant implications depend on the possibility of the Moon having once been habitable [8.47].

Reshaping the boundaries of *Chroococcidiopsis*'s tolerance under extreme conditions has implications in the search for extraterrestrial life since it contributes to defining the habitability of Mars, of icy moons of the solar system and of planets orbiting other stars. When *Chroococcidiopsis*'s capability of surviving under subfreezing temperatures was tested by using sub-eutectic solutions of sodium sulfate, magnesium sulfate and sodium chloride at 258 K, 233 K and 203 K, survivors were scored at 258 K, whereas cell death occurred at 203 K [8.17]. This survival limit was further extended when air-dried cells were exposed to 193 K, suggesting that in potentially habitable icy moons vitrification might allow the survival of microbes transported from niches, which became non-habitable to new habitable ones [8.17].

8.4 The Use of *Chroococcidiopsis*'s Survival Thresholds for Future Astrobiological Experiments

The experiments carried out to date by using ground-based simulations and ESA exposure facilities have offered new insights into the survival of terrestrial life. However, these investigations were restricted to dried, ametabolic biological samples, analyzable only before and after the exposure to space and Mars-like simulations in low Earth orbit. Hence, *Chroococcidiopsis*'s desiccation tolerance represented a key feature prompting its long-term use in the available space exposure facilities. For example, the EXPOSE-R2 space mission lasted over 900 days from sample preparation, launch to the International Space Station and return to the ground [8.45]. A follow up of the BIOMEX results is foreseen in the BIOSIGN project (BIOSIGnatures and habitable Niches), an ESA ILSRA-2014-0019 selected astrobiology experiment to be performed outside the International Space Station [8.24].

The above-reported endurance of *Chroococcidiopsis* sp. CCMEE 029 under space, simulated Mars and icy-moon conditions supports its suitability for future experiments on platforms beyond the low Earth orbit, such as on the Moon's surface or orbiting around it, e.g., the Gateway. The Moon is outside the influence of the Earth's magnetosphere and its surface is subjected to high solar and galactic irradiation; as such it provides an invaluable test platform for astrobiological experiments [8.13] [8.23]. However, the key point here is to expose metabolically active cells that are not eradicated by exposure to deep space radiation. Previous ground-based experiments revealed that desert strains of *Chroococcidiopsis* can cope with ionizing radiation doses as high as 15 kGy [8.7]. Hence it is expected that, as a by-product of their extreme-tolerance, desert strains of *Chroococcidiopsis* have the potential to adapt to the space environment by avoiding and/or repairing the induced damage. Indeed, *Chroococcidiopsis* sp. CCMEE 029 was reported as avoiding oxidative damage and efficiently protect and/or repair its genome under different DNA-damaging conditions, including extreme desiccation [8.4], as well as UVC radiation [8.1] and ionizing radiation [8.26].

Remarkably, the evaluation of cellular responses to the space environment is constrained by the difficulty of performing real-time monitoring of a given cellular activity. The use of *Chroococcidiopsis* sp. CCMEE 029 as a model system is made possible by the availability of a genetic system favoring its manipulation and use of gene expression reporters [8.8]. Indeed, it has been reported that after 18 months of dry storage, intact plasmids could be recovered from dried cells of *Chroococcidiopsis* sp. CCMEE 029, suggesting that the strategies to stabilize its genome upon desiccation were efficient in protecting exogenous plasmid DNA [8.5]. This is a prerequisite to the use of this cyanobacterium in future space missions that might need to launch the biological sample in a dry state, its rehydration in orbit and long-term exposure under deep-space conditions. But this finding has also implications for a future use of *Chroococcidiopsis* as a chassis for space synthetic biology, e.g., endowing it with new abilities, such us producing compounds on demand or by using it to feed synthetic biology-enhanced cells for bioprinting of biomaterials [8.46].

Acknowledgments

Research for this chapter was supported by the Italian Space Agency (BOSS_Cyano grant 053-R.0; BIOMEX_Cyano grant 051-R.0; BIOSIGN_Cyano grant 2018-15-U.0; Life in Space grant 2019-3 U.0) and by the National Antarctic Research Program (MIDAS grant PNRA2016/00101).

References

8.1. Baqué, M., Viaggiu, E., Scalzi, G., Billi, D., Endurance of the endolithic desert cyanobacterium *Chroococcidiopsis* under UVC radiation. *Extremophiles*, 17, 1, 161–169, 2013.

8.2. Baqué, M., Scalzi, G., Rabbow, E., Rettberg, P., Billi, D., Biofilm and planktonic lifestyles differently support the resistance of the desert cyanobacterium *Chroococcidiopsis* under space and Martian simulations. *Orig. Life Evol. Biosph.*, 43, 4-5, 377–389, 2013.

8.3. Bauermeister, A., Moeller, R., Reitz, G., Sommer, S., Rettberg, P., Effect of relative humidity on *Deinococcus radioduran*' resistance to prolonged desiccation, heat, ionizing, germicidal, and environmentally relevant UV radiation. *Microb. Ecol.*, 61, 3, 715–722, 2011.

8.4. Billi, D., Subcellular integrities in Chroococcidiopsis sp. CCMEE 029 survivors after prolonged desiccation revealed by molecular probes and genome stability assays. *Extremophiles*, 13, 1, 49–57, 2009.

8.5. Billi, D., Plasmid stability in dried cells of the desert cyanobacterium *Chroococcidiopsis* and its potential for GFP imaging of survivors on Earth and in space. *Orig. Life Evol. Biosph.*, 42, 2-3, 235–245, 2012.

8.6. Billi, D., Desert cyanobacteria under space and planetary simulations: A tool for searching for life beyond Earth and supporting human space exploration. *Int. J. Astrobiol.*, 1–7, 2018.

8.7. Billi, D., Friedmann, E.I., Hofer, K.G., Grilli Caiola, M., Ocampo-Friedmann, R., Ionizing-radiation resistance in the desiccation-tolerant cyanobacterium *Chroococcidiopsis*. *Appl. Environ. Microbiol.*, 66, 4, 1489–1492, 2000.

8.8. Billi, D., Friedmann, E.I., Helm, R.F., Potts, M., Gene transfer to the desiccation-tolerant cyanobacterium *Chroococcidiopsis*. *J. Bacteriol.*, 183, 7, 2298–2305, 2001.

8.9. Billi, D., Staibano, C., Verseux, C., Fagliarone, C., Mosca, C., Baqué, M., Rabbow, E., Rettberg, P., Dried biofilms of desert strains of *Chroococcidiopsis* survived prolonged exposure to space and Mars-like conditions in low Earth orbit. *Astrobiology*, 19, 8, 1008–1017, 2019.

8.10. Billi, D., Verseux, C., Fagliarone, C., Napoli, A., Baqué, M., de Vera, J.-P., A desert cyanobacterium under simulated Mars-like conditions in low Earth orbit: Implications for the habitability of Mars. *Astrobiology*, 19, 2, 158–169, 2019.
8.11. Billi, D., Mosca, C., Fagliarone, C., Napoli, A., Verseux, C., Baqué, M., de Vera, J.-P., Exposure to low Earth orbit of an extreme-tolerant cyanobacterium as a contribution to lunar astrobiology activities. *Int. J. Astrobiol.*, 19, 1, 53–66, 2020.
8.12. Buitink, J. and Leprince, O., Glass formation in plant anhydrobiotes: Survival in the dry state. *Cryobiology*, 48, 3, 215–28, 2004.
8.13. Carpenter, J.D., Fisackerly, R., De Rosa, D., Houdou, B., Scientific preparations for lunar exploration with the European Lunar lander. *Planet. Space Sci.*, 74, 1, 208–223, 2012.
8.14. Cockell, C.S., Astrobiology – What can we do on the Moon? *Earth Moon Planets*, 107, 1, 3–10, 2010.
8.15. Cockell, C.S., Catling, D.C., Davis, W.L., Snook, K., Kepner, R.L., Lee, P., McKay, C.P., The ultraviolet environment of Mars: Biological implications past, present, and future. *Icarus*, 146, 2, 343–359, 2000.
8.16. Cockell, C.S., Schuerger, A.C., Billi, D., Friedmann, E.I., Panitz, C., Effects of a simulated martian UV Flux on the cyanobacterium, *Chroococcidiopsis* sp. 029. *Astrobiology*, 5, 2, 127–140, 2005.
8.17. Cosciotti, B., Balbi, B., Ceccarelli, A., Fagliarone, C., Mattei, E., Lauro, S.E., Di Paolo, F., Pettinelli, E., Billi, D., Survivability of anhydrobiotic cyanobacteria in salty ice: Implications for the habitability of icy worlds. *Life*, 9, 4, 86, 2019.
8.18. Cottin, H., Kotler, J.M., Billi, D., Cockell, C., Demets, R., Ehrenfreund, P., Elsaesser, A., d'Hendecourt, L., van Loon, J.J.W.A., Martins, Z. *et al.*, Space as a tool for astrobiology: Review and recommendations for experimentations in Earth orbit and beyond. *Space Sci. Rev.*, 209, 1-4, 83–181, 2017.
8.19. Crowe, J.H., Carpenter, J.F., Crowe, L.M., The role of vitrification in anhydrobiosis. *Annu. Rev. Physiol.*, 60, 73–103, 1998.
8.20. Crowe, L.M. and Crowe, J.H., Anhydrobiosis: A strategy for survival. *Adv. Space Res.*, 12, 4, 239–247, 1992.
8.21. Cox, M. and Battista, J.R., *Deinococcus radiodurans* – the consummate survivor. *Nat. Rev. Microbiol.*, 3, 11, 882–892, 2005.
8.22. Daly, M.J., Gaidamakova, E.K., Matrosova, V.Y., Vasilenko, A., Zhai, M., Venkateswaran, A., Hess, M., Omelchenko, M.V., Kostandarithes, H.M., Makarova, K.S. *et al.*, Accumulation of Mn(II) in *Deinococcus radiodurans* facilitates gamma-radiation resistance. *Science*, 306, 5698, 1025–1028, 2004.
8.23. de Vera, J.-P., Boettger, U., de la Torre, R., Sanchez, F.J., Grunow, D., Schmitz, N., Lange, C., Hübers, H.W., Billi, D., Baqué, M. *et al.*, Supporting Mars exploration: BIOMEX in low Earth orbit and further astrobiological studies on the Moon using Raman and PanCam technology. *Planet. Space Sci.*, 74, 1, 103–110, 2012.
8.24. de Vera, J.-P. and The Life Detection Group of BIOMEX/BIOSIGN, A systematic way to life detection: Combining field, lab and space research in low Earth orbit, in: *Biosignatures for Astrobiology. Advances in Astrobiology and Biogeophysics*, B. Cavalazzi and F. Westall (Eds.), pp. 111–122, Springer, Cham, 2019.
8.25. Dose, K., Bieger-Dose, A., Dillmann, R., Gill, M., Kerz, O., Klein, A., Meinert, H., Nawroth, T., Risi, S., Stridde, C. *et al.*, ERA-experiment "Space Biochemistry. *Adv. Space Res.*, 16, 8, 119–129, 1995.
8.26. Fagliarone, C., Mosca, C., Ubaldi, I., Verseux, C., Baqué, M., Wilmotte, A., Billi, D., Avoidance of protein oxidation correlates with the desiccation and radiation resistance of hot and cold desert strains of the cyanobacterium *Chroococcidiopsis*. *Extremophiles*, 21, 6, 981–991, 2017.

8.27. Ferrari, F. and Szuszkiewicz, E., Cosmic rays: A review for astrobiologists. *Astrobiology*, 9, 4, 413–436, 2009.
8.28. Friedmann, E.I., Kappen, L., Meyer, M.A., Nienow, J.A., Long-term productivity in the cryptoendolithic communities of the Ross desert, Antarctica. *Microb. Ecol.*, 25, 1, 51–69, 1993.
8.29. Friedmann, E.I. and Ocampo, R., Endolithic blue-green algae in the dry valleys: Primary producers in the antarctic desert ecosystem. *Science*, 193, 4259, 1247–1249, 1976.
8.30. Grilli Caiola, M., Ocampo-Friedmann, R., Friedmann, E.I., Cytology of long-term desiccation in the desert cyanobacterium *Chroococcidiopsis* (Chroococcales). *Phycologia*, 32, 5, 315–322, 1993.
8.31. Hassler, D.M., Zeitlin, C., Wimmer-Schweingruber, R.F., Ehresmann, B., Rafkin, S., Eigenbrode, J.L., Brinza, D.E., Weigle, G., Böttcher, S., Böhm, E. *et al.*, Mars' surface radiation environment measured with the Mars Science Laboratory's Curiosity rover. *Science*, 343, 6169, 1244797, 2013.
8.32. Hershkovitz, N., Oren, A., Cohen, Y., Accumulation of trehalose and sucrose in cyanobacteria exposed to matric water stress. *Appl. Environ. Microbiol.*, 57, 3, 645–648, 1991.
8.33. Horneck, G., Bücker, H., Reitz, G., Long-term survival of bacterial spores in space. *Adv. Space Res.*, 14, 10, 41–45, 1994.
8.34. Horneck, G., Klaus, D.M., Mancinelli, R.L., Space microbiology. *Microbiol. Mol. Biol. Rev.*, 74, 1, 121–156, 2010.
8.35. Klähn, S. and Hagemann, M., Compatible solute biosynthesis in cyanobacteria. *Environ. Microbiol.*, 13, 3, 551–562, 2001.
8.36. Koeberl, C., The late heavy bombardment in the inner solar system: Is there any connection to Kuiper Belt Objects? *Earth Moon Planets*, 92, 1-4, 79–87, 2003.
8.37. Kumar, A., Tyagi, M.B., Jha, P.N., Evidences showing ultraviolet-B radiation-induced damage of DNA in cyanobacteria and its detection by PCR assay. *Biochem. Biophys. Res. Commun.*, 318, 4, 1025–1030, 2004.
8.38. Mattimore, V. and Battista, J.R., Radioresistance of *Deinococcus radiodurans*: Functions necessary to survive ionizing radiation are also necessary to survive prolonged desiccation. *J. Bacteriol.*, 178, 3, 633–637, 1996.
8.39. Meyer, J.N., QPCR: A tool for analysis of mitochondrial and nuclear DNA damage in ecotoxicology. *Ecotoxicology*, 19, 4, 804–811, 2010.
8.40. Mileikowsky, C., Cucinotta, F.A., Wilson, J.W., Gladman, B., Horneck, G., Lindegren, L., Melosh, J., Rickman, H., Valtonen, M., Zheng, J.Q., Natural transfer of viable microbes in space. *Icarus*, 145, 2, 391–427, 2000.
8.41. Moeller, R., Setlow, P., Reitz, G., Nicholson, W.L., Roles of small, acid-soluble spore proteins and core water content in survival of Bacillus subtilis spores exposed to environmental solar UV radiation. *Appl. Environ. Microbiol.*, 75, 16, 5202–5208, 2009.
8.42. Nicholson, W.L., Munakata, N., Horneck, G., Melosh, H.J., Setlow, P., Resistance of *Bacillus* endospores to extreme terrestrial and extraterrestrial environments. *Microbiol. Mol. Biol. Rev.*, 64, 3, 548–72, 2000.
8.43. Nicholson, W.L., Schuerger, A.C., Setlow, P., The solar UV environment and bacterial spore UV resistance: Considerations for Earth-to-Mars transport by natural processes and human spaceflight. *Mutat. Res.*, 571, 1-2, 249–264, 2005.
8.44. Potts, M., Desiccation tolerance: A simple process? *Trends Microbiol.*, 9, 11, 553–559, 2001.
8.45. Rabbow, E., Rettberg, P., Parpart, A., Panitz, C., Schulte, W., Molter, F., Jaramillo, E., Demets, R., Weiß, P., Willnecker, R., EXPOSE-R2: the astrobiological ESA mission on board of the International Space Station. *Front. Microbiol.*, 8, 1533, 2017.
8.46. Rothschild, L.J., Synthetic biology meets bioprinting: Enabling technologies for humans on Mars (and Earth). *Biochem. Soc. Trans.*, 44, 4, 1158–1164, 2016.

8.47. Schulze-Makuch, D. and Crawford, I.A., Was there an early habitability window for Earth's Moon? *Astrobiology*, 18, 8, 985–988, 2018.
8.48. Slade, D. and Radman, M., Oxidative stress resistance in *Deinococcus radiodurans. Microbiol. Mol. Biol. Rev.*, 75, 1, 133–219, 2011.
8.49. Sun, W.Q. and Leopold, A.C., Cytoplasmic vitrification and survival of anhydrobiotic organisms. *Comp. Biochem. Physiol. Part A Physiol.*, 117, 3, 327–333, 1997.
8.50. Verseux, C., Baqué, M., Cifariello, R., Fagliarone, C., Raguse, M., Moeller, R., Billi, D., Evaluation of the resistance of *Chroococcidiopsis* spp. to sparsely and densely ionizing irradiation. *Astrobiology*, 17, 2, 118–125, 2017.
8.51. Warren-Rhodes, K.A., Rhodes, K.L., Pointing, S.B., Ewing, S.A., Lacap, D.C., Gómez-Silva, B., Amundson, R., Friedmann, E.I., McKay, C.P., Hypolithic cyanobacteria, dry limit of photosynthesis, and microbial ecology in the hyperarid Atacama Desert. *Microb. Ecol.*, 52, 3, 389–398, 2006.
8.52. Westall, F., Loizeau, D., Foucher, F., Bost, N., Betrand, M., Vago, J., Kminek, G., Habitability on Mars from a microbial point of view. *Astrobiology*, 13, 9, 887–897, 2013.
8.53. Wilhelm, M.B., Davila, A.F., Parenteau, M.N., Jahnke, L.L., Abate, M., Cooper, G., Kelly, E.T., García, V.P., Villadangos, M.G., Blanco, Y. *et al.*, Constraints on the metabolic activity of microorganisms in Atacama surface soils inferred from refractory biomarkers: Implications for Martian habitability and biomarker detection. *Astrobiology*, 18, 7, 955–966, 2018.

9

Lichens as Astrobiological Models: Experiments to Fathom the Limits of Life in Extraterrestrial Environments

Rosa de la Torre Noetzel[1]* and Leopoldo García Sancho[2]

[1]*Department of Earth Observation, Instituto Nacional de Técnica Aeroespacial (INTA), Madrid, Spain*
[2]*Department of Plant Biology II, Complutense University of Madrid (UCM), Madrid, Spain*

Abstract
Amongst extremophiles on Earth, lichens are one of the most resistant forms of life to harsh terrestrial environments, as well as some bacteria, cyanobacteria, cryptoendolithic communities, and lithic fungi. Many lichens species are well known as survivors under the most stressful environments, capable of coping with harsh climatic conditions, including extreme ranges of temperature, radiation and humidity. Lichens combine in a single, compact and long-term living system, the heterotrophic and autotrophic paths of life. Consequently, space simulation and space exposure experiments have focused on these complex organisms, demonstrating their high capacities of survival and recovery and becoming valuable models in astrobiological research to fathom the limits and limitations of terrestrial life. These experiments will contribute to the comprehension of the effects of space and planetary environment on biological systems for future human exploration, as well as deepen our knowledge on the limits, distribution and origin of life.

Keywords: Space experiments, Mars habitability, extremophiles, lichens, biomarkers, Lithopanspermia

9.1 Introduction

Astrobiological research using lichens as model systems has an outstanding origin. Selection of lichens for space experiments was a consequence of field studies done both in Antarctica and in high mountains [9.60] [9.13]. The results showed that they belong to the most resistant organisms able to colonize harsh environments, which include arid lands of both hemispheres, high mountains and polar regions. In these environments, lichens have

**Corresponding author*: torrenr@inta.es
Rosa de la Torre Noetzel: www.researchgate.net › profile › Rosa_Noetzel, wikipedia.org › wiki › BIOPAN, scholar.google.com › citations
Leopoldo García Sancho: sancholg@farm.ucm.es

Joseph Seckbach and Helga Stan-Lotter (eds.) Extremophiles as Astrobiological Models, (197–220)

to cope with high intensities of solar radiation [9.36] [9.68], long periods of desiccation, and a broad range of temperatures [9.62] [9.7], which are close to conditions at space or planetary environments, like Mars, Europa, and Enceladus. For instance, in Antarctica's Dry Valleys lichens are able to grow in a range of temperatures between 0 °C to -50° with an annual precipitation below 50 mm. Life that could exist in these conditions must be especially resistant, therefore the adapted organisms are referred to as extremophiles. This capacity is based on a successful life strategy dependent on the lichen symbiosis between a fungus (mycobiont) and an alga or cyanobacteria (photobiont), which allows them to colonize extreme terrestrial habitats where no other multicellular organism is able to survive. Additionally, the secondary metabolites produced in the uppermost layer of the entire lichen thallus function as a shield against UV radiation and act as light filters [9.66] [9.21] [9.23] [9.51] [9.68] [9.43].

Previous space-flight and ground-based experiments have shown that especially micro-colonies of bacteria, meristematic black fungi and symbiotic associations of microorganisms like lichens are able to survive and be reactivated after space experiments [9.30] [9.54] [9.61]. Bacterial strains, such as from *Bacillus subtilis* and *Deinococcus radiodurans*, have shown a certain radiation and vacuum tolerance [9.33]. During the BIOPAN V experiment [9.61], the lichens *Rhizocarpon geographicum* and *Xanthoria elegans* were analyzed after space exposure (15 days). The results led to the conclusion that the tested symbiotic eukaryotic associations of algae and fungi in the lichen were not seriously damaged and 70 to 90% of the tested lichens survived.

A following experiment—the Lithopanspermia experiment (ESA, project Nr. AO-2004-087)—was performed with lichens to test the likelihood of step 2 and step 3 of Lithopanspermia on board Biopan-6 and Stone-6 (September 2007); and after 10 days of exposure to harsh space conditions (solar UV radiation, cosmic radiation, space vacuum and extreme temperatures) showed the high resistance of these organisms (*R. geographicum, X. elegans* and *Circinaria gyrosa*) and their reproduction structures (ascospores). The vagrant lichen *C. gyrosa* showed a maximal recovery with an optimal gas interchange; the ascospores demonstrated a high germination rate (80–95%). In the case of *R. geographicum,* space desiccation had an inducing effect in germination, and *X. elegans* spores showed an earlier germination as observed in the controls. These phenomena were interpreted as "space activation of germination." The resting cells (akinetes) of Anabaena cyanobacteria [9.53] were also indicative of a high resistance to space. The high resistance of this previous lichen species, *R. geographicum,* were confirmed with LIFE (Lichens and Fungi Experiment), a long-term experiment for 18 months (2008–2009) on EXPOSE-E/ISS [9.54]. Results showed a 10–15% survival of these species to real space and simulated Mars conditions. These data suggest that lichens (photobiont and ascospores) and resting cells of cyanobacteria are suitable candidates for testing the concept of Lithopanspermia because they are extremely resistant to the harsh environment of outer space [9.15]. As was shown by analysis on akinetes (resting-state cells of cyanobacteria), cyanobacteria were also able to survive the low-earth orbit and simulated extraterrestrial conditions; while vegetative cells of *Chroococcidiopsis* sp. CCMEE 029 survive prolonged desiccation as well as a few minutes of exposure of an attenuated Martian UV flux and 4 hours of exposure to a Martian UV flux, if shielded under 1 mm of simulant Martian soil or gneiss. Mars simulation tests on archaea consisting of methanogenic species have also shown a remarkable level of survival and even methane production activities during simulated subsurface Mars conditions. The same has been observed for meristematic black fungi during the ground-based experiment in the facilities at DLR Cologne named EVT and during the EXPOSE-E experiment LIFE [9.54].

Other experiments as preparation for the space experiment BIOMEX (launched on the July 24th 2014), which were performed in the ESA ground-based facilities at DLR MUSC, at INTA PASC or at the Mars simulation facility (MSF) at DLR Berlin, have also shown a high survival activity of the lichens *Buelia frigida*, *C. gyrosa* and *Pleopsidium chlorophanum* [9.45] [9.58]. All the results demonstrated that the lichens were physiologically active and were able to germinate and grow. Specifically, the lichen species *C. gyrosa* [9.21] showed high resistance against space and Mars conditions when reproduced with scientific verification tests (SVT) for the EXPOSE-R2 mission which were conducted in the Space Simulation facilities of DLR in Cologne, Germany, which was necessary for the selection of the most promising biological organisms for the BIOMEX-ESA experiment. After the return of the mission in June 2016, the biological analysis showed a quick and complete recovery of the dark space control samples exposed to space vacuum and Mars-like atmosphere, in contrast to the slow and lower recovery of the samples directly exposed to space short-wavelength UV radiation. The experiment demonstrated that UV radiation combined with space vacuum caused the most damage. These findings contribute to the characterization of space- and Mars-resistant organisms relevant to Mars habitability, and will contribute to answering questions on the likelihood of the Lithopanspermia hypothesis [9.72], which will be of relevance for planetary protection issues.

9.2 Survival of Lichens in Outer Space

This chapter is an overview of research conducted over the past 19 years and that is currently being conducted with lichens in the Earth's orbit or upon re-entering the Earth's atmosphere, simulating a meteorite impact. It focuses on the most relevant developments and results for future space exploration programs. It describes the relevance of the use of the space radiation environment and the research in the field of astrobiology for understanding basic biological mechanisms, such as (i) the biological effects of the radiation field in space, and the likelihood of interplanetary transport of microorganisms (lichen, algae, fungi, cyanobacteria, halophilic bacteria) via meteorites, and the application-oriented aspects, such as the detection of biomarkers. The results will contribute to installation of an electronic biosignature data base for the most robust organisms, with application in future exploration projects focused on the search for life in the universe.

The questions related to these experiments were:

1) What are the effects of space environment on these symbiotic organisms and how can they contribute to know more about limits of life in space and on other planets?
2) Do they have any chance to survive and grow under current Mars conditions?
3) Could they survive a hypothetical interplanetary transfer in space (Phase II of Lithopanspermia) on meteorites and a reentry and landing on another planet (Phase III of Lithopanspermia)?

These questions have been answered by exposing lichens to the outer space environment or to selected parameters of it and studying their responses after retrieval.

In this chapter, first the space environment is explored, then the exposure platforms common for both chemistry and biology are presented, and then an overview of current and past astrobiology experiments with lichens is included. Finally, a brief mention is given to studies using terrestrial ground chambers that simulate selected parameters of outer space, such as space vacuum and solar extraterrestrial UV radiation, and Mars-like conditions, such as Mars atmospheric pressure and composition, and Mars UV radiation, as well as the use of heavy accelerators to simulate cosmic rays or polychromatic UV sources.

9.3 Space Environment: Relevance in Space Science

Lichens were selected as ideal organisms for space experiments and were included as part of payloads of the European Space Agency (ESA) space missions due to their ability to survive extreme conditions. Experiments with lichens in space were performed using Earth-orbiting robotic spacecrafts, e.g., the Russian Foton satellites [9.17] and the International Space Station (ISS) [9.2]. Arriving in space without any protection, lichens are confronted with an extremely hostile environment, characterized by an intense radiation field of galactic and solar origin, high vacuum, extreme temperatures, and microgravity. The scenario of an interplanetary transfer phase, that is the exposure to the space environment [9.29] [9.50] during a hypothetical interplanetary transfer, implies an environment characterized by a high vacuum, an intense radiation of galactic and solar origin and extreme temperatures. The pressures in interplanetary transfer prevail down to 10^{-14} but near a body the pressure could be higher due to outgassing. Low Earth Orbit (LEO) has pressures between 10^{-6} to 10^{-4} Pa. Molecular oxygen and nitrogen are part of this environment, as well as highly reactive oxygen and nitrogen atoms [9.33].

9.3.1 Space Radiation

The radiation environment of our solar system is contained by solar and galactic components. The galactic cosmic radiation of our solar system is composed of protons (85%), electrons, α-particles (14%) and heavy ions (1%) of charge $Z>2$, the so-called HZE particles (High charge Z and high E). The solar particle radiation, which is emitted in solar wind and during solar particle events, is composed of 90–95% protons, 5–10% α-particles and a low quantity of heavy ions. In LEO the radiation dose can be higher than in interplanetary space, due to the trapping of protons and electrons by the geomagnetic field.

The spectrum of electromagnetic radiation ranges from short-wavelength X-rays to radio frequencies. Solar irradiance amounts to 1.360 W m^{-2}, the solar constant, at the distance of the Earth to the Sun (1 AU). This radiation is composed of an infrared fraction (45%), a visible fraction (48%) and only 7% of the ultraviolet range. Solar UV radiation can be divided into three spectral ranges: UVC (200 to 280 nm), contributing 0.5% to the whole solar electromagnetic spectrum; UVB (280 to 315 nm), contributing 1.5%; and UVA (315 to 400 nm), contributing 6.3%. Although the UVC and UVB regions make up only 2% of the entire solar extraterrestrial irradiance, they are mainly responsible for the high lethality of extraterrestrial solar radiation to microorganisms that have been exposed to it [9.8], due to the high absorption of those wavelengths by DNA, the decisive target for inactivation and mutation induction within that UV range [9.33].

In order to determine the biological effectiveness of environmental UV radiation, E_{eff}, spectral data are multiplied with an action spectrum of a relevant photobiological reaction. For example, for DNA damage [9.66], the effectiveness of environmental UV radiation follows the equation

$$E_{eff} = \int E_{\lambda}(\lambda) \cdot S_{\lambda}(\lambda) d\lambda \tag{9.1}$$

where $E_{\lambda}(\lambda)$ = the solar spectral irradiance and $S_{\lambda}(\lambda)$ = the spectral sensitivity or action spectrum for a critical biological effect. The biologically effective irradiance E_{eff} is given in the parameter $(W/m^2)_{eff}$.

9.3.2 Temperature

The temperature of a body in space is determined by the absorption and emission of energy, depending also on its position with respect to the Sun and other orbiting bodies, and also depending on its surface, size, mass, and albedo. In Earth's orbit, the energy sources include solar radiation (1.360 W m^{-2}), the Earth's albedo (480 W m^{-2}), and terrestrial radiation (230 W m^{-2}). When an object orbits a planet, the Sun can be shaded due to its passing the planet's night side. Therefore, the temperature of a body in space can be extreme, very high and very low, as experiments in LEO have shown (240 K-320 K). For the ISS, the tolerable temperature limits are defined, with +120 °C as the highest value and-120 °C as the lowest value [9.2].

9.3.3 Microgravity

Even in low Earth orbit, the force of gravity is not actually removed. Rather, without the significant equal and opposite resistance needed to impart weight, an orbiting object simply experiences a continuous state of gravity-induced free fall (i.e., accelerated motion) around the Earth. This state is what is commonly referred to as "weightlessness" or "microgravity," as attributed to a relative frame of reference in which an object appears to "float" inside the spacecraft [9.38] [9.70].

9.4 Biological Effects of Space

9.4.1 Extraterrestrial Solar UV Radiation

One of the most harmful parameters of space is solar UV radiation, as results showed with viruses, bacterial and fungal spores at desiccated status [9.28]. Among all organisms tested, only lichens (*Rhizocarpon geographicum, Xanthoria elegans, and Circinaria gyrosa*) maintained full viability after 2 weeks in outer space, whereas all other test systems were inactivated by orders of magnitude [9.61] [9.15].

Using optical filters and spores of *Bacillus subtilis* as a biological UV dosimeter, it was found that the current ozone layer reduces the biological effectiveness of solar UV by 3 orders of magnitude. If shielded against solar UV, spores of *B. subtilis* were capable of surviving in space for up to 6 years, especially if embedded in clay or meteorite powder (artificial meteorites) [9.34]. The data support the likelihood of interplanetary transfer of microorganisms within meteorites, the so-called Lithopanspermia hypothesis.

In experiments, when spores of *B. subtilis* were exposed to the complete spectrum of extraterrestrial UV radiation they were killed within seconds [9.29], which is about a thousand times more efficient than the UV radiation at the Earth surface due to the high protection of our biosphere by the stratospheric ozone layer. The cause of this high biological efficiency is the high energetic UV-C (190–280 nm) and vacuum UV region (<190 nm) that is directly absorbed by the DNA. This high biological efficiency is increased if bacterial spores or other organisms, such as lichens, are simultaneously exposed to solar UV radiation and space vacuum. Alterations in DNA will appear, generating photoproducts in extremophile organisms, like spores of *B. subtilis:* (i) thymidine decomposition products (*cis-syn* and *trans-syn*) cyclobutadithymine; (ii) 5,6-dihydro-5-thymine (TDHT); (iii) DNA protein crosslinking [9.28] [9.42].

9.4.2 Cosmic Radiation

The effect of γ-rays on lichens and other astrobiological models has rarely been tested. As a part of primary cosmic radiation [9.3] [9.40], cosmic γ-ray bursts [9.39] [9.71], and solar flares [9.1], ionizing γ-radiation is a crucial factor of non-terrestrial environments such as space, Mars and other planetary bodies. Concerning space exposure conditions of LEO, corpuscular ionizing radiation is effectively shielded by the Earth's magnetic field. On the one hand, recent long duration exposure experiments in the EXPOSE-E facility in LEO reported doses of ionizing radiation of ca. 0.2 Gy/a that do not affect lichen viability [9.3] [9.5]. On the other hand, it is difficult to reproduce and apply ionizing radiation in ground-based space simulation facilities, especially in combination with other space parameters.

9.4.3 Space Vacuum

Space vacuum has a high dehydrating effect, therefore it has been included as one of the factors that restrain interplanetary transfer of life [9.52]. Nevertheless, 70% of bacterial and fungal spores have actually survived short-term experiments (e.g., 14 days) without protection [9.29].

Vacuum exposure in space produces damage based on the extreme desiccation. Exposure of cells in vacuum that are not protected by internal or external substances suffer dramatic changes in lipids, carbohydrates, proteins and nucleic acids. Functional changes, such as altered membrane permeability, enzyme activity, decreased energy production, alteration of genetic information, as well as structural changes, such as deformation of bilayers, Maillard reactions, i.e., amino-carbonyl reactions giving irreversible polymerization of the biomolecules, are common. Vacuum also produces lethal or mutagenic damage to DNA, first demonstrated with spores of *B. subtilis* during the Spacelab-1 mission [9.28]. Additionally, tandem double-base changes at specific sites on DNA have been observed [9.49]. Accumulation of this damage would take place during long-term exposure to space vacuum, due to the fact that DNA is not able to repair anhydrobiosis. Survival then depends on the efficiency of the repair systems after germination.

9.4.4 Microgravity

Microgravity is another space parameter that cells can escape only if hosted in an on-board 1×g reference centrifuge. The responses of microorganisms to microgravity at the molecular and cellular levels are discussed in "Role of Gravity in Basic Biological Processes" in [9.35]. It should be noted that microgravity mostly interferes with growing or metabolizing cells; however, in the exposure experiments, dry layers of cells, algal cells, fungal conidia, or bacterial spores were used that were not affected by the gravity environment.

9.5 Current and Past Astrobiological Facilities for Experiments with Lichens

To expose lichens and other extremophile organisms to outer space or to Mars simulated conditions, the European Space Agency (ESA) invented different facilities for integration on the outer part of the spacecraft. For short-duration exposure experiments (10 to 12 days), BIOPAN (Figure 9.1), a pan-shaped facility with the experimental hardware integrated in its lid and bottom, mounted on the Russian Earth-orbiting satellite Foton was developed in 1992. When reaching the correct orbit, BIOPAN opens its lids 180° to expose the experiments in the bottom and in the lid to space. Space parameters were registered by solar-UV, radiation and temperature sensors. During ascent and re-entry, the lid is hermetically closed and the whole facility is covered by an ablative heat shield. The first experiment with lichens, called Lichens was launched in May 2005 (Table 9.1) with the BIOPAN-5 facility (Figure 9.2) on the Foton-M2 satellite mission. The next opportunity for experiments with lichens in outer space was provided in 2007 by the Foton-M3 mission (September 2007). The experiment was called LITHOPANSPERMIA, which studied the interplanetary transfer and entry processes of epi- and endolithic microbial communities.

For long-duration experiments in outer space, ESA developed the last exposure units called EXPOSE facilities (Figures 9.3 and 9.4), which consist of three trays, each housing four compartments similar to previous exposure devices, such as ERA. The first one was

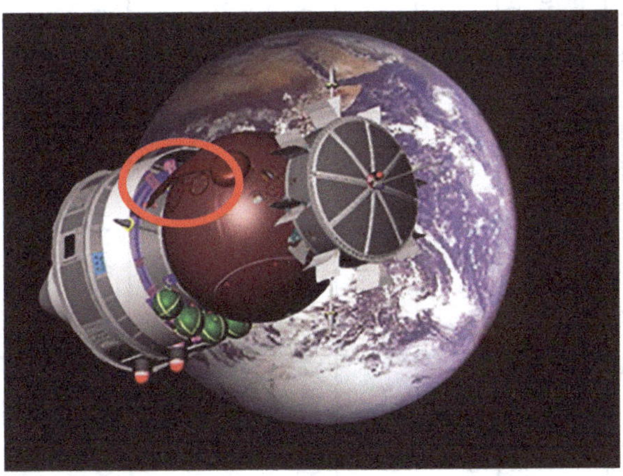

Figure 9.1 Foton satellite with BIOPAN-5 facility orbiting the Earth. (Courtesy of ESA).

Table 9.1 Exposure conditions during the BIOPAN and during the EXPOSE experiments.

Mission	Flight date day/ month/year	Exposure period (lid open) days	Solar UV [kJ m^{-2}]	Space vacuum simulated Mars atm. [Pa] (+ 95% CO2)	Temperature variations [°C]	Cosmic radiation [mGy]
BIOPAN 5	31/05 to 15/05/2005	15 days	>170 nm 22 473	In orbit: 0.133×10^{36}	−21.7 to +21.8	3.16[b]
BIOPAN 6	14/09 to 26/09/2007	10 days	>170 nm 11 800	In orbit: 0.133×10^{-3}	−23.0 to +27.0	100 ± 20[a] 4 ± 0.5[b]]
EXPOSE E	07/02/2008 to 02/02/2009	540 days	Biologically active range >200 nm to <400 nm 634 MJm^{-2} (100% transmission insolated)	Space vacuum: 10^{-4} Mars atm. pressure: 980	-21.5 to +59.6	190
EXPOSE R2	23/07/2014 to 02/03/2016	469 days	Biologically active range >200 nm to <400 nm: 536 ± 116 MJm^{-2}	Space vacuum: 1.33×10^{-3} to 1.33×10^{-4} Mars atm. pressure: 980	-20.9 to + 57.98 °C	1000

Figure 9.2 Opened Biopan-5 facility with experiment LICHENS surrounded in red. (Courtesy of ESA).

Figure 9.3 EXPOSE-E on the International Space Station with experiment LIFE. (Courtesy of ESA).

Figure 9.4 EXPOSE-R2 on the International Space Station with experiment BIOMEX. (Courtesy of ESA).

EXPOSE-E, which was launched with the STS-122 mission in 2008 and mounted by extravehicular activity to the European Technology Exposure Facility (EuTEF) platform.

One tray of EXPOSE-E was preserved for experiments on prebiotic chemical evolution, the second tray provides outer space conditions (space vacuum and a solar UV spectrum of >110 nm), and the third tray provides conditions that simulate the Mars surface climate (600 Pa of pressure, 95% CO_2, and solar UV of >200 nm).

EXPOSE-E remained in space for more than 1 year. The second EXPOSE facility, EXPOSE-R, was launched in November 2008 and remained attached to the URM-D platform, an external ISS facility at the Russian Zvezda module, for about 1 year. Before being launched into space, all EXPOSE experiments were tested in carefully designed ground simulation experiments and in an experiment sequence test, using the Planetary and Space Simulation Facilities (PSI) at the German Aerospace Center (DLR) [9.55]. EXPOSE-E and EXPOSE-R housed a total of 13 different experiments that were performed through international cooperation [9.2] [9.32]. On July 23, 2014, the Progress cargo spacecraft 56P was launched from Baikonur to the International Space Station (ISS), carrying EXPOSE-R2, the third ESA (European Space Agency) EXPOSE facility. It was the second EXPOSE on the outside platform of the Russian Zvezda module, with six international space experiments. Two experiments with lichens were exposed to space and to Mars-like conditions, the experiment LIFE at EXPOSE-E and the experiment BIOMEX (Figure 9.4) at EXPOSE-R2 [9.21].

9.6 Space Experiments with Lichens

9.6.1 The BIOPAN Experiments: Exposure to Outer Space Conditions

Lichens Experiment

Since the beginning of space science many organisms have been exposed to outer space to assess the limits of life (1966, Gemini IX and XII missions) [9.44]. The first experiments with lichens were conducted on the Foton-M2 satellite in 2005 [9.61]. Various species of epilithic lichens, *Rhizocarpon geographicum* and *Xanthoria elegans* (Figure 9.5), that colonize high mountain regions (South and Central Mountain Systems of Spain and the Alps) were exposed to LEO parameters, extraterrestrial radiation (different ranges of solar-UV radiation, and cosmic radiation) and space vacuum, as well as to temperature cycles (Table 9.1), during a 2-week flight, with other extremophile organisms integrated at BIOPAN. So far, only a few microbial systems are known that cope with the full space environment; these were the lichens *R. geographicum* and *X. elegans* (Figure 9.5), which had completely restored their vitality after retrieval from the 2-week BIOPAN-5 flight (Figure 9.11) [9.15].

The lichens experiment [9.61] showed for the first time, the high survival capacity by photosynthetic performance (chlorophyll *a* fluorescence) of both epilithic lichens to low

Figure 9.5 Lichens (a) *Rhizocarpon geographicum* and (b) *Xanthoria elegans*. (Image credit: the authors).

Earth orbit conditions. Further multimicroscopic analyses revealed no detectable ultrastructural changes in most of the algal and fungal cells of the lichens (Figure 9.6). Both systems—the lichens and the halophilic cyanobacteria—possess UV-screening capacities, namely, the thick and dense cortex, with UV-screening pigments rhizocarpic and parietin phenolic acids as an upper layer and as indigenous protection, for the lichen symbiotic system [9.18] [9.23].

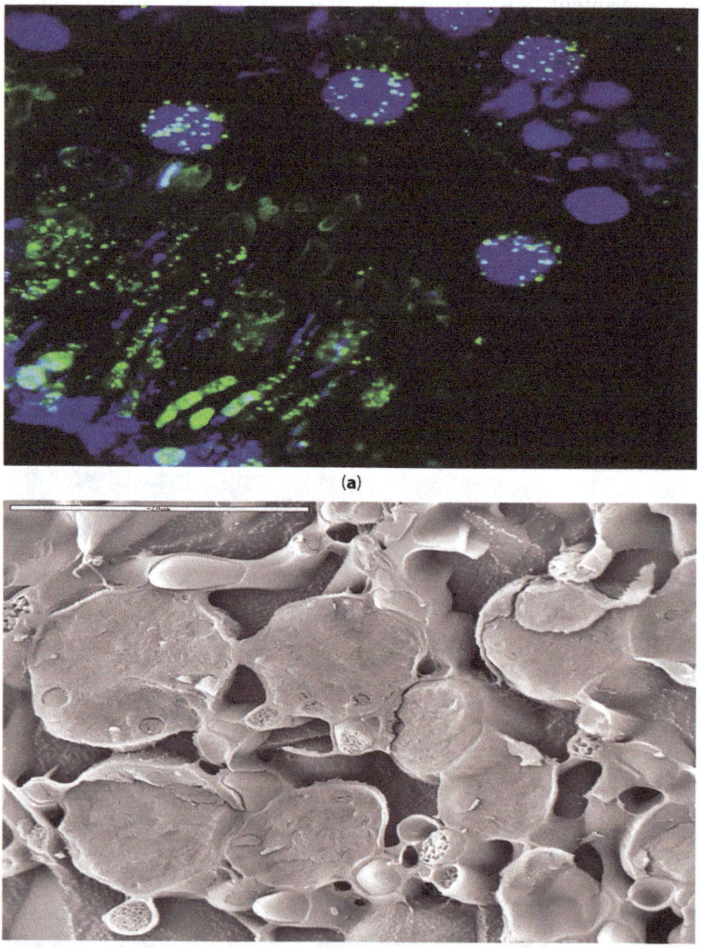

(a)

(b)

Figure 9.6 (a) Confocal scanning laser microscopy image showing high vitality of algal and fungal cells; (b) Transmission electronic microscopy image showing cryofractured intact fungal and algal cells. (Courtesy of Wierzchos, Ascaso, and De los Ríos).

LITHOPANSPERMIA and LITHOPANSPERMIA/STONE Experiments

Studies were conducted on the interplanetary transfer and entry processes of epi- and endolithic microbial communities.

After one successful experiment, lichens were exposed again to space conditions in the LITHOPANSPERMIA experiment on ESA's Foton-M3 mission in 2007.

The LITHOPANSPERMIA experiment, as part of the payload of the BIOPAN-6 mission, included the same rock-colonizing lichen species, *Rhizocarpon geographicum* and *Xanthoria elegans,* a new vagrant lichen species called *Circinaria gyrosa* (from *Aspicilia fruticulosa,* ren. see Sohrabi [9.67] (Figure 9.7)), originally from the steppic highlands in Spain, and endolithic and endoevaporitic communities of cyanobacteria and bacteria with their natural rock substrate. In order to assess the shielding effect of the pigments, samples with intact thallus as well as those without pigments, which were extracted by use of pure acetone 100%, according to Gauslaa and Ustvedt [9.22], were used for the flight experiment. Also, samples of *R. geographicum* and *X. elegans* with fruiting bodies (containing ascospores), collected at the highest insolated areas of the Alps, were integrated. The samples were exposed to selected space conditions, that is, to space vacuum, cosmic radiation, and different spectral ranges of solar UV radiation (>110, >200, >290, or >400 nm, respectively). After 10 days at space, the samples were analyzed, determining the survival capacity by photosynthetic activity, the ultrastructure, the vitality by live/dead staining dye and CO_2 exchange rates, and the germination capacity of the ascospores [9.15]. The results showed optimal rates of net photosynthesis (Figure 9.11) and dark respiration with the gas exchange method, and a general high and fast recovery of the biological activity (chlorophyll *a* fluorescence) of lichens after 72 h of reactivation. Although the ultrastructure showed probable stress-induced changes, the conclusion was that these lichens were able to repair all space-induced

Figure 9.7 The vagrant lichen species *Circinaria gyrosa.* (Courtesy of L. García Sancho).

damages. Additionaly, germination and growth capacity of ascospores of *R. geographicum* and *X. elegans* showed a germination index between 75% and 90% for *X. elegans* (compared to 97% of the ground control) and an even higher germination index (81–100%) for *R. geographicum* flight samples, with an indirect correlation to the UV solar radiation wavelengths. An earlier start of germination of 1–2 days of the space-exposed samples in relation to the ground controls was observed.

But the cyanobacterial communities within their natural habitat, such as halite or limestone/sandstone, were less resistant in comparison with the lichens, probably due to the extraction from their natural environment. However, resting-state cyanobacteria are likely to confer greater resistance to space conditions. Akinetes are resting-state cyanobacteria that can survive decades of desiccating conditions [9.53], do not require sunlight, and are not limited to the upper surface of the rock.

This experiment provided the opportunity to learn that organisms that adapted to tolerate extreme conditions on our planet, like epilitihic and vagrant lichens, and resting-state phototrophic organisms of microbial communities, could resist a "hypothetical" interplanetary travel through space. Ascospores in the fruiting bodies and photobiont cells from deep layers of the lichen thallus, as well as akinetes, or similar resting-state phototrophic organisms [9.10] [9.11], are the best protected cells and candidates for surviving natural transfer between planets [9.48]. Together they comprise all prerequisites for generating a new lichen thallus on the new planet, if conditions are favorable for life.

Another experiment took part in the Foton-M3 mission in 2007 (Figure 9.8). The aim of the LITHOPANSPERMIA/STONE experiment [9.15] was to identify the physical, chemical and biological modifications caused by atmospheric entry in meteorites and to their possibly embedded microorganisms.

We used the lichen *R. geographicum* as biological rock-inhabiting model system to test its ability to survive during re-entry into the Earth's atmosphere. After atmospheric re-entry, the granite of the Stone sample was transformed into a glassy, nearly homogeneous material, with several friction striae. None of the lichen cells survived this re-entry process. The data suggest that lichens are suitable candidates for testing the concept of lithopanspermia

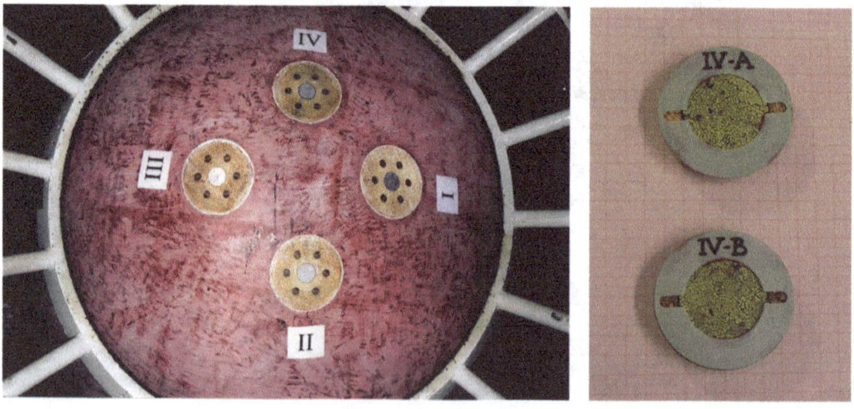

Figure 9.8 LITHOPANSPERMIA/STONE Experiment: (left) Location at the heat shield of the Foton-M4 capsule; (right) Hardware of glass textolite on back of lichen sample. (Photo courtesy of ESA).

because they are extremely resistant to the harsh environment of outer space. The more critical event is the atmospheric re-entry after being captured by a planet. Experiments simulating the re-entry process of a microbe-carrying meteoroid did not show any survivors.

The space experiments reported here have shown that lichens can survive short-term, i.e., 10–14 days, exposure to outer space. From laboratory experiments mimicking impact ejection at shock pressures up to 50 GPa it is known that photobiont cells and particularly ascospores of *X. elegans* are able to survive impact ejection [9.69] [9.34].

9.6.2 EXPOSE Facility On Board the ISS: Experiments at Space- and Mars-Like Conditions

The high resistance of lichens demonstrated by the previous experiments have led us to test the next objective: their survival after long-term exposure to environmental conditions in space and on Mars.

LIFE Experiment

The 2008 Lichens and Fungi Experiment (LIFE) [9.54] [9.63] was a long-term experiment integrated on the EXPOSE-E facility (Columbus laboratory) of the ISS. This experiment was relevant to demonstrate for the first time, the limits of the survival capacity of rock-dwelling organisms and microbial communities from harsh climatic areas: Antarctic cryptoendolithic (dwelling inside rocks) communities in their natural sandstone, microcolonial black cryptoendolithic fungi (*Cryomyces antarcticus* and *Cryomyces minteri*) isolated from Antarctic sandstone [9.64] [9.65], and high mountain epilithic lichens (*R. geographicum* and *X. elegans*) (Figure 9.9). The biological samples were accommodated in wells

Figure 9.9 LIFE experiment: Lichens thalli, lichens mycobionts cryptoendolithic communities in rocks within the wells of carrier integrated in tray 1 of EXPOSE-E, and fungi samples located in the hardware wells. (Photo courtesy of ESA [9.12]).

(1.4 cm in diameter) as shown in Figure 9.9 [9.12]. During the space mission they were exposed either to the full space environment (vacuum from 10^{-7} to 10^{-4} Pa, fluctuations of temperature between −21.5 and +59.6 °C, cosmic ionizing radiation up to 190 mGy, and solar extraterrestrial electromagnetic radiation up to 6.34×10^8 Jm^{-2}) or they were shielded from insolation. After 1.5 years in space, the samples were retrieved and their viability was investigated. During the mission the sun-exposed LIFE samples had been exposed to 1,879 eSCh (estimated Solar Constant hours) [9.56]. The samples were also kept in simulated Mars atmosphere and pressure (1.6% Ar, 0.15% O_2, 2.7% N_2, 370 ppm H_2O, in CO_2 at a pressure of 103 Pa), exposed to simulated Mars UV radiation, cutting-off the spectrum of solar extraterrestrial electromagnetic radiation at a wavelength of $\lambda = 200$ nm by use of optical filters. Some samples were insolated with a reduced irradiance by three orders of magnitude by using neutral density filters, with fluences of 9.19×105 Jm^{-2}, below a 0.1% transmission neutral density filter. In addition, dark flight samples were located beneath the insolated ones.

The viability techniques used were: (a) photosynthetic activity of the lichen photobiont, (b) colony-forming ability of *C. antarcticus* and *C. minteri* and isolation of microorganisms from rock fragments, (c) fraction of DNA amplified from cells with undamaged membrane (PMA, propidium-monoazide assay) of *C. antarcticus* and *C. minteri* and of cryptoendolithic communities inside sandstone fragments, and (d) viability of *X. elegans* and the fungus of the lichen (mycobiont, cultured without the algal symbiont and dried) by means of vital staining.

After flight, the circumpolar and alpine red lichen *X. elegans* was able to retain its photosynthetic activity almost completely after 14 days in space [9.54] [9.5] [9.6], as well as after 565 days (this study). Particularly the mycobiont seems to play a fundamental role in maintenance of viability of the entire lichen system, because 565 days in space also appears to not have any effect on its physiological activity (Figure 9.11). However, at EXPOSE-E the anhydrobiotic (anabiotic) state of desiccated lichens [9.26] and the potential to avoid damage caused by insolation with wavelengths >110 nm (space vacuum) and >200 nm (Mars analogue conditions), respectively, were focused on. But surprising was the demonstration of the higher viability rates Fv/Fm of the Mars analogue samples after shorter reactivation than the space-exposed samples. Depending on the length of the desiccation event the re-activation time needed was more prolonged, as indicated by the time-dependent increase of Fv/Fm in the (space) vacuum samples. The LIVE/DEAD staining results for the PB indicated less difference between the viability of Mars analogue and space vacuum samples than the chlorophyll *a* fluorescence data pointed to, indicating that Fv/Fm measurements should not be the only test for viability of the photobiont in the lichen thallus.

Lichens, as small and self-supporting ecosystems, are considered as possible earliest colonizers of terrestrial environments [9.27]. As autotrophic systems, they could theoretically be able to colonize any environment, terrestrial or extraterrestrial, if some moisture and a few elements are present. Up till now, the observed high survival rate of these pioneer organisms with such peculiar ecology raises new possibilities for unexpected and surprising scenarios in the discussion of the colonization of other planets through natural processes.

The experiment showed that lichens are one of the most resistant organisms to space and Mars conditions, and therefore suitable candidates for life experiments (Lithopanspermia theory), as well as for planetary protection efforts.

BIOMEX Experiment

The Biology and Mars Experiment (BIOMEX) was launched on board a Progress spacecraft on July 2014 from Baikonur, which was one of four experiments that made up the EXPOSE-R2 facility attached to the external part of the ISS platform on the Zvezda module (Figure 9.10) (http://www.astrobio.net/topic/origins/extreme-life/biomex-exploring-mars-low-earth-orbit/#sthash.KlrD8vy7.dpuf).

The main aim of BIOMEX (Figure 9.10) is to search for information about which type of basic traces should be searched for when looking for signs of life on extraterrestrial habitats—in particular on the Moon and on Mars. The second aim, secondary to the main goal, is to identify space-resistant organisms that may be relevant for life support systems on spacecraft, lunar and Mars bases.

Additional objectives are:

1) Expose organisms (bacteria, cyanobacteria, biofilms, archea, lichens, black fungi and mosses) and biomolecules to different parameters of the space environment, including analogue minerals and the atmospheric conditions on Mars.
2) Compare the exposed biological material with ground controls, focusing on the most basic features (e.g., vitality, viability, tissue and cellular morphology), on DNA integrity and on the level of dehydration (H_2O content), as well as the composition or alteration of the organic and inorganic material used in the experiments by Raman spectroscopy.

The data obtained by the investigation on biomarkers after being exposed to Mars-like conditions will support the analysis of data obtained during future instrumental detection operations in future space missions on Mars (i.e., ExoMars).

The results of the present BIOMEX experiment demonstrate that, after a long exposure period (18 months) at the EXPOSE-R2 facility on the ISS, it remains possible for *C. gyrosa* (Figure 9.7) to recover PSII activity [9.37]. Although PSII activity, as determined by measurement of chlorophyll *a* fluorescence, was further reduced in Sun-exposed samples at space and Mars conditions (Figure 9.11), with a maximal mean reduction of 21% in comparison of pre-flight values under vacuum, at Mars-like conditions activity decreased to a mean of 44.25%; however, dark-control samples increased their activity from an average

Figure 9.10 (left) Zvezda module at International Space Station; (right) EXPOSE-R2 facility with experiment BIOMEX. (Photo courtesy of ESA).

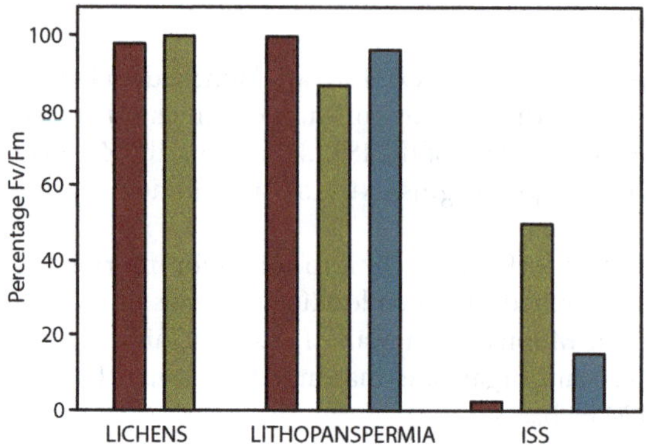

Figure 9.11 Percentage of activity of photosystem II of lichen species measured after exposure to space conditions in relation to activity measured before flight in three different experiments. Red, *Rhizocarpon geographicum*; yellow, *Xanthoria elegans*; cyan, *Circinaria gyrosa*. (Image credit: the authors).

45.41% under space vacuum to an average of 85.71% under Mars atmospheric composition and pressure. These data indicate a lower recovery tendency of the photosynthetic system of lichens following exposure to space conditions, including UV irradiation, compared with earlier reports of simulation experiments [9.58] [9.24] [9.25], including the Lichens experiment on board BIOPAN-5 [9.61] [9.62], the LITHOPANSPERMIA experiment [9.15] on board BIOPAN-6, and the LIFE experiment on the EXPOSE-E of the ISS [9.46].

Electron and fluorescence microscopy techniques also revealed that the viability of *C. gyrosa* exposed to space conditions decreased in comparison to those exposed to Mars-like environment [9.9]. Moreover, differences were observed between samples positioned at level 1 and level 2. In general, TEM and FESEM observations [9.16] showed that samples at level 2 (basal samples) were slightly affected in their morphology/ultrastructure by the exposure conditions. In contrast, cellular ultrastructure alterations were clearly evident for samples exposed to space radiation, which led to a shrinkage process. The cell walls were irregularly shaped and debris of the major organelles were visible.

Biomolecular analyses (PCR and sequencing) also revealed that the viability of *C. gyrosa* exposed to space conditions decreased in comparison to those exposed to the Mars-like environment [9.58]. In contrast to these studies, the biogeochemical variations have been examined with Raman spectroscopic analyses to look for possible degradation of cell surfaces and pigments which were in contact with terrestrial rocks, and martian analogue regolith. The identification of a biomarker which is called whewellite (calcium oxalate) and other organic compounds and mineral products of the biological activity of the astrobiology model system *C. gyrosa*, were done [9.4].

The results will contribute to the installation of an electronic biosignature data base for the most robust organisms, with application in future exploration projects focused on the search for life in the universe.

9.7 Simulation Studies

Simulation tests are necessary to check the suitability of a new organism to extraterrestrial conditions and also as a fit check before space flight. Studies performed with lichens at simulated planetary conditions and at ionizing campaigns (simulated space- and Mars-like conditions), have demonstrated high capacities of resistance and recovery [9.15] [9.18] [9.19] [9.20] [9.58] [9.59] [9.5] [9.47] which have identified them as valuable model organisms in astrobiological research, contributing to the better understanding of the adaptability of extremophile organisms to harsh environments, as well as to estimate the habitability of a planet, like Mars, surface.

But, up to date, space and planetary simulation chambers are not able to provide a variety of environmental extreme parameters that exist in space. In particular the combination of low pressure, temperature oscillations, short wavelength UV (ultraviolet) and the complex cosmic ionizing radiation is only available in space and provides a research environment for astrobiology that resembles other solar system bodies and moons closer than laboratories on Earth can reproduce [9.57].

9.8 Summary and Conclusions

One remarkable advantage of lichens is that a single thallus contains the two main metabolic pathways of life (heterotrophic and autotrophic) linked to two very different phylogenetic lines.

- Lichens are ideal astrobiological model systems to test the limits of survival under extraterrestrial conditions.
- Epilithic and endolithic lichens offer opportunities to contribute to the Lithopanspermia theory showing how life can survive interplanetary transfer. If sufficiently shielded by meteorite-like material, endolithic microorganisms may survive the journey through space and finally be captured, with their host rock, by another planet. If the planet has an atmosphere, which might make it habitable [9.41], the outer shell of the rock is subjected to very high temperatures during atmospheric re-entry. So far, microorganisms that were embedded in 2-cm-thick rocks were fitted at the outer surface of a re-entry capsule (Stone facilities of Foton missions), thereby simulating the entry of a meteorite, did not survive this entry process [9.73]. In order to perform a more realistic simulation of the entry of meteorites carrying endolithic microorganisms, rocks of different composition, e.g., basalts, should be tested and the technical concept of Stone needs to be modified so that larger, and in particular thicker rock samples, can be accommodated. However, for safety reasons samples thicker than 2 cm cannot be installed on a Foton capsule. The very thick and resistant heat shield of the Foton spacecraft will safely do its job even if equipped with recesses to accommodate Stone as it is now, but a further deepening of the recesses in the heat shield to make room for thicker Stone samples has justifiably been rejected by the Foton engineers and safety people [9.15].
- The experiments (LICHENS, Foton M2-2005 and LITHOPANSPERMIA, Foton M3-2007) [9.61] [9.15] have shown that short-term exposures to LEO (low earth orbit) conditions do not have a significant deleterious effect on their vitality; meanwhile, long-term experiments show more relevant changes

(LIFE experiment, EXPOSE-E, ISS, 2008-2009) [9.54] (Figure 9.11). This could be explained by the lichen's structural and physiological features, which allowed them to survive the harsh environmental conditions on Earth (high UV radiation, extreme temperatures and desiccation), as well as in space (high UV space radiation, UV vacuum, extreme temperatures and microgravity).
- Both symbionts have showed different survival levels, which have been proved by aposymbiotic laboratory cultures.
- Some lichen species after space flight still retain the ability to reproduce sexually.

These experiments have shown that lichens are one of the most resistant organisms to space and Mars conditions, and therefore are suitable candidates for life experiments (Lithopanspermia theory), as well as for planetary protection efforts.

9.9 Future Possibilities and Recommendations

Recommendations will focus on experimentation with suitable extremophile lichens for long-term exposures to space conditions. A deeper study of the survival capacity on molecular level with metagenomics will help to decipher how space- and Mars-like conditions affect the genomic expression and protein synthesis.

These experiments will contribute to the comprehension of the effects of space and planetary environment on biological systems for future human exploration, as well as deepen our knowledge on the limits, distribution and origin of life.

References

9.1. Ajello, M. *et al.*, Impulsive and long duration high-energy gamma-ray emission from the very bright 2012 March 7 solar flares. *Astrophys. J.*, 789, 1, 20, 2014.
9.2. Bagnoli, P., Sabbatini, M., Horneck, G., Astrobiology experiments in low Earth orbit: Facilities, instrumentation, and results, in: *Complete Course in Astrobiology*, G. Horneck and P. Rettberg (Eds.), pp. 273–319, Wiley-VCH, New York, NY, USA, 2007.
9.3. Berger, T., Hajek, M., Bilski, P., Körner, C., Vanhavere, P., Reitz, G., Cosmic radiation exposure of biological test systems during the EXPOSE-E mission. *Astrobiology*, 12, 5, 387–392, 2012.
9.4. Böttger, U., Meessen, J., Martinez-Frias, J., Hübers, H.-W., Rull, F., Sánchez, F.J., de la Torre, R., de Vera, J.P., Raman spectroscopic analysis of the calcium oxalate producing extremotolerant lichen Circinaria gyrosa. *Int. J. Astrobiol.*, 13, 1, 19–27, 2013.
9.5. Brandt, A., de Vera, J.P., Onofri, S., Ott, S., Viability of the lichen *Xanthoria elegans* and its symbionts after 18 months of space exposure and simulated Mars conditions on the ISS. *Int. J. Astrobiol.*, 14, 411–425, 2015.
9.6. Brandt, A., Meeßen, J., Jänicke, R.U., Raguse, M., Ott, S., Simulated space radiation: Impact of four different types of high-dose ionizing radiation on the lichen *Xanthoria elegans*. *Astrobiology*, 17, 136–144, 2017.

9.7. Büdel, B., Darienko, T., Deutschwitz, K., Dojani, S., Friedl, T., Mohr, K.L., Salisch, M., Reisser, W., Weber, B., Southern African biological soil crusts are ubiquitous and highly diverse in drylands, being restricted by rainfall frequency. *Microb. Ecol.*, 57, 229–247, 2009.
9.8. Cockell, C.S. and Horneck, G., The history of the UV radiation climate of the Earth-theoretical and space-based observations. *Photochem. Photobiol.*, 73, 447–451, 2001.
9.9. Cockell, C.S., Schuerger, A., Billi, D., Friedmann, E.I., Panitz, C., Effects of a simulated Martian UV flux on the Cyanobacterium, *Chroococcidiopsis* sp.029. *Astrobiology*, 5, 127–140, 2005.
9.10. Cockell, C.S., Brack, A., Wynn-Williams, D.D., Baglioni, P., Brandstätter, F., Demets, R., Edwards, H.G., Gronstal, A.L., Kurat, G., Lee, P. *et al.*, Interplanetary transfer of photosynthesis: An experimental demonstration of a selective dispersal filter in planetary island biogeography. *Astrobiology*, 7, 1–9, 2007.
9.11. Cockell, C.S., The interplanetary exchange of photosynthesis. *Orig. Life Evol. Biosph.*, 38, 1, 87–104, 2008.
9.12. Cottin, H., Kotler, J.M., Billi, D., Cockell, C.S., Demets, R., Ehrenfreund, P., Elsaesser, A., d'Hendecourt, L., van Loon, J.W.A., Martins, Z. *et al.*, Space as a tool for astrobiology: Review and recommendations for experimentations in Earth orbit and beyond. *Space Sci. Rev.*, 209, 83, 2017.
9.13. de la Torre, R., Horneck, G., Sancho, L.G., Scherer, K., Facius, R., Urlings, T., Rettberg, P., Reina, M., Pintado, A., Photoecological characterisation of an epilithic ecosystem at a high mountain locality (Central Spain), in: *Proceedings of the Second European Workshop on Exo/Astrobiology*, ESA SP-518 (Eds.), pp. 443–445, 2002.
9.14. de la Torre, R., Garcia-Sancho, L., Horneck, G., Adaptation of the lichen *Rhizocarpon geographicum* to harsh high-altitude conditions: Relevance to a habitable Mars, in: *ROME: Response of Organisms to the Martian Environment*, C.S. Cockell, G. Horneck, ESA SP.1299 (Eds.), pp. 145–150, European Space Agency, ESTEC, Noordwijk, The Netherlands, 2007.
9.15. de la Torre, R., Sancho, L.G., Horneck, G., de los Ríos, A., Wierzchos, J., Olsson-Francis, K., Ott, S., Survival of lichens and bacteria exposed to outer space conditions–results of the Lithopanspermia experiments. *Icarus*, 208, 735–748, 2010.
9.16. de la Torre, R., Ortega, M.V., López-Ramirez, M.R., Miller, A.Z., Bassy, O., Granja, C., Cubero, B., Jordão, L., Martínez Frías, J., Rabbow, E. *et al.*, The BIOMEX Experiment on-board the International Space Station: Biomolecular- and bio-geochemical changes of lichens exposed to space- and to Mars-like conditions. *Conference: 42nd COSPAR Scientific Assembly, Conference Paper*, 2016.
9.17. Demets, R., Schulte, W., Baglioni, P., The past, present and future of Biopan. *Adv. Space Res.*, 36, 311–316, 2005.
9.18. de Vera, J.P., Horneck, G., Rettberg, P., Ott, S., The potential of the lichen symbiosis to cope with the extreme conditions of outer space. I: Influence of UV radiation and space vacuum on the vitality of lichen symbiosis and germination capacity. *Int. J. Astrobiol.*, 1, 285–293, 2003.
9.19. de Vera, J.P., Möhlmann, D., Butina, F., Lorek, A., Wernecke, R., Ott, S., Survival potential and photosynthetic activity of lichens under Mars-like conditions: A laboratory study. *Astrobiology*, 10, 2004.
9.20. de Vera, J.P., Horneck, G., Rettberg, P., Ott, S., The potential of the lichen symbiosis to cope with the extreme conditions of outer space. II: Germination capacity of lichen ascospores in response to simulated space conditions. *Adv. Space Res.*, 33, 1236–1243, 2004.
9.21. de Vera, J.P., Alawi, M., Backhaus, T., Baqué, M., Billi, D., Böttger, U., Berger, T., Bohmeier, M., Cockell, C., Demets, R. *et al.*, Limits of life and the habitability of Mars: The ESA Space Experiment BIOMEX on the ISS, 2019.

9.22. Gauslaa, Y. and Ustvedt, E.M., Is parietin a UV-B or a blue-light screening pigment in the lichen *Xanthoria parietina*? *Photochem. Photobiol. Sci.*, 2, 424–432, 2003.
9.23. Gauslaa, Y. and Solhaug, K.A., Photoinhibition in lichens depends on cortical characteristics and hydration. *Lichenologist*, 36, 133–143, 2004.
9.24. Gómez, F., Mateo-Martí, E., Prieto-Ballesteros, O., Martín-Gago, J., Amils, R., Protection of chemolithoautotrophic bacteria exposed to simulated Mars environmental conditions. *Icarus*, 209, 482–487, 2010.
9.25. Gómez-Gómez, J.M., Estébanez, B., Sanz-Arranz, A., Mateo-Martí, E., Medina, J., Survival of moss reproductive structures under simulated Martian environmental conditions and extreme thermal stress: Vibrational spectroscopy study and astrobiological implications. *J. Astrobiol. Outreach*, 4, 2, 2016.
9.26. Green, T.G.A., Sancho, L.G., Pintado, A., Ecophysiology of dessication/rehydration cycles in mosses and lichens, in: *Plant Dessication Tolerance, Ecological Studies*, U. Lüttge, E. Beck, D. Bartels (Eds.), pp. 89–120, Springer, Berlin, 2011.
9.27. Heckman, D.S., Geiser, D.M., Brooke, R., Eidell, B.R., Stauffer, R.L., Kardos, N.L., Hedges, S.B., Molecular evidence for the early colonization of land by fungi and plants. *Science*, 293, 1129–1133, 2001.
9.28. Horneck, G., Bücker, H., Reitz, G., Requardt, H., Dose, K., Martens, K.D., Mennigmann, H.D., Weber, P., Microorganisms in the space environment. *Science*, 225, 226–228, 1984.
9.29. Horneck, G., Responses of *Bacillus subtilis* spores to space environment: Results from experiments in space. *Orig. Life Evol. Biosph.*, 23, 37–52, 1993.
9.30. Horneck, G., Bücker, H., Reitz, G., Long-term survival of bacterial spores in space. *Adv. Space Res.*, 14, 41–45, 1994.
9.31. Horneck, G., Exobiological experiments in Earth orbit. *Adv. Space Res.*, 22, 317–326, 1998.
9.32. Horneck, G., Impact of microgravity on radiobiological processes and efficiency of DNA repair. *Mutat. Res.*, 430, 221–228, 1999.
9.33. Horneck, G. and Baumstark-Khan, C. (Eds.), *Astrobiology. The Quest for the Conditions of Life*, Springer-Verlag, Berlin Heidelberg New York, 2002.
9.34. Horneck, G., Stöffler, D., Ott, S., Hornemann, U., Cockell, C.S., Moeller, R., Meyer, C., de Vera, J.P., Fritz, J., Schade, S., Artemieva, N.A., Microbial rock inhabitants survive hypervelocity impacts on Mars-like host planets: First phase of lithopanspermia experimentally tested. *Astrobiology*, 8, 17–44, 2008.
9.35. Horneck, G., Klaus, D.M., Mancinelli, R.L., Space microbiology. *Microbiol. Mol. Biol. Rev.*, 74, 1, 121, 2010.
9.36. Hughes, K.A. and Lawley, B., A novel Antarctic microbial endolithic community within gypsum crusts. *Environ. Microbiol.*, 5, 555–565, 2003.
9.37. Kappen, L. and Valladares, F., Opportunistic growth and desiccation tolerance, the ecological success of the poikilohydrous strategy, in: *Functional Plant Ecology*, F. Pugnaire and F. Valladares (Eds.), pp. 8–65, Taylor & Francis, New York, 2007.
9.38. Klaus, D.M., Gravitational influence on biomolecular engineering processes. *Gravit. Space Biol. Bull.*, 17, 51–65, 2004.
9.39. Klebesadel, R.W., Strong, I.B., Olson, R.A., Observations of gamma-ray bursts of cosmic origin. *Astrophys. J.*, 182, L85–L83, 1973.
9.40. Kraushaar, W.L., Clark, G.W., Garmire, G.P., Borken, R., Higbie, P., Leong, V., High-energy cosmic gamma-ray observations from the OSO-3 satellite. *Astrophys. J.*, 177, 341–348, 1972.
9.41. Lammer, H., Bredehöft, J.H., Coustenis, A., Khodachenko, M.L., Kaltenegger, L., Grasset, O., Prieur, D., Raulin, F., Ehrenfreund, P., Yamauchi, M. *et al.*, What makes a planet habitable. *Astron. Astrophys. Rev.*, 17, 181–249, 2009.

9.42. Lindberg, C. and Horneck, G., Action spectra for survival and spore photoproduct formation of *Bacillus subtilis* irradiated with short wavelength (200–300 nm) UV at atmospheric pressure and in vacuo. *J. Photochem. Photobiol. B: Biol.*, 11, 69–80, 1991.
9.43. Lohézic-Le Dévéhat, F., Legouin, B., Couteau, C., Boustie, J., Coiffard, L., Lichenic extracts and metabolites as UV filters. *J. Photochem. Photobiol. B: Biol.*, 120, 17–28, 2013.
9.44. Mancinelli, R.L., White, M.R., Rothschild, L.J., Biopan-survival. I. Exposure of the osmophilic *Synechococcus* sp. (Nageli) and *Haloarcula* sp. to the space environment. *Adv. Space Res.*, 22, 3, 327–334, 1998.
9.45. Meeßen, J., Sánchez, F.J., Brandt, A., Balzer, E.M., de la Torre, R., Sancho, L.G., de Vera, J.P., Ott, S., Extremotolerance and resistance of lichens: Comparative studies on five species used in astrobiological research I. Morphological and anatomical characteristics. *Orig. Life Evol. Biosph.*, 43, 283–30, 2013.
9.46. Meeßen, J., Sánchez, F.J., Sadowsky, A., de la Torre, R., Ott, S., de Vera, J.P., Extremotolerance and resistance of lichens: Comparative studies of fives species use in astrobiological research II. Secondary lichen compounds. *Orig. Life Evol. Biosph.*, 43, 501–526, 2014.
9.47. Meeßen, J., Wuthenow, P., Schille, P., Rabbow, E., de Vera, J.P., Ott, S., Resistance of the lichen *Buellia frigida* to simulated space conditions during the preflight tests for BIOMEX – Viability assay and morphological stability. *Astrobiology*, 15, 8, 601–615, 2015.
9.48. Mileikowsky, C., Cucinotta, F., Wilson, J.W., Gladman, B., Horneck, G., Lindegren, L., Melosh, J., Rickman, H., Valtonen, M., Zheng, J.Q., Natural transfer of viable microbes in space, part 1: From Mars to Earth and Earth to Mars. *Icarus*, 145, 391–427, 2000.
9.49. Munakata, N., Saitou, N., Takahashi, N., Hieda, K., Morohoshi, F., Induction of unique tandem-base change mutations in bacterial spores exposed to extreme dryness. *Mutat. Res.*, 390, 189–195, 1997.
9.50. Nicholson, W.L., Munakata, N., Horneck, G., Melosh, H.J., Setlow, P., Resistance of *Bacillus* endospores to extreme terrestrial and extraterrestrial environments. *Microbiol. Mol. Biol. Rev.*, 64, 548–572, 2000.
9.51. Nybakken, L., Solhaug, K.A., Bilger, W., Gauslaa, Y., The lichens *Xanthoria elegans* and *Cetraria islandia* maintain a high protection against UV-B radiation in Arctic habitats. *Oecologia*, 140, 211–216, 2004.
9.52. Nussinov, M.D. and Lysenko, S.V., Cosmic vacuum prevents radiopanspermia. *Orig. Life Evol. Biosph.*, 13, 2, 153–164, 1983.
9.53. Olsson-Francis, K., de la Torre, R., Towner, M.C., Cockell, C.S., Survival of akinetes (resting-state cells of cyanobacteria) in low Earth orbit and simulated extraterrestrial conditions. *Orig. Life Evol. Biosph.*, 39, 565–579, 2009.
9.54. Onofri, S., de la Torre, R., de Vera, J.P., Ott, S., Zucconi, L., Selbmann, L., Scalzi, G., Venkateswaran, K.J., Rabbow, E., Sánchez Iñigo, F.J., Horneck, G., Survival of rock-colonizing organisms after 1.5 years in outer space. *Astrobiology*, 12, 508–516, 2012.
9.55. Rabbow, E., Rettberg, P., Panitz, C., Drescher, J., Horneck, G., Reitz, G., SSIOUX–space simulation for investigating organics, evolution and exobiology. *Adv. Space Res.*, 36, 297–302, 2005.
9.56. Rabbow, E., Rettberg, P., Barczyk, S., Bohmeier, M., Parpart, A., Panitz, C., Horneck, G., von Heise-Rotenburg, R., Hoppenbrouwers, T., Willnecker, *et al.*, EXPOSE-E: An ESA astrobiology mission 1.5 years in space. *Astrobiology*, 12, 374–386, 2012.
9.57. Rabbow, E., Rettberg, P., Parpart, A., Panitz, C., Schulte, W., Molter, F., Jaramillo, E., Demets, R., Weiss P. and Willnecker, R., EXPOSE-R2: The Astrobiological ESA mission on board of the International Space Station. *Front. Microbiol.*, 8, 1533, 2017.
9.58. Sanchez, F.J., Mateo-Martí, E., Raggio, J., Meessen, J., Martinez-Frias, J., Sancho, L.G., de la Torre, R., The resistance of the lichen *Circinaria gyrosa* (nom. provis.) towards simulated

Mars conditions - a model test for the survival capacity of an eukaryotic extremophyle. *Planet. Space Sci.*, 72, 102–110, 2012.

9.59. Sánchez, F.J., Meeßen, J., Ruiz, M.C., Leopoldo, G., Ott, S., Vílchez, C., Horneck, G., Sadowsky, A., de la Torre, R., UV-C tolerance of symbiotic *Trebouxia* sp. in the space-tested lichen species *Rhizocarpon geographicum* and *Circinaria gyrosa*: role of the hydration state and cortex/screening substances. *Int. J. Astrobiol.*, 13, 1–18, 2014.

9.60. Sancho, L.G., Schroeter, B., del Prado, R., Ecophysiology and morphology of the globular erratic lichen *Aspicilia fruticulosa* (Eversm.) FLAGEY from Central Spain. New aspects in cryptogamic research. *Bibl. Lichenol.*, 75, 137–147, 2000.

9.61. Sancho, L.G., de la Torre, R., Horneck, G., Ascaso, C., de los Rios, A., Pintado, A., Schuster, M., Lichens survive in space: Results from the 2005 LICHENS experiment. *Astrobiology*, 7, 443–454, 2007.

9.62. Sancho, L.G., de la Torre, R., Pintado, A., Lichens, new and promising material from experiments in Astrobiology. *Fungal Biol. Rev.*, 22, 103–109, 2008.

9.63. Scalzi, G., Selbmann, L., Zucconi, L., Rabbow, E., Horneck, G., Albertano, P., Onofri, S., LIFE experiment: isolation of cryptoendolithic organisms from Antarctic colonized sandstone exposed to space and simulated Mars conditions on the international space station. *Orig. Life Evol. Biosph.*, 42, 2-3, 253–62, 2012.

9.64. Selbmann, L., de Hoog, G.S., Mazzaglia, A., Friedmann, E.I., Onofri, S., Fungi at the edge of life, cryptoendolithic black fungi from Antarctic deserts. *Stud. Mycol.*, 51, 1–32, 2005.

9.65. Setlow, R.B., The wavelengths in sunlight effective in producing skin cancer: A Theoretical analysis. *Proc. Natl. Acad. Sci.*, 71, 9, 3363–6, 1974.

9.66. Solhaug, K.A., Gauslaa, Y., Nybakken, L., Bilger, W., UV-induction of sun-screening pigments in lichens. *New Phytol.*, 158, 91–100, 2003.

9.67. Sohrabi, M., *Taxonomy and phylogeny of the manna lichens and allied species (Megasporaceae)*, PhD thesis, Publications in Botany from the University of Helsinki, 2012.

9.68. Stivaletta, N. and Barbieri, R., Endoliths in terrestrial arid environments: Implications for astrobiology, in: *From Fossils to Astrobiology*, pp. 319–333, Springer, Dordrecht, 2009.

9.69. Stöffler, D., Horneck, G., Ott, S., Hornemann, U., Cockell, C.S., Moeller, R., Meyer, C., de Vera, J.P., Fritz, J., Artemieva, N.A., Experimental evidence for the potential impact ejection of viable microorganisms from Mars and Mars-like planets. *Icarus*, 186, 585–588, 2007.

9.70. Van Loon, J.J.W.A., The gravity environment in space experiments, in: *Biology in Space and Life on Earth, Effects of Spaceflight on Biological Systems*, E. Brinckmann (Ed.), pp. 17–32, Wiley-VCH, Weinheim, Germany, 2007.

9.71. Verter, F., Cosmic gamma-ray bursts. *Phys. Rep.*, 81, 293–349, 1982.

9.72. Westall, F. and de la Torre Noetzel, R., Meteorites: Stones with stowaways?, in: *Looking up: Europe´s Quiet Revolution in Microgravity Research*, pp. 8–15, Scientific American, 2008.

10

Resistance of the Archaeon *Halococcus morrhuae* and the Biofilm-Forming Bacterium *Halomonas muralis* to Exposure to Low Earth Orbit for 534 Days

Stefan Leuko[1]*, Helga Stan-Lotter[2], Greta Lamers[3], Sebastian Sjöström[4], Elke Rabbow[1], Andre Parpart[1] and Petra Rettberg[1]

[1]German Aerospace Centre, Radiation Biology Department, Köln, Germany
[2]University of Salzburg, Department of Biosciences, Salzburg, Austria
[3]ETH Zürich, Department of Health Sciences and Technology, Schwerzenbach, Switzerland
[4]ETH Zürich, Department of Chemistry and Applied Biosciences, Zürich, Switzerland

Abstract

The halophilic archaeon *Halococcus morrhuae* and the biofilm-forming bacterium *Halomonas muralis* were exposed to space conditions during the EXPOSE-R2 mission. Evidence for both strains co-existing on a mural in the castle Herberstein (Austria) has been found and here we tested the theory that the biofilm produced by *Hlm. muralis* may act as a protective layer for *Hcc. morrhuae* during exposure to extreme conditions. During Pre-flight tests it was shown that *Hcc. morrhuae* is significantly more resistant to extreme conditions compared to *Hlm. muralis*. A mixture of both strains proved advantageous only for the survival of *Hcc. morrhuae*. *Hcc. morrhuae* was exposed independently and in combination with *Hlm. muralis* to space conditions outside of the International Space Station (ISS) for 534 days. Survival of *Hcc. morrhuae* was investigated by most probable number test and colony-forming unit assay; damage inflicted to genomic DNA was evaluated by random amplified polymorphic DNA-PCR and the integrity of the 16S rRNA by real-time PCR. Exposure to extreme conditions on Earth as well as in outer space had a strong detrimental effect on the survival and genomic stability of *Hcc. morrhuae*; however, we were able to re-cultivate *Hcc. morrhuae* from samples exposed outside of the ISS.

Keywords: Halophiles, biofilm, prolonged outer space exposure, BOSS, ISS, EXPOSE-R2

**Corresponding author*: stefan.leuko@dlr.de
Stefan Leuko: https://www.dlr.de/me/en/desktopdefault.aspx/tabid-2016/
Helga Stan-Lotter: https://www.researchgate.net/scientific-contributions/38431685_Helga_Stan-Lotter
Greta Lamers: https://energymetab.ethz.ch/about-us/people/person-detail.MjU1OTU2.TGlzdC8yMTksODczNTMzMzQ2.html, https://www.researchgate.net/profile/Greta_Lamers
Sebastian Sjöström: https://www.researchgate.net/profile/Sebastian_Sjoestroem, https://scholar.google.com/citations?user=gUFlSdsAAAAJ&hl=en

Joseph Seckbach and Helga Stan-Lotter (eds.) Extremophiles as Astrobiological Models, (221–236)

10.1 Introduction

The universe is a hostile place and currently we are only aware of one planet that harbors life: Earth. Most of the existing life on Earth thrives in moderate environmental conditions, however, over the last decades extreme environments have been shown to harbor a great diversity of life [10.43]. The quest to understand modern life in extreme environments addresses some profound questions. How are these organisms able to cope with or even flourish in these hostile environmental conditions? Could this imply they might withstand other extreme conditions such as those prevalent in outer space or on foreign planets such as Mars? To answer these questions, experiments with different parameters are conducted in ground-based laboratories and increasingly in outer space on board the ISS and in particular on the outside of the ISS, e.g., the EXPOSE missions [10.38] [10.39]. Here, we investigated the resistance and survival of the halophilic archaeon *Halococcus morrhuae* in association with the extracellular polymeric substances (EPS)-producing and biofilm-forming bacterium *Halomonas muralis*. Both strains were identified on a mural of the Saint-Catherine chapel in Castle Herberstein, Austria [10.15] [10.36].

Previous research has uncovered that haloarchaea are extremely versatile and able to survive detrimental environmental conditions such as desiccation [10.28] [10.46], fluxes in external osmotic conditions [10.5] [10.22], microgravity [10.9], UV radiation [10.23] [10.31] and even exposure to outer space [10.27] [10.29]. Of particular astrobiological interest is the resistance against desiccation and radiation. Both desiccation and radiation result in DNA double-strand breaks and the connection between resistances to these stresses has been previously established [10.30] [10.6]. An elegant explanation for the radiation and desiccation resistance is the presence of multiple genome copies and the ability to regenerate damaged DNA by utilizing scattered DNA fragments [10.44] [10.45]. Several studies have shown, for example, that the halophilic archaeon *Halobacterium salinarum* NRC-1, is highly resistant to ionizing radiation [10.7] [10.20] [10.26] [10.48] as well as to simulated solar radiation [10.24], which makes haloarchaea excellent candidates for possible survival on planets with a higher radiation regime. Here we tested whether the association with a biofilm-forming bacterium is beneficial for the survival of a candidate halophilic archaeon when exposed to simulated and actual space conditions.

Biofilms have been defined as aggregates of microorganisms in which cells are frequently embedded in a self-produced matrix of EPS that are adherent to each other and/or a surface [10.11] [10.12]. These structures are complex systems that have high cell densities (10^8 to 10^{11} cells per g-1 wet weight) and typically comprise many different species [10.2]. The formation of biofilms is believed to be the most dominant mode of life in nature, and offers several advantages, including increased nutrient supply and better protection against environmental stress factors such as desiccation or UV radiation [10.3] [10.13] [10.35]. Not only do biofilms occur in various environments on Earth, they have been documented on board the Russian Space Station MIR as well as the International Space Station (ISS) [10.19] [10.34]. Furthermore, Kim and colleagues [10.18] have reported evidence that the exposure to microgravity inside the ISS improves biofilm formation by *Pseudomonas aeruginosa*, with increased number of viable cells, biomass, and thickness in spaceflight biofilms compared to normal gravity controls.

Here we report the results of the long-term space exposure experiment BOSS (Biofilm Organisms Surfing Space) as part of the EXPOSE-R2 missions. The effects of long-term storage and exposure to space conditions were shown to be highly detrimental to the survival of the cells as well as the DNA integrity; however, successful re-cultivation of *Hcc. morrhuae* was achieved. These experiments unveiled the ability of *Hcc. morrhuae*, when protected by a biofilm, to withstand exposure to space conditions for a prolonged time. They furthermore add to our understanding of potential life, past or present, on other planets such as Mars.

10.2 Material and Methods

10.2.1 Cultivation

Halococcus morrhuae DSM 1307^T and *Halomonas muralis* DSM 14789^T were obtained from the DSMZ (Deutsche Sammlung von Mikroorganismen und Zellkulturen, Braunschweig, Germany). *Hcc. morrhuae* was routinely cultured in DSM97 medium containing 250 g/L NaCl at 37 °C, shaking at 150 rpm for 14 days. The pH of the medium was adjusted to 7.4 with 10 M NaOH. *Hlm. muralis* was cultivated in DSM514 medium for 3 days at 30 °C without shaking to promote biofilm formation. For solid medium, 1,5% (w/v) agar was added to the media.

10.2.2 UV-C Radiation and Desiccation Resistance

To get a notion of the resistance of both strains, organisms were initially exposed in liquid to UV-C radiation. Cells of *Hlm. muralis* and *Hcc. morrhuae* were exposed to germicidal UV-C radiation (245 nm) up to fluences of 1.000 J/m^2 and the relative survival was determined by analyzing colony-forming units (CFU) before and after exposure.

10.2.3 Sample Preparation for Exposure

Starting cultures for both organisms were prepared from glycerol stocks and grown until late exponential phase. *Hlm. muralis* was cultivated without shaking to promote biofilm formation. Cell density for *Hlm. muralis* was evaluated by counting the cells in a Thoma counting chamber and the bacteria were diluted in 1× PBS to a final concentration of 5×10^7/40 μL. This solution was spotted onto quartz discs (ø 11 mm). Due to the strong clustering of *Hcc. morrhuae*, the determination of the cell concentration with the Thoma cell chamber was not possible. To obtain better accuracy, CFU counts were determined at a set OD (OD_{600nm}=1), which in turn was used to determine the concentration of *Hcc. morrhuae*. One mL of *Hcc. morrhuae* was centrifuged for 5 min at 10.000 × g and the resulting cell pellet was washed three times in 25% TN-buffer (containing 250 g/L NaCl and 6.055 g/L Tris-HCl, pH 7.4). Twenty μL containing 5×10^5 cells were spotted in the middle of the previously spotted samples of *Hlm. muralis* and gently mixed with the tip of the pipette. Quartz discs were left at ambient temperature (~22 °C, 30% RH) in a safety workbench until completely dried (2 days). For the exposure of only *Hcc. morrhuae*, 40 μL containing 10^6 washed cells were spotted onto quartz discs and dried as described above. All experiments conducted during this investigation were conducted with samples prepared in that way.

10.2.4 Pre-Flight Ground Tests

To determine the feasibility of the proposed experiments in space and to optimize sample preparation and analysis, three ground experiments were conducted before the mission. An experiment verification test (EVT) was designed to evaluate if the tested organisms survive extreme environments, such as radiation, desiccation and vacuum, as expected in the space experiment. The single stress factors tested during the EVT are given in Table 10.1. A science verification test (SVT) was designed to simulate the complete mission in a shorter timeframe on Earth, to evaluate if the chosen organism has the potential of survival and as a rehearsal for the procedures and logistics. The tested parameters are given in Table 10.2. To further simulate the space experiment conditions, samples were placed in three different compartments which were stacked as: top (t) compartment, where samples were exposed to the full solar radiation spectrum, middle (m) and bottom (b) where no solar radiation could penetrate the samples. This arrangement was also employed for the flight and a detailed description of the hardware is described by Rabbow *et al.* [10.41].

10.2.5 Flight Preparation and Mission Ground Reference (MGR) Preparation

Flight, MGR, and laboratory control samples were prepared from the same starting cultures. The flight and MGR samples were mounted in the sample trays as described under "*Pre-flight ground tests*" for the SVT test and the laboratory controls were stored in ambient conditions (~22 °C and ~40% relative humidity) and darkness until the flight sample returned. For flight samples, trays were mounted in Tray I and Tray II of the EXPOSE-R2 mission. Tray I was covered with MgF_2 glass windows evacuated and flooded with nitrogen gas (Figure 10.1a,b). Once installed on the outside of the ISS (Figure 10.1c), the valves of the tray were opened and the tray evacuated to expose the samples to space vacuum. Tray II was evacuated and flooded with 980 Pa pressure Mars gas (described in

Table 10.1 Single stress factors tested during EVT (environmental verification tests).

Stress conditions	Description	Duration
Vacuum	10^{-5} Pa	7 d
Mars atmosphere	CO_2 gas composition*, 10^3 Pa	7 d
Temperature cycle	-10 to +45 °C	40 cycles
Temperature min	-25 °C	1 h
Temperature max	+60 °C	1 h
Irradiation	UV-C (254 nm) up to 10.000 J/m^2 with 60 μWcm^{-2}	variable

*Exact composition: 95% CO_2, 2,7% N_2, 1.6% Ar, 0.15% O_2 and 370 ppm H_2O.

Table 10.2 *Hcc. morrhuae* and *Hlm. muralis* were exposed to a combination of space-relevant stress conditions (SVT test), combining the parameters given below.

Setup	Stress factors	Value	Duration
Set 1	Vacuum	10^{-5} Pa	
	Irradiation $_{(200\text{-}400nm)}$	$5x10^5$ kJm^2	
	Temperature max	+10 °C	
	Temperature min	-25 °C	18 d
Set 2	Mars atmosphere	*	
	Irradiation$_{(200\text{-}400nm)}$	$5x10^5$ KJm^2	
	Temperature max	+10 °C	
	Temperature min	-25 °C	18 d
Set 3	Control experiment, 10^3 Pa air pressure, darkness, room temperature		30 d

*Exact composition: 95% CO_2, 2,7% N_2, 1.6% Ar, 0.15% O_2 and 370 ppm H_2O.

Table 10.1), hermetically closed, and covered by a quartz window with transmission of solar electromagnetic radiation of $\lambda > 200$ nm, similar to the spectrum on Mars [10.42]. To attenuate the total mission UV fluence, samples of both trays were additionally covered with quartz neutral density filter (0.001% transmission). These filters were employed due to the results obtained during preliminary experiments (detailed in the results section). An identical arrangement of samples was used for the MGR experiment.

10.2.6 Mission EXPOSE-R2 – Flight Parameters and Conditions

An exhaustive description about the parameters of the mission is given by Rabbow *et al.* [10.41]. Briefly, samples were transported to Baikonur and launched on 23 August 2014 with the Progress 56P rocket. EXPOSE-R2 was installed on the URM-D platform on Zvezda module on the ISS on 18 August 2014 during the EVA-39 by Aleksandr Skvortsov and Oleg Artemyev. The cover protecting the samples from solar radiation prior to the start of the exposure period was removed on 22 October 2014. Samples were exposed on the outside of the ISS for 534 days and exposed to solar radiation for a total of 469 days. During the EXPOSE-R2 experiment, the altitude of the ISS varied between 398 and 417 km above Earth.[1] Over the duration of the mission, the upper surface of the sample trays was exposed to a total of 549.88 MJ/m^2 solar radiation (200–400 nm) but due to the neutral density filters, samples on the top layer were exposed to a total fluence of 5.54 kJ/m^2 over 469 days. The trays were returned to the interior of the ISS on 03 February 2016 and arrived back on Earth on 18 June 2016 with the Soyuz 45S rocket. The total mission duration was

[1]http://www.heavens-above.com/ISSheight.aspx

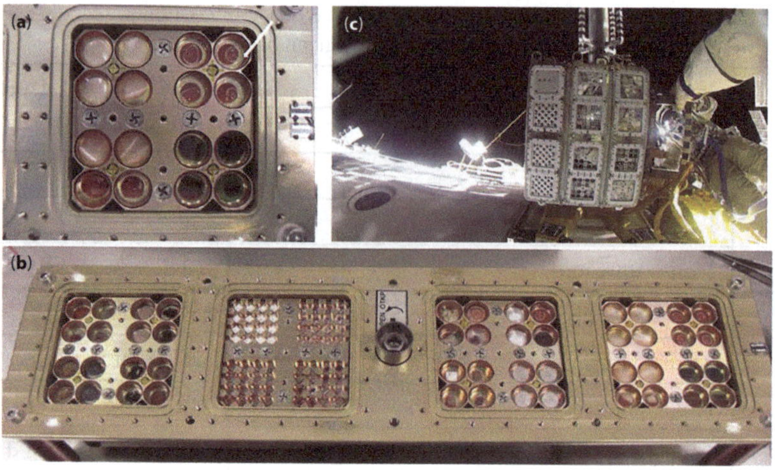

Figure 10.1 Setup of the EXPOSE-R2 mission tray. (a) Close-up of a single-tray compartment (white arrow indicating the spot of the BOSS samples; the dimensions of the carrier are 77×77 mm); (b) a complete tray (with a length of 466 mm, 120 mm width and a height of 88,5 mm; and (c) the complete EXPOSE-R2 setup being installed on the outside of the ISS. The monoblock has a length of 480 mm, a width of 390 mm, a height of 140 mm and it weighs 44 kg. (Image credit: Photos A & B, the authors; C, ESA).

696 days from launch to return to ground. An identical arrangement of samples was used for the MGR experiment. During the MGR, the environmental data of the flight unit (e.g., temperature profile and UV exposure) were monitored, with a time delay of 2 months, mirrored within DLR's Planetary and Space simulation facilities [10.40]. Additional laboratory control samples from the same batch of bacteria were stored in a desiccator, shielded from light, on a laboratory bench until the end of the mission. They were then analyzed in parallel to the flight samples upon arrival back to Earth.

10.2.7 Survival

Survival was determined by analyzing CFU. Quartz discs containing the cells were placed in 48-well plates and covered with 500 μL DSM97 medium. To avoid evaporation, the plates were put into plastic bags containing two water-soaked paper towels within a petri dish. The plates were incubated for 2 months at 37 °C with shaking (150 rpm). Following this initial incubation, 100 μL of serial diluted cells (undiluted, 10^{-1} and 10^{-2}) were spotted onto DSM97 plates and incubated at 37 °C. This procedure was only implemented for the flight and MGR samples. For the pre-flight tests, the discs were dissolved in 500 μL DSM97 medium overnight at room temperature (RT) with shaking at 150 rpm. One hundred μL of serial diluted cells were then spotted onto DSM97 plates and the survival was evaluated by counting CFU every week for up to 2 months. For mission and ground control samples, survival was evaluated once a week for up to 6 months. The rational for these long incubation times is that it has been previously shown that halophilic archaea show a delay in growth following exposure to stress situations such as exposure to pulsed electrons or HZE particles [10.7] [10.26]. To protect the plates from drying, an open petri dish containing two water-soaked paper towels were added to the top of the incubation bag.

10.2.8 Stability of Genomic DNA

Genomic integrity was investigated by RAPD–PCR, which allows for the detection of DNA damage and mutations [10.1]. DNA from cells exposed to space conditions, laboratory references and MGR samples was extracted using the XS-buffer method as preciously described by Tillett and Neilan [10.47], optimized for halophilic archaea by Leuko *et al.* [10.21]. The concentration of the extracted DNA was determined with a NanoDrop spectrophotometer and 25 ng were used as template. PCR reactions were performed with 1 U Platinum *Taq* Polymerase (ThermoFisher Scientific), 1X polymerase buffer, 3 mM $MgCl_2$, 0.2 mM dNTPs, and 0.2 pM of primer XD4 (5'AGCAGCGTGG'3) [10.32]. Thermal cycling conditions were: Initial denaturation 94 °C for 1 min., followed by 40 cycles of denaturation at 94 °C for 5 min, primer annealing at 31 °C for 1 min, strand elongation at 72 °C for 2 min, and a final extension step at 72 °C for 7 min. PCR products were separated by gel electrophoresis on 2% (w/v) agarose-TBE gels at 7 V cm^{-1} for 3 h. Gels were stained with SYBR Safe and visualized on an ImageQuant LAS4000 (GE Healthcare).

10.2.9 Integrity of 16S rRNA Gene

The integrity of the highly conserved 16S rRNA gene following space flight was investigated by quantitative real-time PCR (qRT-PCR). For this analysis, 10 ng were routinely used as template for the analysis. qRT-PCR was performed following exposure to space conditions and simulated Mars conditions by employing the forward primer Halo5F (5'ATTCGGTTGATCCTGCCGGA'3) and the reverse primer Halo1461R (5'CAGATTCCCCTACGGCTACCTT'3), resulting in a 1456 bp product. Primers were designed based on a sequence alignment with sequences obtained from the NCBI database of different halophilic archaea. Amplifications were performed in an Opticon 2 system (Bio-Rad) using the PeqLab KAPA Sybr FAST Kit. Thermal cycling program included: Initial denaturation at 94 °C for 2 min followed by 40 cycles of denaturation at 94 °C for 30 s, primer annealing at 60 °C for 20 s, and strand extension at 72 °C for 90 s. The standard curve was calculated based on a 16S rRNA gene product obtained from serial diluted (1:10; 1:100; 1:1.000; and 1:10.000), isolated DNA, obtained from a freshly grown culture. Melting curve analysis (0.2 °C s^{-1}) and agarose gel electrophoresis (2% agarose) revealed single amplicons for all samples (data not shown). The amplification efficiency (E) was calculated from the slope of the standard curve using the formula: $E(\%) = 10^{(-1/slope)}-1$ [10.4]. PCR efficiency and correlation coefficient for the standard curve were 81% and $r^2 = 0.998$ respectively. Resulting values from the qPCR runs were converted into relative lesion frequencies per 1.501 kB (~1.5 kB) DNA by application of the Poisson distribution (lesions/amplicon = $-\ln(A_t/A_0)$; where A_t represents the amplification of treated samples and A_0 is the amplification of untreated controls) as previously described by Hunter *et al.* [10.16].

10.3 Results

Cells of exponentially grown *Hcc. morrhuae* were embedded in a biofilm of *Hlm. muralis*, dried at RT and exposed to a variety of simulated space conditions and actual space conditions. To get an idea of their resistance, both strains were exposed to UV-C radiation and desiccation. *Hlm. muralis* showed decent survival to prolonged exposure (up to 3 months), with only a two-fold reduction in survival (data not shown); however, germicidal UV-C radiation was detrimental to the survival of *Hlm. muralis* even at low doses (< 100 J/m^2) (Figure 10.2a). In contrast, UV-C radiation had little effect on the survival of *Hcc. morrhuae* (Figure 10.2b) and the resistance to desiccation of halophilic archaea is well documented in the literature [10.44]. Based on these results, we focused on the survival and genetic integrity of *Hcc. morrhuae* for the ensuing experiments.

10.3.1 Survival EVT/SVT

Members of the genus *Halococcus* spp. are generally known to be resistant to detrimental environmental conditions; however, to assess the candidacy of this organism for a long-term space exposure experiment, several preliminary tests were conducted. For the EVTs (Table 10.1), single stress conditions were applied to desiccated samples and the survival of *Hcc. morrhuae* was evaluated, and the results are given in Figure 10.3. It became clear again that the most detrimental effect on the survival of *Hcc. morrhuae* is radiation. Although survival was observed at a fluence of up to 10.000 J/m^2, there was a significant loss in viability. All remaining factors, such as high temperature, vacuum and exposure to a simulated Martian atmosphere, had little effect on *Hcc. morrhuae* (Figure 10.3).

The next logical step was to simulate a trip to outer space similar to the proposed setup for the EXPOSE-R2 mission. The samples were placed in similar trays as used for the flight experiment (for an in-depth description of the trays and the mission, please refer to Rabbow *et al.* [10.41]), exposed to a combination of relevant space conditions as shown in Table 10.2, and the results are given in Figure 10.4. A 3-fold reduction in survival was observed for *Hcc. morrhuae*, but we were not able to recover *Hlm. muralis*. Interestingly, there was no difference in survival for samples that were exposed to solar radiation compared to samples that were kept in the dark.

10.3.2 Survival and Genetic Integrity Following Space Flight

Members of the genus *Halococcus* spp. are known to be notoriously difficult to cultivate. This was further complicated after exposure to stressful conditions. We were able to recover some cells of *Hcc. morrhuae* from 3 discs exposed to space conditions and 3 discs used for the MGR. All the recovered cells had been exposed to space vacuum (Tray 1). No survival was found in Tray 2. The discs were located at the bottom (2 CFU), the middle (1 CFU) and the top (1 CFU) of Tray 1 of the space samples. MGR samples showed growth in discs situated at the bottom (1CFU), the middle (1CFU) and the top (4 CFU). For each setup, 27 plates were exposed. The recovered colonies showed the typical clustering of *Hcc. morrhuae* and analyzing the 16S rRNA gene confirmed that it was *Hcc. morrhuae* (data not shown).

Following exposure to space conditions for a prolonged period of time, we extracted genomic DNA and analyzed the integrity of the 16S rRNA gene as well as the whole genome

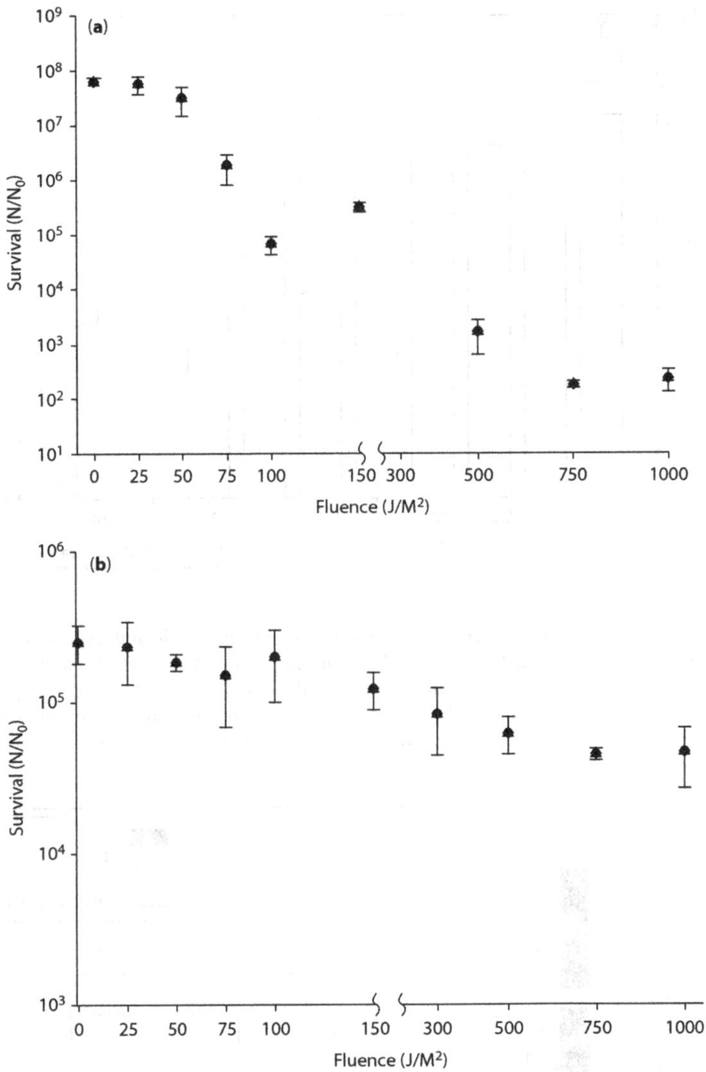

Figure 10.2 Survival of desiccated *Hlm. muralis* (a) and *Hcc. morrhuae* (b) cells to UV-C radiation up to 1.000 J/m^2. (Image credit: the authors).

integrity. Table 10.3 shows the results of the 16S rRNA integrity and it is clear that the DNA was heavily damaged even without the exposure to any detrimental environment. Even the laboratory control accumulated more damage over time, compared to an exposed sample. A similar picture can be drawn for whole genome integrity, analyzed by RAPD-PCR (Figure 10.5). However, the analysis indicates that the presence of a biofilm-forming bacterium provides some protection, as some parts of the genome are not as heavily damaged compared to the unprotected *Hcc. morrhuae*.

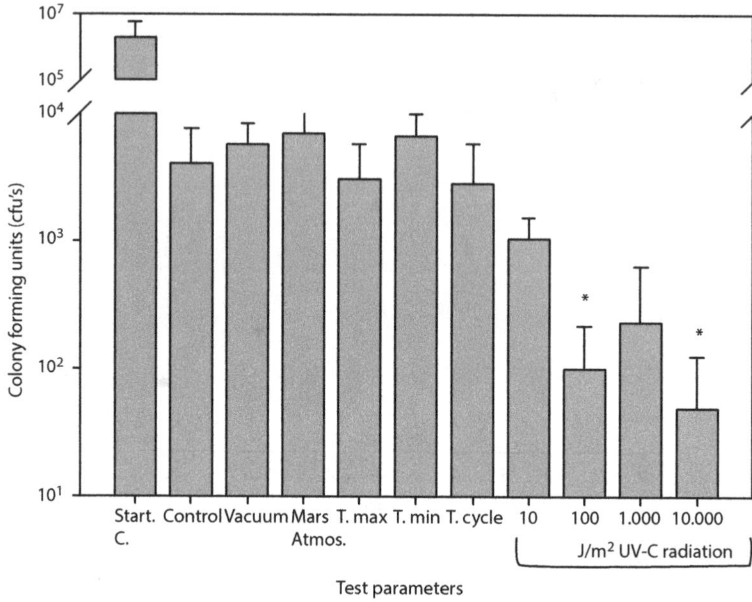

Figure 10.3 Survival of *Hcc. morrhuae*, embedded in a biofilm of *Hlm. muralis* and, exposed to single stress situations conditions as described in Table 10.1. The starting culture (Start. C.) was the culture used to dry onto quartz discs and the control denotes a culture dried for the duration of stress exposure kept in the laboratory. Standard deviations of biological triplicates are given as error bars and statistically significant changes compared to control are given marked with asterisk (*, $p < 0.05$). (Image credit: the authors).

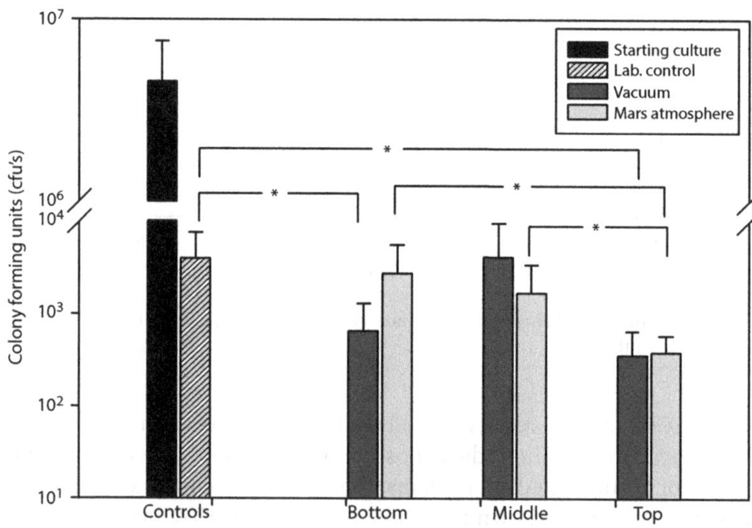

Figure 10.4 Survival of *Hcc. morrhuae* (embedded in the biofilm of *Hlm. muralis*) following the SVT. Standard deviation is shown as error bars and statistically significant changes are given (* indicates $p < 0.05$). (Image credit: the authors).

Table 10.3 Relative lesion frequency of *Hcc. morrhuae* 16S rRNA gene following exposure to simulated space conditions and real space conditions. Standard deviation is given in brackets. LC: Lab control (stored for the duration of the mission); MGR: Mission ground reference; Flight: Samples exposed outside the ISS.

Sample position	**Relative lesion frequency (%)**
LC - (*Hcc.* alone)	78 (± 2)
LC - (*Hcc.* + *Hlm.*)	89 (± 2)
MGR - Top (*Hcc.* + *Hlm.*)	83 (± 2)
MGR - Bottom (*Hcc.* + *Hlm.*)	77 (± 3)
Flight - Top (*Hcc.* + *Hlm.*)	85 (± 2)
Flight - Bottom (*Hcc.* + *Hlm.*)	77 (± 6)
MGR - Top (*Hcc.* alone)	89 (± 3)
MGR - Bottom (*Hcc.* alone)	78 (± 3)
Flight - Top (*Hcc.* alone)	92 (± 8)
Flight - Bottom (*Hcc.* + *Hlm.*)	74 (± 4)

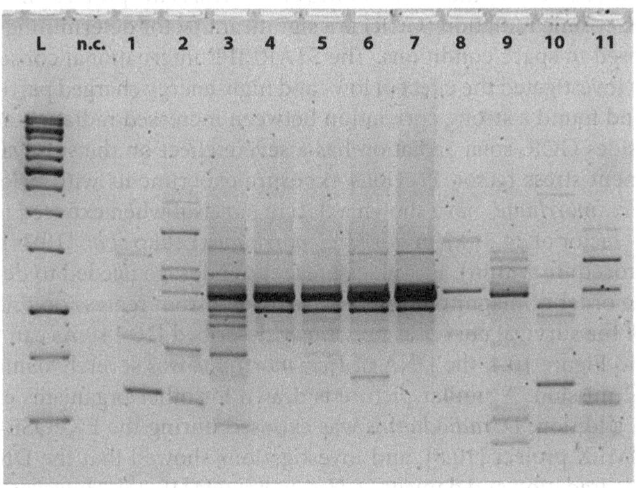

Figure 10.5 RAPD-PCR to visualize the damages accumulated in DNA of *Hcc. morrhuae* during exposure to simulated and real space conditions for the duration of the EXPOSE-R2 mission. L: DNA molecular ladder (1Kb); n.c.: negative control; 1: Extracted DNA of *Hcc. morrhuae* stored at 4 °C in dH_2O for the duration of the mission; 2: *Hcc. morrhuae* alone; 3: *Hcc. morrhuae* and *Hlm. muralis* combined; 4: Mission ground reference (MGR) – *Hcc. morrhuae* and *Hlm. muralis* combined - top layer; 5: MGR - *Hcc. morrhuae* and *Hlm. muralis* combined - bottom layer; 6: Flight samples - *Hcc. morrhuae* and *Hlm. muralis* combined - top layer; 7: Flight sample - *Hcc. morrhuae* and *Hlm. muralis* combined - bottom layer; 8: MGR – *Hcc. morrhuae* alone – top layer; 9: MGR – *Hcc. morrhuae* alone – bottom layer; 10: Flight sample - *Hcc. morrhuae* alone – top layer; 11: Flight sample – *Hcc. morrhuae* alone – bottom layer. (Image credit: the authors).

10.4 Discussion

In this study, we investigated the response of the halophilic archaeon *Hcc. morrhuae* and the bacterium *Hlm. muralis* to space environment for a prolonged time. They were both reported to be associated with a mural. *Hlm. muralis* is known to form biofilms; therefore, it was of interest to investigate whether this biofilm would offer some protection for *Hcc. morrhuae* during the exposure to space environment. Previous exposure experiments, such as the BIOPAN experiment in 1998 by Mancinelli and colleagues [10.27] or the EXPOSE-R mission in 2009 [10.29], have shown that halophilic archaea are able to survive exposure to space environment even for a prolonged time. It was shown that 14 days of exposure (BIOPAN) moderately decreases survival yet the prolonged exposure during the EXPOSE-R mission had a dramatic effect on the survival with only a few cells recovered. A similar scenario was observed in this experiment where only very few CFU were recovered. Interestingly, the lab control samples were also highly affected by long-term storage and showed almost no survival. A possible explanation for this phenomenon may be changes in relative humidity over the course of storage. It is well established that the rewetting of air-dried cells imposes physiological constraints on organisms [10.37], which may also be achieved by elevated humidity. Previous tests (EVT and SVT) also showed good survival for the laboratory controls. However, these tests were conducted in a much shorter time frame; therefore, no previous expertise was available about the survival of *Hcc. morrhuae* during desiccation for more than 2 years. Many factors contribute to the resistance of an organism to stress situations, and one of the cornerstones of survival is the protection of genetic information. Galactic cosmic radiation (GCR) is a significant factor determining the survival of organisms exposed to space conditions. The STARLIFE international consortium, (led by Dr. Ralf Möller) investigated the effect of low- and high-energy charged particles on the stability of DNA and found a strong correlation between increased radiation and DNA damage [10.33]. Besides GCR, solar radiation has a severe effect on the survival of organisms and is a prominent stress factor. Previous exposure experiments with halophilic archaea (in particular *Hcc. morrhuae*) have shown a decent survival when exposed in a dried state, resulting in a F_{10} factor of 24.31 kJ/m^2 for *Hcc. morrhuae* (Leuko *et al.* [10.24]; details on the experimental procedure within). F_{10} values present the dosage needed to decrease the survival rate by one order of magnitude and are calculated from regression lines of the exponential slopes of the survival curves as previously described [10.14]. As can be clearly seen in Table 10.3 and Figure 10.4, the DNA of *Hcc. morrhuae* was severely damaged following the EXPOSE-R2 mission. A similar picture is drawn for other organisms exposed during this mission. In addition, *D. radiodurans* was exposed during the EXPOSE-R2 mission as part of the BIOMEX project [10.8], and investigations showed that the DNA displayed a similar state of degradation and damage as *Hcc. morrhuae* [10.25]. However, like the results from the survival studies, the laboratory controls were severely damaged as well. It has been previously reported by Dose *et al.* [10.10] that prolonged desiccation of *Deinococcus radiodurans* without protective substances, such as glucose and storage in a non-oxidizing atmosphere such as argon, leads to a lethal amount of DNA double-strand breaks over time.

The results of this study reveal that even though halophilic archaea are considered highly resistant to a wide range of extreme conditions, 534 days of exposure to space without any significant protection is lethal to most organisms. Although embedded in a biofilm, the survival was not significantly higher than cells without protective layer. It has been previously shown that several layers of cells increase the survival of organisms when exposed to radiation [10.17]; however, the naturally occurring clustering of *Hcc. morrhuae* appears to have non-beneficial effects on survival or DNA protection during long-term exposure.

Despite *Hcc. morrhuae* showing little survival following this exposure experiment, the possibilities of finding remnants of previous life on other planets is still intriguing. For example, the surface of Mars is too hostile for life as we know it. However, below the surface the picture may be different. Protected from the most detrimental environmental source, radiation, remnants of early life may be preserved. It is therefore of high scientific and human interest to continue exploring the universe for possible signs of life, may it be extinct or extant.

Acknowledgments

The authors thank the European Space Agency for the flight opportunity and would like to acknowledge the support of ESA (special thanks to René Demets) and the MUSC team.

References

10.1. Atienzar, F.A., Venier, P., Jha, A.N., Depledge, M.H., Evaluation of the random amplified polymorphic DNA (RAPD) assay for the detection of DNA damage and mutations. *Mutat. Res.*, 521, 151–163, 2002.

10.2. Balzer, M., Witt, N., Flemming, H.C., Wingender, J., Accumulation of fecal indicator bacteria in river biofilms. *Water Sci. Technol.*, 61, 1105–1111, 2010.

10.3. Billi, D., Staibano, C., Verseux, C., Fagliarone, C., Mosca, C., Baqué, M., Rabbow, E., Rettberg, P., Dried biofilms of desert strains of *Chroococcidiopsis* survived prolonged exposure to space and Mars-like conditions in low earth orbit. *Astrobiology*, 19, 2019.

10.4. Bustin, S.A., Benes, V., Garson, J.A., Hellemans, J., Huggett, J., Kubista, M., Mueller, R., Nolan, T., Pfaffl, M.W., Shipley, G.L. *et al.*, The MIQE guidelines: Minimum information for publication of quantitative real-time PCR experiments. *Clin. Chem.*, 55, 611–622, 2009.

10.5. Coker, J.A., DasSarma, P., Kumar, J., Müller, J.A., DasSarma, S., Transcriptional profiling of the model archaeon *Halobacterium* sp. NRC-1: Responses to changes in salinity and temperature. *Saline Syst.*, 3, 6, 2007.

10.6. Daly, M.J., A new perspective on radiation resistance based on *Deinococcus radiodurans*. *Nat. Rev. Microbiol.*, 7, 237–245, 2009.

10.7. DeVeaux, L.C., Müller, J.A., Smith, J., Petrisko, J., Wells, D.P., DasSarma, S., Extremely radiation-resistant mutants of a halophilic archaeon with increased single-stranded DNA-binding protein (RPA) gene expression. *Radiat. Res.*, 168, 507–514, 2007.

10.8. De Vera, J.P., Alawi, M., Backhaus, T., Baqué, M., Billi, D., Böttger, U., the BIOMEX team, Limits of Life and the Habitability of Mars: The ESA Space Experiment BIOMEX on the ISS. *Astrobiology*, 19, 145–157, 2019.

10.9. Dornmayr-Pfaffenhuemer, M., Legat, A., Schwimbersky, K., Fendrihan, S., Stan-Lotter, H., Responses of haloarchaea to simulated microgravity. *Astrobiology*, 11, 199–205, 2011.

10.10. Dose, K., Bieger-Dose, A., Dillmann, R., Gill, M., Kerz, O., Klein, A., Meinert, H., Nawroth, T., Risi, S., Stridde, C., ERA-experiment "Space biochemistry. *Adv. Space Res.*, 16, 119–129, 1995.

10.11. Flemming, H.C., Wingender, J., Szewzyk, U., Steinberg, P., Rice, S.A., Kjelleberg, S., Biofilms: an emergent form of bacterial life. *Nat. Rev.*, 14, 563–575, 2016.

10.12. Flemming, H.C. and Wuertz, S., Bacteria and archaea on Earth and their abundance in biofilms. *Nat. Rev. Microbiol.*, 17, 247–260, 2019.

10.13. Fröls, S., Dyall-Smith, M., Pfeifer, F., Biofilm formation by haloarchaea. *Environ. Microbiol.*, 14, 3159–3174, 2012.

10.14. Harm, W., *Biological Effects of Ultraviolet Radiation*, Cambridge University Press, Cambridge, UK, 1980.

10.15. Heyrman, J., Balcaen, A., De Vos, P., Swings, J., *Halomonas muralis* sp. nov., isolated from microbial biofilms colonizing the walls and murals of the Saint-Catherine chapel (Castle Herberstein, Austria). *Int. J. Syst. Evol. Microbiol.*, 52, 2049–2054, 2002.

10.16. Hunter, S.E., Jung, D., Di Giulio, R.T., Meyer, J.N., The QPCR assay for analysis of mitochondrial DNA damage, repair, and relative copy number. *Methods*, 51, 444–451, 2010.

10.17. Kawaguchi, Y., Yang, Y., Kawashiri, N., Shiraishi, K., Takasu, M., Narumi, I., Satoh, K., Hashimoto, H., Nakagawa, K., Tanigawa, Y. *et al.*, The possible interplanetary transfer of microbes: assessing the viability of *Deinococcus* spp. under the ISS environmental conditions for performing exposure experiments of microbes in the Tanpopo mission. *Orig. Life Evol. Biosph.*, 43, 411–428, 2013.

10.18. Kim, W., Tengra, F.K., Young, Z., Shong, J., Marchand, N., Chan, H.K., Pangule, R.C., Parra, M., Dordick, J.S., Plawsky, J.L., Collins, C.H., Spaceflight promotes biofilm formation by *Pseudomonas aeruginosa*. *PLoS ONE*, 8, 4, e62437, 2013.

10.19. Klintworth, R., Reher, H.J., Viktorov, A.N., Bohle, D., Biological induced corrosion of materials II: new test methods and experiences from MIR station. *Acta Astronaut.*, 44, 7–12, 1999.

10.20. Kottemann, M., Kish, A., Iloanusi, C., Bjork, S., DiRuggiero, J., Physiological responses of the halophilic archaeon *Halobacterium* sp. strain NRC1 to desiccation and gamma irradiation. *Extremophiles*, 9, 219–227, 2005.

10.21. Leuko, S., Goh, F., Ibáñez-Peral, R., Burns, B.P., Walter, M.R., Neilan, B.A., Lysis efficiency of standard DNA extraction methods for *Halococcus* spp. in an organic rich environment. *Extremophiles*, 12, 301–308, 2008.

10.22. Leuko, S., Raftery, M.J., Burns, B.P., Walter, M.R., Neilan, B.A., Global protein-level responses of *Halobacterium salinarum* NRC-1 to prolonged changes in external sodium chloride concentrations. *J. Proteome Res.*, 8, 2218–2225, 2009.

10.23. Leuko, S., Neilan, B.A., Burns, B.P., Walter, M.R., Rothschild, L.J., Molecular assessment of UVC radiation-induced DNA damage repair in the stromatolitic halophilic archaeon, *Halococcus hamelinensis*. *J. Photochem. Photobiol. B*, 102, 140–145, 2011.

10.24. Leuko, S., Domingos, C., Parpart, A., Reitz, G., Rettberg, P., The survival and resistance of *Halobacterium salinarum* NRC-1, *Halococcus hamelinensis*, and *Halococcus morrhuae* to simulated outer space solar radiation. *Astrobiology*, 15, 987–997, 2015.

10.25. Leuko, S., Bohmeier, M., Hanke, F., Böttger, U., Rabbow, E., Parpart, A., Rettberg, P., de Vera, J.P.P., On the stability of Deinoxanthin exposed to Mars conditions during a long-term space mission and implications for Biomarker detection on other Planets. *Front. Microbiol.*, 8, 1680, 2017.

10.26. Leuko, S. and Rettberg, P., The effects of HZE particles, γ and X-ray radiation on the survival and genetic integrity of *Halobacterium salinarum* NRC-1, *Halococcus hamelinensis*, and *Halococcus morrhuae*. *Astrobiology*, 17, 110–117, 2017.

10.27. Mancinelli, R.L., White, M.R., Rothschild, L.J., Biopan-survival I: Exposure of the osmophiles *Synechococcus* sp. (Nägeli) and *Haloarcula* sp. to the space environment. *Adv. Space Res.*, 22, 327–334, 1998.

10.28. Mancinelli, R.L., Fahlen, T.F., Landheim, R., Klovstad, M.R., Brines and evaporites: Analogs for Martian life. *Adv. Space Res.*, 33, 1244–1246, 2004.
10.29. Mancinelli, R.L., The affect of the space environment on the survival of *Halorubrum chaoviator* and *Synechococcus* (Nägeli): data from the space experiment OSMO on EXPOSE-R. *Int. J. Astrobiol.*, 14, 123–128, 2015.
10.30. Mattimore, V. and Battista, J.R., Radioresistance of *Deinococcus radiodurans*: Functions necessary to survive ionizing radiation are also necessary to survive prolonged desiccation. *J. Bacteriol.*, 178, 633–637, 1995.
10.31. McCready, S., Mueller, J.A., Boubriak, I., Berquist, B.R., Ng, W.L., DasSarma, S., UV irradiation induces homologous recombination genes in the model archaeon, *Halobacterium* sp. NRC-1. *Saline Syst.*, 1, 3, 2005.
10.32. Moschetti, G., Aponte, M., Blaiotta, G., Casaburi, A., Chiurazzi, M., Ventorino, V., Villani, F., Characterization of halophilic Archaea isolated from different hypersaline ecosystems. *Ann. Microbiol.*, 56, 119–127, 2006.
10.33. Möller, R., Raduse, M., Leuko, S., Berger, T., Hellweg, C.E., Fujimoi, A., Okayasu, R., Horneck, G., the STARLIFE Research group, STARLIFE – An international campaign to study the role of galactic cosmic radiation in astrobiological model systems. *Astrobiology*, 17, 101–109, 2017.
10.34. Novikova, N., De Boever, P., Poddubko, S., Deshevaya, E., Polikarpov, N., Rakova, N., Coninx, I., Mergeay, M., Survey of environmental biocontamination on board the International Space Station. *Res. Microbiol.*, 157, 5–12, 2006.
10.35. Panitz, C., Frösler, J., Wingender, J., Flemming, H.C., Rettberg, P., Tolerances of *Deinococcus geothermalis* biofilms and planktonic cells exposed to Space and simulated Martian conditions in low earth orbit for almost two years. *Astrobiology*, 19, 2019.
10.36. Piñar, G., Saiz-Jimenez, C., Schabereiter-Gurtner, C., Blanco-Varela, M.T., Lubitz, W., Rölleke, S., Archaeal communities in two disparate deteriorated ancient wall paintings: detection, identification and temporal monitoring by denaturating gradient gel electrophoresis. *FEMS Microbiol. Ecol.*, 37, 45–54, 2001.
10.37. Potts, M., Desiccation tolerance of Prokaryotes. *Microbiol. Rev.*, 58, 755–805, 1994.
10.38. Rabbow, E., Horneck, G., Rettberg, P., Schott, J.U., Panitz, C., L'Afflitto, A., von Heise-Rotenburg, R., Willnecker, R., Baglioni, P., Hatton, J. *et al.*, EXPOSE, an astrobiological exposure facility on the International Space Station – from proposal to flight. *Orig. Life Evol. Biosph.*, 39, 581–598, 2009.
10.39. Rabbow, E., Rettberg, P., Barczyk, S., Bohmeier, M., Parpart, A., Panitz, C., von Heise-Rotenburg, R., Hoppenbrouwers, T., Willnecker, *et al.*, EXPOSE-E: an ESA astrobiology mission 1.5 years in space. *Astrobiology*, 12, 374–386, 2012.
10.40. Rabbow, E., Parpart, A., Reitz, G., The planetary and space simulation facilities at DLR Cologne. *Microgravity Sci. Technol.*, 28, 215–229, 2016.
10.41. Rabbow, E., Rettberg, P., Parpart, A., Panitz, C., Schulte, W., Molter, F., Jaramillo, E., Demets, R., Weiß, P., Willnecker, R., EXPOSE-R2: The astrobiological ESA mission on board of the International Space Station. *Front. Microbiol.*, 8, 1533, 2017.
10.42. Rontó, G., Bérces, A., Lammer, H., Cockell, C.S., Molina-Cuberos, G.J., Patel, M.R., Selsis, F., Solar UV irradiation conditions on the surface of Mars. *Photochem. Photobiol.*, 77, 34–40, 2003.
10.43. Rothschild, L.J. and Mancinelli, R.L., Life in extreme environments. *Nature*, 409, 1092–1101, 2001.
10.44. Stan-Lotter, H. and Fendrihan, S., Halophilic archaea: Life with desiccation, radiation and oligotrophy over geological times. *Life*, 5, 1487–1496, 2015.

10.45. Soppa, J., Evolutionary advantages of polyploidy in halophilic archaea. *Biochem. Soc. Trans.*, 41, 339–343, 2013.

10.46. Soppa, J., Polyploidy in archaea and bacteria: About desiccation resistance, giant cell size, long-term survival, enforcement by a eukaryotic host and additional aspects. *J. Mol. Microbiol. Biotechnol.*, 24, 409–419, 2014.

10.47. Tillett, D. and Neilan, B.A., Xanthogenate nucleic acid isolation from cultured and environmental cyanobacteria. *J. Phycol.*, 36, 251–258, 2000.

10.48. Whitehead, K., Kish, A., Pan, M., Kaur, A., Reiss, D.J., King, N., Hohmann, L., DiRuggiero, J., Baliga, N.S., An integrated systems approach for understanding cellular responses to gamma radiation. *Mol. Syst. Biol.*, 2006.

11

The Amazing Journey of *Cryomyces antarcticus* from Antarctica to Space

Silvano Onofri[1], Claudia Pacelli[1,2]*, Laura Selbmann[1] and Laura Zucconi[1]

[1]*Department of Ecological and Biological Sciences, University of Tuscia, Viterbo, Italy*
[2]*Italian Space Agency (ASI), Rome, Italy*

Abstract

The cryptoendolithic endemic black fungus *Cryomyces antarcticus* was first isolated from sandstone collected at Linnaeus Terrace, McMurdo Dry Valleys, Southern Victoria Land, Antarctica, by E.I. Friedmann in 1982, but published as a new species and genus in 2005. The McMurdo Dry Valleys are the coldest hyperarid desert on Earth and one of the best terrestrial analogues for Mars. There, microbes last chance for survival is to find refuge inside the porous rocks. The ability of *C. antarcticus* to withstand stresses has been widely demonstrated, and its stunning resistance is nowadays regarded as one of the best eukaryotic test organisms for astrobiological researches. Its extreme tolerance has been widely reported in about 10 years of space and ground experiments. It maintained survival, DNA integrity, ultrastructure stability and rapid metabolic activity recovery after 18 months exposure to space and Mars-like conditions in low Earth orbit (LEO) in the frame of ESA's LIFE (LIchens and Fungi Experiment) and ESA-ASI's BIOMEX (BIOlogy and Mars EXperiment) projects. Further investigations of space simulations on Earth within the STARLIFE project, aimed at characterizing the effect of different kinds of ionizing radiation (gamma radiation and α-particle), have shown its bewildering ability to resist stress radiation beyond all expectations. The resistance of fungal biomolecules will be further investigated within the recently approved ESA-ASI BIOSIGN (BIOSIGnature and habitable Niches) project.

Keywords: Black fungi, cryptoendolithic, melanin, Mars, space, international space station, resistance, radiation

**Corresponding author*: claudia.pacelli@asi.it; pacelli@unitus.it
Silvano Onofri: https://scholar.google.it/citations?user=H-9KA5cAAAAJ&hl=it&oi=ao, https://www.researchgate.net/profile /Silvano_Onofri
Caludia Pacelli: https://scholar.google.com/scholar?hl=it&as_sdt=0%2C5&q=Pacelli+C+Cryomyces&btnG=, https://www.researchgate.net/profile/Claudia_Pacelli
Laura Selbmann: https://www.researchgate.net/profile/Laura_Selbmann, https://scholar.google.it/citations?user=FwefdmkAAAAJ&hl=en
Laura Zucconi: ORCID ID: orcid.org/0000-0001-9793-230

Joseph Seckbach and Helga Stan-Lotter (eds.) Extremophiles as Astrobiological Models, (237–254)

11.1 Introduction

Astrobiology is an emerging science that deals with the origin, evolution and distribution of life in the Universe. A first step in searching for life on other planets is to understand "what" life is and how it originated on our planet, and which are the biophysical limits for life. For this purpose, the study of organisms living in extreme environments, called extremophiles, can help us to understand if and how life might be able to exist in other planetary bodies. Examples of extreme environments hosting extremophiles are hot and cold deserts, acidic or basic soils, salt soils, deep see hydrothermal vents, and hot springs.

Antarctica is one of the best terrestrial models for the search for life on other planets. It is the largest and least explored continent, and one of the most inhospitable places on Earth. It is the coldest continent, with the lowest air temperature record set on June 2018 at -98 °C on the East Antarctic Plateau [11.42]. Antarctica can be divided into three major areas—West Antarctica, East Antarctica and the Antarctic Peninsula—which differ from a climatic point of view. The Antarctic Peninsula has the most moderate climate, while more prohibitive climatic conditions can be found in Continental Antarctica. The Transantarctic Mountains, a 4,000 km long mountain chain extending across the continent, separates the larger East Antarctica ice sheet from the one of West Antarctica; the ice sheet covering East Antarctica, up to 4 km thick, is home to some of the coldest and driest conditions on Earth.

11.2 The McMurdo Dry Valleys

In Southern Victoria Land, between the East Antarctic Ice Sheet and the seasonally open water of the Ross Sea, are located the McMurdo Dry Valleys, characterized by extremely low humidity and lack of snow and ice coverage. They are the largest ice-free area on the continent (4,800 square kilometers, around 0.03% of the continent, and about 15% of the ice-free surfaces in Antarctica) and are considered one of the best models for astrobiological studies as terrestrial environment closest to Mars. The Dry Valley ecosystem is characterized by low temperatures (air temperatures average about -17 °C, with the winter minimum about -40 °C and the summer maximum a few degrees above freezing) [11.14], frequent freezing-thawing cycles, low precipitation levels and biologically available water (less than 100 mm water equivalent per year), low nutrient availability, strong winds, and high incidence of solar and UV radiation; high sublimation/evaporation exceeds precipitation and salts, deposited by precipitation of snow, become concentrated and accumulated in the soils.

Norm Horowitz, former head of the Biology Division of NASA's Jet Propulsion Laboratory (JPL), was the first to consider the Dry Valley soils as suitable models for astrobiological studies, based on the first images returned from Mariner 4 mission that showed Mars as a cratered, cold and dry planet; the soils of the Dry Valleys were considered sterile for a long time [11.21], until the record of viable microbes by Wolf Vishniac of Yale Medical School. Later on, Imre Friedmann, a microbiologist from Florida State University, studying some sandstone samples collected for him by Vishniac, discovered and described the presence of indigenous microflora colonizing the inside of Antarctic desert rocks, opening a new frontier in exobiological research [11.16] [11.15] [11.10] (Figure 11.1a). The presence of endolithic microbial ecosystems in such stressful environmental conditions suggested that life is more

Figure 11.1 (a) Sandstone slopes in the McMurdo Dry Valleys (Southern Victoria Land) landscape; (b) the micro-colonial black yeast-like fungus *C. antarcticus* CCFEE 515, scale bar=20 μm; (c) section of a dried sandstone fragment, colonized by a stratified cryptoendolithic microbial community. The community develops 10 mm deep below the rock crust and is organized into different colored and biologically distinct bands. Black fungi occupy the dark zone of the community, scale bar=10 mm; (d) the EXPOSE facility that has been attached to the Columbus module of the ISS in 2009, accommodating the LIFE experiment. (Image credit: Photos a, b & c, the authors; Photo d, ESA).

robust than previously thought and pointed to rocks being a suitable habitat for the search for extant or extinct life forms on the surface of Mars.

11.3 Cryptoendolithic Communities

Two types of endolithic communities were described on the basis of the primary producers: the lichen- and cyanobacteria-dominated communities, the former being the most prevalent in sandstone in the McMurdo Dry Valleys [11.24]. Here, lichens do not show epilithic development, but both mycobiont and photobiont colonize the pore spaces of exposed rocks, protected from the desiccating environmental conditions of the surface and from an excessive solar radiation, in spatial association with bacteria, cyanobacteria, non-lichenized filamentous fungi and black meristematic fungi, disposed to form different bands (Figure 11.1c). Microbes of these communities are very attractive because they evolved and adapted in the conditions that best mimic the ones on the Red Planet, and if a putative life has ever evolved on Mars, rocks may have been the ultimate habitat for life on its subsurface [11.48]. Being that the conditions on the surface of Mars are inhospitable for life as we know it, due to heavy oxidation and exposition to harmful UV and ionizing radiations [11.20] [11.50], the subsurface may offer refuge and may have extended the window of habitability. Indeed, the upcoming exploration mission on Mars, ESA-Roscosmos ExoMars 2020, looking for signs of life, will investigate the subsurface and within outcrops surface, by drilling down to 2 meters.

In the austral summer 1996-97, E.I. Friedmann and S. Onofri were collecting colonized samples in Antarctica; a few years later, in 2001, Friedmann donated a number of strains isolated from Antarctic rock samples to the University of Tuscia, representing the initial core of the CCFEE collection (Culture Collection of Fungi from Extreme Environments), housing today a significant number of strains from the McMurdo Dry Valleys [11.47].

11.4 The Black Microcolonial Yeast-like Fungus *Cryomyces antarcticus*

Since the discovery of the cryptoendolithic communities, their distribution in Northern Victoria Land has been mapped [11.53], and a quite large number of rock black fungi has been isolated and described by molecular tools for their scarcely differentiated morphology [11.25] [11.43] [11.44] [11.46] [11.12]. They share similar morpho-physiological characteristics, such as a thick and highly melanized cell wall, an extra-cellular polymeric layer, a slow growth and a cauliflower-like shape of the colony, all together ensuring survival in stressful environmental conditions. Among them, *Cryomyces antarcticus* was described in 2005 [11.43], based on a strain (CCFEE 534) isolated by R. Ocampo-Friedmann from weathered sandstone samples collected at Linnaeus Terrace (Dry Valleys) by E.I. Friedmann during the Antarctic expedition 1981–1982. It has slow-growing cerebriform colonies, black and thick-walled torulose hyphae, and dark and very thick-walled conidia; it grows in the range from 0 °C to 20 °C, tolerates strong thermal fluctuations and produces extracellular polysaccharides in response to extreme conditions, protecting against cycles of desiccation, and freeze/thawing [11.43] (Figure 11.1b).

Its ability to survive in one of the most extreme environments on Earth, and withstand different stressing conditions in laboratory tests, has made it one of the best eukaryotic models for many astrobiological studies, such as to undertake *an amazing journey from Antarctica to Space.*

11.5 The Polyextremotolerance of *Cryomyces antarcticus*

From the description of *C. antarcticus* onwards, a series of studies have been carried out to understand the amplitude of its polyextremotolerance, in addition to Experiment Verification Tests at the German Aerospace Center (DLR, Cologne, Germany), to support its reliability as a valid eukaryotic model.

The temperature was the first parameter analyzed among the abiotic factors. The ability to grow in a temperature range from 0 °C (minimum value tested) up to 20 °C, with an optimum at 15 °C, makes it well adapted to the Antarctic environment, which is characterized by temperatures up to 15 °C or higher at the rock surface and by strong thermal variations [11.43]. Its ability to germinate and form colonies remained practically unaffected after up to seven repeated cycles of freezing (up to -25 °C) and thawing (15 °C) [11.16], in accordance with endolithic life in the Antarctic desert where, during the austral summer, the temperature at the rock surface can cross the freezing point even 14 times within only 42 min [11.24]. This ability was further stressed later on, and dried isolates demonstrated an ability to survive, despite a strong reduction in growth, also after 50 cycles of heating (+20 °C) and cooling (-20 °C) within 2 weeks, which are ranges compatible with freeze and thaw cycles in space [11.26].

Despite its psychrophilic behavior, unexpectedly it proved to survive, when dehydrated, also at temperatures up to 90 °C for 60 min, an important skill in the light of the possibility of experimenting with conditions, albeit temporary, of temperatures up to 90 °C during a space flight [11.17].

The combination of low precipitation and high evaporation leads to salts accumulation in the Dry Valleys that represents a further stress for microbial life. *C. antarcticus* showed high tolerance against NaCl, growing up to 24% NaCl amended medium, with a higher production of melanin, covering the thick multilayered cell wall, and the occurrence of intracellular globules possibly related to the production of osmoregulators, as trehalose and glycerol [11.25].

Regarding the response of *C. antarcticus* to different types and intensities of radiation, in a first experiment it was shown to survive increasing UV-B (280–360 nm) irradiation doses, corresponding to over five to eight times the Antarctic terrestrial UV-B irradiance, with some ultrastructural damage and a lethal effect on almost 50% of propagules after doses ranging between 21.6 and 43.2 kJ/m^2 [11.26]. The highly pigmented thick cell walls and the presence of EPSs are reputed to be the winning features in a natural habitat suffering from increasing UV-B radiation levels because of the stratospheric ozone depletion. The melanin shielding ability probably avoids the occurrence of DNA damage, as proved by RAPD and rDNA amplification PCR-based assays on two Antarctic strains of *Cryomyces: C. antarcticus* CCFEE 534 and *C. minteri* CCFEE 5187. The template activity of these strains was maintained even after 240 min (the higher dose tested) of UV-B exposure, while it was drastically inhibited in *Saccharomyces pastorianus*, a hyaline yeast strain used as comparison, already after 30 min of exposure [11.45].

The effects on propagules' germination of UV-C (245 nm at 10, 100, and 1000 Jm^2), total UV (200–400 nm, at 1.5, 1.5×10^3, and 1.5×10^5 kJm^2), and total UV irradiation in combination once with vacuum and once with simulated CO_2 Mars atmosphere, were later examined in ground-based experiments carried out at the German Aerospace Center, from the perspective of sending the fungus into space. *C. antarcticus* colonies remained practically unaffected by the lower doses, while the total UV irradiation, once combined with vacuum and once with simulated Mars atmosphere, caused a substantial reduction, up to a complete inhibition under vacuum condition [11.27]. Finally, although analysis of the effect of the vacuum alone proved that it was capable of negatively influencing the ability of *C. antarcticus* to grow, it still maintained a detectable ability to grow after 22 days of exposure [11.27].

Since hypothetical life on other planets, or during travel through space (Lithopanspermia), encounters radiation conditions different from those on Earth, *C. antarcticus* was tested for a wide set of radiations.

Along with the presence of melanin, *C. antarcticus* resistance to radiation is possibly related to its ability to tolerate prolonged desiccation: colonies dried under laminar flow in a sterile cabinet, perfectly preserve their ultrastructure, with intact cell walls and membranes and a well-organized cytoplasm with a nucleus, vacuoles and lipid bodies. Fungal cells, upon losing their water content, enter a complete a-metabolic state, facing and overcoming environmental stresses, to reanimate when water is available again [11.32].

11.6 *Cryomyces antarcticus* and its Resistance to Radiation in Ground-Based Simulated Studies

One of the major concerns about astrobiological research is the radiation environment outside the Earth, which is very complex, and different from any highly irradiated environment on Earth. Indeed, beyond the shielding influence of Earth's magnetic field and atmosphere, the space radiation environment is dominated by high-energy electromagnetic waves from our Sun (ultraviolet, gamma and X-rays) and subatomic particles from the cosmos (electrons, protons, neutrons, and heavy ions). The high energy of ionizing radiation is able to remove electrons from molecules, resulting in cellular DNA and protein damage either directly, or indirectly, through the interaction with water and the following production of reactive oxygen species (ROS). Particles in space radiation flux have different masses, energies, and charges and thus damage organisms in different ways [11.20] [11.7]. Particularly dangerous are the so-called HZE particles (particles of High charge Z and high Energy), that are defined as cosmic ray primaries with charges $Z > 2$ and with *e*nergies high enough to penetrate at least 1 mm of shielding [11.20]. Even if they contribute only about 1% of the galactic cosmic rays (GCR) flux, they are of special interest to radiobiologists because of the inefficiency of an adequate shielding; they can even generate harmful secondary radiation as they interact with materials (i.e., spacesuit, metal walls of a spacecraft).

Studying the resistance of black fungi to ionizing radiation provides relevant information regarding astrobiology-related topics, including the search for life beyond Earth, planetary protection issues, and development of an adequate shielding material for long-term manned missions. As part of the STARLIFE irradiation campaign, which aims to test the resistance of microorganisms to space-relevant radiation [11.22], *C. antarcticus* was exposed to different kinds of ionizing radiation, namely gamma rays (up to 117.07 Gy) and alpha particles (He nuclei, up to 1000 Gy). The aim was to assess the resistance of dried colonies to photon and ion radiation, with various LET values.

Interestingly, in an exposure experiment performed at the German Aerospace Center in Cologne, Germany, with increasing doses of gamma radiation (up to 117.07 kGy, ^{60}Cobalt), colonies still preserved an astonishing ability to grow. After irradiation, the dried colonies were re-hydrated to determine their survivability, by counting the number of colony-forming units (CFU). Results, reported in Pacelli *et al.* [11.33], clearly indicated a very high resistance to gamma radiation, with a 12% survival rate recorded even at the dose of 55.81 kGy (Figure 11.2a). These results were corroborated in the analyses of the ultrastructural integrity by transmission electron microscopy analyses and the DNA integrity. These results are extremely important in the context of searching for life on Mars. In fact, considering that NASA's Mars Science Laboratory rover has recently reported the dose rate of 76 mGy year^{-1} on the Martian surface at Gale Crater [11.18], a hypothetical colony of *C. antarcticus* on the Martian surface could survive in the radiation environment for around 700,000 years. In this experiment, the fungus was reported to have the best radio-resistance, compared to the other eukaryotic test models, a green alga and a non-melanized lichen. Therefore, it

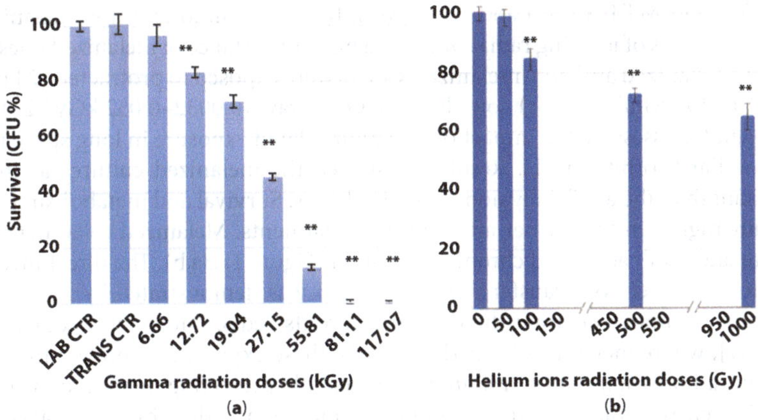

Figure 11.2 Resistance to space-relevant radiation in dried condition. Cultivation test of dried colonies of *C. antarcticus* after (a) γ-radiation from a ^{60}Co source (γ-rays at 1.17 MeV, low LET of 0.3 keV/μm; (b) accelerated helium ion irradiation (150 MeV/n, LET 2.2 keV/μm), ranging from 50 to 1,000 Gy. Data were normalized with the non-irradiated samples (control and transport samples). Significant differences were calculated by Tukey test with * = $p < 0.05$ and ** = $p < 0.001$. (Image credit: the authors).

was chosen for the following experiment in order to understand which mechanisms (repair or protection) are involved in fungal resistance against radiation. The biological effect of helium ion radiation (up to 1,000 Gy) was assessed by different tests, and all proved the maintenance of high survival and metabolic activity, even after the highest dose [11.34]. Cultivation test showed a progressive increase of mortality from 100 Gy onward, up to 65% of vitality maintained at the highest dose applied (Figure 11.2b). Similarly, the metabolic activity (XTT assay) started to be reduced at 100 Gy, but 66% was still maintained at the dose 1,000 Gy. Congruently, almost 63% of cells maintained their membrane integrity, assessed by using qPCR after treatments with propidium monoazide, after the highest dose; this integrity of cell membrane was also confirmed by TEM analyses. The DNA integrity was investigated by the amplification of ITS and ITS-LSU fragments (700, 1,600, and 2,000 bp): those genes were perfectly maintained and detectable in all the irradiation conditions. The integrity of the whole genome was further investigated, demonstrating a good maintenance of the fingerprinting profiles up to the highest irradiation dose [11.34].

Taken together, these studies have widely shown that the adaptation ability of this terrestrial microorganism enables its survival in conditions that never occur in its natural habitat (ionizing radiation and alpha particles, for example). Microorganisms limit radiation damage by a variety of microbial protection mechanisms, and the production of specialized organic pigments, including scytonemin, carotenoids and melanins [11.52], has already been reported. A crucial structural mechanism in the adaptation of *C. antarcticus* to radiation is the presence of a very highly melanized thick cell wall [11.28].

Considering the high survival reported with no detectable DNA damage, we focused our study on the role of melanin in photoprotection, by comparing the radiation resistance of melanized and non-melanized strains of *C. antarcticus*. To this scope, strains were irradiated with densely ionizing deuterons (^2H up 1,500 Gy) and sparsely ionizing X-rays (up to 300 Gy) radiation in physiological conditions, and the effects assessed by measuring their metabolic activity (XTT and MTT assays) and ATP levels. Both melanized and non-melanized cells survived acute doses of ionizing radiation, with a higher resistance of melanized ones [11.35].

The same melanized and non-melanized strains were exposed to protracted (24 h, potentially allowing for cellular repair), low-dose rates of X-rays (0.0032–0.052 kGy h21) [11.36], more relevant for assessing the impact of radiation-related exposure in long space missions beyond low Earth orbit [11.11]. Results confirmed the melanized cultures as invariably more resistant than the artificially non-melanized ones. Survival and metabolism were both substantially higher in the former after chronic treatments. Melanized cells increase their metabolism after both acute and chronic irradiation (Figure 11.3a,b). The direct involvement of melanin in the electron-transfer post-irradiation was demonstrated by the XTT assay. This substance cannot penetrate the cell, therefore its reduction might occur at the cell surface [11.1], where melanin is located. Together, these experiments proved that melanin has a role in radioresistance of *C. antarcticus*, efficiently protecting the fungus against high acute and low chronic radiation doses of ionizing radiation under physiological conditions [11.35] [11.36].

Melanins are unique multifunctional pigments ubiquitous on Earth that have been consistently associated with the extreme tolerance of black microorganisms to different kinds of stressors. The ubiquity of melanins indicates early evolution, conservation along the tree of life and functional importance; they emerged early in the course of evolution, when the basic background radiation was higher than today, which may have allowed the settlement of life on Earth. Even if the exact mechanism involved is still unknown, the melanin-related response to irradiation occurs in various species, and in response to different forms of radiation as well, as reported in [11.23]. For fungal melanin it has been speculated that the combination of physical shielding [11.6] and quenching of free radicals [11.49] is a persuasive mechanism of photoprotective action. Our observations also have important implications for basic space radiobiology and for creating melanin-based radioprotectors for humans. In the context of long-term deep space expeditions and human settlement on

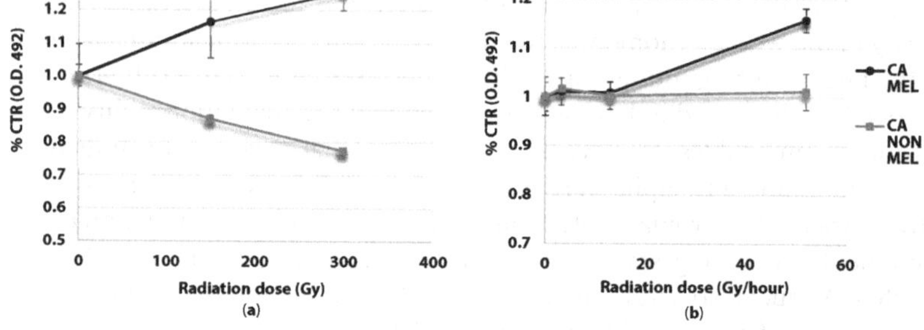

Figure 11.3 Understanding the role of melanin: Metabolic activity recovery after irradiation in physiological condition, a comparison between melanized and non-melanized strains of *C. antarcticus*. Effects of irradiation on *C. antarcticus* XTT reducing activity after (a) acute exposure to acute X-rays; (b) exposure to protracted X-rays. Data were normalized with the controls. Significant differences were calculated by Tukey test with * = $p < 0.05$ and ** = $p < 0.001$. (Figures were modified from [11.34] [11.36], open access or with permission from Wiley, respectively).

Mars, microbial radioprotection application could greatly decrease the cost of human space exploration and increase the success of missions by synthetizing, for instance, structural materials for radiation protection.

In fact, since the beginning of human space exploration, radiation environment has been recognized as the major risk, potentially limiting the duration of human sojourns in space. Far from the Earth's protective gas atmosphere and magnetosphere, astronauts and vehicles are exposed to harmful levels of ionizing radiation that are extremely dangerous for any form of life and biomolecules.

Melanin is a biopolymer derived from natural sources, so it has good biocompatibility and biostability. In this regard, natural melanins are excellent candidates with unique properties for biotechnological applications. Therefore, further work with melanized strains should focus on understanding the molecular mechanisms involved in the radioprotection and the hypothetical utilization of fungal melanin as radiation shielding material, both for astronaut protection and for the vehicular materials for long-term human space exploration [11.3].

One of the main efforts of space research is also related to the development of a comprehensive framework to interpret potential biomarkers (signs of extant or recently extinct life, abiotic biomarkers, false positives, and false negatives), and to increase confidence and consensus in interpretations, during the exploration mission. In this context, a putative microorganism on Mars or traveling in space could benefit from a melanization status. In addition, melanin itself can be considered a good biomarker; since it is ancient and distributed in all kingdoms of life on Earth, it is quite stable under space conditions and after irradiation, and is exceptionally preserved in fossils [11.17]. Melanin-like molecules, therefore, can be searched for as signs of life on a planet such as Mars.

11.7 *C. antarcticus* and its Resistance to Actual Space Exposure in Low Earth Orbit

Despite the big increase in knowledge achieved by the previous experiments, it must be taken into account that in ground-based simulations only one heavy ion with a specific energy coming from one direction is available at the same time. In space, the radiation field is composed of different ions with different energies and the radiation is coming from all directions. Environmental conditions that encompass space vacuum, galactic cosmic rays and solar particle events are only possible by direct exposure in space [11.4], therefore space experiments are necessary.

By taking advantage of the possibility to allocate samples at the exterior of the International Space Station (ISS) by the ESA exposure facility EXPOSE, the fungus has been exposed twice, for more than one year, to the space conditions experimented in the low Earth orbit, in the frame of two ESA experiments; LIFE (LIchen and Fungi Experiment) and BIOMEX (BIOlogy and Mars EXperiment).

11.7.1 LIFE Experiment on EXPOSE-E on Board the International Space Station

The aim of the LIFE experiment was to investigate the resistance of the cryptoendolithic black fungus *C. antarcticus* during long-term travel in space on board the ESA EXPOSE-E facility (Figure 11.1d) [11.28] [11.29.] [11.30] [11.41]. During space exposure, the fungal samples, accommodated in wells (1.4 cm in diameter), experienced either the full space environment (vacuum from 10^{-7} to 10^{-4} Pa, fluctuations of temperature between -21.5 and +59.6°C, cosmic ionizing radiation up to 190 mGy, and solar extraterrestrial electromagnetic radiation up to $6.34 \times 10^{8} Jm^{-2}$); or they were shielded from insolation. During the mission the Sun-exposed LIFE samples were exposed to 1,879 eSCh (estimated Solar Constant hours) [11.38]. The samples were also kept in simulated Mars atmosphere and pressure (1.6% Ar, 0.15% O_2, 2.7% N_2, 370 ppm H_2O, in CO_2 at a pressure of 10^3 Pa), and simulated Mars UV radiation, cutting off the spectrum of solar extraterrestrial electromagnetic radiation at a wavelength of $\lambda = 200$ nm by use of optical filters. Some samples were insolated with a reduced irradiance by three orders of magnitude by using neutral density filters, with fluencies of $9.19 \times 10^{5} Jm^{-2}$, below a 0.1% transmission neutral-density filter. In addition, dark flight samples were located beneath the insolated ones.

After 1.5 years in space, the samples were retrieved, and different assays were applied: (a) colony-forming ability; (b) amplification of DNA fractions from cells with undamaged membrane (PMA, propidium monoazide assay); (c) integrity of cellular ultrastructure.

After exposure to full insolation ($\lambda > 110$ nm; 100% insolated samples), *C. antarcticus* lost colony-forming ability, but the percentage of culturable cells was still 12.5 (±4.11) when 0.1% insolated. The PMA assay showed about 80% of DNA amplified from cells with undamaged membranes (100% insolated), in comparison to total extracted DNA [11.28]. The survival of cryptoendolithic organisms of long-term space exposure is of special interest in terms of Lithopanspermia, because rocks may supply an additional external protection to face the impact-driven ejection into space [11.19] and transfer from one planet to another.

In Mars conditions simulated in space (100% solar electromagnetic radiation at $\lambda > 200$ nm), the viability of black Antarctic fungi cells was 0.8% (±0.18) (Figure 11.4a). In the PMA assay, amplified DNA from cells with undamaged membranes was 66.32% (±6.75) [11.28] [11.29]. Results from PCR-based assays and TEM approaches to detect DNA and ultrastructural damage in *C. antarcticus* showed that solar radiation in space was the most threatening factor in terms of cellular and macromolecular (DNA) damage [11.30].

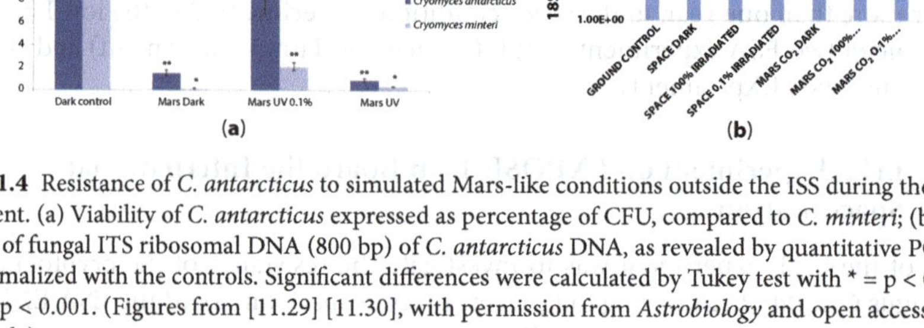

Figure 11.4 Resistance of *C. antarcticus* to simulated Mars-like conditions outside the ISS during the LIFE experiment. (a) Viability of *C. antarcticus* expressed as percentage of CFU, compared to *C. minteri*; (b) Integrity of fungal ITS ribosomal DNA (800 bp) of *C. antarcticus* DNA, as revealed by quantitative PCR. Data were normalized with the controls. Significant differences were calculated by Tukey test with * = $p < 0.05$ and ** = $p < 0.001$. (Figures from [11.29] [11.30], with permission from *Astrobiology* and open access *Life*, respectively).

Results showed that: (i) in the single-gene amplification assay, it was always possible to get products, even for long genes (up to 2000 bp); (ii) the mutational load never exceeded 5%, even at the highest treatment; (iii) according to qPCR assays, any decrease of the DNA copy number yielded in all samples (Figure 11.4b); and (iv) RAPD profiles were perfectly maintained in all treatments. Simulated Mars conditions also inflicted considerable damage to samples. *C. antarcticus* demonstrated to be considerably resilient to space conditions and Mars conditions simulated in space. Its resistance is possibly related to the thick layers of melanin protecting the cells, but its surprising resistance must be more deeply investigated.

11.7.2 BIOMEX Experiment on EXPOSE-R2 on Board the International Space Station

Taking advantage of the combined harsh conditions present in LEO (UV and ionizing radiation, temperature extremes, space vacuum and microgravity), the EXPOSE-R2 space mission was successfully launched to the ISS on July 24, 2014, on board the space cargo Progress 56P, whereas on August 18th it was installed outside the ISS on the Russian Zvezda module. The experiment ended in June 2016, for a total of external exposure of 531 days [11.39]. Among the four experiments allocated in the facility, for 16 months the BIOMEX Experiment exposed selected extremophiles not only to space conditions, but also to a simulated Mars-like environment (CO_2 atmosphere and UV > 200 nm). A set of microorganisms, representing species of all three main branches of the tree of life were chosen, including microorganisms that are known to be relevant for Mars (e.g., methane producing archaea, cyanobacteria and iron bacteria) and to corroborate the Lithopanspermia theory. The main aims of the project were to investigate the survival of selected extremophiles and the stability/degradation of their biological components (pigments, cell wall components, etc.) [11.8] [11.9], to be proposed as biomarkers for searching for life on Mars. To mimic real planetary conditions, microorganisms were grown on lunar regolith analogue rocks like anorthosite and on two Mars regolith analogue mixtures, namely phyllosilicatic Mars regolith simulant (P-MRS) and sulfatic Mars regolith simulant (S-MRS), that feature igneous rocks and an analogue for a more acidic environment with sulfate deposits, respectively [11.2]. The fungus *C. antarcticus* was chosen among the selected extremophiles. Before exposure to space, laboratory simulations were performed to select the most suitable extremophiles and to test the experimental setup, including the optimization of the samples' preparation, before the space mission. Laboratory simulations were performed at the Planetary and Space Simulation Facilities (PSI) at the DLR Institute of Aerospace Medicine in Cologne, named the Experiment Verification Tests (EVTs) and Scientific Verification Tests (SVTs). Mission ground reference (MGR) experiments were carried out in parallel to the space mission, to reproduce the environmental parameters of the space mission and using the same flight samples allocation; results were used as comparison for the space results interpretation.

Ground-based simulations, EVT1 and EVT2, were preliminary experiments to validate samples for the space mission. Although a reduced survival was observed after exposure to the most stressing parameters, *C. antarcticus* retained colony-forming ability after UV irradiation as well as after simulated treatments with other outer space stressors, which included temperature cycles, vacuum, and temperature extremes (60 °C and -25 °C). Accordingly,

TEM observations showed that, in addition to cells showing ultrastructural damage after the highest irradiation dose, colonies were still in good condition. Horneck *et al.* [11.20] reported the UV radiation as the most dangerous factor in space for microorganisms in low Earth orbit; in our experiments, the main damage was observed in cells exposed to increasing doses of UV radiation during the EVT2 test [11.32].

The last set of ground-based experiments, namely the SVT, were a dress rehearsal of the planned mission, providing a combination of simulated space vacuum and UV radiation and simulated Martian conditions. The post-exposure viability was tested by both cultivation and molecular methods, as previously and successfully applied on samples of the EVTs. The results on SVT tests demonstrated that *C. antarcticus* was able to tolerate the conditions of the SVT experiment, regardless of the substratum where it was grown, including the vibration test. DNA demonstrated high stability after treatments, and it was confirmed as a possible biomarker molecule, while melanin was chosen as a new target for biomarker detection [11.37]. Being that EVTs and SVT successfully performed, samples have been launched into space in 2014.

The 531 days of space mission exposed fungal samples to different stressors, which included temperature cycles, vacuum, UV and ionizing radiation. Due to the constantly changing attitude of the orbiting EXPOSE facility towards the Sun, samples were exposed to temperature cycles: the highest recorded temperature was +47 °C, the lowest -20 °C. According to ground-based results, this stress does not represent a threat to fungal colonies. Concerning the radiation environment, samples on the EXPOSE-R2 mission were exposed to about 0.5 Gy of ionizing radiation [11.5], while the mean polychromatic UV radiation fluence for trays 1 and 2 was about 458 and 492 MJ m^{-2}, respectively, both attenuated with the 0.1% neutral-density filter [11.39]. Redshift calculation reported that each sample was exposed to a different UV dose depending on its position [11.40]. Hence, *C. antarcticus* samples were exposed to polychromatic (UV and visible light) radiation doses ranging from about 2×10^3 to 4×10^3 kJm^{-2}.

Once back on Earth, samples were investigated with different approaches in order to i) understand the resistance of the fungus grown on lunar and Martian rock analogues and ii) investigate the resilience of its biomolecules, to be accounted as biomarkers. Survival and metabolic activity recovery, with no detectable DNA and cell membrane damage, were reported for *C. antarcticus* colonies regardless of the substratum in which the fungus was grown [11.31]. The fungal ultrastructure was perfectly maintained despite the prolonged desiccation and the radiation environment during the mission, to which the fungus has been exposed (Figure 11.5). Overall, these results have implications for the hypothesis that desiccation-tolerant life forms could have survived for long periods in protected habitable niches during the climatic history of Mars. The resilience of fungal biomolecules to the space environment and to simulated Martian conditions has been further investigated by using Raman spectroscopy, a technique planned for the upcoming ESA ExoMars mission which aims to nondestructively identify both the mineral context and putative signs of life on Mars [11.51]. Melanin spectra have been acquired from exposed samples by this technique; being that melanin pigment is ubiquitous in all kingdoms of life, it has been counted among the most resistant biomolecules to help in the search for life on other planets.

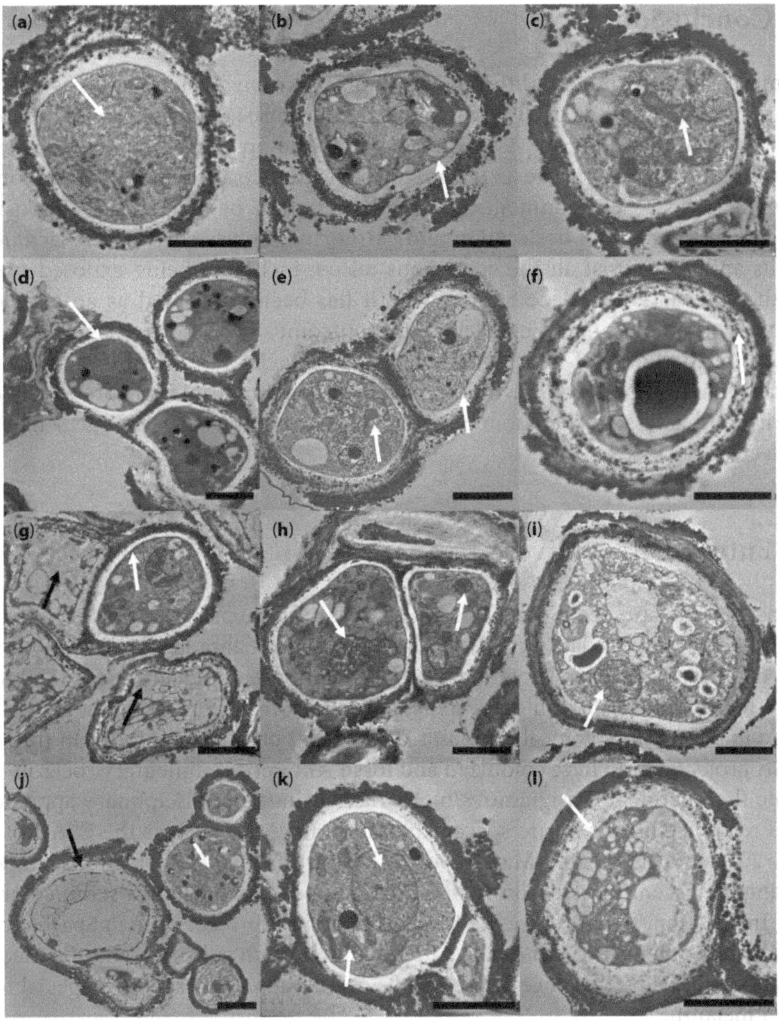

Figure 11.5 BIOMEX experiment outside the ISS: TEM micrographs of cells of *C. antarcticus* grown on Antarctic sandstone (a, b, c); Lunar analogue (d, e, f); Phyllosilicatic Mars Regolith Simulant (P-MRS) (g, h, i); Sulfatic Mars Regolith Simulant (S-MRS) (j, k, l), after 15 months of low Earth orbit exposure. Conditions in space: Irradiated (>120 nm) in vacuum (a, d), irradiated (>200 nm) in Mars-like atmosphere (g, j), vacuum in the dark (b, e), Mars-like atmosphere in the dark (h, k) and lab controls (c, f, i, l). White arrows indicate intact structures; black arrows indicate damage. Bars 2 μm. Description: Cells from cultures on Antarctic sandstone showed high preservation of ultrastructural features of nuclei, membranes and organelles in space conditions, both irradiated and in-flight dark samples, and in lab controls (a, b, c; white arrows); the same was observed in Lunar analogues (d, e, f). Some damage to membranes and organelles can be noticed in S-MRS and P-MRS samples exposed to Mars-like radiation (>200 nm) and Mars-like simulated atmosphere (g, j; black arrows), but ultrastructure was preserved in in-flight dark and lab control samples (h, i, k, l). (Figure from [11.31], with permission from *Astrobiology*).

11.8 Conclusion

To summarize, the cryptoendolithic black fungus *Cryomyces antarcticus* is remarkably resistant to a large number of diverse chemical and physical stresses, even different and more harmful than the ones of its natural habitat. The previously reported experiments were carried out to investigate extremophiles' resistance to some stressor conditions encountered in space environments, i.e., wide variation of temperature, exposition to UV and ionizing radiation, desiccation and extremely low-pressure values. *C. antarcticus* showed its ability to resist all the conditions tested, including being exposed two times outside the International Space Station, and it has been confirmed as an excellent test model for astrobiological studies. Hence, the outcomes of these studies have stretched the concept of the limits of microbial life and provide clues for defining the boundaries for habitable environments on Earth, discovering life in environments previously considered sterile, and assessing the habitability of Mars or other planets, if Earth-like life has ever evolved.

11.9 Future Perspectives

For its past climate and geological features, Mars is still the prime target for searching for life beyond Earth. Recent data from NASA's Mars Science Laboratory Curiosity rover have strengthened models on Mars' potential past habitability: sedimentary rocks were found to have been recently (78 ± 30 million years) exposed to surface conditions, hence furnishing protection to potential biological remnants or dormant organisms [11.13]. In this context, a new ESA project will analyze colonized and fossil Antarctic sedimentary rocks in order to investigate the microbial biosignatures/biomarkers using multidisciplinary approaches, to provide a reference list for future space exploration missions such as the ESA-Roscosmos ExoMars 2020 and the NASA Mars rovers. These studies will be of utmost importance considering the recent decision on ExoMars landing sites to include a sedimentary rocks location. In addition, as part of the Italian astrobiological project "Life in Space," the resistance of the test organism *C. antarcticus* to perchlorate-rich substrata and the mechanism of adaptation will be further investigated with "omics" approaches, with a special focus on the role of melanin.

Acknowledgments

The authors wish to thank the Italian National Antarctic Research Program (PNRA) for funding research in Antarctica (2013/AZ1.17, PNRA14_00132, PNRA16_00006), the Italian National Antarctic Museum (MNA) for preserving the CCFEE (Culture Collection of Fungi from Extreme Environments), ASI for co-funding the BIOMEX-MCF Experiment (ASI N. 2013-063-R.0) and for co-funding the BIOSIGN-Microfossils project (ASI N. 2018-6-U.0), and ESA for supporting and organizing LIFE and BIOMEX experiments in space on the International Space Station.

References

11.1. Berridge, M.V., Herst, P.M., Tan, A.S., Tetrazolium dyes as tools in cell biology: New insights into their cellular reduction. *Biotechnol. Annu. Rev.*, 11, 127–152, 2005.

11.2. Böttger, U., de Vera, J.P., Fritz, J., Weber, I., Hübers, H.W., Schulze-Makuch, D., Optimizing the detection of carotene in cyanobacteria in a Martian regolith analogue with a Raman spectrometer for the ExoMars mission. *Planet. Space Sci.*, 60, 1, 356–362, 2012.

11.3. Cordero, R.J., Melanin for space travel radioprotection. *Environ. Microbiol.*, 19, 7, 2529–2532, 2017.

11.4. Cottin, H., Kotler, J.M., Billi, D., Cockell, C., Demets, R., Ehrenfreund, P., Elsaesser, A., d'Hendecourt, L., van Loon, J.J.W.A., Martins, Z. *et al.*, Space as a tool for astrobiology: Review and recommendations for experimentations in Earth orbit and beyond. *Space Sci. Rev.*, 209, 83–181, 2017.

11.5. Dachev, T.P., Bankov, N.G., Tomov, B.T., Matviichuk, Y.N., Dimitrov, P.G., Häder, D.P., Horneck, G., Overview of the ISS Radiation Environment Observed during the ESA EXPOSE-R2 Mission in 2014–2016. *Space Weather*, 15, 1475–1489, 2017.

11.6. Dadachova, E., Bryan, R.A., Howell, R.C., Schweitzer, A.D., Aisen, P., Nosanchuk, J.D., Casadevall, A., The radioprotective properties of fungal melanin are a function of its chemical composition, stable radical presence and spatial arrangement. *Pigm. Cell Melanoma Res.*, 21, 2, 192–199, 2008.

11.7. Dartnell, L., Biological constraints on habitability. *Astron. Geophys.*, 52, 1, 1–25, 2011.

11.8. de Vera, J.P., Boettger, U., de la Torre Noetzel, R., Sánchez, F.J., Grunow, D., Schmitz, N., Lange, C., Hübers, H.-W., Billi, D., Baqué, M. *et al.*, Supporting Mars exploration: BIOMEX in Low Earth Orbit and further astrobiological studies on the Moon using Raman and PanCam technology. *Planet. Space Sci.*, 74, 1, 103–110, 2012.

11.9. de Vera, J.P., Alawi, M., Backhaus, T., Baqué, M., Billi, D., Böttger, U., Berger, T., Bohmeier, M., Cockell, C.S., Demets, R. *et al.*, Limits of life and the habitability of Mars: the ESA space experiment BIOMEX on the ISS. *Astrobiology*, 19, 2, 145–157, 2019.

11.10. Doran, P.T., Lyons, W.B., McKnight, D.M. (Eds.), *Life in Antarctic Deserts and Other Cold Dry Environments: Astrobiological Analogs*, vol. 5, Cambridge, United Kingdom, Cambridge University Press, 2010.

11.11. Durante, M., New challenges in high-energy particle radiobiology. *Br. J. Radiol.*, 87, 20130626, 2014.

11.12. Egidi, E., de Hoog, G.S., Isola, D., Onofri, S., Quaedvlieg, W., de Vries, M., Verkley, G.J.M., Stielow, J.B., Zucconi, L., Selbmann, L., Phylogeny and taxonomy of meristematic rock-inhabiting black fungi in the dothidemycetes based on multi-locus phylogenies. *Fungal Divers.*, 65, 1, 127–165, 2014.

11.13. Farley, K.A., Malespin, C., Mahaffy, P., Grotzinger, J.P., Vasconcelos, P.M., Milliken, R.E., Malin, M., Edgett, K.S., Pavlov, A.A., Hurowitz, J.A. *et al.*, *In situ* radiometric and exposure age dating of the Martian surface. *Science*, 343, 6169, 1247166, 2014.

11.14. Fountain, A.G., Nylen, T.H., Monaghan, A., Basagic, H.J., Bromwich, D., Snow in the McMurdo Dry Valleys, Antarctica. *Int. J. Climatol.: A Journal of the Royal Meteorological Society*, 30, 5, 633–642, 2010.

11.15. Friedmann, E.I., Endolithic microorganisms in the Antarctic cold desert. *Science*, 215, 1045–1053, 1982.

11.16. Friedmann, E.I. and Ocampo-Friedmann, R., Endolithic blue-green algae in the dry valleys: Primary producers in the Antarctic desert ecosystem. *Science*, 103, 1247–1249, 1976.

11.17. Glass, K., Ito, S., Wilby, P.R., Sota, T., Nakamura, A., Bowers, C.R., Vintherf, J., Duttag, S., Summonsg, R., Briggs, D.E.G. *et al.*, Direct chemical evidence for eumelanin pigment from the Jurassic period. *Proc. Natl. Acad. Sci.*, 109, 26, 10218–10223, 2012.
11.18. Hassler, D.M., Zeitlin, C., Wimmer-Schweingruber, R.F., Ehresmann, B., Rafkin, S., Eigenbrode, J.L., Brinza, D.E., Weigle, G., Böttcher, S., Böhm, E. *et al.*, Mars' surface radiation environment measured with the Mars Science Laboratory's Curiosity rover. *Science*, 343, 6169, 1244797, 2014.
11.19. Horneck, G., Stöffler, D., Ott, S., Hornemann, U., Cockell, C.S., Moeller, R., Meyer, C., de Vera, J.-P., Fritz, J., Schade, S., Artemieva, N.A., Microbial rock inhabitants survive impact and ejection from host planet: First phase of Lithopanspermia experimentally tested. *Astrobiology*, 8, 17–44, 2008.
11.20. Horneck, G., Klaus, D.M., Mancinelli, R.L., Space microbiology. *Microbiol. Mol. Biol. Rev.*, 74, 1, 121–156, 2010.
11.21. Horowitz, N.H., Cameron, R.E., Hubbard, J.S., Microbiology of the dry valleys of Antarctica. *Science*, 176, 4032, 242–245, 1972.
11.22. Moeller, R., Raguse, M., Leuko, S., Berger, T., Hellweg, C.E., Fujimori, A., Okayasu, R., Horneck, G., the STARLIFE Research Group, STARLIFE—An international campaign to study the role of galactic cosmic radiation in astrobiological model systems. *Astrobiology*, 17, 101–109, 2017.
11.23. Malo, M.E. and Dadachova, E., Melanin as an energy transducer and a radioprotector in black fungi, in: *Fungi in Extreme Environments: Ecological Role and Biotechnological Significance*, pp. 175–184, Springer, Cham, 2019.
11.24. Nienow, J.A. and Friedmann, E.I., Terrestrial lithophytic (rock) communities, in: *Antarctic Microbiology*, E.I. Friedmann (Ed.), pp. 343–412, Wiley-Liss, New York, 1993.
11.25. Onofri, S., Pagano, S., Zucconi, L., Tosi, S., *Friedmanniomyces endolithicus* (Fungi, Hyphomycetes), anam. - gen. and sp. nov., from continental Antarctica. *Nova Hedwigia*, 68, 175–181, 1999.
11.26. Onofri, S., Selbmann, L., de Hoog, G.S., Grube, M., Barreca, D., Ruisi, S., Zucconi, L., Evolution and adaptation of fungi at boundaries of life. *Adv. Space Res.*, 40, 1657–1664, 2007.
11.27. Onofri, S., Barreca, D., Selbmann, L., Isola, D., Rabbow, E., Horneck, G., de Vera, J.-P.P., Hatton, J., Zucconi, L., Resistance of Antarctic black fungi and cryptoendolithic communities to simulated space and Martian conditions. *Stud. Mycol.*, 61, 99–109, 2008.
11.28. Onofri, S., de la Torre, R., de Vera, J.P., Ott, S., Zucconi, L., Selbmann, L., Isola, D., Horneck, G., Survival of rock-colonizing organisms after 1.5 years in outer space. *Astrobiology*, 12, 508–516, 2012.
11.29. Onofri, S., de Vera, J.P., Zucconi, L., Selbmann, L., Scalzi, G., Venkateswaran, K.J., Rabbow, E., de la Torre, R., Horneck, G., Survival of Antarctic cryptoendolithic fungi in simulated Martian conditions on board the International Space Station. *Astrobiology*, 15, 1052–1059, 2015.
11.30. Onofri, S., Selbmann, L., Pacelli, C., de Vera, J.P., Horneck, G., Hallsworth, J.E., Zucconi, L., Integrity of the DNA and cellular ultrastructure of cryptoendolithic fungi in space or Mars conditions: A 1.5-year study at the International Space Station. *Life*, 8, 23, 2018.
11.31. Onofri, S., Selbmann, L., Pacelli, C., Zucconi, L., Rabbow, E., de Vera, J.P., Survival, DNA, and ultrastructural integrity of a cryptoendolithic Antarctic fungus in Mars and lunar rock analogs exposed outside the International Space Station. *Astrobiology*, 19, 170–182, 2019.
11.32. Pacelli, C., Selbmann, L., Zucconi, L., De Vera, J.P., Rabbow, E., Horneck, G., de la Torre, R., Onofri, S., BIOMEX experiment: Ultrastructural alterations, molecular damage and survival of the fungus *Cryomyces antarcticus* after the experiment verification tests. *Orig. Life Evol. Biosph.*, 47, 187–202, 2017.

11.33. Pacelli, C., Selbmann, L., Zucconi, L., Raguse, M., Moeller, R., Shuryak, I., Onofri, S., Survival, DNA integrity, and ultrastructural damage in Antarctic cryptoendolithic eukaryotic microorganisms exposed to ionizing radiation. *Astrobiology*, 17, 126–135, 2017.
11.34. Pacelli, C., Selbmann, L., Moeller, R., Zucconi, L., Fujimori, A., Onofri, S., Cryptoendolithic Antarctic black fungus *Cryomyces antarcticus* CCFEE 515 irradiated with accelerated helium ions: survival, DNA and ultrastructural damage. *Front. Microbiol.*, 8, 1–8, 2017.
11.35. Pacelli, C., Bryan, R.A., Onofri, S., Selbmann, L., Shuryak, I., Dadachova, E., Melanin is effective in protecting fast and slow growing fungi from various types of ionizing radiation. *Environ. Microbiol.*, 19, 1612–1624, 2017.
11.36. Pacelli, C., Bryan, R.A., Onofri, S., Selbmann, L., Zucconi, L., Shuryak, I., Dadachova, E., The effect of protracted X-rays exposure on cell survival and metabolic activity of fast and slow growing fungi capable of melanogenesis. *Environ. Microbiol. Rep.*, 10, 3, 255–263, 2018.
11.37. Pacelli, C., Selbmann, L., Zucconi, L., Coleine, C., de Vera, J.P., Rabbow, E., Boettger, U., Dadachova, E., Onofri, S., Responses of the black fungus *Cryomyces antarcticus* to simulated Mars and space conditions on rock analogs. *Astrobiology*, 19, 209–220, 2019.
11.38. Rabbow, E., Rettberg, P., Barczyk, S., Bohmeier, M., Parpart, A., Panitz, C., Horneck, G., von Heise-Rotenburg, R., Hoppenbrouwers, T., Willnecker, R. *et al.*, EXPOSE-E: An ESA astrobiology mission 1.5 years in space. *Astrobiology*, 12, 374–386, 2012.
11.39. Rabbow, E., Rettberg, P., Parpart, A., Panitz, C., Schulte, W., Molter, F., Jaramillo, E., Demets, R., Weiss, P., Willnecker, R., EXPOSE-R2: The astrobiological ESA mission on board of the International Space Station. *Front. Microbiol.*, 8, 1533, 2017.
11.40. Beuselinck, T., van Bavinchove, C., Hatton, J., and Demets, R. RedShift BVBA, Solar Irradiation by Calculation - Expose-R2 Simulation Results EXPR2-RP-002-RS Issue A *Redshift Design and Engineering BVBA*, 1, 2018.
11.41. Scalzi, G., Selbmann, L., Zucconi, L., Rabbow, E., Horneck, G., Albertano, P., Onofri, S., LIFE Experiment: isolation of cryptoendolithic organisms from Antarctic colonized sandstone exposed to space and simulated Mars conditions on the International Space Station. *Orig. Life Evol. Biosph.*, 42, 253–262, 2012.
11.42. Scambos, T.A., Campbell, G.C., Pope, A., Haran, T., Muto, A., Lazzara, M., Reijmer, C.H., van den Broeke, M.R., Ultra-low surface temperatures in East Antarctica from satellite thermal infrared mapping: The coldest places on Earth. *Geophys. Res. Lett.*, 45, 12, 6124–6133, 2018.
11.43. Selbmann, L., de Hoog, G.S., Mazzaglia, A., Friedmann, E.I., Onofri, S., Fungi at the edge of life: Cryptoendolithic black fungi from Antarctic deserts. *Stud. Mycol.*, 51, 1–32, 2005.
11.44. Selbmann, L., de Hoog, G.S., Zucconi, L., Isola, D., Ruisi, S., van den Ende, G., Ruibal, C., De Leo, F., Urzì, C., Onofri, S., Drought meets acid: Three new genera in a dothidealean clade of extremotolerant Fungi. *Stud. Mycol.*, 61, 1-20, 2008.
11.45. Selbmann, L., Isola, D., Zucconi, L., Onofri, S., Resistance to UV-B induced DNA damage in extreme-tolerant cryptoendolithic Antarctic fungi: Detection by PCR assays. *Fungal Biol.*, 115, 937–944, 2011.
11.46. Selbmann, L., Isola, D., Egidi, E., Zucconi, L., Gueidan, C., de Hoog, G.S., Onofri, S., Mountain tips as reservoirs for new rock-fungal entities: *Saxomyces* gen. nov. and four new species from the Alps. *Fungal Divers.*, 65, 167–182, 2014.
11.47. Selbmann, L., Zucconi, L., Isola, D., Onofri, S., Rock black fungi: Excellence in the extremes, from the Antarctic to space. *Curr. Genet.*, 61, 3, 335–345, 2015.
11.48. Selbmann, L., Pacelli, C., Zucconi, L., Dadachova, E., Moeller, R., de Vera, J.P., Onofri, S., Resistance of an Antarctic cryptoendolithic black fungus to radiation gives new insights of astrobiological relevance. *Fungal Biol.*, 122, 6, 546–554, 2018.

11.49. Schweitzer, A.D., Howell, R.C., Jiang, Z., Bryan, R.A., Gerfen, G., Chen, C.-C., Mah, D., Cahill, S., Casadevall, A., Dadachova, E., Physico-chemical evaluation of rationally designed melanins as novel nature-inspired radioprotectors. *PLoS One*, 4, 9, e7229, 2009.

11.50. Vago, J.L., Witasse, O., Svedhem, H., Baglioni, P., Haldemann, A., Gianfiglio, G., Blancquaert, T., McCoy, D., de Groot, R., ESA ExoMars program: The next step in exploring Mars. *Sol. Syst. Res.*, 49, 7, 518–528, 2015.

11.51. Vago, J.L., Westall, F., Pasteur Instrument Teams, Landing Site Selection Working Group, Other Contributors, Habitability on early Mars and the search for biosignatures with the ExoMars Rover. *Astrobiology*, 17, 6-7, 471–510, 2017.

11.52. Wynn-Williams, D.D., Edwards, H.G.M., Newton, E.M., Holder, J.M., Pigmentation as a survival strategy for ancient and modern photosynthetic microbes under high ultraviolet stress on planetary surfaces. *Int. J. Astrobiol.*, 1, 1, 39–49, 2002.

11.53. Zucconi, L., Onofri, S., Cecchini, C., Isola, D., Ripa, C., Fenice, M., Madonna, S., Reboleiro-Rivas, P., Selbmann, L., Mapping the lithic colonization at the boundaries of life in Northern Victoria Land, Antarctica. *Polar Biol.*, 39, 1, 91–102, 2016.

Part III

REVIEWS OF EXTREMOPHILES ON EARTH AND IN SPACE

12

Tardigrades – Evolutionary Explorers in Extreme Environments

K. Ingemar Jönsson

Department of Environmental Science and Bioscience, Kristianstad University, Kristianstad, Sweden

Abstract

Tardigrades have long been recognized for their ability to survive the most extreme environmental conditions. In particular, the widespread ability of tardigrades to dehydrate and enter a state of arrested metabolism (cryptobiosis) has allowed these aquatic animals to persist in habitats regularly deprived of water. The evolutionary strategy to allow metabolic arrest during periods of hostile conditions has several advantages, including arrest of the biological clock and energy consumption, but since all cryptobiotic animal taxa have very small body size there must be physical or physiological constraints preventing this strategy to evolve in larger animals. Our understanding of the molecular mechanisms of cryptobiotic tardigrades has improved considerably over the last decade, much due to the genome and transcriptome sequencing of some species. Although we are still far from a comprehensive understanding of how tardigrades maintain cellular structure under complete dehydration, it is clear that multiple molecular mechanisms with carbohydrates and proteins are involved, and that different adaptive mechanisms have evolved within tardigrades. Based on their documented tolerances, tardigrades made their entrance into astrobiology with the FOTON-M3 mission in 2007, and were the first animals to survive a combined exposure to space vacuum, cosmic radiation and UV radiation. With their ancient adaptations for coping with environmental extremes, tardigrades are excellent animals for studies of how early metazoan life evolved, and for understanding the constraints of cryptobiotic adaptations. Most tardigrades do not fit within the traditional definition of extremophiles, but rather belong to the category of extremotolerant organisms, surviving extreme conditions but with optimal performance under more normal environmental conditions. However, tardigrades have been reported from some permanently extreme environments, such as deep sea and hot and radioactive springs, and more studies of these populations would be of great interest.

Keywords: Tardigrada, cryptobiosis, anhydrobiosis, extremotolerants, extremophiles, evolutionary adaptations, desiccation tolerance

Email: ingemar.jonsson@hkr.se
K. Ingemar Jönsson: https://www.researchgate.net/profile/K_I_Joensson, https://www.hkr.se/en/research/staff/k.-ingemar-jonsson/

Joseph Seckbach and Helga Stan-Lotter (eds.) Extremophiles as Astrobiological Models, (257–274)

12.1 Introduction

Tardigrades (phylum Tardigrada) represent one of the animal taxa that has managed to evolve adaptations that allow persistence in some of the most challenging environments on Earth. They belong to our smallest metazoans, with a normal adult body length of 0.25–0.50 mm but in some species reaching 1 mm, and inhabit a variety of freshwater, limnoterrestrial and marine ecosystems. Despite being essentially aquatic organisms, the tardigrade phylum is well-known for containing many species that tolerate complete desiccation and freezing, conditions that force them into a state of metabolic arrest called cryptobiosis [12.44] [12.86] (Figure 12.1). The remarkable tolerance of tardigrades was recognized and studied soon after this animal group was discovered in the late 18th century [12.19]. Among animals they share their cryptobiotic abilities with many species of rotifers and nematodes and with a few insects and crustaceans, but while tardigrades, rotifers and nematodes are able to enter cryptobiosis in all life stages ("holo-cryptobiosis" [12.36]), cases of cryptobiosis in insects and crustaceans are restricted to early developmental stages (embryo or larval stages). In addition to these tolerances that may be derived from adaptations to natural environmental conditions, animals with cryptobiotic capacities also show high tolerance to more unnatural agents, in particular ionizing radiation. A number of studies have now shown that adult tardigrades survive ionizing radiation (both low linear energy transfer and high linear energy transfer radiation) at doses of several kiloGray, but as expected developmental stages (embryos) are more sensitive (see review in Jönsson *et al.* [12.43]). The background to this cross-tolerance is not clear, but probably represents a by-product of adaptations for the natural environmental agents, or some other common factor that results in functional correlations between, e.g., desiccation and radiation tolerance. The tolerance to multiple environmental extremes makes tardigrades excellent organisms for studies of how metazoan life has evolved in extreme environments, and tardigrades are currently the only animals that have survived under near open space conditions [12.41]. In this chapter I discuss some conceptual and evolutionary perspectives of tardigrades and their ability to enter into cryptobiotic life stages, with a main focus on desiccation-induced cryptobiosis (anhydrobiosis).

Figure 12.1 Tardigrades in active hydrated state (left) and in the desiccated anhydrobiotic state (right). The yellow animals (active + desiccated) are of the species *Richtersius* cf. *coronifer* and the white animals are of the species complex *Macrobiotus* cf. *hufelandi*. (Photos by K.I. Jönsson).

12.2 The Evolutionary Transition Towards Cryptobiotic Adaptations in Tardigrades

The evolutionary history behind the evolution of cryptobiotic adaptations in tardigrades is still unknown. The origin of the Tardigrada phylum has been dated by molecular methods to Precambrium (>541 mya), which suggests an evolutionary origin in a marine environment [12.21]. This estimate is based on molecular clock analyses and there are no tardigrade fossil records older than 90 myr [12.6], although fossils in ca. 500 myr Siberian orsten has been reported as possible records of a stem tardigrade [12.55]. Colonization of the land environments (terrestrialization) by tardigrades has been estimated to have occurred in the Late Silurian–Early Devonian, following the Cambrian radiation of arthropods and mirroring the evolution of land plants [12.72]. Since tardigrades evolved in the sea, the initial selection for anhydrobiosis (cryptobiosis induced by desiccation) and probably also other forms of cryptobiosis (in particular anoxybiosis) likely took place in shorelines and intertidal environments, characterized by frequent changes between dry and wet conditions, and creation of temporary pools that dried out. Little [12.54] also emphasized marine interstitial habitats as routes for the transition of arthropods towards terrestrial life, and such shoreline habitats may also have imposed selection for anhydrobiotic adaptations. Thus, the first evolutionary steps towards cryptobiosis in tardigrades may well have taken place before the terrestrialization of this group. Some tardigrades in such marine intertidal habitats also show high tolerance to desiccation (e.g., *Echiniscoides sigismundi* [12.32] and *Styraconyx haploceros* [12.34]). Once desiccation tolerance had evolved in these border habitats between the sea and land the possibility to disperse further into terrestrial habitats was prepared for, by the tolerance adaptations themselves, but also by the wind dispersal opportunity that the dry state gave rise to. However, based on phylogenetic data and information on the presence of cryptobiotic species within different tardigrade taxa, Guidetti, Altiero and Rebecchi [12.22] suggested that cryptobiosis should not be considered a plesiomorphic (phylogenetically common) trait within the Tardigrada phylum. Rather, cryptobiosis may have evolved in several evolutionary lines within the Eutardigrada and Heterotardigrada classes, respectively. They also suggested that adaptative evolution in response to environmental conditions, rather than phylogenetic relationships, explains the distribution of cryptobiotic abilities among tardigrade species within higher taxa.

Theoretical analyses on the evolution of cryptobiosis have suggested that even at low frequency (probability) of exposure to desiccating conditions, natural selection may promote adaptive mechanisms for desiccation tolerance, if the survival consequences (i.e., mortality) of such exposure are very high [12.39]. And if both the frequency and survival consequences of exposure to dry conditions are high, selection will be even stronger. All known species of desiccation-tolerant tardigrades are holo-anhydrobiotic, showing tolerance to desiccation in all life stages, but this was not necessarily the case in the early forms. Early developmental stages (eggs) in tardigrades may have evolved tolerance to desiccation before similar tolerance evolved in juveniles and adults. However, Olofsson, Ripa and Lundberg [12.60] have shown that the simultaneous evolution of multiple-stage cryptobiosis (e.g., holo-anhydrobiosis) is promoted by low serial autocorrelation in environmental conditions, i.e., unstable and unpredictable water conditions. In many semiterrestrial microhabitats where we often find desiccation-tolerant tardigrades today this condition is usually fulfilled due to stochasticity of weather conditions, but if this was the case also in the early habitats where cryptobiotic forms of tardigrades started to emerge is less clear.

12.3 Cryptobiosis as an Evolutionary Adaptive Strategy

What are the advantages and disadvantages from an evolutionary perspective of entering into an ametabolic cryptobiotic state, induced by, e.g., desiccation? Given that an organism cannot leave a habitat that turns into a hostile condition, there are two options: either to resist the conditions or tolerate them. In the case of desiccating environmental condition these two options are represented by partial dehydration with sustained metabolism (e.g., diapause), and more complete dehydration with arrested metabolism (i.e., anhydrobiosis), respectively. Although tardigrades are mainly known for the latter option, there are also species within this group that use diapause (encystment), but the environmental conditions inducing this state is not desiccation, but more predictable factors such as temperature and possibly oxygen level [12.22].

A number of advantages of cryptobiosis have been proposed, including the opportunity to colonize new terrestrial niches where competitors and predators may also be fewer due to harsh conditions, and improved dispersal ability in the anhydrobiotic state [12.22]. However, these advantages could also be valid for an organism that evolves a survival strategy with reduced but maintained metabolism, requiring adaptations that, in the case of dehydrating environmental conditions, maintain enough water in the cells so that metabolism will not be arrested. The advantage of sustained metabolism is that physiological functions such as cellular repair can be maintained, while these come to a halt in the anhydrobiotic state. The main difference between the metabolic and ametabolic dormancy strategies is directly related to the presence/absence of metabolism. First, with ongoing metabolism processes related to senescence will continue, while they will be halted in the anhydrobiotic state. Therefore, an important advantage of anhydrobiosis is that the biological time of the organism is reduced or halted, and decoupled from the chronological time spent in the dry state [12.28]. This "Sleeping Beauty" effect has been shown in both tardigrades and rotifers [12.25] [12.71]. Second, since metabolic activity consumes energy and no intake of food is possible in the dormant state, there is a definitive limit of the time that the dormant but metabolically active animal can stay in this state. In contrast, entrance into the anhydrobiotic state removes energy as a time limiting factor, since no energy is expended in this state. This is a considerable advantage in environments where the length of dry periods is unpredictable. Third, expending energy with the only purpose to overcome a period of hostile conditions and without possibilities to grow or reproduce is an energetically wasteful option compared to a complete shutdown of energy consumption.

But anhydrobiosis is not without costs. Physiological processes related to adjustment of the body for the dry anhydrobiotic state, and for the readjustment at rehydration, requires energy. Such costs have been indicated by reduction in the size of cells that carry the energy stores in tardigrades ("storage cells") over a single anhydrobiotic cycle [12.38]. Since no cellular repair mechanisms are active, any damage to cellular components such as DNA, proteins, and membranes (imposed by oxidative reactions with surrounding air), cannot be repaired until the animal has returned to the active hydrated state. For instance, increased damage to DNA with time spent in anhydrobiosis has been reported in tardigrades [12.59]. Increased repair of accumulated cellular damage will carry increased energy expenditure, but it also has another important effect. In a similar way as energy imposes a limited time frame in the case of metabolic dormancy, the accumulation of cellular damage imposes a time limit for animals in the anhydrobiotic state. When the accumulated damage exceeds the repair capacity at rehydration, the animal will be unable to recover from the dry state. This explains why tardigrades and other invertebrates capable of anhydrobiosis cannot stay indefinitely in the dry state, but show reduced recovery rate with time spent in the anhydrobiotic state [12.67]. The current record of anhydrobiotic survival in tardigrades under ambient laboratory conditions is 20 years [12.35], and under frozen conditions 30 years [12.79].

Despite the apparent advantages of entering an anhydrobiotic state in response to desiccating environmental conditions, in animals this evolutionary strategy is restricted to a few invertebrate taxa and to life forms with a maximum body size of about 5 millimeters. Given that anhydrobiotic animals evolved at an early stage of the terrestrialization of animals, this suggests that there are some fundamental physical or physiological constraints connected with a complete dehydration and arrest of metabolism [12.2].

12.4 Defining Life in Cryptobiotic Animals

The apparent lifelessness of desiccated tardigrades, rotifers and nematodes in their cryptobiotic state has fascinated zoologists for several centuries, representing a condition that provides some problems to define with respect to the concept of life. In the 19th century this became an issue of considerable concern when zoologists claimed that these animals were able to dry out completely, a condition that suggested cessation of life activities [12.44]. A commission at the Biological Society of France, led by Paul Broca, set out to clarify if these animals were in fact able to revive after complete desiccation, arranged by using vacuum pumps and heating to 100 °C. In their 1860 report they concluded that these animals, when "brought to the most complete degree of drying we can achieve under these conditions and within the current state of the science, may still retain the ability to revive in contact with water." ([12.10], p. 139).

Given that desiccated anhydrobionts are truly deprived of metabolism, Clegg [12.14] concluded that there in fact exist three states of biological organization—alive, dead, and cryptobiotic—where cryptobiosis occupies a position *in between* the alive and dead condition. Since animals in cryptobiosis must also be considered alive since they have the potential to return to an active life, I would rather use the expression "metabolically alive" for active animals. With respect to metabolism, the cryptobiotic animal has reached an endpoint, but obviously this does not lead to the state of death. Thus, while metabolic arrest is a necessary condition for death, it is not a sufficient condition. The transition from a cryptobiotic animal to a dead animal requires a change in the biological organization of cells and tissues such that a return to the metabolically active state is not possible. Therefore, cell structure can be considered a more constraining and fundamental condition of life than metabolic state in cryptobiotic animals such as tardigrades. This is consistent with the proposition of the Nomenclature Committee on Cell Death (NCCD) that "a cell should be considered dead when any one of the following molecular or morphological criteria is met: (1) the cell has lost the integrity of its plasma membrane, as defined by the incorporation of vital dyes (e.g., PI) *in vitro*; (2) the cell, including its nucleus, has undergone complete fragmentation into discrete bodies (which are frequently referred to as 'apoptotic bodies'); and/or (3) its corpse (or its fragments) has been engulfed by an adjacent cell *in vivo*." ([12.49], p. 5). The first two criteria are clearly based on structural characteristics. Since the state of death is generally recognized as an irreversible condition, the state of life can involve both metabolic and ametabolic conditions, as long as the structural integrity of the cells are maintained. The importance of structural integrity in explaining the revivability of desiccation-tolerant invertebrates was recognized already in the report by Broca [12.10], when he concluded that "the whole of the facts known so far makes it possible to regard the revivability as a property inherent in the organized matter, and as permanent neither more nor less as the molecular state on which it depends." ([12.10], p. 135). This conclusion simultaneously rejected the alternative view that the animals maintained a minimal metabolic activity in the dry state. Few studies have evaluated the metabolic status of anhydrobiotic animals, including tardigrades, and measurements of desiccated tardigrades made by Pigòn and Węglarska [12.62] [12.63] were interpreted as indicating very low (400–600 times below the active level) but yet sustained metabolism. This conclusion has, however, been questioned since estimates of such low oxygen consumption levels may represent non-metabolic oxygen reactions rather than metabolism [12.44] [12.86]. Also, studies on desiccation-tolerant *Artemia* cysts have indicated that metabolic processes cannot be sustained below water levels of about 10% (0.1 g H_2O/g dry mass) [12.13], which is well above the dehydration levels that animals in anhydrobiosis may experience and survive. Even if providing direct evidence of absent metabolism is connected with methodological challenges in organisms with naturally very low metabolic rates, the fact that tardigrades can survive long periods in a dry state under very low temperatures [12.77] and without oxygen (e.g., under space vacuum) [12.41] should provide sufficient evidence that these animals are able to arrest their metabolism completely.

12.5 A Resilience Approach to the Cryptobiotic State

Cryptobiosis may be considered the most resilient biological system at the individual organism level. The concept of resilience captures a system's ability to remain within the boundaries of a defined structural or functional state when exposed to an external stress [12.29]. Cryptobiotic organisms extend the boundaries of life to their most extreme level, a level at which the biological structure is the only remaining characteristic defining the organism as being alive. This structure also maintains the ability to react to an external environmental stimulus, and to reorganize the body and return to its active metabolic state. In the field of ecological resilience, a common metaphor is the "ball and cup" [12.50], where the ball represents the current state of the system (which in our case represents an individual organism), and the cup-shaped landscape represents possible states of the system (Figure 12.2).

Each cup represents a "state regime," and in Figure 12.2 the regimes of life and death of an organism are visualized. From the normal state within the life regime (Figure 12.2a) an environmental agent, e.g., desiccation, may push the organism towards and over the border of the death regime. This is the case in non-desiccation-tolerant organisms. Cryptobiotic organisms manage to create a pocket of a stable state (or temporary "basin of attraction") near the threshold between the life and death regimes, beyond which it would fall irreversibly into the death regime (Figure 12.2b). The existence of this pocket relies on the stable structural organization of the body and its cell components, discussed below. As we know from empirical evidence, the wall of this pocket will erode over time towards the death regime, due to oxidative processes, and eventually result in an inability to maintain structure and reorganize the body for a return to the active metabolic state. Revealing the processes involved in this

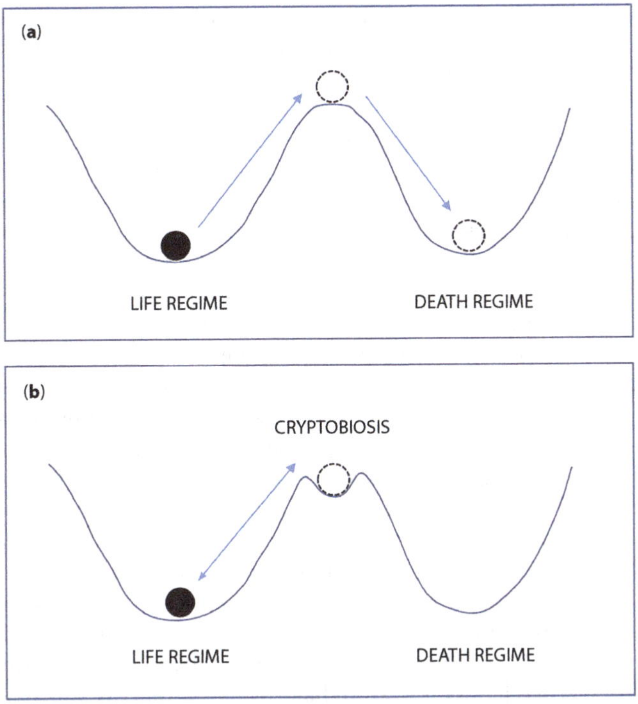

Figure 12.2 The ball and cup heuristics as used in resilience science to illustrate that systems may be in different states (regimes) in a system landscape but strive towards attractors (the lowest point) within each regime. Strong enough stress agents may, however, push the system over the threshold into a new regime. The ball represents the current state of the system. In Figure 12.2a the system landscape consists of a life regime and a death regime, which may represent the alternative of an individual organism. From a normal active life state within the life regime (black ball) stress agents may push the organism over the threshold and into the death regime. This is what happens when a desiccation-sensitive organism is exposed to severe desiccation. Figure 12.2b depicts the system change in a desiccation-tolerant organism such as tardigrades, where desiccation has pushed the organism towards the death regime, but adaptations for structural protection have created a stable pocket in between the life and death regimes, allowing it to maintain vitality in the dry state and return to the active life regime when water conditions change. (Image credit: K.I. Jönsson).

erosion, and the adaptations evolved to prevent erosions and rebuilding cellular structure is a major goal in understanding the evolution of cryptobiotic animals such as tardigrades. The molecular mechanisms reviewed in the next section represent such adaptations.

12.6 Molecular Mechanisms for Structural Stability in the Dry State

The conclusion already established in the Broca report [12.10] (mentioned above in Section 12.4), suggesting metabolic arrest in the desiccated state and a crucial importance of structural integrity, still holds and represents a foundation for current research on anhydrobiotic animals. Thus, understanding the mechanisms that support structural integrity of cells and tissues in anhydrobiotic animals has been a main focus of modern studies of cryptobiosis, with a particular interest in revealing the "desiccome," the adaptive set of genes, proteins and metabolites behind the ability to survive complete desiccation [12.52]. This research has developed along several lines, two of which are the complementary hypotheses of "water replacement" and "vitrification" [12.16] [12.74], in which the disaccharide trehalose has played a central role. Maintaining the structural integrity of cell membranes is an essential aspect of desiccation tolerance, and starting from physiological studies on desiccation-tolerant nematodes and tardigrades in the early 1970s, John Crowe and his collaborators developed seminal investigations on membrane systems under dehydration over the next 40 years. The role of the disaccharide trehalose in replacing water in membranes under dehydration was one of the main areas of study for this research [12.17]. Trehalose is synthesized to high levels (15–20% dw) in several groups of invertebrates capable of surviving complete dehydration, such as nematodes [12.57], insects [12.82] and *Artemia* [12.12], and has proved to protect and preserve membranes, liposomes and proteins in model systems against damage in the dry state [12.15]. Trehalose as well as other disaccharides also have the property of forming a glassy state (vitrification) at low water levels, creating a stable solid-like matrix that protects membranes and intracellular structure in the dry state. Sakurai *et al.* [12.73] have provided evidence for such protective mechanism of trehalose in desiccation-tolerant larvae of the chironomid *Polypedilum vanderplanki*. In tardigrades, trehalose is induced by dehydration in some desiccation-tolerant species but not in others, but the levels of trehalose are very low (<3% dw in all species examined so far and in most species <1%) [12.26] [12.40] [12.83]. Also, desiccation-tolerant rotifers have been reported to have no trehalose [12.51]. Therefore, if trehalose has a protective role related to desiccation tolerance in some species, other biochemical mechanisms must have evolved in tardigrade species with absent or very low levels of trehalose. Hengherr *et al.* [12.27] reported evidence of glass transition and vitrified states in two tardigrade species of the tardigrade family Macrobiotidae, which also is the family where the highest levels of trehalose have been found. Interestingly, vitrified states have been documented also in tardigrades without trehalose, and Boothby *et al.* [12.9] provided evidence that "tardigrade-specific intrinsically disordered proteins" (TDPs) may act in a similar way as trehalose by creating a glassy state in dry tardigrades of the species *Hypsibius exemplaris.*

Over the last decade, our knowledge of the molecular biology of tardigrades has improved considerably, especially related to genes and molecular components involved in tolerance to desiccation and survival in the anhydrobiotic state [12.74]. The genome sequencing of two tardigrade species (*Hypsibius exemplaris, Ramazzottius varieornatus*) [12.5] [12.24] [12.46] and transcriptome sequencing of a third species (*Milnesium tardigradum*) [12.81] has contributed considerably to this. Several tardigrade-specific genes and proteins with proposed roles in desiccation tolerance have been identified (e.g., Dsup [12.24], TDPs [12.9]). Although we still do not have a clear understanding of the molecular mechanisms behind desiccation tolerance in tardigrades, the available data strongly suggest two things: (i) that multiple molecular components (sugars, proteins) are involved and likely act in synergistic ways, and (ii) that multiple adaptive mechanisms have evolved within the tardigrade phylum as a result of natural selection for desiccation tolerance. Tardigrades therefore represent a very interesting group of animals for studies on convergent evolution of adaptive mechanisms for stress responses and survival in extreme environments.

12.7 Tardigrades as Astrobiological Models

On September 14, 2007, around 3000 desiccated tardigrades of four different species entered the low Earth orbit in the FOTON-M3 mission, launched at Baikonur in Kazakhstan. The tardigrades were kept in a sample container within the BIOPAN 6 experimental facility provided by the European Space Agency. About half of the samples were protected from the sunlight, the other half received UV radiation of either the full UV spectrum (117–400 nm) or the UV_A/UV_B spectrum (280–400 nm). All were exposed to space vacuum and cosmic radiation. For ten days the animals were orbiting around the Earth, until they were navigated back through the atmosphere and landed in Kazakhstan. This was the first time in history that an animal was exposed to the combined exposure of space vacuum, cosmic radiation, and UV radiation in an experiment in space. A year later the results were published, reporting two astonishing findings: that ten days under space vacuum and cosmic radiation had no apparent effect on the animals, and that some animals of one of the tardigrade species (*M. tardigradum*) exposed to very high doses of the UV_A/UV_B spectrum had survived and managed to reproduce [12.41] (Figure 12.3). Tardigrades became the first animals in history to survive conditions close to "open space." A later publication reported no evidence of reduced fitness in the F1 and F3 offspring generations of the *M. tardigradum* animals exposed in the TARDIS experiment [12.42].

The idea that tardigrades may be able to survive in space was not new, however. Already in 1923, the German zoologist Gilbert Rahm wrote, "The theory of panspermia represented and spread by the Swedish scientist Svante Arrhenius does not encounter any problem from a biological standpoint, even if his doctrine sounds fantastic and lacks evidence. The cold condition in space, the ultraviolet radiation, and even the lack of oxygen, can these animals, tardigrades, rotifers and nematodes withstand without damage" [12.64]. Later on, the French biologist Raoul May and collaborators wrote that "Tardigrades are a material of choice for biological studies on interplanetary spaces" ([12.58], p. 365).

Although the TARDIS experiment was the first study where tardigrades were exposed to open space conditions at low Earth orbit, there were also other experiment with tardigrades included in the FOTON-M3 mission (RoTaRad [12.61]; TARSE [12.68] [12.70]). These did not expose the animals to either space vacuum or UV radiation, only cosmic radiation. In the RoTaRad experiment adult tardigrades and embryos of the species *R. coronifer* were exposed to an estimated dose of cosmic radiation of 4 Gy, and in contrast to the TARDIS experiment, survival for both groups was lower compared to controls [12.60]. The TARSE experiment included both desiccated and (for the first time) active hydrated tardigrades of the species *Paramacrobiotus richtersi* that were exposed to low doses (1.9 mGy) of cosmic ionizing radiation [12.70]. In both hydrated and dehydrated tardigrades no effects were observed on survival or DNA damage from space exposure [12.68] [12.70], but the space exposed tardigrades had increased levels of heat shock proteins [12.70].

In the last flight of the space shuttle Endeavour to the International Space Station, desiccated samples of tardigrades were carried inside the aircraft in the TARDIKISS project. The cosmic radiation dose after the 16-day flight was only 300–400 uSv, and no significant effects on survival of the tardigrades were seen [12.80]. To my knowledge, there have been no subsequent experiments with tardigrades under real space conditions. Tardigrades were also contained in the Phobos LIFE experiment [12.8], which was part of the interplanetary

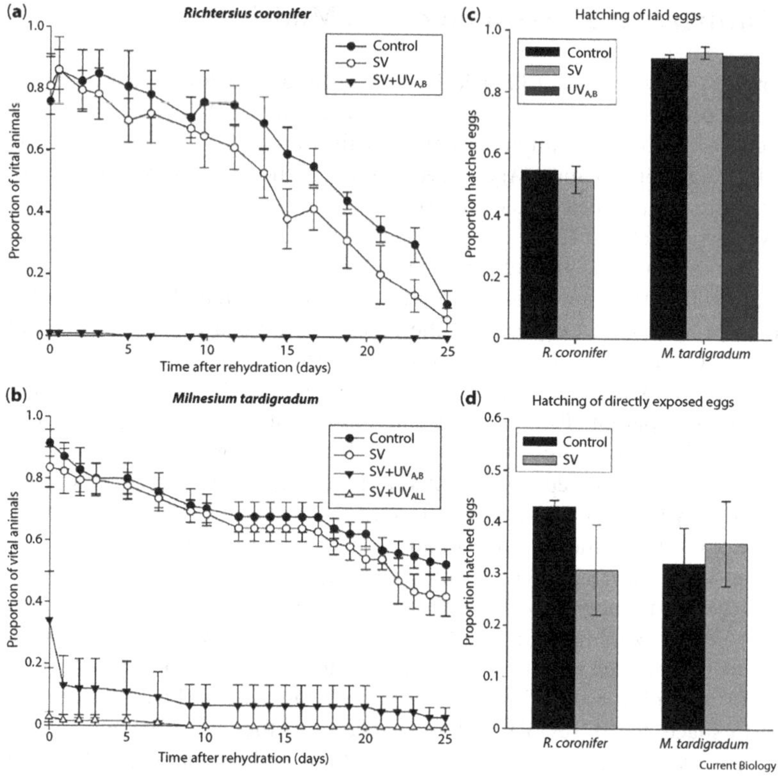

Figure 12.3 The main results from the TARDIS experiment performed within the BIOPAN 6 platform (provided by the European Space Agency) on the FOTON-M3 mission launched in 2007. The figure shows data on adult vitality (survival, a,b) and egg hatching (c,d) from two of the four tardigrade species included in the study, *Richtersius coronifer* and *Milnesium tardigradum*. Vitality data show that space vacuum (SV) had negligible effects on the animals, while tardigrades exposed to both vacuum and UV radiation (SV+UV) were much more affected. Hatching rate of eggs laid by tardigrades after the exposure in space (c) did not show any effect of space vacuum or UV radiation, while eggs exposed directly in space (d) did not hatch after exposure to UV but were not significantly affected by space vacuum. (Reprinted with permission from Jönsson *et al.* [12.41]).

Phobos Grunt mission launched in November 2011 and destined for a 34 month return trip to Phobos, one of the moons of Mars. The samples in the Phobos LIFE module were planned to be exposed to space radiation, but not to space vacuum or UV. Unfortunately, control of the spacecraft was lost soon after launch and it returned to Earth and fell into the Pacific Ocean on January 2012.

Given the multiple tolerances that some tardigrades, along with some rotifers, nematodes and a few arthropods, have to desiccation, high and low temperature, and radiation, they are exceptionally well suited as metazoan model organisms in astrobiological research [12.7] [12.23] [12.30] [12.37]. Such research can provide important knowledge on a variety of issues related to exposure of life in space, including, e.g., extraterrestrial life, interplanetary transfer of life forms, ecological life support systems in space, and space safety [12.1].

Unfortunately, opportunities to expose organisms under real space conditions during short-term periods (a few weeks), similar to those provided by the FOTON recoverable spacecraft, are rare nowadays.

12.8 Tardigrades - Extremotolerants or Extremophiles?

Tardigrades are sometimes considered as extremophiles (e.g., [12.75]), but for several reasons this is a characterization that needs clarification. First, it is important to remember that the phylum Tardigrada is a diverse taxonomic group, currently including more than 1300 different species [12.18] inhabiting a wide variety of habitats, and with some species being very tolerant to environmental extremes while other species are not very tolerant. This variation in tolerance most likely originates from adaptive evolution in response to selection from habitat conditions. Thus, tardigrades inhabiting so-called limnoterrestrial habitat (e.g., terrestrial mosses and lichens) generally show high tolerance to desiccation, while more limnic species are more sensitive to desiccation [12.31] [12.33]. Also, in marine tardigrades, differences in desiccation tolerance related to habitat and natural exposure to desiccation have been reported. Species inhabiting intertidal areas where they are regularly exposed to periods of desiccation also show tolerance to desiccation, while species inhabiting marine habitats that do not dry out are not desiccation tolerant [12.34]. The most desiccation-tolerant marine species is *Echiniscoides sigismundi* which lives on barnacles in the intertidal zone [12.32].

Second, whether one should classify tardigrades with cryptobiotic capabilities as extremophiles or not depends on how we define the term extremophile. In the original formulation by MacElroy [12.56], an extremophile was described as an "organism able to populate environments hostile to mesophiles, or organisms which grow only in intermediate environments." Such definition, where the criterium is based on the *ability* of an organism to populate an extreme environment, would definitely allow tardigrades to be considered extremophiles. Tardigrade species with cryptobiotic adaptations are able to persist in many habitats where other aquatic invertebrates (apart from cryptobiotic nematodes, rotifers, and arthropods), cannot survive due to temporary hostile conditions that prevent maintained metabolism.

Wharton [12.85] made a distinction between extremophiles and extremotolerants, where the former are organisms that have their physical life space and optimal physiological performance outside the life space of what we consider "normal" conditions, while extremotolerants do not deviate from mesophiles in the conditions under which they show optimal physiological performance in their active state, but which can tolerate and survive extreme physical conditions by entering specific survival states.

Rampelotto [12.66] also made an attempt to find a terminological distinction between the two groups discussed by Wharton [12.85], but in a different way. He divided extremophiles into two categories: "extremophilic organisms which require one or more extreme conditions in order to grow, and extremotolerant organisms which can tolerate extreme values of one or more physicochemical parameters though growing optimally at 'normal' conditions" ([12.66], p. 482). I don't see this conceptual proposal as a progress compared to the extremophile–extremotolerant distinction by Wharton [12.85], since the *extremophilic* term is more or less identical to the *extremophile* term. A new higher level term incorporating both extremophiles and extremotolerant organisms, or a new term (replacing *extremophiles*) for organisms requiring extreme conditions while keeping the extremophile term as the overarching concept, would be a better conceptual alternative.

While many tardigrade species qualify to be called extremotolerants, there are so far few examples of tardigrade species that belong to the traditional extremophile categories (thermophiles, halophiles, etc.). Most tardigrades have their optimal performance in the active state under quite normal conditions. With respect to water temperature, available experimental studies suggest that hydrated eutardigrades do not survive above 36–37 °C [12.53] [12.69] and temperatures for optimal performance under culturing conditions is reported to be in the range between 10 and 25 °C, depending on species [12.3] [12.4] [12.53]. In the desiccated anhydrobiotic state, temperature tolerance of some tardigrades is considerably higher (air temperature up to 100 °C) [12.27], but at this state the animal cannot grow or reproduce, only survive until hydrated conditions return.

There are, however, some reports of tardigrades from more permanently extreme environmental conditions. This includes the report of a taxonomically unique tardigrade from hot springs in Japan by Rahm [12.65], where water temperature was estimated at ca. 40 °C [12.20]. This record has received much attention since it represents the only report of tardigrades in the proposed class Mesotardigrada, an intermediate taxon between the main classes Heterotardigrada and Eutardigrada. The type specimens described by Rahm [12.65] are lost and recent attempts to find the type locality and new specimens of this unique tardigrade were unsuccessful [12.20]. This finding and the description of the Mesotardigrada class is therefore currently considered a *nomen dubium* [12.20]. Grothman *et al.* [12.20] also mention some other old reports from the 1920s and 1930s of tardigrades from hot springs (32–38 °C) in Japan, but so far there are no more recent verifications of these findings. Tardigrades have also been reported from homothermic (<20 °C) and radioactive springs in Greenland [12.47] [12.48]. Another extreme environment where tardigrades have been found are the deep-sea manganese nodules [12.11], at a depth of around 4000 m. The environmental conditions at that depth are characterized by constantly low temperature at 2–3 °C and a pressure of about 400 atmospheres (or 400 times that at sea level) [12.78]. Finally, marine tardigrades (*Dipodarctus subterraneus, Tanarctus ramazzottii*) have also been reported from benthic environment in the Black Sea, at a depth of 88–250 m where the level of oxygen was estimated at 0.12–0.16 PPM, suggesting that these populations live under permanent anoxic conditions [12.45].

It is most likely that the tardigrade populations living under these extreme environmental conditions also have evolved biochemical and physiological adaptations that prevent them from persisting under more normal conditions, and that they therefore may be considered true thermophiles (organisms with optimal performance under high temperature), psychrophiles (organisms with optimal performance under low temperature), or barophiles/piezophiles (organisms with optimal performance under high hydrostatic pressure). However, evidence of this is currently lacking. I have not found any evidence of halophilic or acidophilic/alkaliphilic tardigrades.

An interesting question related to the extremotolerant vs. extremophile status of tardigrades is whether species inhabiting environments that regularly dry out also have evolved a *requirement* of entrance into this cryptobiotic state for optimal performance. For instance, Westh and Kristensen ([12.84], p. 694) claimed that the eutardigrade *Richtersius coronifer* (previously *Adorybiotus coronifer*) "has to dry out several times before it can reproduce." I am not aware of any experimental study evaluating this particular proposition, but there are behavioral observations that tardigrades in culture may actively search for opportunities to dry out. In a life-history study on *Milnesium tardigradum* Suzuki ([12.76], p. 56) reported that "In cases where the side-walls of the culture dishes were coated with agar, tardigrades repeatedly crawled out of the water and entered a state of anhydrobiosis." While culturing of desiccation-tolerant tardigrades like *M. tardigradum* under permanently hydrated conditions show that regular dehydration is certainly not obligatory for persistence of the cultured populations, fitness could still be influenced. However, in a comparison between cultures of *M. tardigradum* kept under continuous hydration and cultures dehydrated repeatedly at 7-day intervals, Hengherr *et al.* [12.25] did not find any difference in total life span (dry period of the dehydrated group disregarded). No statistical analysis was made of reproductive parameters in this study, but both treatment groups were reported to show normal life history and reproduction. This suggests that regular dehydration has no direct positive effect on fitness, but more studies evaluating this are needed.

If a positive fitness effect of regular entrance into anhydrobiosis would indeed be verified, this would qualify anhydrobiotic tardigrades not only to be called xerotolerant, but also xerophilic, showing optimal performance only when exposed to desiccation. The possible mechanism underlying such a positive link between desiccation and animal performance remains to be elucidated.

Acknowledgments

I am grateful to Prof. Reinhardt M. Kristensen for information about tardigrades in extremophilic environments, and to Prof. Andrew Gooday for information about the deep-sea manganese nodule environment.

References

12.1. Alekseev, V.R., Sychev, V.N., Novikova, N.D., Studying the phenomenon of dormancy – why it is important for space exploration, in: *Diapause in Aquatic Invertebrates*, V.R. Alekseev, B. De Stasio, J.J. Gilbert (Eds.), pp. 207–214, Springer, Dordrecht, 2007.

12.2. Alpert, P., Constraints of tolerance: Why are desiccation-tolerant organisms so small or rare? *J. Exp. Biol.*, 209, 1575–1584, 2006.

12.3. Altiero, T. and Rebecchi, L., Rearing tardigrades: results and problems. *Zool. Anz.*, 240, 217–221, 2001.

12.4. Altiero, T., Suzuki, A.C., Rebecchi, L., Reproduction, development and life cycles, in: *Water Bears: The Biology of Tardigrades*, R.O. Schill (Ed.), Zoological Monographs vol. 2, pp. 211–247, Springer, Cham, 2018.

12.5. Arakawa, K., No evidence for extensive horizontal gene transfer from the draft genome of a tardigrade. *Proc. Natl. Acad. Sci. (USA)*, 113, 22, e3057, 2016.

12.6. Bertolani, R. and Grimaldi, D., A new Eutardigrade (Tardigrada: Milnesiidae) in amber from the upper cretaceous (Turonian) of New Jersey, in: *Studies on Fossils in Amber, with Particular Reference to the Cretaceous of New Jersey*, D. Grimaldi (Ed.), pp. 103–110, Backhuys, Leiden, 2000.

12.7. Bertolani, R., Rebecchi, L., Jönsson, K.I., Borsari, S., Guidetti, R., Altiero, T., Tardigrades as a model for experiences of animal survival in the space. *Microgravity and Space Station Utility*, 2, 211–212, 2001.
12.8. Betts, B.H., Warmflash, D., Frazc, R.E., Friedman, L., Vorobyova, E., Lilburn, T.G., Smith, A., Rettberg, P., Jönsson, K.I., Ciftcioglu, N. *et al.*, Phobos LIFE (Living Interplanetary Flight Experiment). *Astrobiology*, 19, 9, 1177–1185, 2019.
12.9. Boothby, T.C., Tapia, H., Brozena, A.H., Piszkiewicz, S., Smith, A.E., Giovannini, I., Rebecchi, L., Pielak, G.J., Koshland, D., Goldstein, B., Tardigrades use intrinsically disordered proteins to survive desiccation. *Mol. Cell*, 65, 975–984, 2017.
12.10. Broca, P., Rapport sur la question soumise à la Société de Biologie au suject de la reviviscence des animaux desséchés. *Mém. Soc. Biol. Paris*, 2, 1–140, 1860.
12.11. Bussau, C., New deep-sea Tardigrada (Arthrotardigrada, Halechiniscidae) from a manganese nodule area of eastern South Pacific. *Zool. Scr.*, 21, 1, 79–91, 1992.
12.12. Clegg, J.S., The origin of trehalose and its significance during the formation of encysted dormant embryos of *Artemia salina*. *Comp. Biochem. Physiol.*, 14, 135–143, 1965.
12.13. Clegg, J.S., The physical properties and metabolic status of *Artemia* cysts at low water contents: The 'water replacement hypothesis', in: *Membranes, Metabolism and Dry Organisms*, A.C. Leopold (Ed.), pp. 169–187, Cornell University Press, Ithaca, NY, 1986.
12.14. Clegg, J.S., Cryptobiosis: A peculiar state of biological organization. *Comp. Biochem. Physiol. B*, 128, 613–624, 2001.
12.15. Crowe, J.H., Trehalose as a "chemical chaperone": Fact and fantasy. *Adv. Exp. Med. Biol.*, 594, 143–158, 2008.
12.16. Crowe, J.H., Anhydrobiosis: An unsolved problem with applications in human welfare, in: *Membrane Hydration: The Role of Water in the Structure and Function of Biological Membranes*, E.A. Disalvo (Ed.), Subcellular Biochemistry 71, pp. 263–280, Springer, Cham, 2015.
12.17. Crowe, J.H., Crowe, L.M., Oliver, A.E., Tsvetkova, N., Wolkers, W., Tablin, F., The trehalose myth revisited: Introduction to a symposium on stabilization of cells in the dry state. *Cryobiology*, 43, 89–105, 2001.
12.18. Degma, P., Bertolani, R. & Guidetti, R. Actual checklist of Tardigrada species 2009-2020, 38th Edition, 2020, http://dx.doi.org/10.25431/11380_1178608.
12.19. Greven, H., From Johann August Ephraim Goeze to Ernst Marcus: A ramble through the history of early tardigrade research (1773 until 1929), in: *Water Bears: The Biology of Tardigrades*, R.O. Schill (Ed.), Zoological Monographs vol. 2, pp. 1–55, Springer, Cham, 2018.
12.20. Grothman, G.T., Johansson, C., Chilton, G., Kagoshima, H., Tsujimoto, M., Suzuki, A.C., Gilbert Rahm and the status of Mesotardigrada Rahm, 1937. *Zool. Sci.*, 34, 5–10, 2017.
12.21. Guidetti, R. and Bertolani, R., Paleontology and molecular dating, in: *Water Bears: The Biology of Tardigrades*, R.O. Schill (Ed.), Zoological Monographs vol. 2, pp. 131–143, Springer, Cham, 2018.
12.22. Guidetti, R., Altiero, T., Rebecchi, L., On dormancy strategies in tardigrades. *J. Insect Physiol.*, 57, 567–576, 2011.
12.23. Guidetti, R., Rizzo, A.M., Altiero, T., Rebecchi, L., What can we learn from the toughest animals of the Earth? Waterbears (tardigrades) as multicellular model organisms in order to perform scientific preparations for lunar exploration. *Planet. Space Sci.*, 74, 97–102, 2012.
12.24. Hashimoto, T., Horikawa, D.D., Saito, Y., Kuwahara, H., Kozuka-Hata, H., Shin-I, T., Minakuchi, Y., Ohishi, K., Motoyama, A., Aizu, T. *et al.*, Extremotolerant tardigrade genome and improved radiotolerance of human cultured cells by tardigrade-unique protein. *Nat. Commun.*, 7, 12808, 2016.

12.25. Hengherr, S., Brümmer, F., Schill, R.O., Anhydrobiosis in tardigrades and its effects on longevity. *J. Zool.*, 275, 216–220, 2008.
12.26. Hengherr, S., Heyer, A.G., Köhler, H.R., Schill, R.O., Trehalose and anhydrobiosis in tardigrades - evidence for divergence in responses to dehydration. *FEBS J.*, 275, 2, 281–288, 2008.
12.27. Hengherr, S., Worland, M.R., Reuner, A., Brümmer, F., Schill, R.O., High temperature tolerance in anhydrobiotic tardigrades is limited by glass transition. *Physiol. Biochem. Zool.*, 82, 749–755, 2009.
12.28. Hochachka, P.W. and Guppy, M., *Metabolic Arrest and the Control of Biological Time*, Harvard University Press, Cambridge, Mass, 1987.
12.29. Holling, C.S., Resilience and stability of ecological systems. *Annu. Rev. Ecol. Syst.*, 4, 1–23, 1973.
12.30. Horikawa, D.D., The tardigrade *Ramazzottius varieornatus* as a model animal for astrobiological studies. *Biol. Sci. Space*, 22, 3, 93–98, 2008.
12.31. Horikawa, D.D., Cumbers, J., Sakakibara, I., Rogoff, D., Leuko, S., Harnoto, R., Arakawa, K., Katayama, T., Kunieda, T., Toyoda, A. *et al.*, Analysis of DNA repair and protection in the tardigrade *Ramazzottius varieornatus* and *Hypsibius dujardini* after exposure to UVC radiation. *PLoS ONE*, 8, 6, e64793, 2013.
12.32. Hygum, T.L., Clausen, L.K.B., Halberg, K.A., Jørgensen, A., Møbjerg, N., Tun formation is not a prerequisite for desiccation tolerance in the marine tidal tardigrade *Echiniscoides sigismundi*. *Zool. J. Linn. Soc.*, 178, 4, 907–911, 2016.
12.33. Ito, M., Saigo, T., Abe, W., Kubo, T., Kunieda, T., Establishment of an isogenic strain of the desiccation-sensitive tardigrade *Isohypsibius myrops* (Parachela, Eutardigrada) and its life history traits. *Zool. J. Linn. Soc.*, 178, 863–870, 2016.
12.34. Jørgensen, A. and Møbjerg, N., Notes on the cryptobiotic capability of the marine arthrotardigrades *Styraconyx haploceros* (Halechiniscidae) and *Batillipes pennaki* (Batillipedidae) from the tidal zone in Roscoff, France. *Mar. Biol. Res.*, 11, 214–217, 2015.
12.35. Jørgensen, A., Møbjerg, N., Kristensen, R.M., A molecular study of the tardigrade *Echiniscus testudo* (Echiniscidae) reveals low DNA sequence diversity over a large geographical area. *J. Limnol.*, 66, 77–83, 2007.
12.36. Jönsson, K.I., The evolution of life-histories in holo-anhydrobiotic animals: A first approach. *Integr. Comp. Biol.*, 45, 764–770, 2005.
12.37. Jönsson, K.I., Tardigrades as a potential model organism in space research. *Astrobiology*, 7, 757–766, 2007.
12.38. Jönsson, K.I. and Rebecchi, L., Experimentally induced anhydrobiosis in the tardigrade *Richtersius coronifer*: Phenotypic factors affecting survival. *J. Exp. Zool.*, 293, 578–584, 2002.
12.39. Jönsson, K.I. and Järemo, J., A model on the evolution of cryptobiosis. *Ann. Zool. Fenn.*, 40, 331–340, 2003.
12.40. Jönsson, K.I. and Persson, O., Trehalose in three species of desiccation tolerant tardigrades. *Open Zool. J.*, 3, 1–5, 2010.
12.41. Jönsson, K.I., Rabbow, E., Schill, R.O., Harms-Ringdahl, M., Rettberg, P., Tardigrades survive exposure to space in low Earth orbit. *Curr. Biol.*, 18, 17, R729–R731, 2008.
12.42. Jönsson, K.I., Schill, R.O., Rabbow, E., Rettberg, P., Harms-Ringdahl, M., The fate of the TARDIS offspring: No intergenerational effects of space exposure. *Zool. J. Linn. Soc.*, 178, 924–930, 2016.
12.43. Jönsson, K.I., Levine, E.B., Wojcik, A., Haghdoost, S., Harms-Ringdahl, M., Environmental adaptations: Radiation, in: *Water Bears: The Biology of Tardigrades*, R.O. Schill (Ed.), Zoological Monographs vol. 2, pp. 311–330, Springer, Cham, 2018.
12.44. Keilin, D.K., The Leeuwenhoek Lecture. The problem of anabiosis or latent life: history and current concept. *Proc. R. Soc. Lond., Ser. B*, 150, 149–191, 1959.

12.45. Kharkevych, Kh.O. and Sergeeva, N.G., Deep-water Tardigrada of the Istanbul Strait's (Bosporus) outlet area of the Black Sea. *Vestn. Zool.*, 47, 3, 17–27, 2013.
12.46. Koutsovoulos, G., Kumar, S., Laetsch, D.R., Stevens, L., Daub, J., Conlon, C., Maroon, H., Thomas, F., Aboobaker, A.A., Blaxter, M., No evidence for extensive horizontal gene transfer in the genome of the tardigrade *Hypsibius dujardini*. *Proc. Natl. Acad. Sci. USA*, 113, 18, 5053–5058, 2016.
12.47. Kristensen, R.M., New aberrant eutardigrades from homothermic springs on Disko Island, West Greenland, in: *Proceedings of the Third International Symposium on Tardigrada, Tennessee 1980*, D.R. Nelson (Ed.), pp. 203–220, 1982.
12.48. Kristensen, R.M., The "southern" flora and the "marine" fauna elements in the homothermic springs on Disko Island, West Greenland, in: *Grönland Exkursion 2.- 25. August 1987*, Institut für Polarökologie (Kiel), pp. 202–226, 1988.
12.49. Kroemer, G., Galluzzi, L., Vandenabeele, P., Abrams, J., Alnemri, E.S., Baehrecke, E.H., Blagosklonny, M.V., El-Deiry, W.S., Golstein, P., Green, D.R. *et al.*, Classification of cell death: Recommendations of the Nomenclature Committee on Cell Death 2009. *Cell Death Differ.*, 16, 3–11, 2009.
12.50. Lamothe, K.A., Somers, K.M., Jackson, D.A., Linking the ball-and-cup analogy and ordination trajectories to describe ecosystem stability, resistance, and resilience. *Ecosphere*, 10, 3, e02629, 2019.
12.51. Lapinski, J. and Tunnacliffe, A., Anhydrobiosis without trehalose in bdelloid rotifers. *FEBS Lett.*, 553, 387–390, 2003.
12.52. Leprince, O. and Buitink, J., Desiccation tolerance: From genomics to the field. *Plant Sci.*, 179, 554–564, 2010.
12.53. Li, X. and Wang, L., Effect of thermal acclimation on preferred temperature, avoidance temperature and lethal thermal maximum of *Macrobiotus harmsworthi* Murray (Tardigrada, Macrobiotidae). *J. Therm. Biol.*, 30, 443–448, 2005.
12.54. Little, C., *The Terrestrial Invasion - An Ecophysiological Approach to the Origins of Land Animals*, Cambridge University Press, Cambridge, 1990.
12.55. Maas, A. and Waloszek, D., Cambrian derivatives of the early arthropod stem lineage, Pentastomids, Tardigrades and Lobopodians – an 'Orsten' perspective. *Zool. Anz.*, 240, 451–459, 2001.
12.56. MacElroy, R.D., Some comments on the evolution of extremophiles. *Biosystems*, 6, 1, 74–75, 1974.
12.57. Madin, K.A.C. and Crowe, J.H., Anhydrobiosis in nematodes: Carbohydrate and lipid metabolism during drying. *J. Exp. Zool.*, 193, 335–342, 1975.
12.58. May, R.M., Maria, M., Guimard, J., Actions différentielles des rayons ex et ultraviolets sur le tardigrade *Macrobiotus areolatus*, à l'état actif et desséché. *Bull. Biol. Fr. Belg.*, 98, 349–367, 1964.
12.59. Neumann, S., Reuner, A., Brümmer, F., Schill, R.O., DNA damage in storage cells of anhydrobiotic tardigrades. *Comp. Biochem. Physiol. Part A: Mol. Integr. Physiol.*, 153, 425–429, 2009.
12.60. Olofsson, H., Ripa, J., Lundberg, P., Evolution of multi-stage dormancy in temporally autocorrelated environments. *Evol. Ecol. Res.*, 7, 8, 1125–1137, 2005.
12.61. Persson, D., Halberg, K.A., Jørgensen, A., Ricci, C., Møbjerg, N., Kristensen, R.M., Extreme stress tolerance in tardigrades: Surviving space conditions in low earth orbit. *J. Zool. Syst. Evol. Res.*, 49, Suppl. 1, 90–97, 2011.
12.62. Pigòn, A. and Węglarska, B., The respiration of Tardigrada: A study in animal anabiosis. *Bull. Acad. Polon. Sci., CL. II*, 1, 69–72, 1953.

12.63. Pigòn, A. and Węglarska, B., Rate of metabolism in tardigrades during active life and anabiosis. *Nature*, 176, 121–122, 1955.
12.64. Rahm, P.G., Gibt es einen Stillstand des Lebens, ohne daß der Tod eintritt. *Die Umschau*, 28, 51, 987–992, 1923.
12.65. Rahm, G., Eine neue Tardigraden-Ordnung aus den heißen Quellen von Unzen, Insel Kyushu, Japan. *Zool. Anz.*, 120, 65–71, 1937.
12.66. Rampelotto, P.H., Extremophiles and extreme environments. *Life (Basel)*, 3, 3, 482–485, 2013.
12.67. Rebecchi, L., Guidetti, R., Borsari, S., Altiero, T., Bertolani, R., Dynamics of long-term anhydrobiotic survival of lichen-dwelling tardigrades. *Hydrobiologia*, 558, 23–30, 2006.
12.68. Rebecchi, L., Altiero, T., Guidetti, R., Cesari, M., Bertolani, R., Negroni, M., Rizzo, A.M., Tardigrade resistance to space effects: first results of experiments on the LIFE-TARSE mission on FOTON-M3 (September 2007). *Astrobiology*, 9, 6, 581–591, 2009.
12.69. Rebecchi, L., Boschini, D., Cesari, M., Lencioni, V., Bertolani, R., Guidetti, R., Stress response of a boreo-alpine species of tardigrade, *Borealibius zetlandicus* (Eutardigrada, Hypsibiidae). *J. Limnol.*, 68, 1, 64–70, 2009.
12.70. Rebecchi, L., Altiero, T., Guidetti, R., Caselli, V., Cesari, M., Resistance of the anhydrobiotic eutardigrade *Paramacrobiotus richtersi* to space flight (LIFE–TARSE mission on FOTON-M3). *J. Zool. Syst. Evol. Res.*, 49, Suppl. 1, 98–103, 2011.
12.71. Ricci, C. and Covino, C., Anhydrobiosis of *Adineta ricciae*: Costs and benefits. *Hydrobiologia*, 546, 307–314, 2005.
12.72. Rota-Stabelli, O., Daley, A.C., Pisani, D., Molecular timetrees reveal a Cambrian colonization of land and a new scenario for ecdysozoan evolution. *Curr. Biol.*, 23, 5, 392–398, 2013.
12.73. Sakurai, M., Furuki, T., Akao, K., Tanaka, D., Nakahara, Y., Kikawada, T., Watanabe, M., Okuda, T., Vitrification is essential for anhydrobiosis in an african chironomid, *Polypedilum vanderplanki*. *Proc. Natl. Acad. Sci.*, 105, 5093–5098, 2008.
12.74. Schill, R.O. and Hengherr, S., Environmental adaptations: desiccation tolerance, in: *Water Bears: The Biology of Tardigrades*, R.O. Schill (Ed.), Zoological Monographs vol. 2, pp. 273–293, Springer, Cham, 2018.
12.75. Schulze-Makuch, D. and Seckbach, J., Tardigrades: an example of multicellular extremophiles, in: *Polyextremophiles. Cellular Origin, Life in Extreme Habitats and Astrobiology*, vol. 27, J. Seckbach, A. Oren, H. Stan-Lotter (Eds.), pp. 599–607, Springer, Dordrecht, 2013.
12.76. Suzuki, A.C., Life history of *Milnesium tardigradum* Doyère (Tardigrada) under a rearing environment. *Zool. Sci.*, 20, 49–57, 2003.
12.77. Sømme, L. and Meier, T., Cold tolerance in Tardigrada from Dronning Maud Land, Antarctica. *Polar Biol.*, 15, 221–224, 1995.
12.78. Thistle, D., The deep-sea floor: An overview, in: *Ecosystems of the World 28*, P.A. Tyler (Ed.), pp. 5–37, Elsevier Science, New York, 2003.
12.79. Tsujimoto, M., Imura, S., Kanda, H., Recovery and reproduction of an Antarctic tardigrade retrieved from a moss sample frozen for over 30 years. *Cryobiology*, 72, 1, 78–81, 2016.
12.80. Vukich, M., Ganga, P.L., Cavalieri, D., Rizzetto, L., Rivero, D., Pollastri, S., Mugnai, S., Mancuso, S., Pastorelli, S., Lambreva, M. *et al.*, BIOKIS: A model payload for multisciplinary experiments in microgravity. *Microgravity Sci. Technol.*, 24, 397–409, 2012.
12.81. Wang, C., Grohme, M.A., Mali, B., Schill, R.O., Frohme, M., Towards decrypting cryptobiosis – analyzing anhydrobiosis in the tardigrade *Milnesium tardigradum* using transcriptome sequencing. *PLoS ONE*, 9, 3, e92663, 2014.

12.82. Watanabe, M., Kikawada, T., Minagawa, N., Yukuhiro, F., Okuda, T., Mechanism allowing an insect to survive complete dehydration and extreme temperatures. *J. Exp. Biol.*, 205, 2799–2802, 2002.
12.83. Westh, P. and Ramløv, H., Trehalose accumulation in the tardigrade *Adorybiotus coronifer* during anhydrobiosis. *J. Exp. Zool.*, 258, 303–311, 1991.
12.84. Westh, P. and Kristensen, R.M., Ice formation in the freeze-tolerant eutardigrades *Adorybiotus coronifer* and *Amphibolus nebulosus* studied by differential scanning calorimetry. *Polar Biol.*, 12, 693–699, 1992.
12.85. Wharton, D.A., *Life at the Limits: Organisms in Extreme Environments*, Cambridge University Press, Cambridge, 2002.
12.86. Wright, J.C., Cryptobiosis 300 years from van Leuwenhoek: What have we learned about tardigrades? *Zool. Anz.*, 240, 563–582, 2001.

13

Spore-Forming Bacteria as Model Organisms for Studies in Astrobiology

Wayne L. Nicholson

Department of Microbiology and Cell Science, University of Florida, Merritt Island, Florida, USA

Abstract

Spore-forming bacteria have a long history of use as model organisms for studying astrobiology, including the origin of life, evolution and survival in extreme environments, and interplanetary transfer of life via natural processes and human activities. This chapter will discuss the various aspects of astrobiological studies utilizing bacterial spore formers.

Keywords: Astrobiology, *Bacillus*, endospore, extreme environments, lithopanspermia, Mars, panspermia, spore

13.1 Introduction

Bacterial endospores (hereafter referred to simply as spores) are the dormant resting forms of numerous Firmicutes in the Classes Bacilli and Clostridia [13.18]. While most spore-forming species may not be considered extremophiles in their vegetative growing form, in their spore form they are considered among the hardiest and longest-lived types of cells known [13.52] [13.56]. They can resist harsh physical and chemical treatments and can survive for extended (up to geological) periods of time. Furthermore, because spores are readily and widely distributed by wind and water, they can be found in virtually every conceivable ecological niche on Earth. Because of their extreme hardiness and longevity, spores have been considered superb experimental models to test several aspects of astrobiology. In this chapter I will review the history of the topic and report on recent advances in this area. This chapter expands and updates earlier reviews to which the reader is referred for background information [13.18] [13.50] [13.51] [13.52] [13.53] [13.54] [13.55] [13.56] [13.58] [13.65] [13.88].

Email: WLN@ufl.edu
Wayne L. Nicholson: https://scholar.google.com/citations?hl=en&user=3YxR-s0AAAAJ, http://microcell.ufl.edu/people/faculty-directory/nicholson/

Joseph Seckbach and Helga Stan-Lotter (eds.) Extremophiles as Astrobiological Models, (275–290)

13.2 Historical Beginnings

The modern beginnings of this story can largely be traced to the state of scientific thought concerning evolution, the origin and distribution of life, and microbiology, which were being formulated during the 19th century (reviewed in [13.10] [13.53] [13.63]).

13.2.1 Evolution and the Origin of Life

Charles Darwin published his groundbreaking work *On the Origin of Species* in 1859 [13.9], in which he put forth the theory now known as Evolution, that advanced forms of life arose from simpler forms through natural selection operating on naturally occurring heritable variations among individuals (in his words, "descent with modification"). The theory immediately presented a problem: if all modern life had descended from a primordial form, this implied that the first living form must have arisen from nonliving matter—a problem that Darwin avoided addressing in *Origin of Species.* Coincidentally, this very question was occupying some of the finest scientific minds of the time, who were wrestling over the notion that microbes could arise spontaneously from non-living material. This idea of spontaneous generation or abiogenesis was first coherently synthesized by Aristotle (384–322 BCE) and endured for the next ~2,000 years. Experiments conducted during the 17th and 18th centuries essentially put an end to the idea of spontaneous generation for macroscopic creatures such as maggots, mice, or scorpions. However, in the late 1600s, Antonie van Leeuwenhoek invented the microscope and described a hitherto unknown world of microscopic single-celled organisms. His discovery triggered a revival of the notion that these simple cells could arise by spontaneous generation; indeed, the presence of spores resistant to being killed by the sterilization techniques of the time were often the cause of spurious claims of spontaneous generation. Finally, in the mid-19th century the spontaneous generation of microbes was disproved through seminal experiments conducted by Louis Pasteur and John Tyndall, which indicated that even microbial life arose only from pre-existing life (i.e., biogenesis). Although Darwin was well aware of his contemporaries and their evidence against spontaneous generation, he realized that the origin of life must have involved an abiogenic event. He even went so far as to make his famous "warm little pond" conjecture in a letter to his friend Joseph Hooker in 1871 [13.63]. Darwin also correctly surmised that 19th century technology was not advanced enough to make significant inroads into understanding the origin of life. Indeed, research into the origin of life did not gain scientific traction until the early 20th century. The origin of life has since become a major topic of scientific investigation in several fields, including Astrobiology, but despite many theoretical insights gained, to date the problem still has not been solved experimentally (for a recent review, see [13.68]).

13.2.2 Panspermia

At the same time as the debates surrounding the origin of life and evolution were brewing, the 19th century also saw advances in the modern concept of panspermia, the idea that life could arise wherever conditions are favorable and be distributed throughout the universe. This idea also had ancient roots, being originally conceived by the pre-Socratic philosopher Anaxagoras (c. 510–c. 428 BCE). But a possible mechanism for panspermia was lacking until the 19th century, when H.E. Richter in 1865, and later Hermann von Helmholtz and William Thomson (Lord Kelvin) in 1871 and Ferdinand Cohn in 1872, each postulated a mechanism for transfer of life through space involving meteorites (reviewed in [13.10] [13.53]). In the version dubbed lithopanspermia ("panspermia in rocks"), comets or meteorites impacting a living planet like Earth could blast life-bearing rocks into space, and similar meteorites blasted off other living worlds might have inoculated the early Earth with life. This idea met with several criticisms, not the least of which was that panspermia merely kicked the origin-of-life question to another planet. In addition, at the time it was thought that living organisms could not possibly survive ejection by impact, transit through space, and entry onto another planet. These views began to change only in the late 20th century, as we shall see below.

13.2.3 Microbiology

To understand why spore-forming bacteria have become popular organisms for astrobiological studies also requires a foray back into the 19th century. From his microscopic studies of the life forms found inhabiting infusions of hay, in 1838 Ehrenberg reported the presence of small optically refractile bodies (now known as spores) located within rod-shaped bacterial cells [13.13]. Later, in the 1870s, Ferdinand Cohn, Robert Koch, and John Tyndall independently reported the existence of forms of bacteria from various plant infusions which were resistant to being killed by boiling [13.8] [13.31] [13.85], and these resistant forms were soon associated with the optically refractile spore bodies. While Cohn was concerned with the general understanding of the life history of the "hay bacillus" (now recognized as *Bacillus subtilis*), Koch was more concerned with understanding the causative agent of the disease anthrax, *Bacillus anthracis*. Both scientists observed that spores developed in nutrient-depleted cultures, and that they could rapidly germinate and resume growth when immersed in fresh nutrient solutions, thus establishing the fundamental life cycle of spore-forming bacteria.

The 19th century would also see the first attempts to organize and classify the bacterial world along the lines of the binomial nomenclature system of taxonomy, previously established for macroscopic life by Linnaeus in 1753. Early taxonomic schemes relied on microscopic descriptions of easily distinguishable morphological features such as cell shape. Thus the ability to produce optically refractile spores soon became a major defining characteristic in bacterial taxonomy. Over the next century it became evident that the vast majority of spore-forming bacteria were Gram-positive rod-shaped cells belonging to the taxonomic Classes Bacilli (aerobic) and Clostridia (anaerobic).

13.2.4 Bacterial Spores

From the early beginnings of bacteriology, scientists have worked to understand: (i) how bacterial cells could sporulate, i.e., sense nutrient depletion from their environment and differentiate into dormant, metabolically inactive spores; (ii) how spores could withstand harsh physical and chemical treatments known to be lethal to other forms of life; and (iii) how spores could germinate, i.e., sense the return of nutrients to their environment and rapidly resume growth. Through the first half of the 20th century, microbiological, physiological, and biochemical studies utilized a wide variety of spore-forming species of *Bacillus*

(*B. anthracis, B. cereus, B. coagulans, B. licheniformis, B. megaterium,* and *B. subtilis*, among others) as well as various *Clostridium* spp., to unravel the molecular and biochemical details underlying these developmental processes. However, discoveries and developments in the 1950s and 1960s led to the rise of *B. subtilis*, particularly descendants of a single strain designated 168, as the dominant experimental organism for the study of spore biology. First among these was the discovery by John Spizizen in 1958 of DNA-mediated transformation in *B. subtilis* strain 168 [13.80], followed soon afterward by the independent discoveries reported in 1961 and 1962 by Iwao Takahashi and Curtis Thorne of phage-mediated transduction in *B. subtilis* [13.82] [13.83]. At about the same time, Pierre Schaeffer and coworkers were creating and characterizing mutants of *B. subtilis* blocked at various stages of the sporulation process (these early studies are summarized in [13.67]). Application of these newly discovered genetic tools, as well as subsequent advances in molecular biology and genomics, opened the door for an explosion in sporulation studies using mainly *B. subtilis* strain 168 and its derivatives, which continue to the present day. For collections of detailed reviews on various aspect of *B. subtilis* spore biology, the reader is directed to recent volumes on the topic [13.12] [13.64] [13.79].

Due to its emergence as the best-characterized spore-forming bacterium, *B. subtilis* also emerged as the premier model organism for astrobiology studies using bacterial spores. Spores of *B. subtilis* have been measured to be orders of magnitude more resistant than their vegetative counterparts to various extreme treatments such as: wet and dry heat; repeated freeze/thaw cycles; extreme desiccation resulting from ultrahigh vacuum; various organic chemical disinfectants (e.g., chloroform, phenol, formaldehyde); acids and strong oxidants (e.g., hydrogen peroxide, nitrous acid, sodium perchlorate); and energetic radiation (UV and ionizing radiation), among others. Many of these harsh environments can be found in space and on other celestial bodies. As a result of intensive research over the past 70 years, investigators have explored and uncovered myriad molecular mechanisms that underlie the amazing resistance properties of bacterial spores. These mechanisms have been reviewed elsewhere in exhaustive detail; for a recent comprehensive review, the reader is referred to [13.73] and references therein.

13.3 Revival of Lithopanspermia

As mentioned earlier, a major argument against the lithopanspermia theory which lingered until the late 20th century was that microbial cells could not possibly survive the forces prevailing during impact-mediated launch, transit through space, and entry onto another planet. This view began to change in the early 1980s as a number of meteorites in collections were identified as having originated from Mars. Analyses of the isotopic composition of gases trapped within certain meteorites revealed that they closely matched the composition of the Mars atmosphere, as measured directly by the Viking landers in 1976 [13.3]. Since the 1980s, more than 200 confirmed Martian meteorites have been found on Earth (for details, see the Meteoritical Bulletin Database at https://www.lpi.usra.edu/meteor/). Detailed petrologic examinations of Martian meteorites revealed that they had experienced shock pressures of 5–45 GPa and corresponding temperature spikes of 10–800 °C [13.21] [13.60]. From a microbiological standpoint, the lower end of these pressure-temperature conditions would be too mild to sterilize the meteorite.

Strengthening this notion were results from magnetic analysis of Martian meteorite ALH84001, indicating that it had not been heated above 40 °C during ejection and transit from Mars to Earth [13.87].

How could rocks be ejected from the surface of a planet under such mild pressure-temperature conditions? Shortly after the discovery of Martian meteorites, planetary geophysicist Jay Melosh proposed a mechanism which explained how rocks could be launched off the surface of a planet into space by nearby hypervelocity impacts without suffering high shock pressures and temperatures. Around the margin of an impact site is located a spallation zone (spalls are fragments or chips from a piece of stone) where the reflected shock wave of the impact would create a very large pressure gradient near the surface crust. Because the pressure above the crustal surface is effectively zero, most of the energy of the reflected shock wave would be translated directly into acceleration of surface rocks upwards, rather than causing their compression and resultant heating [13.42] [13.43]. This spallation model has since been confirmed experimentally [13.23].

Because of their resistance to environmental extremes, bacterial spores are ubiquitous on Earth and have been isolated and characterized from nearly every environment, including the interiors of rocks (reviewed in [13.18] [13.52] [13.76] [13.78]). Of particular interest to lithopanspermia is whether spores can be found in the interiors of rocks situated near the surface where they could be ejected by spallation. We sampled the interiors of two common igneous rock types found at the Earth's surface, basalt and granite. We found that basalt, a low-porosity rock, harbored cultivable spores of *B. pumilus* and *B. subtilis* strains on the order of ~10^4 per kg [13.2]. In contrast granite, a highly porous rock, was found to contain on the order of ~10^5-10^6 cultivable spores per kg and a larger variety of *Bacillus* spp., including strains related to *B. megaterium, B. simplex, B drentensis, B. niacinii,* and *B. barbaricus* [13.19]. The finding of relatively large numbers of viable spores inside rocks at the Earth's surface, in a location most likely to be ejected by impacts, suggested that spores would be good experimental models to test the lithopanspermia hypothesis.

13.4 Testing Lithopanspermia Experimentally

For purposes of experimental investigation, lithopanspermia has been divided into three stages: (i) impact-mediated launch from the donor planet; (ii) transit through interplanetary space; and (iii) entry and deposition onto the recipient planet [13.46] [13.56]. A putative interplanetary traveler must survive all three stages, each fraught with its own particular hazards.

13.4.1 Impact-Mediated Launch

As discussed above, upon impact a microbe-bearing rock is subjected to a strong shock pressure wave which heats the rock. Survival of *B. subtilis* spores to pressure and thermal shocks simulating impact-mediated launch has been measured using layers of spores sandwiched between two disks of gabbro (an igneous rock) subjected to explosive compression at various pressures [13.29]. These experiments showed that spores were inactivated with first-order kinetics with respect to shock pressure; at 5 GPa, ~50% of spores survived, and at 40 GPa, survival averaged ~10^{-3} to 10^{-5} [13.29].

In addition to shock and heating, in order to escape into space upon impact a rock must also be subjected to acceleration from rest to escape velocity (5.03 and 11.2 km/sec for Mars and Earth, respectively) with very short rise times, on the order of 0.5 msec (so-called "jerk"). In an impact-mediated launch from Mars, a rock and its associated microbes would be subjected to a calculated acceleration of 3×10^6 m/sec^2 (~300,000 xg) and a corresponding jerk of 6×10^9 m/sec^3 [13.40]. By firing spores loaded into the backs of lead pellets into chilled plasticine targets we were able to exceed these values, and we showed that spores survived such extreme accelerations (actually, decelerations) at high frequencies (~90%). Thus it appeared that shock and heating, but not acceleration and jerk, are the major factors limiting spore survival to impact-mediated launch.

To test all launch parameters simultaneously, we devised an experiment in which we recreated an impact-mediated launch scenario using the NASA Ames Vertical Gas Gun [13.16]. In this experiment, *B. subtilis* spores were infused into a flat slab of granite to simulate the surface of a planet. We then impacted the granite target with an aluminum projectile fired at 5.4 km/sec, and captured the spall fragments in layers of polyurethane foam. Spall fragments were recovered aseptically from the foam and assayed for viable spores. Although the peak shock pressure at the impact site itself was greater than 50 GPa, the spall fragments were only lightly shocked (5–7 GPa), and we recovered spore survivors at a frequency of $\sim 10^{-4}$ [13.16]. These results indicated that rocks ejected by an impact could contain upwards of ~100 viable spores per kg. While this may not sound like much, keep in mind that a single impact of the magnitude of the 180-km diameter Chicxulub impact (which contributed to the Cretaceous-Tertiary mass extinctions 65 million years ago) would have ejected many billions of meter-sized terrestrial surface rocks from Earth's gravity well with relatively little shock damage [13.56]. The slowest of these ejecta would fall back to Earth or enter Earth's orbit; the fastest of these ejecta would escape Earth's orbit and enter heliocentric orbits that could eventually be captured by another celestial body.

13.4.2 Transit Through Interplanetary Space

Shortly after ejection from a planetary surface, spores would face an entirely new set of challenges to their survival—exposure to the extreme environment of space—characterized by: ultrahigh vacuum; extreme desiccation; high and low temperature extremes; and radiation from solar and galactic sources such as high-energy photons (UV, X-rays, γ-rays), electrons, protons, and heavy nuclei from He ($z = 2$) to U ($z = 92$) [13.56] [13.58]. Since the 1980s, spores of *B. subtilis* have been exposed to the space environment in low Earth orbit (LEO) on several missions ranging from a few days up to nearly 6 years (reviewed in [13.25] [13.27] [13.28] [13.56]). The results from these experiments can be summarized as follows. Unshielded monolayers of spores were inactivated within a few minutes exposure to solar UV [13.26], but minimal shielding provided by regolith, rock, or even the upper layers of spores in multilayers, could effectively block UV [13.56] [13.58]. When spores were shielded from solar UV but otherwise unprotected, they survived exposure to space in LEO for 1.5 years on the PROTECT experiment outside the International Space Station (55% survival) [13.28] and 5.7 years on the Long Duration Exposure Facility (LDEF) satellite (0.3% survival) [13.25]. Survival of spores on the LDEF satellite was considerably enhanced by prior air-drying of spores in the presence of buffer salts (11% survival) or glucose (28% survival). Because both the PROTECT and LDEF experiments were performed in LEO, the spacecraft were periodically shaded from the sun, thus experienced temperature cycling ranging from -20 to +59 (PROTECT) and +9 to +29 (LDEF). In addition, both spacecraft flew inside the Earth's magnetic field, so were subjected to an attenuated ionizing radiation field, resulting in rather mild exposure to ionizing radiation (0.14 and 4.8 Gy, respectively) relative to interplanetary space. In aggregate, the results from these long-duration experiments indicated that spores shielded from solar UV could survive for several years in the space environment.

One factor potentially limiting spore survival is the time spent during the interplanetary journey through space. As mentioned above, upon ejection, rocks enter heliocentric orbits that are altered by the gravitational effects of the planets, which upon close passage can "slingshot" them into new orbits either closer to, or farther away from, the Sun. The time that a rock has spent in interplanetary space can be determined by analysis of how exposure to cosmic rays has changed the ratios of certain isotopes, the so-called cosmic-ray exposure (CRE) age. Measurements of the CRE ages of 16 Martian meteorites indicated relatively long transit times, on the order of 0.6–14.7 million years [13.14] [13.60]. Calculations of survival times of spores based on thermal inactivation kinetics indicated that spores of mesophilic species such as *B. subtilis* would be unlikely to survive a journey of these long durations, but that spores of thermophiles might persist for such lengths of time [13.48]. However, it should be noted that such a small sampling of CRE ages reveals only the typical or bulk-average transfer times of Mars meteorites, and would likely miss rarer fast-transfer events. Numerical simulations have indicated that a small but appreciable fraction of rocks ejected from Mars (up to 0.1%) would reach Earth in less than 100,000 years [13.46]. Similar calculations suggested that ejecta from Earth could reach Venus (0.1%) and Mars (0.001%) within the relatively short time span of 30,000 years or less [13.22]. Spore longevity calculations based on thermal inactivation kinetics [13.48] indicated that even modest populations of spores from mesophilic species like *B. subtilis* could survive these shorter sojourns through space. Extremely short transit times of ejecta ranging from a few months to a few years are theoretically possible (just as modern spacecraft are launched on trajectories taking them from Earth to Mars in 5–10 months), but numerical simulations suggest that the fraction of impact-ejected rocks entering such direct orbits becomes exceedingly low [13.44]. Again, these low probabilities are offset by the large numbers of rocks ejected per impact and the large number of impacts suffered by planets since their origin [13.46].

13.4.3 Atmospheric Entry and Deposition

The final stage of lithopanspermia concerns the capture of a rock from space by the gravity field of the recipient planet, its entry through the atmosphere, and its deposition by impact into the planet's surface (discussed in detail in [13.17] [13.56]). Natural meteoroids enter the recipient planet's atmosphere at speeds higher than escape velocity and are subjected to large aerodynamic stresses and strong frictional heating. This stress and heating commonly breaks the meteor into smaller fragments [13.62] [13.81] and causes melting at the surface of the rocks [13.6]. The melted material is ablated away in an incandescent trail visible from the ground, and the ablation process leaves a blackened surface crust on the resulting meteorite. Despite this fiery display, below a few millimeters the interior of a meteorite never receives strong heating, as the total duration of entry is at most a few tens of seconds [13.17].

Relatively few experimental attempts have been made to measure spore survival to simulated hypervelocity atmospheric entry. We first showed that *B. subtilis* spores, embedded in artificial granite meteorites attached to the exterior of a sounding rocket, could survive atmospheric entry at 1.2 km/sec and heating to a peak temperature of 145 °C [13.17]. However, this speed is much below the typical entry speed of an interplanetary meteor, >11.2 and >5.03 km/sec for Earth and Mars, respectively. In a later experiment called STONE-5, *B. subtilis* spores in dolerite (a basaltic rock) were subjected to a higher entry velocity of 7.6 km/sec on the leading face of the Russian FOTON-2 capsule as it returned from orbiting the Earth, but no viable spores were recovered [13.5]. However, in a more recent experiment, spores of the thermophilic anaerobic bacterium *Thermoanaerobacter siderophilus* embedded in basalt were demonstrated to survive entry from Earth's orbit at 7.6 km/sec on the leading face of the FOTON-M4 capsule [13.77].

Concerning deposition onto the surface of the recipient planet, it turns out that planets with atmospheres are nearly ideal recipients of potentially life-bearing rock fragments (reviewed in [13.56]). Entry through the atmosphere serves to disaggregate meteorites and slow the resulting fragments to atmospheric terminal velocities of only ~112 and ~550 m/sec for Earth and Mars, respectively. As discussed above, our ballistics experiments have demonstrated that *B. subtilis* spores readily survived impacts at 100 and 300 m/sec at high frequencies [13.40]. Impact with the surface would serve to further shatter the rock fragments and mix them with the soil, regolith, or water, thus effectively inoculating the recipient planet's surface and introducing life into environments potentially conducive to growth.

13.5 Lithopanspermia, Spores, and the Origin of Life

As we have seen in the discussion above, spores can survive the lethal forces prevailing during impact-mediated launch from a donor planet, transit through space, and entry onto a recipient planet. Although the probability of any single spore surviving is rather low, this constraint is amply compensated for by (i) the large numbers of spores in rocks and (ii) the vast numbers of rocks that have been exchanged among the inner planets during the past ~3–4 billion years of life on Earth. In fact, during the early history of the solar system, the impact flux was likely 2–3 orders of magnitude higher than it is today [13.4]. In particular, a spike in the frequency of impacts striking the inner terrestrial planets, known as the Late Heavy Bombardment (LHB), occurred between ~4.2 and ~3.5 Ga [13.4]. Coincidentally, from dating of the earliest microbial fossils at ~3.4–3.8 Ga [13.11] [13.59] [13.69], life on Earth—morphologically indistinguishable from present-day microbes—apparently had already been established by the end of the LHB. At this early time in solar system evolution it is also thought that the environments of Venus, Earth, and Mars were much more similar to each other than they are today [13.32]. Thus, a scenario has been envisioned in which life may have arisen on any of those three planets and rapidly been transferred to the other two via impacts, where it may have encountered clement environments for subsequent propagation and evolution.

This scenario raises the prediction that putative extraterrestrial life would share a common ancestor with, thus be demonstrably related to, extant Earth life. Coupling of phylogenetic and physiologic analyses indicates that the earliest known ancestors of Earth life were bacterial and archaeal autotrophs (acetogens and methanogens) inhabiting hot anaerobic submarine hydrothermal vents, whose carbon and energy needs were supplied by reduction of CO_2 using electrons from H_2 produced abiotically by serpentinization of minerals (reviewed in [13.38] and references therein). Among these earliest-branching acetogenic bacteria were identified members of the genus *Clostridium*, which are anaerobic spore-forming Gram-positive bacteria. These results imply that sporulation may have arisen very early in the history of life. This notion is strengthened by a recent combination of phylogenetic and electron cryotomographic studies suggesting that sporulation is an extremely ancient process and furthermore that the earliest common ancestors of life formed spores [13.84].

Current theories concerning the evolution of the inner planets posit that during the LHB both the frequency and the size of impactors was larger than today, and that early life may have arisen only to be extinguished again by a large impactor that delivered enough energy to boil away the oceans and heat the crust, thereby sterilizing the planet [13.75]. It has been calculated that it would take up to a few thousand years for the impact energy to dissipate to space and for the planet to cool sufficiently for the oceanic vapor to condense and rain out [13.74]. In this scenario, such "sterilizing impacts" could have created one or more early evolutionary bottlenecks, and what we think of as the earliest common ancestors of life on Earth may actually be the survivors of the last sterilizing impact [13.74]. An impactor large enough to cause such a catastrophe would certainly launch a vast amount of life-bearing crustal material into space, including endolithic spores. Much of this material would be launched at less than escape velocity and would enter a variety of Earth-centric orbits that would decay and cause the rocks to return to Earth over tens, hundreds, or thousands of years. Such rocks would serve both as orbiting refuges for microbes while the Earth was uninhabitable, and sources of re-inoculation once the Earth had cooled sufficiently [13.74].

13.6 Interstellar Lithopanspermia

The possibility that life could be transported within the solar system has prompted the larger question of whether life could be spread between star systems and ultimately throughout the galaxy. This notion hearkens back to Anaxagoras, who posited that the "seeds" (σπέρματα or spermata) of life are present throughout the universe. Due to a nearly total lack of evidence, consideration of interstellar lithopanspermia has been by necessity theoretical. Studies—by the same groups that found interplanetary transport of life to be likely—concluded an essentially zero probability of interstellar transport of life from one terrestrial planet to another, with one exception: life-bearing rocks may have been transferred between stars during the early formation of our solar system, which presumably occurred within a dense cluster of stars packed together at 100–1000 times the present-day density [13.1] [13.45] [13.86]. A similar argument has been invoked for interstellar panspermia within the dense cluster of stars near the galactic core [13.7]. As with interplanetary panspermia (see above), extremely long transit times tend to disfavor spore or microbe survival.

One argument against interstellar panspermia was that astronomers had never detected an object from another star entering our solar system [13.45]. Such objects would be predicted to pass through the solar system at high speeds on hyperbolic trajectories. In 2017, an oblong interstellar object dubbed 1I/2017 U1 "Oumuamua" was detected as it rapidly passed through our solar system on a hyperbolic path, the first such object of its kind [13.41]. Then in 2019, a second interstellar object, comet 2I (2019 Q4) Borisov, was detected passing through our solar system [13.89]. These two events demonstrate that interstellar objects are indeed entering our solar system at a higher frequency than previously thought, renewing interest in the notion of interstellar lithopanspermia.

13.7 Humans as Agents of Panspermia

Up until now we have considered the natural transport of life between planets by impacts. But for the past 60 years, since the first launch of the Soviet spacecraft Luna 1 to the Moon in 1959, humans have been launching artificial spacecraft (and themselves with NASA's Apollo program) outside of Earth's orbit into the solar system. Because bacterial spores are ubiquitous and highly resistant to being killed, it is a virtual certainty that spores have also traveled to various solar system destinations and beyond as unintended passengers on spacecraft. Indeed, microbial contamination of spacecraft is sufficient cause for concern that each space-faring nation supports its own Planetary Protection program aimed to prevent the spread of Earth microbes to other bodies in the solar system (so-called "forward contamination") or the return of putative extraterrestrial microbes to Earth ("back contamination"). Although detailed discussion of Planetary Protection is beyond the scope of this review, the fields of Planetary Protection and Astrobiology do share some common themes, particularly the understanding of how Earth microbes might survive or grow in extraterrestrial environments (see below). For some recent reviews of Planetary Protection as it relates to Astrobiology, the reader is referred to [13.15] [13.34] [13.50] [13.65].

13.8 Survival and Growth of Spores in the Mars Environment

Interplanetary transfer of life is ecologically relevant only if the transported microbe finds itself in an environment conducive to growth [13.20]. As discussed above, in the early days of the solar system the environments of the inner terrestrial planets Venus, Earth, and Mars were likely similar, but since that time the environments of the three planets have embarked on markedly distinct trajectories.

Liquid water is thought to be a prerequisite for life as we know it. Although direct evidence is scant, early in its history Venus is thought to have embarked on a "runaway greenhouse" pathway leading to its present-day hot (>450 °C!), dry, high-pressure, and likely uninhabitable surface [13.30]. In contrast to our limited measurements of the surface of Venus, we possess a wealth of information about the environment on the surface of Mars from numerous flybys, orbiters, landers, and rovers that have been exploring the Red Planet since 1965. The portrait of Mars built up from this data indicates that Mars apparently started its history as a warmer, wetter planet with a thicker atmosphere that subsequently evolved into its current cold, low-pressure, dry state [13.32]. Present-day Mars is characterized by extreme cold, with a near-surface air temperature averaging -61 °C and ranging from -113 °C to -39 °C [13.39]. However, at local noon the ground temperature can transiently climb as high as +17 to +28 °C, as measured by the MER-1 and MER-2 rovers, respectively [13.39]. The atmosphere is dominated by CO_2 (96%), with lesser amounts of N_2 (1.9%), Ar (1.9%), O_2 (0.15%), and traces of other gases [13.35]. The Mars atmosphere is quite dry, with a water volume mixing ratio of 30–75 ppm [13.24]. The atmospheric pressure of Mars averages ~600 Pa at the mean altitude, but varies seasonally between ~400 and ~870 Pa due to freezing and sublimation of CO_2 at the polar caps. (By way of comparison, Earth's atmospheric pressure at sea level is 168 times greater at 101,300 Pa.) Pressure varies inversely with altitude, and can be as low as 72 Pa at the top of Olympus Mons and as high as 1,240 Pa at the bottom of Hellas Planitia. Under most of the pressure-temperature conditions on present-day Mars, water is locked up as ice either at the polar caps or in the regolith, and a small and variable amount sublimates into the atmosphere as vapor. However, abundant evidence exists indicating that liquid water once flowed on the Martian surface [13.36], pointing to an ancient potentially habitable Mars. Moreover, the recent detection by radar of a large (20-km) lake of liquid water, probably hypersaline, located 1.5 km below Mars' southern ice cap [13.61] revives the possibility that present-day microbial life may exist there, in a manner similar to that seen in subglacial Arctic and Antarctic lakes [13.33] [13.66].

How do spore-forming bacteria fit into this picture? Because spores would most likely survive interplanetary transit on exploration spacecraft headed to Mars, there is interest in understanding whether these accidental passengers could survive and/or grow if deposited into the Martian environment. Spore survival has been tested experimentally in a Mars simulation chamber (MSC) that replicates the temperature, pressure, atmospheric gas composition, and UV radiation environment of Mars using spores of the model laboratory strain *B. subtilis* 168 as well as *Bacillus* spp. spores isolated from Mars-bound spacecraft and from Spacecraft Assembly Facilities (SAFs) (reviewed in [13.20]). From these experiments it rapidly became apparent that spores exposed to simulated Mars solar UV were rapidly inactivated within minutes to hours [13.47] [13.70] [13.71], but that minimal amounts of shielding, such as dust or regolith [13.37], or even the upper layers of spores in multilayers [13.57], could provide almost complete protection from UV (reviewed in [13.58]). When protected from UV, spores of *B. subtilis* were shown to survive at levels approaching 100% for up to 21 days of continuous exposure in the MSC [13.49] [13.57].

Over the years, numerous attempts have been made to cultivate various bacteria, including spore-forming *Bacillus* spp., under atmospheric conditions closely mimicking those on Mars (anoxic CO_2-dominated atmosphere, 0 °C, 700 Pa). Literally tens of thousands of microbes have been tested, and only a small minority of species, mostly isolated from Siberian permafrost, Arctic and Antarctic soils, have been demonstrated to grow under these atmospheric conditions, including species of spore-forming bacteria (*Bacillus, Clostridium, Paenibacillus,* and *Streptomyces*) [13.72]. It should be noted here that although these microbes could grow under Mars atmospheric composition, pressure, and temperature conditions, they were cultivated on hydrated nutrient agar media, a conditions certainly not found on present-day Mars. However, these environmental isolates capable of survival and growth in simulations of the Mars atmosphere are excellent candidates for studies to understand what attributes putative Martian microbes might possess, to inform upcoming life-detection missions to the Red Planet.

Acknowledgments

I wish to thank the NASA Exobiology, Astrobiology, and Mars Fundamental Research programs for their generous funding of this research. I especially wish to thank Patricia Fajardo-Cavazos both for our close and productive research collaboration spanning the past 30 years and for critical reading of this manuscript.

References

13.1. Adams, F.C. and Spergel, D.N., Lithopanspermia in star-forming clusters. *Astrobiology*, 5, 4, 497–514, 2005.
13.2. Benardini, J.N., Sawyer, J., Venkateswaran, K., Nicholson, W.L., Spore UV and acceleration resistance of endolithic *Bacillus pumilus* and *Bacillus subtilis* isolates obtained from Sonoran desert basalt: Implications for lithopanspermia. *Astrobiology*, 3, 4, 709–717, 2003.
13.3. Bogard, D.D. and Johnson, P., Martian gases in an Antarctic meteorite. *Science*, 221, 4611, 651–654, 1983.
13.4. Bottke, W.F. and Norman, M.D., The Late Heavy Bombardment, in: *Annual Review of Earth and Planetary Sciences*, vol. 45, K.H. Freeman and R. Jeanloz (Eds.), pp. 619–647, Annual Reviews, Palo Alto, 2017.
13.5. Brandstätter, F., Brack, A., Baglioni, P., Cockell, C.S., Demets, R., Edwards, H.G.M., Kurat, G., Osinski, G.R., Pillinger, J.M., Roten, C.A., Sancisi-Frey, S., Mineralogical alteration of artificial meteorites during atmospheric entry. The STONE-5 experiment. *Planet. Space Sci.*, 56, 7, 976–984, 2008.
13.6. Bronshten, V.A., *Physics of Meteoric Phenomena*, D. Reidel Publishing Co., Dordrecht, The Netherlands, 1983.
13.7. Chen, H., Forbes, J.C., Loeb, A., Habitable evaporated cores and the occurrence of panspermia near the galactic center. *Astrophys. J. Lett.*, 855, 1, 7, #L1, 2018.
13.8. Cohn, F., Untersuchungen über Bacterien. I. *Beitr. Biol. Pflanz.*, 1, 2, 127–224, 1872.
13.9. Darwin, C., *On the Origin of Species*, John Murray, London, UK, 1859.
13.10. Demets, R., Darwin's contribution to the development of the panspermia theory. *Astrobiology*, 12, 10, 946–950, 2012.
13.11. Dodd, M.S., Papineau, D., Grenne, T., Slack, J.F., Rittner, M., Pirajno, F., O'Neil, J., Little, C.T.S., Evidence for early life in Earth's oldest hydrothermal vent precipitates. *Nature*, 543, 7643, 60–64, 2017.
13.12. Driks, A. and Eichenberger, P. (Eds.), *The Bacterial Spore: From Molecules to Systems*, ASM Press, Washington, DC, 2016.
13.13. Ehrenberg, C.G., *Die infusionsthierchen als vollkommene organismen. Ein blick in das tiefere organische leben der natur*, L. Voss, Leipzig, 1838.
13.14. Eugster, O., Busemann, H., Lorenzetti, S., Terrebilini, D., Ejection ages from krypton-81-krypton-83 dating and pre-atmospheric sizes of martian meteorites. *Meteorit. Planet. Sci.*, 37, 10, 1345–1360, 2002.
13.15. Fairen, A.G., Schulze-Makuch, D., Whyte, L., Parro, V., Pavlov, A., Gomez-Elvira, J., Azua-Bustos, A., Fink, W., Baker, V., Planetary Protection and the astrobiological exploration of Mars: Proactive steps in moving forward. *Adv. Space Res.*, 63, 5, 1491–1497, 2019.
13.16. Fajardo-Cavazos, P., Langenhorst, F., Melosh, H.J., Nicholson, W.L., Bacterial spores in granite survive hypervelocity launch by spallation: Implications for lithopanspermia. *Astrobiology*, 9, 7, 647–657, 2009.

13.17. Fajardo-Cavazos, P., Link, L., Melosh, H.J., Nicholson, W.L., *Bacillus subtilis* spores on artificial meteorites survive hypervelocity atmospheric entry: Implications for lithopanspermia. *Astrobiology*, 5, 6, 726–736, 2005.
13.18. Fajardo-Cavazos, P., Maughan, H., Nicholson, W.L., Evolution in the Bacillaceae, in: *The Bacterial Spore: From Molecules to Systems*, A. Driks and P. Eichenberger (Eds.), pp. 21–58, ASM Press, Washington, DC, 2016.
13.19. Fajardo-Cavazos, P. and Nicholson, W.L., *Bacillus* endospores isolated from granite: Close molecular relationships to globally distributed *Bacillus* spp. from endolithic and extreme environments. *Appl. Environ. Microbiol.*, 72, 4, 2856–2863, 2006.
13.20. Fajardo-Cavazos, P., Schuerger, A., Nicholson, W., Testing interplanetary transfer of bacteria between Earth and Mars as a result of natural impact phenomena and human spaceflight activities. *Acta Astronaut.*, 60, 4–7, 534–540, 2007.
13.21. Fritz, J., Artemieva, N., Greshake, A., Ejection of Martian meteorites. *Meteorit. Planet. Sci.*, 40, 9–10, 1393–1411, 2005.
13.22. Gladman, B., Dones, L., Levison, H.F., Burns, J.A., Impact seeding and reseeding in the inner solar system. *Astrobiology*, 5, 4, 483–496, 2005.
13.23. Gratz, A.J., Nellis, W.J., Hinsey, N.A., Observations of high-velocity, weakly shocked ejecta from experimental impacts. *Nature*, 363, 6429, 522–524, 1993.
13.24. Harri, A.M., Genzer, M., Kemppinen, O., Gomez-Elvira, J., Haberle, R., Polkko, J., Savijarvi, H., Renno, N., Rodriguez-Manfredi, J.A., Schmidt, W. *et al.*, Mars Science Laboratory relative humidity observations: Initial results. *J. Geophys. Res.-Planets*, 119, 9, 2132–2147, 2014.
13.25. Horneck, G., Responses of *Bacillus subtilis* spores to the space environment: results from experiments in space. *Orig. Life Evol. Biosph.*, 23, 1, 37–52, 1993.
13.26. Horneck, G., Bücker, H., Reitz, G., Requardt, H., Dose, K., Martens, K.D., Mennigmann, H.D., Weber, P., Microorganisms in the space environment. *Science*, 225, 4658, 226–228, 1984.
13.27. Horneck, G., Klaus, D.M., Mancinelli, R.L., Space microbiology. *Microbiol. Mol. Biol. Rev.*, 74, 1, 121–156, 2010.
13.28. Horneck, G., Moeller, R., Cadet, J., Douki, T., Mancinelli, R.L., Nicholson, W.L., Panitz, C., Rabbow, E., Rettberg, P., Spry, A. *et al.*, Resistance of bacterial endospores to outer space for planetary protection purposes–experiment PROTECT of the EXPOSE-E mission. *Astrobiology*, 12, 5, 445–456, 2012.
13.29. Horneck, G., Stoffler, D., Ott, S., Hornemann, U., Cockell, C., Moeller, R., Meyer, C., De Vera, J.P., Fritz, J., Schade, S., Artemieva, N., Microbial rock inhabitants survive hypervelocity impacts on Mars-like host planets: First phase of lithopanspermia experimentally tested. *Astrobiology*, 8, 1, 17–44, 2008.
13.30. Ingersoll, A.P., Runaway greenhouse - a history of water on Venus. *J. Atmos. Sci.*, 26, 6, 1191–1198, 1969.
13.31. Koch, R., Die Ätiologie der Milzbrand-Krankheit, begrïdet auf die Entwicklungsgeschichte des *Bacillus anthracis*. *Beitr. Biol. Pflanz.*, 2, 2, 277–310, 1876.
13.32. Lammer, H., Zerkle, A.L., Gebauer, S., Tosi, N., Noack, L., Scherf, M., Pilat-Lohinger, E., Gudel, M., Grenfell, J.L., Godolt, M., Nikolaou, A., Origin and evolution of the atmospheres of early Venus, Earth and Mars. *Astron. Astrophys. Rev.*, 26, 72, #2, 2018.
13.33. Laybourn-Parry, J. and Pearce, D., Heterotrophic bacteria in Antarctic lacustrine and glacial environments. *Polar Biol.*, 39, 12, 2207–2225, 2016.
13.34. Lupisella, M.L. and Race, M.S., Low-latency teleoperations, planetary protection, and astrobiology. *Int. J. Astrobiol.*, 17, 3, 239–246, 2018.
13.35. Mahaffy, P.R., Webster, C.R., Atreya, S.K., Franz, H., Wong, M., Conrad, P.G., Harpold, D., Jones, J.J., Leshin, L.A., Manning, H. *et al.*, Abundance and isotopic composition of gases in the Martian atmosphere from the Curiosity rover. *Science*, 341, 6143, 263–266, 2013.

13.36. Malin, M.C. and Edgett, K.S., Evidence for persistent flow and aqueous sedimentation on early Mars. *Science*, 302, 5652, 1931–1934, 2003.
13.37. Mancinelli, R. and Klovstad, M., Martian soil and UV radiation: Microbial viability assessment on spacecraft surfaces. *Planet. Space Sci.*, 48, 11, 1093–1097, 2000.
13.38. Martin, W.F., Weiss, M.C., Neukirchen, S., Nelson-Sathi, S., Sousa, F.L., Physiology, phylogeny, and LUCA. *Microb. Cell*, 3, 12, 451–456, 2016.
13.39. Martinez, G.M., Newman, C.N., De Vicente-Retortillo, A., Fischer, E., Renno, N.O., Richardson, M.I., Fairen, A.G., Genzer, M., Guzewich, S.D., Haberle, R.M. *et al.*, The modern near-surface Martian climate: A review of *in-situ* meteorological data from Viking to Curiosity. *Space Sci. Rev.*, 212, 1–2, 295–338, 2017.
13.40. Mastrapa, R., Glanzberg, H., Head, J., Melosh, H., Nicholson, W., Survival of bacteria exposed to extreme acceleration: Implications for panspermia. *Earth Planet. Sci. Lett.*, 189, 1–2, 1–8, 2001.
13.41. Meech, K.J., Weryk, R., Micheli, M., Kleyna, J.T., Hainaut, O.R., Jedicke, R., Wainscoat, R.J., Chambers, K.C., Keane, J.V., Petric, A. *et al.*, A brief visit from a red and extremely elongated interstellar asteroid. *Nature*, 552, 7685, 378–381, 2017.
13.42. Melosh, H., Impact ejection, spallation, and the origin of meteorites. *Icarus*, 59, 2, 234–260, 1984.
13.43. Melosh, H., Ejection of rock fragments from planetary bodies. *Geology*, 13, 2, 144–148, 1985.
13.44. Melosh, H., The rocky road to panspermia. *Nature*, 332, 6166, 687–688, 1988.
13.45. Melosh, H.J., Exchange of meteorites (and life)? between stellar systems. *Astrobiology*, 3, 1, 207–215, 2003.
13.46. Mileikowsky, C., Cucinotta, F., Wilson, J., Gladman, B., Horneck, G., Lindegren, L., Melosh, H.J., Rickman, H., Valtonen, M., Zheng, J.Q., Natural transfer of viable microbes in space - 1. From Mars to Earth and Earth to Mars. *Icarus* 145, 2, 391–427, 2000.
13.47. Newcombe, D.A., Schuerger, A.C., Benardini, J.N., Dickinson, D., Tanner, R., Venkateswaran, K., Survival of spacecraft-associated microorganisms under simulated martian UV irradiation. *Appl. Environ. Microbiol.*, 71, 8147–8156, 2005.
13.48. Nicholson, W., Using thermal inactivation kinetics to calculate the probability of extreme spore longevity: Implications for paleomicrobiology and lithopanspermia. *Orig. Life Evol. Biosph.*, 33, 6, 621–631, 2003.
13.49. Nicholson, W. and Schuerger, A., *Bacillus subtilis* spore survival and expression of germination-induced bioluminescence after prolonged incubation under simulated Mars atmospheric pressure and composition: Implications for planetary protection and lithopanspermia. *Astrobiology*, 5, 4, 536–544, 2005.
13.50. Nicholson, W., Schuerger, A., Race, M., Migrating microbes and planetary protection. *Trends Microbiol.*, 17, 9, 389–392, 2009.
13.51. Nicholson, W.L., Roles of *Bacillus* endospores in the environment. *Cell. Mol. Life Sci.*, 59, 3, 410–416, 2002.
13.52. Nicholson, W.L., Ubiquity, longevity, and ecological roles of *Bacillus* spores, in: *Bacterial Spore Formers: Probiotics and Emerging Applications*, E. Ricca, A.O. Henriques, S.M. Cutting (Eds.), pp. 1–15, Horizon Scientific Press, Norfolk, UK, 2004.
13.53. Nicholson, W.L., Ancient micronauts: Interplanetary transport of microbes by cosmic impacts. *Trends Microbiol.*, 17, 6, 243–250, 2009.
13.54. Nicholson, W.L., Experimental evolution to explore adaptation of terrestrial bacteria to the Martian environment, in: *Molecular Mechanisms of Microbial Evolution*, P.H. Rampelotto (Ed.), pp. 241–265, Springer Nature, Cham, Switzerland, 2018.
13.55. Nicholson, W.L., Fajardo-Cavazos, P., Rebeil, R., Slieman, T.A., Riesenman, P.J., Law, J.F., Xue, Y.M., Bacterial endospores and their significance in stress resistance. *Anton. Leeuw. Int. J. Gen. Mol. Microbiol.*, 81, 1–4, 27–32, 2002.

13.56. Nicholson, W.L., Munakata, N., Horneck, G., Melosh, H.J., Setlow, P., Resistance of *Bacillus* endospores to extreme terrestrial and extraterrestrial environments. *Microbiol. Mol. Biol. Rev.*, 64, 3, 548–572, 2000.
13.57. Nicholson, W.L., Schuerger, A.C., Douki, T., The photochemistry of unprotected DNA and DNA inside *Bacillus subtilis* spores exposed to simulated Martian surface conditions of atmospheric composition, temperature, pressure, and solar radiation. *Astrobiology*, 18, 4, 393–402, 2018.
13.58. Nicholson, W.L., Schuerger, A.C., Setlow, P., The solar UV environment and bacterial spore UV resistance: Considerations for Earth-to-Mars transport by natural processes and human spaceflight. *Mutat. Res.*, 571, 1–2, 249–264, 2005.
13.59. Nutman, A.P., Bennett, V.C., Friend, C.R.L., Van Kranendonk, M.J., Chivas, A.R., Rapid emergence of life shown by discovery of 3,700-million-year-old microbial structures. *Nature*, 537, 7621, 535–538, 2016.
13.60. Nyquist, L., Bogard, D., Shih, C., Greshake, A., Stoffler, D., Eugster, O., Ages and geologic histories of Martian meteorites. *Space Sci. Rev.*, 96, 1–4, 105–164, 2001.
13.61. Orosei, R., Lauro, S.E., Pettinelli, E., Cicchetti, A., Coradini, M., Cosciotti, B., Di Paolo, F., Flamini, E., Mattei, E., Pajola, M. *et al.*, Radar evidence of subglacial liquid water on Mars. *Science*, 361, 6401, 490–493, 2018.
13.62. Passey, Q.R. and Melosh, H.J., The effects of atmospheric breakup on crater field formation. *Icarus*, 42, 211–233, 1989.
13.63. Peretó, J., Bada, J.L., Lazcano, A., Charles Darwin and the origin of life. *Orig. Life Evol. Biosph.*, 39, 5, 395–406, 2009.
13.64. Ricca, E., Henriques, A.O., Cutting, S.M. (Eds.), *Bacterial Spore Formers: Probiotics and Emerging Applications*, Horizon Bioscience, Trowbridge, UK, 2004.
13.65. Rummel, J.D., Beaty, D.W., Jones, M.A., Bakermans, C., Barlow, N.G., Boston, P.J., Chevrier, V.F., Clark, B.C., de Vera, J.P., Gough, R.V. *et al.*, A new analysis of Mars "Special Regions": Findings of the second MEPAG Special Regions Science Analysis Group (SR-SAG2). *Astrobiology*, 14, 11, 887–968, 2014.
13.66. Rutishauser, A., Blankenship, D.D., Sharp, M., Skidmore, M.L., Greenbaum, J.S., Grima, C., Schroeder, D.M., Dowdeswell, J.A., Young, D.A., Discovery of a hypersaline subglacial lake complex beneath Devon Ice Cap, Canadian Arctic. *Sci. Adv.*, 4, 4, #eaar4353, 2018.
13.67. Schaeffer, P., Asporogenous mutants of *Bacillus subtilis* Marburg. *Folia Microbiol.*, 12, 3, 291–296, 1967.
13.68. Scharf, C., Virgo, N., Cleaves, H.J., Aono, M., Aubert-Kato, N., Aydinoglu, A., Barahona, A., Barge, L.M., Menner, S.A., Biehl, M. *et al.*, A strategy for origins of life research. *Astrobiology*, 15, 12, 1031–1042, 2015.
13.69. Schopf, J.W., Kitajima, K., Spicuzza, M.J., Kudryavtsev, A.B., Valley, J.W., SIMS analyses of the oldest known assemblage of microfossils document their taxon-correlated carbon isotope compositions. *Proc. Natl. Acad. Sci. U.S.A.*, 115, 1, 53–58, 2018.
13.70. Schuerger, A., Mancinelli, R., Kern, R., Rothschild, L., McKay, C., Survival of endospores of *Bacillus subtilis* on spacecraft surfaces under simulated martian environments: Implications for the forward contamination of Mars. *Icarus*, 165, 2, 253–276, 2003.
13.71. Schuerger, A., Richards, J., Newcombe, D., Venkateswaran, K., Rapid inactivation of seven *Bacillus* spp. under simulated Mars UV irradiation. *Icarus*, 181, 1, 52–62, 2006.
13.72. Schuerger, A.C. and Nicholson, W.L., Twenty species of hypobarophilic bacteria recovered from diverse soils exhibit growth under simulated martian conditions at 0.7 kPa. *Astrobiology*, 16, 12, 964–976, 2016.
13.73. Setlow, P., Spore resistance properties, in: *The Bacterial Spore: From Molecules to Systems*, A. Driks and P. Eichenberger (Eds.), pp. 201–215, ASM Press, Washington, DC, 2016.

13.74. Sleep, N. and Zahnle, K., Refugia from asteroid impacts on early Mars and the early Earth. *J. Geophys. Res.-Planets*, 103, E12, 28529–28544, 1998.
13.75. Sleep, N.H., Zahnle, K.J., Kasting, J.F., Morowitz, H.J., Annihilation of ecosystems by large asteroid impacts on the early Earth. *Nature*, 342, 6246, 139–142, 1989.
13.76. Slepecky, R.A. and Leadbetter, E.R., Ecology and relationships of endospore-forming bacteria: Changing perspectives, in: *Regulation of bacterial differentiation*, P.J. Piggot, J. Moran, C.P., P. Youngman (Eds.), pp. 195–206, American Society for Microbiology, Washington, D.C, 1994.
13.77. Slobodkin, A., Gavrilov, S., Ionov, V., Iliyin, V., Spore-forming thermophilic bacterium within artificial meteorite survives entry into the Earth's atmosphere on FOTON-M4 satellite landing module. *PLoS One*, 10, 7, 13, #e0132611, 2015.
13.78. Sonenshein, A.L., Endospore-forming bacteria: An overview, in: *Prokaryotic Development*, Y. Brun and L. Shimkets (Eds.), pp. 133–150, ASM Press, Washington, DC, 2000.
13.79. Sonenshein, A.L., Hoch, J.A., Losick, R. (Eds.), *Bacillus subtilis and its Closest Relatives: From Genes to Cells*, ASM Press, Washington, DC, 2002.
13.80. Spizizen, J., Transformation of biochemically deficient strains of *Bacillus subtilis* by deoxyribonucleate. *Proc. Natl. Acad. Sci. U.S.A.*, 44, 10, 1072–1078, 1958.
13.81. Tabetah, M.E. and Melosh, H.J., Air penetration enhances fragmentation of entering meteoroids. *Meteorit. Planet. Sci.*, 53, 3, 493–504, 2018.
13.82. Takahashi, I., Genetic transduction in *Bacillus subtilis*. *Biochem. Biophys. Res. Commun.*, 5, 3, 171–175, 1961.
13.83. Thorne, C.B., Transduction in *Bacillus subtilis*. *J. Bacteriol.*, 83, 106–111, 1962.
13.84. Tocheva, E.I., Ortega, D.R., Jensen, G.J., Sporulation, bacterial cell envelopes and the origin of life. *Nat. Rev. Microbiol.*, 14, 8, 535–542, 2016.
13.85. Tyndall, J., Further researches on the department and vital persistence of putrefactive and infective organisms from a physical point of view. *Philos. Trans. R. Soc.*, 167, 149–206, 1877.
13.86. Valtonen, M., Nurmi, P., Zheng, J., Cucinotta, F., Wilson, J., Horneck, G., Lindegren, L., Melosh, J., Rickman, H., Mileikowsky, C., Natural transfer of viable microbes in space from planets in extra-solar systems to a planet in our solar system and vice versa. *Astrophys. J.*, 690, 1, 210–215, 2009.
13.87. Weiss, B., Kirschvink, J., Baudenbacher, F., Vali, H., Peters, N., Macdonald, F., Wikswo, J., A low temperature transfer of ALH84001 from Mars to Earth. *Science*, 290, 5492, 791–795, 2000.
13.88. Zeigler, D.R. and Nicholson, W.L., Experimental evolution of *Bacillus subtilis*. *Environ. Microbiol.*, 19, 9, 3415–3422, 2017.
13.89. Gibbs, W.W. Alien comets may be common, analysis of 2I/Borisov suggests. *Science*, 366, 558–558, 2019.

14

Potential Energy Production and Utilization Pathways of the Martian Subsurface: Clues from Extremophilic Microorganisms on Earth

Varun G. Paul[1*] and Melanie R. Mormile[2]

[1]*Department of Geosciences, Mississippi State University, Mississippi, USA*
[2]*Department of Biology, Missouri University of Science and Technology, Rolla, Missouri, USA*

Abstract

Life on Mars, if present, is suggested to exist only in the subsurface as the surface conditions are extremely oxidizing for known life forms to survive. Past knowledge from meteorites and rover expeditions have expanded our understanding of the conditions present on Mars. To enhance our approach to narrow down the search for life, two key aspects need to be considered: explore the available energy sources on Mars, and gather clues from the type of microbial life on Earth that exploit those resources. The best candidates to investigate on Earth are microorganisms that can survive in similar extreme conditions (extremophiles), and explore mechanisms and pathways on how they utilize the energy sources, while adapting themselves to stressful conditions. In this review, we explore the potential energy-generating electron donors and acceptors that are available on Mars, and consider the extremophiles on Earth capable of metabolizing these energy sources. Hydrogen and methane are two essential molecules that serve as ideal electron donors, and are produced by a variety of processes on Mars, including serpentinization, and hydration of the ultramafic rock units. Hydrogen is a key electron donor that is mostly discussed in this review because it is plentifully available and a wide variety of extreme microorganisms can utilize this energetically favorable source. Methane is another important electron donor, and the complex process of methane utilization is orchestrated by a broad group of uniquely adapted microbial populations. Apart from these two sources, iron-, sulfur- and nitrogen-based compounds, carbon monoxide and manganese are other electron donors that can be potentially exploited. A thorough understanding of the clues that are created by these microbial activities, including biomarkers, isotopic signatures, and morphological traits, can advise us on future explorative missions to Mars and other life-supporting heavenly bodies.

Keywords: Extremophiles, energy production, electron donors, Martian subsurface

**Corresponding author*: vgp25@msstate.edu
Varun Paul: https://www.geosciences.msstate.edu/people/professors/varun-paul/, https://www.researchgate.net/profile/Varun_Paul, https://scholar.google.com/citations?user=NDNLlH0AAAAJ&hl=en
Melanie Mormile: https://people.mst.edu/faculty/mmormile/index.html, https://www.researchgate.net/profile/Melanie_Mormile, https://scholar.google.com/citations?user=mtamEzwAAAAJ&hl=en

Joseph Seckbach and Helga Stan-Lotter (eds.) Extremophiles as Astrobiological Models, (291–316)

14.1 Introduction

Microorganisms have inhabited the surface and subsurface of Earth at least 80% of the time since Earth's formation about 4.56 billion years ago [14.146]. The origins of life most likely took place under grueling conditions considered typical of the early atmospheres, aquatic settings and terrestrial systems of the rocky planets of our solar system [14.17] [14.56] [14.76] [14.113]. The study of the chemistry and mineralogy of Martian meteorites has provided evidence for the presence of necessary geochemical conditions that could have supported previous surface life [14.13] [14.68] [14.94]. However, current conditions on the surface of Mars are highly oxidizing and subject to intense insolation by UV radiation, with a low atmospheric pressure consisting of mainly carbon dioxide (96%), nitrogen (1.9%), argon (1.9%), oxygen (0.15%) and carbon monoxide (<0.1%), and a temperature range of approximately -120 to +10 °C [14.73]. There is general consensus that current life on the Martian surface would be highly unlikely and recent interest has been focused on the subsurface realm and icy waters of the planet [14.11] [14.34] [14.82] [14.116] [14.142] [14.145]. While suitable electron donors are suspected to be available for biological life in Martian subsurface, the more determining factor is the presence of water. Since surficial processes that can transfer electron donors to the deeper regions are restricted, the life that likely exists in the subsurface would have to make use of components already present in soil, rock or water (in the form of ice that extends to 2,000 meters thickness).

The study of present-day microorganisms that can survive under extreme conditions is an obvious choice to understand the origins and evolution of life on Earth. Several authors have already established the link between extremophiles and astrobiological applications [14.41] [14.103] [14.105] [14.135]. With the past, current and future rover explorations to Mars, it has become increasingly evident that Earth's microbial extremophiles can provide crucial clues and advise scientists on how to approach exploration of life on Mars. The more recent expedition, Curiosity, has provided a "robotic, hands-on" knowledge of the surface mineralogy and produced high-definition images of the surface, revolutionizing our understanding of the physical and chemical processes on Mars [14.68] [14.74] [14.125]. The largest revelation thus far, from an astrobiological perspective, was the confirmation of the presence of water and that of surface and subsurface hydrological processes [14.68] [14.98] [14.107] [14.115]. Furthermore, presence of organic material and Earth-like mineral morphology indicative of possible biological origins have piqued the interest of scientists searching for clues to indicate life [14.21] [14.36] [14.68] [14.87].

The Mars 2020 rover mission, named Perseverance, is explicitly focused on detecting biosignatures and will have the capacity to bring rock cores and soil back to the Earth for more thorough analyses. Similarly, the Water Ice Subsurface Deposit Observation on Mars (WISDOM) ground-penetrating radar instrument intended to study the Martian subsurface from 3 to 10 meters is part of the payload for the ESA-Roscosmos 2020 ExoMars Rover mission [14.22]. With these missions in mind, interdisciplinary researchers are ever more interested in studying Earth proxies. For example, biomarkers, isotopic signatures, and mineralogy of the oldest fossil evidence for life, namely stromatolites, and their modern analogues have been suggested as a potential tool that could reveal information about the environmental conditions and metabolic reactions that occurred in the past [14.15] [14.90] [14.101] [14.111]. This information can be used to extrapolate possible conditions on Mars that could have favored past and possibly present life in the subsurface. An ideal approach is to explore potential sources of energy of Mars' subsurface that could support life, especially extremophiles.

The main goals of this chapter are to review potential energy sources available and suitable for microbial metabolism in the Martian subsurface conditions, and the extremophiles on Earth that could potentially thrive and survive in such settings. Since we are considering subsurface life, photosynthetic capabilities and resistance to direct UV radiation [14.82] are generally considered to be insignificant in Martian context and will not be discussed in this review.

14.2 Energy Sources

Autotrophic energy production is a primary foundation mechanism in any living ecosystem. On Earth's surface, it is dominated by photosynthesis. However, several microbial groups use alternative sources to obtain energy. These include chemolithoautotrophs, which obtain energy by oxidizing inorganic compounds through carbon dioxide fixation (e.g., sulfur oxidizing and iron oxidizing bacteria); chemolithoheterotrophs, which are similar to the previous category, except for lack of carbon dioxide fixation (e.g., certain sulfate-reducing bacteria (SRB), *Beggiatoa*); and the more common chemoorganoheterotrophs, which obtain energy from organic compounds (e.g., *E. coli*, *Actinobacteria*).

Adopting the famous quote by Nobel Prize winning biochemist, Albert Szent-Györgyi, "Life is nothing but an electron looking for a place to rest," in our context, identifying suitable electron donors and acceptors are key to exploring potential energy sources in underground Mars. Subsurface lithoautotrophic microbial ecosystems, or SLiMEs [14.130], are communities of microorganisms that exist in subsurface environments of Earth and likely Mars, where the energy is derived solely from reactions taking place in the rock units. This energy serves as the primary trophic level, with by-products being utilized by other community members, creating a self-sustaining subsurface ecosystem. On Earth, electron donors and acceptors can be either organic or inorganic. The end point in the flow of electrons is the terminal electron accepting reaction and is indicative of the microbial community within a given biome. Determining the starting and final step of this electron transport chain is critical in understanding the diversity of the microbial populations and functional structure. Aerobic reactions on Earth's surface have the highest energy yield (-2,880 kJ/mol glucose), whereas reactions dominant in subsurface yield much lower energies. For example, methanogenesis and sulfate reduction generate low energy, -492 and -428 kJ/mol glucose, respectively [14.108].

The subsurface conditions on Mars have all the essential ingredients for supporting life. It is important to note that just the presence of these elements and compounds are not sufficient; a flux or diffusion to move these components is also critical [14.38]. Examples of extremophiles that have features of interest in the context of Martian environments are provided in Table 14.1. While discussing the energy sources, such as electron donor-acceptor pairs potentially present in the Martian subsurface, similar Earth conditions can be used to estimate the amount of energy that could be available for use by putative Martian microorganisms. It is also important to understand that many of the electron donors and acceptors discussed below have multiple roles and are able to support more than one group of microorganisms.

Table 14.1 Electron donors available in the Martian subsurface and examples of extremophiles on Earth capable of using each that can be studied as putative Martian microbial proxies.

Electron donors present in the Martian subsurface	Examples of Extremophilic Microorganisms present on Earth	Type of Extremophile	References
Hydrogen	*Methanopyrus* sp.	Hyperthermophile	[14.130]
	Methanosarcinales	Psychrophile/ Thermophile	[14.25] [14.59] [14.70]
	Aquifex pyrophilus	Hyperthermophile/ Halophile	[14.51]
	Hydrogenobacter subterraneus	Thermophile	[14.131]
	Marinitoga piezophila	Thermopiezophile	[14.2]
	Pyrococcus yayanosii CH1	Piezophile	[14.83]
	Clostridium caminithermale	Halophile	[14.14]
	Acidithiobacillus ferridurans	Acidophile	[14.47]
	Strain AHO 1	Alkaliphile	[14.122]
Methane	ANME and sulfate-reducing bacteria	Hyperthermophile/ Piezophile/ Psychrophile	[14.91]
	Methylohalobius crimeensis	Halophile	[14.49]
	Kam1 (Verrucomicrobia)	Acidophile, Thermophile	[14.55]
Iron	*Tepidibacillus decaturensis*	Halothermophile/ Piezophile	[14.28]
	Leptospirillum ferrooxidans and *Acidithiobacillus ferrooxidans*	Acidophile	[14.99]
Sulfur and sulfur-based compounds	*Pyrococcus yayanosii* CH1	Piezophile	[14.83]
	Thioalkalivibrio thiocyanoxidans	Alkaliphile	[14.120]
	Thiohalospira halophila	Haloalkaliphile	[14.119]
Nitrogen and related compounds	Crenarchaeota	Thermophile/Halophile	[14.143]
	Methylomicrobium sp.	Haloalkaliphile	[14.121]

(*Continued*)

Table 14.1 Electron donors available in the Martian subsurface and examples of extremophiles on Earth capable of using each that can be studied as putative Martian microbial proxies. (*Continued*)

Electron donors present in the Martian subsurface	Examples of Extremophilic Microorganisms present on Earth	Type of Extremophile	References
Carbon Monoxide	*Alkalilimnicola ehrlichii* MLHE-1 and *Halorubrum* str. BV1	Halophile	[14.60]
Manganese	*Bacillus* sp.	Psychrophile	[14.132]
	Bacillus subterraneus	Thermophile	[14.58]

14.2.1 Hydrogen

Hydrogen is a key element involved in almost all biological functions and is considered to be the main energy driver that is expected to be on Mars [14.89]. In catabolic reactions, hydrogen is used as a redox-active component, transferring reducing equivalents during oxidation of carbon. It has a very low reduction potential and is therefore a high energy value electron donor when coupled with CO_2, sulfate and Fe(III) reduction. Hydrogen can be generated either abiotically or through microbial metabolism. Examples of abiotic processes include graphitization (conversion of methane to graphite), radiolysis of water, serpentinization, cataclasis (hydrogen produced in fault zones due to hydromechanical processes), and degassing of hydrogen from the deep mantle [14.43]. Amongst these, radiolysis [14.129], serpentinization [14.112], and cataclasis [14.80] are well known to occur on Mars. Radiolysis was suggested to be an important process of generating hydrogen in the subsurface of Noachian Mars (3.7–4.1 Gyr ago). Serpentinization has also received special attention owing to the mafic-rich rock units on Mars [14.5]. Magnesium and iron-rich silicates (e.g., olivine and pyroxenes) that make up the ultramafic rocks, upon hydration form serpentine and molecular hydrogen, along with brucite and magnetite, as shown by the equation: $(Mg, Fe)_2\ SiO_4+H_2O \rightarrow Mg_3Si_2O_5(OH)_4+Mg(OH)_2+Fe_3O_4+H_{2(aq)}$. Olivine has been reported to make up to 88% of the Martian meteorite Chassigny, among other meteorites, and is expected to be a major component in the interior of the planet [14.38]. Mars also has a basaltic crust, where hydrogen production is expected to be common due to reactions such as serpentinization [14.81].

Microorganisms themselves are able to generate hydrogen, primarily through the enzyme hydrogenase and when coupled with fermentation, nitrogen fixation, anaerobic CO oxidation, and phosphite and acetate oxidation [14.43]. Molecular hydrogen can also be photochemically produced in the atmosphere along with carbon monoxide and diffuse into the top few tens of meters of the planet's surface [14.144].

Extremophiles Using Hydrogen

High pressures and temperatures characterize the terrestrial subsurface along with limited exposure to the atmosphere on Earth. The biomass in Earth's subsurface is expected to be several magnitudes larger than that of the surface, further elucidating that similar biodiversity could exist in Mars' subsurface [14.8] [14.72].

Geothermal areas in Iceland, New Zealand, Japan, Italy, the former Soviet Union, and North America, such as Yellowstone National Park, have been studied extensively for hydrothermal systems that are suitable analogues for Mars. A good example of a geogenic process to produce hydrogen is when the CO_2-bearing, NaCl-rich hydrothermal fluids interact with minerals, such as olivine, at high temperatures and pressures [14.9] [14.38]. Hyperthermophilic, hydrogen-utilizing bacteria such as *Methanopyrus* and members of the order *Methanococcales* can process hydrogen along with CO_2 and produce methane [14.130]. One of the predominant mechanisms by which microbes in the Martian subsurface are expected to function is through the coupling of hydrogen oxidation and CO_2 reduction [14.11] [14.18]. The resulting process will lead to the production of methane and two molecules of water as given by the equation, $4H_2 + CO_2 \rightarrow CH_4 + 2H_2O$. *Methanosarcinales*, an order of methanogens, can survive temperatures ranging from 1 to 70 °C and utilize a variety of sources, including hydrogen+CO_2, acetate, and CO [14.25] [14.59] [14.70]. Geologically produced hydrogen could thus be a very important source of energy for chemosynthetic microorganisms. Hydrothermal waters from igneous rocks in Idaho studied for their microbial population showed that the subsurface environment was dominated by methanogens using hydrogen as their primary source [14.18]. As Mars has a basaltic crust [14.81], deep basalt aquifers on Earth are ideal locations to investigate the functioning of the microbial populations hosted within. Methane-producing, hydrogen-utilizing community dominance in deep basalt aquifers has been shown earlier, where the environment was also poor in organic material [14.18]. Apart from methanogenesis, sulfate reduction and acetogenesis are the two other important processes for utilizing hydrogen.

The terminal electron accepting reactions when using hydrogen are varied, and ranges from aerobic to completely anaerobic reactions (Figure 14.1). Krumholz *et al.* [14.64], characterized two novel species, *Desulfomicrobium hypogeium* sp. nov. and *Acetobacterium*

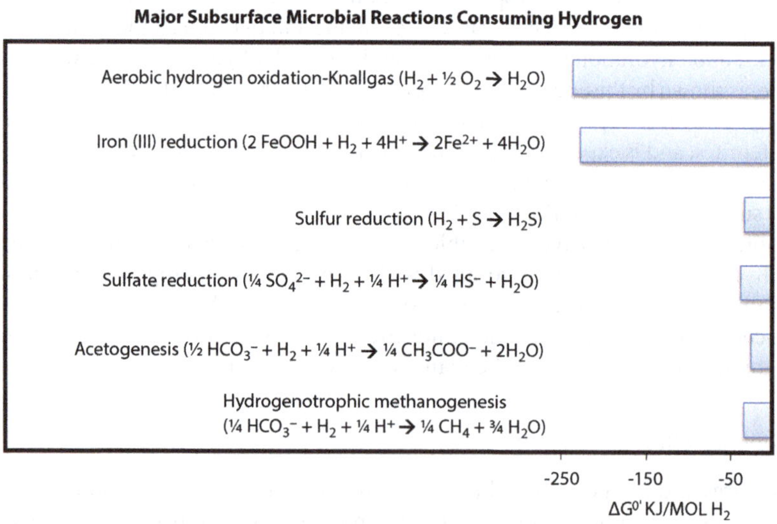

Figure 14.1 Energetics of common hydrogen-utilizing microbial reactions in the subsurface. (Image credit: the authors).

psammolithicum sp. nov., from cultures isolated from Cretaceous sandstone core material from drilling sites in central New Mexico. In the thermopiezophile, *Marinitoga piezophila*, hydrogen use was inhibited with any electron acceptor; however, the presence of elemental sulfur seemed to remove the inhibition [14.2]. Another example of a novel species from the subsurface is reported by Takai *et al.* [14.131]. These authors were able to isolate *Hydrogenobacter subterraneus* sp. nov., a thermophilic microorganism (optimum temperature for growth 78 °C) from a geothermal water pool at a depth of 1500 m.

Piezophiles are microorganisms that can survive under high-pressure environments and concentrate fatty acids within their cell membrane to withstand the stress [14.117]. Deep-sea hydrothermal vents are another excellent proxy to study adaptations to extreme environments, where several piezophiles, hyperthermophiles and psychrophiles have been studied and pure species isolated [14.24] [14.83]. A vast number of thermophilic isolates from deep-sea environments are chemoautotrophic anaerobes [14.106]. Many thermophiles found in both shallow and deep-sea environments metabolize elemental sulfur and oxidize hydrogen. Symbionts growing on deep-sea hydrothermal vent mussels, *Bathymodiolus* from the Mid-Atlantic Ridge, were found to contain the hupL gene that produces hydrogenase and is responsible for oxidation of hydrogen [14.102]. Because of the availability of hydrogen and CO in the Martian atmosphere and near subsurface, hydrogenase will be an important enzyme to consider [14.140]. Hydrogenase can also function in a reverse fashion, where in addition to catalyzing hydrogen activation, it can also induce its production. Many thermophiles have been found to contain this vital enzyme [14.1], a more thermostable version than that produced by their mesophilic counterparts. Additionally, the mesophilic hydrogenases contain a Ni-S or Fe-S catalytic site. However, it is suggested that for the enzyme to function at higher temperatures, these sites should be absent. Thermophilic hydrogenases are of two types; membrane-bound and soluble. The former oxidizes hydrogen, donating the resulting electrons to elemental sulfur, while the latter activates hydrogen and catalyzes CO_2 reduction [14.1].

The bacterium, *Pyrococcus yayanosii* CH1 can be considered as an example to elucidate energy conservation strategies under stressful conditions. This well-studied obligate piezophile is also a hyperthermophile due to its ability to survive in high temperatures [14.83]. The bacterium was isolated from a depth of 4,100 meters at the Ashadze deep-sea hydrothermal vent. The tryptophan aromatic amino acid biosynthesis pathway was not detected in this strain, similar to other piezophiles. The production of tryptophan, the largest amino acid, involves a high-energy cost, which the bacteria cannot afford. The genes involved in energy production and conversion (ATP/ADP synthase (V-ATPase)) and also genes coding for hydrogenases and ferredoxin oxidoreductases were found twice as frequently than in other similar strains. The hydrogenases, especially, are of particular interest under stressful conditions. When tested under higher pressures, the *P. yayanosii* strain indicated that hydrogenases coupled with formate metabolism is one of the more regulated energy pathways when compared to other electron donors. One of the hydrogenase protein clusters, Mbh, represents a basic respiratory process, where the energy generated through hydrogen production is used to create an ionic gradient by accepting electrons from ferredoxin, an iron-sulfur protein.

Several of the subsurface environments of Mars are suggested to be briny aquifers [14.60]. Liquid water is expected to be present in the form of brines associated with features known as recurrent slope lineae (RSL). Investigations from the nakhlite Martian meteorites revealed that the properties of altered meteorites are similar to diluted terrestrial brines found on Earth, with a high pH [14.13]. Such alkaline environments are suggested to be more habitable when compared to other similar, but acidic environments. Halophiles in briny aquifers and lakes on Earth are efficiently adapted to the stress due to excess salt by accumulating osmotic organic solutes (organic "compatible" solutes) or developing salt-tolerant proteins and enzymes (salt-in strategy) [14.69] [14.109]. In haloalkaline environments, the cells have to adapt themselves to not only the high salt but also the high pH. The energy metabolism to maintain osmolarity and pH homeostasis is balanced by a combined process of inward ion transfer and ATP synthesis. Several hydrogen-oxidizing halophilic members have been isolated from a variety of salty environments [14.14] [14.51]. As an example, a strictly anaerobic, slightly halophilic, hydrogen-oxidizing, and moderately thermophilic species, *Clostridium caminithermale,* was isolated from deep-sea hydrothermal vents at depths of 800 m in the Atlantic Ocean Ridge [14.14].

Knallgas bacteria are a group of microaerophilic, hydrogen-oxidizing members that are found in several extreme environments. While oxygen is limited in the Martian subsurface, this particular microbial metabolism with respect to hydrogen utilization can be further explored in extreme settings. For example, a hyperthermophilic and mildly halophilic bacterium, *Aquifex pyrophilus,* was isolated from deep-sea hydrothermal vent, and was found to be capable of using molecular hydrogen, thiosulfate and elemental sulfur as electron donors [14.51]. These are strict chemolithoautotrophs deriving energy from chemical breakdown of minerals from rocks.

Microbes that can utilize CO and hydrogen have also been reported [14.60]. Two extremely halophilic species of microorganisms isolated from salt brines, *Halorubrum* str. BV1 and *Alkalilimnicola ehrlichii* MLHE-1, were shown to oxidize CO at low water potentials and anoxic conditions. Similar metabolisms of hydrogen consumption along with CO were found to occur in Hawaiian volcanic deposits [14.61]. Additionally, under deep-sea conditions, carbon monoxide dehydrogenases (CODH; the *cox* genes) were found to be abundant, suggesting that CO oxidation could be an important energy source [14.50] [14.60] [14.77].

The presence of molecular hydrogen in the polar ice caps of Mars is difficult to study. Remote neutron measurements have shown the presence of hydrogen atoms in the ice caps; however, these have been almost exclusively associated with water and methane clathrates [14.20] [14.35]. However, the presence of molecular hydrogen cannot be ruled out owing to the presence of clathrates, and thus it is pertinent to look at psychrophilic (cold-loving) metabolisms that can utilize hydrogen. Terrestrial permafrost of Earth's ice caps and glaciers has been studied extensively and a wide variety of microbial metabolisms have been reported [14.86] [14.141]. Methanogenic archaea that can utilize hydrogen as their sole source of energy along with CO_2, are ideal candidates for studies related to Martian subsurface. Additionally, several subsurface water deposits are briny and alkaline or acidic in nature both on Earth and Mars, and warrants investigation of metabolic functions [14.41] [14.127]. *Acidithiobacillus ferridurans*, for example, was identified as a novel species that is capable of oxidizing iron(II), hydrogen and sulfur, and is classified as an acidophilic and chemolithotrophic, Gammaproteobacterium [14.47]. On the other hand, alkaliphilic members have also been identified from soda lakes. An example of this would be the strain AHO 1, closely related to alkaliphilic bacterium *Roseinatronobacter thiooxidans* [14.122].

14.2.2 Methane

Methane is considered to be a significant source of energy capable of supporting life on Mars. Several studies have reported the release of methane into the atmosphere from the terrestrial environment [14.5] [14.42]. This methane release, as observed at the Gale crater, is suggested to be seasonal, owing to absorption of the gas into the regolith and subsequent release into the atmosphere [14.84]. Potential sources and sinks of methane on Mars are given in Figure 14.2. Microseepage of methane adds to the amount of the gas released back, especially at the Gale location. A big challenge has been to differentiate between abiotically produced methane (Fischer-Tropsch-type (FTT), where transition metals catalyze methane production reported in hydrothermal environments) versus a biotic origin (i.e., through methanogenic bacteria) [14.67] [14.151]. Proposed areas for methane discharge in Mars include mud-volcano-like mounds, ancient springs and rims of large impact craters, faults and fractures. Abiotic subsurface processes that can produce methane in the above areas include alteration of basaltic rocks, hydration of ultramafic units (serpentinization), and hydrothermal changes at high temperature and pressure (Figure 14.2 [14.95]). Apart from methanogenesis, acetogenesis is also a likely process that can make use of the abundant CO_2 and use it for production of acetic acid. It is more probable that these processes are coupled with redox reactions involving Fe(II), Fe(III), Mn(II) or subsurface gases, including hydrogen and hydrogen sulfide.

Another possible source of methane is in the hydrated form of the clathrates. Calculations based on thermodynamics revealed that the gas clathrate hydrate deposits might be present at Martian polar caps and in some of the subsurface environments [14.20]. Under certain low temperature and high-pressure conditions, solidification of liquid water along with gases into a crystalline structure called gas clathrate hydrate takes place [14.78].

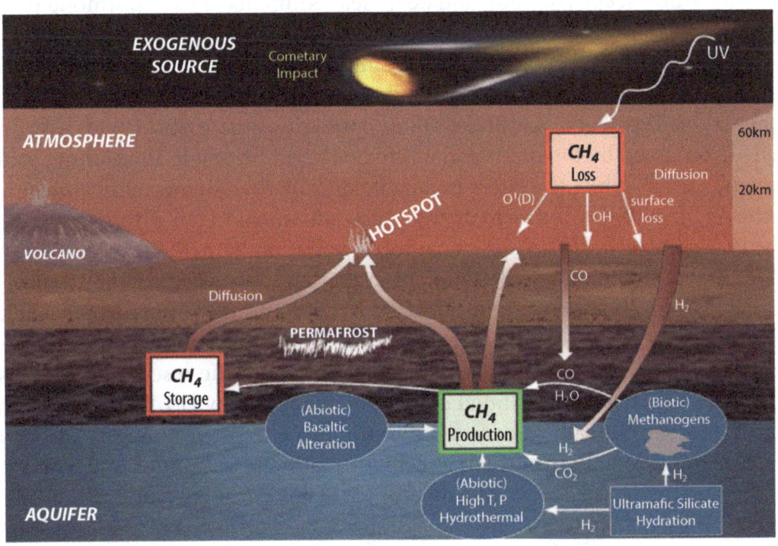

Figure 14.2 Possible sources and sinks for methane on Mars. (Reprinted with permission from Atreya *et al.*, 2007 [14.5]).

Extremophiles Using Methane

On Earth, methane is generated both biotically and abiotically, as determined by using isotope fractionation to detect the origins of the gas. Similar strategies are been adopted to classify Martian methane [14.126]. Methane produced from hydrogen oxidation and CO_2 reduction (discussed earlier) can be utilized by methanogens. Some of these methanogens can also reverse the pathway and can oxidize methane but only during net methane production [14.134]. The generated methane can be oxidized by other anaerobic bacteria through anaerobic oxidation of methane (AOB). Thus, methane could simultaneously serve as an electron donor and support different microbial groups simultaneously. These members possess unique terminal reductase enzymes that help to biochemically facilitate energy conservation during complete methane oxidation to CO_2 [14.79]. This process is mediated by a mixed population of SRB and archaeal members of anaerobic methanotrophs (ANME), where the electrons generated through methane oxidation by ANME are transferred to SRB, which in turn reduce the sulfate as per the equation, $CH_4 + SO_4^{2-} \rightarrow HCO_3^- + HS^- + H_2O$, G°'= -16.6 kJ/mol.

Yamagishi *et al.* [14.149] suggested the methane oxidizers could be present in the near surface environment of Mars and could derive their energy from methane coupled with oxidative compounds such as ferric oxide, sulfate and perchloride. ANME were isolated from a terrestrial methane seep in an Arctic hypersaline (~24% salinity), subzero (-5 °C), perennial spring, arising through thick permafrost [14.91], indicating their halophilic and psychrophilic properties. Thermogenic methane was the primary gas being released into this environment, and the functioning of the community, which included 84% bacteria and 3.8% archaea (majority were ANME), was portrayed as a model to simulate Martian methane plumes. ANME has several end-members that are tolerant to a variety of extreme environments, including acidic soils [14.54], hot springs [14.63], deep granitic rock units [14.53], hydrocarbon seeps [14.100], and deep-sea sediments [14.123]. Among these, many are halophilic, thermophilic and psychrophilic members.

In addition to the production of methane in permafrost environments, the produced methane can be oxidized. Three types of methanotrophs, I, II and X, are capable of oxidizing methane. Both Type I and X utilize ribulose monophosphate (RuMP) for assimilation of formaldehyde obtained by the oxidation of methane and belong to the gamma subdivision of the Proteobacteria, while the latter is capable of growth at higher temperatures. Type II employs the serine pathway for formaldehyde assimilation and fall within the beta subdivision of the Proteobacteria [14.46]. The anaerobic oxidation of methane was also reported in deep gas clathrates found at 600–800 meters in the Hydrate Ridge off the coast of Oregon in the USA [14.136]. The association between ANME and SRB was further elucidated in this case, and was accompanied by the production of copious amounts of hydrogen sulfide.

While anaerobic methanotrophy has dominated most of the research questions related to Mars, recent focus has also included aerobic methanotrophy [14.114]. Aerobic methane oxidation should theoretically generate the minimum energy production rate required to support endogenous metabolism, such as cellular maintenance, under Martian conditions. Aerobic methane oxidation occurs in a diverse set of environments on Earth. Some methanotrophs have the capacity to oxidize and assimilate methane even at subzero temperatures [14.137]. However, aerobic oxidation of methane has been shown to be lacking in halophilic environments [14.23].

14.2.3 Iron

Another significant biogeochemical mechanism catalyzed by microorganisms on Earth that is pertinent to Mars involves iron redox reactions. Mafic and ultramafic rocks, and weathered deposits are all suggested to be rich in iron and iron-bearing minerals on both planets. On Mars, jarosite, goethite and hematite are examples of weathered minerals found in the surficial sediments and rocks. The surface sediments are rich in iron oxides, contributing to the red color of the soil, while reduced iron, Fe(II), can exist as low as a few centimeters below the surface [14.138].

Layered basaltic sediments in the Sinus Meridiani and Aram Chaos contain 10–15% of crystalline gray hematite [14.21]. Hematite formation is a complex phenomenon involving fluid transport of reduced iron accelerated by acidification of orthopyroxene volcaniclastic source rocks, change in redox conditions, and eventual precipitation [14.97]. The famous "blueberries," hematite spherules found on the surface of Meridiani Planum, have been linked to the presence of water [14.84]. Similarly, iron deposits in Terra Meridiani were suggested to have been formed by thermal oxidation of volcanic deposits and fluid circulation with the layered deposits [14.52]. Both iron and sulfur were predominantly present in many of the terrains and led to the proposal that extant metabolisms exploiting these two elements were highly probable [14.92]. The identification of the iron sulfate salt, jarosite, suggests that an acidic, sulfate-rich brine must have existed, which eventually froze or evaporated [14.75]. On the other hand, the alteration of mafic and ultramafic rocks will result in alkaline lakes with high concentrations of dissolved salts; thus, alkaline environments on Earth are also of interest in this aspect [14.10]. Analysis of Martian meteorites showed that in addition to the reduced Fe(II), zero-valent iron is also present in the Martian smectites [14.19].

Extremophiles Using Iron

Several iron-reducing bacteria (IRB) and iron-oxidizing bacteria (IOB) have members in a variety of extreme conditions, and possess the ability to reduce/oxidize Fe(III) to Fe(II) depending on the geochemical environment [14.45]. Under subsurface conditions, energy is generated by using Fe(II) as a source for chemical energy and Fe(III) as electron acceptor (Figure 14.3

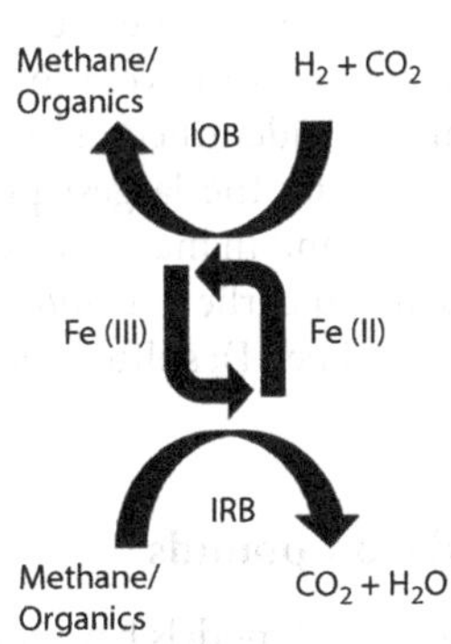

Figure 14.3 Microbial iron redox cycling showing the co-dependent operation of iron-reducing bacteria (IRB) and iron-oxidizing bacteria (IOB). (Image credit: the authors).

[14.110]). In cases where organic matter is limited, methane is typically used as a source of carbon. As discussed earlier, many methanogens function closely with IRB and SRB. The diversity of the IRB and IOB is fascinating, with novel species found in extreme salt, high and low pH, and at deep locations in the subsurface. For example, *Tepidibacillus decaturensis,* a moderate halothermophile IRB found at depths of 1.7 km in the Mount Simon Sandstone groundwater system in the Illinois Basin, is suggested as a model organism for what could survive on Mars [14.28]. The presence of zero-valent iron on Mars can also be exploited by methanogens, which can use the native form of iron as an electron donor to produce methane [14.19].

Iron-oxidizing bacteria isolated from psychrophilic, deep-sea hydrothermal vents near Juan de Fuca were able to grow in pyrite, basalt glass (~10 wt% FeO), siderite, and Fe substrates as their sole energy sources [14.31]. These cultures grew optimally at 3 to 10 °C, are obligate chemolithoautotrophs that could grow in the presence of oxygen or nitrogen, could not use organic substrates, and were classified under the gamma and alpha class of Proteobacteria. Similar organisms were found in the basaltic ocean crust, where aerobic or anaerobic Fe(II) oxidation and Fe(III) reduction accompanied by hydrogen consumption were reported [14.7]. Interestingly, it was observed that at constant temperature, increasing pressure decreases the overall energy yield $\left(\Delta G_r^\circ\right)$ of all redox-related reactions. The only exception were those reactions involving Fe(III) reduction, where ΔG_r° seemed to increase with pressure, suggesting the important role of minerals associated with iron [14.33]. In terrestrial environments, iron reduction by halophiles and thermophiles has been extensively reported [14.28] [14.29] [14.118].

Iron-oxidizing bacteria also has several acidophilic members capable of surviving in pH as low as 0.5 [14.32]. *Leptospirillum ferrooxidans* (aerobic) are uniquely adapted to such acidic environments, and have been used as models to study biomarkers for future Mars missions [14.99]. One of the most commonly studied sites for acid-tolerant microorganisms is Rio Tinto Basin (Iberian Pyrite Belt), which has high concentration of ferric iron and sulfates found in its waters. The ferric iron is produced as a result of pyrite biooxidation. The iron- and sulfide-rich subsurface rocks at Rio Tinto and waters in the system host unique anaerobic, acidophilic organisms. Studying the adaptability of these organisms in terms of their ability to use iron as a source of their energy and their resistance to radiation, oxidative and acidic stress has revealed probable clues into metabolisms that could be exhibited by potential life on Mars. Furthermore, the adaptations of these microorganisms to acidic waters could be applicable as proxies to understand life on Europa or Enceladus, where the subsurface water is considered to have a low pH [14.128]. A wide variety of chemolithotrophic microorganisms were identified in this Rio Tinto system, with iron- and sulfur-metabolizing organisms occupying the largest proportion [14.4]. Iron and sulfur metabolisms are very closely associated and in many cases are dependent on each other. The acid-tolerant chemolithotroph discussed earlier, *Acidithiobacillus ferrooxidans,* can obtain energy by oxidizing both reduced S and Fe(II); sulfur oxidation reaction has been shown to be energetically more favorable [14.4].

14.2.4 Sulfur and Associated Compounds

Meteorite data, remote sensing analysis and models have shown that Mars contains significant amounts of sulfur and sulfur-rich compounds. Mantle melts degassed volatile sulfur, leading to a significant greenhouse effect in the past. It is suspected that just as CO_2 is critical for Earth's climate, sulfur is (was) critical for influencing Martian climate. Sulfate has been routinely identified in landing missions with some sites, such as the Meridiani Planum, having more than 30% of Mg sulfates with minerals, including kieserite, gypsum and polyhydrated sulfates [14.57]. Sulfate salts have been detected even in the polar ice caps [14.65]. In environments where iron was found, sulfur and its compounds were generally identified in close association. As mentioned earlier, the identification of the iron-sulfate mineral, jarosite, is indicative of acidic conditions indicative of a past lake or sea. The source of the acid is suspected to be pyrite; when subjected to weathering and oxidative conditions, pyrite transforms to jarosite. Hydrogen sulfide is another gas that can get released along with methane from the hydrate reservoirs [14.71].

Extremophiles Using Sulfur

Just like iron oxidation and reduction, electron transfer occurs between sulfide and sulfate. Both sulfur and sulfide serve as sources of energy (electron donors), with sulfate serving as a terminal electron acceptor. Sulfate reduction is one of the most ubiquitous processes on Earth. Glacial ice caps, permafrost, deep-sea environments, terrestrial subsurface, hydrothermal vents, basaltic rocks, etc., are examples of locations where SRB are commonly found. In the subsurface, these organisms dominate metabolic processes along with iron reduction and methanogens. The process of sulfate reduction is closely associated with ANME as discussed earlier.

In the presence of oxygen, elemental sulfur can be oxidized by acidophilic bacteria and archaea, generating sulfuric acid, $S^0 + H_2O + 1.5\ O_2 \rightarrow H_2SO_4$. "Solfatara" are natural sulfur-rich systems near terrestrial geothermal sites and can reach temperatures up to 80 °C. Most of the sulfur-metabolizing organisms are bacteria, with some hyperthermophiles belonging to the archaeal phylum, Crenarchaeota [14.30]. A number of psychrotolerant acidophiles that can metabolize sulfur and iron have also been reported [14.44]. The metabolisms of sulfur oxidizers are quite varied and include obligate autotrophs, heterotrophs, or facultative autotrophs. Chemolithotrophic sulfur-oxidizing bacteria are capable of utilizing ferrous iron and/or hydrogen as electron donors. Reduced inorganic sulfur compounds include thiosulfates and polysulfides, and oxidation of these compounds can take place either aerobically or anaerobically [14.30]. Several series of genes have been shown to be uniquely upregulated in both cases.

The iron- and sulfur-oxidizing bacteria in Rio Tinto provide excellent examples of microbial activities that can take place in acidic environments [14.4]. Atacama Desert in Chile is one of the analogue sites for Mars that has been intensively studied. The dry soils help simulate the highly oxidized surface conditions of Mars. Gypsum is one of the sulfate minerals found in the Atacama soil. Analysis of the gypsum crust revealed abundant colonization, suggesting that potential microbes in the Martian past could have found refuge in such evaporite minerals during development of arid conditions [14.148].

14.2.5 Other Potential Energy Sources

Nitrogen

The basaltic rocks on Earth have very limited amount of nitrogen and its compounds (0.5 ppm) but even these low quantities support a plethora of microbial life [14.38].

Nitrate has been confirmed in modern aeolian and ancient sediment deposits on Mars [14.104] [14.125]. Nitrogen was detected in the form of nitric oxide and could be released from the breakdown of nitrates during heating. This essential element was estimated to be able to support ~10^{15} kg of biomass based on its current atmospheric composition in Mars [14.12]. These authors suggested that enough energy is available to fix nitrogen in the soils and ice caps. Just like photolytically produced atmospheric hydrogen can seep through the regolith, nitrogen can also diffuse into the subsurface in the form of transient thin liquid films. Ammonia synthesis is also considered to be a possibility on Mars [14.152].

Nitrate-dependent iron oxidation is suggested as a theoretical possibility under Martian conditions [14.104]. Ammonia oxidation, along with denitrification and dissimilatory nitrate reduction, has been shown to occur under thermophilic conditions at two hot springs, with archaea capable of ammonia oxidation being a dominant community member [14.27]. Certain haloalkaliphilic methanotrophs can oxidize ammonia to nitrate at pH 10 [14.121]. Similarly, the phylum Crenarchaeota was reported to have members capable of oxidizing ammonia in subterranean thermal Alpine springs with high amounts of radiation [14.143]. Anaerobic ammonia oxidation bacteria from Arctic fjord sediments were found to be adapted to low temperatures [14.16].

Manganese

Manganese is the third most abundant element in the crust of both Earth and Mars and is commonly present in its +2 oxidation state in a majority of igneous rocks. Several rocks in Gale Crater were found to contain an abundance of manganese through the ChemCam instrument on the *Curiosity* rover [14.66]. The evidence of manganese-rich minerals is indicative of the presence of water in the near surface environments of Mars as water is needed to precipitate the minerals with manganese (II) sourced from weathered igneous rocks.

Manganese (II)-oxidizing and manganese (IV)-reducing bacteria have been reported in a wide variety of extreme environments on Earth [14.88] [14.133]. For example, manganese-oxidizing bacteria were isolated from the deep-sea hydrothermal vents of Guaymas Basin [14.26] and oxidize pillow basalts from the cold, outer portions of Loihi Seamount [14.132]. A rapid oxidation of Mn(II) was mediated by a *Bacillus* species using a mnxG multicopper oxidase gene. Since manganese and iron are so closely related, several of the extremophilic iron-oxidizing bacteria mentioned earlier can also oxidize manganese [14.45]. Manganese- and iron-reducing bacteria were found in terrestrial thermal aquifers [14.58]. The bacterial group in this system was also able to use nitrate, nitrite and fumarate as their electron acceptors. Bacterial isolates from Antarctica were able to catalyze both oxidation and reduction of manganese forms and the oxidation reactions were influenced by cobalt, nickel and, only to a lesser extent, iron [14.62].

Perchlorate

Perchlorate has been previously reported at high levels for two locations on Mars' surface [14.87]. A strongly oxidized compound, perchlorate can be reduced by bacteria both in *in-situ* subsurface and laboratory conditions [14.48] [14.150]. Hydrogen is one of the common electron donors used for the reduction of perchlorate [14.150]. Several halophilic members are able to use perchlorate as an electron acceptor under anaerobic conditions [14.96]. Interestingly, the hydrogen produced from zero-valent iron corrosion can also be used as electron donor for the reduction of perchlorate [14.150].

Organics

The presence of organic material on Mars has been conclusively proven with the identification of chlorinated hydrocarbons and sulfur-containing organics [14.40]. The origin of these compounds is still debated, with meteorites and interplanetary dust particles suspected of delivering as much as 2.4×10^6 kg/year [14.39]. Nevertheless, pristine organic material is expected to be found on the surface and probably the subsurface as well. Simple organic compounds are excellent electron donors and pretty much all domains and species of life on Earth can metabolize organic material to obtain energy. The likely association of organics in the near-surface environment of Mars presents a variety of opportunities for exploration. One of the best proxies to study microbial life on dry, highly oxidizing, near-surface environment is the Atacama Desert, Chile. The amount of organic material and cultivable bacteria found in this environment was low, with most of the organic material being transported by wind and water, and native organics occupying a smaller proportion [14.6]. Interestingly, endolithic cyanobacteria and other microbial communities were found colonizing halite and Ca-sulfate evaporite crystals a few centimeters below the surface [14.147] [14.148]. Similar halophilic bacteria have been isolated from a halite core obtained from a 186 meter subsurface salt pan, and were at least 10,000 or more years old [14.85]. Other bacteria found in fluid inclusions of salts like gypsum and halite have been shown to survive on trace organics from original depositional conditions and have an extremely slow metabolism [14.93].

Special Mention: Oxygen

Oxygen is the most energetically favorable common electron acceptor. While Mars has limited free oxygen (0.13% or 0.01 mbar), new research has shown that oxygen could very well be present and useful as an electron acceptor [14.124]. The source of oxygen is expected to be photolysis of CO_2 present in the dissolved form found in near-surface brines of Mars [14.37]. Therefore, halophilic microaerophiles on Earth are good counterparts that can be considered. The dissolved oxygen availability in saline and hypersaline environments are much lower compared to non-saline systems. Both bacterial and archaeal populations have been reported that utilize oxygen and can obtain their electrons from organic and inorganic sources [14.139]. Typically, any of the electron donors discussed in the review can be exploited by such microaerophilic bacteria. Hydrogen-oxidizing, hyperthermophilic bacteria of the genus *Aquifex* can use oxygen concentrations as low as 7.5 ppm, i.e., less than 0.01 mbar of the gas [14.38]. Similarly, obligately chemolithoautotrophic, halophilic, sulfur-oxidizing bacteria, *Thiohalospira halophila* gen. nov., sp. nov. and *Thiohalospira alkaliphila* sp. nov., were found to use 2% oxygen [14.119].

14.3 Conclusion

Autotrophic energy production on Earth takes place in a wide variety of extreme environments ranging from terrestrial to deep-sea settings. With the current advances in metagenomics, metabolomics, and analytical instrumentation, we are presented with an excellent opportunity to obtain clarity in the biogeochemical functioning of these enigmatic ecosystems. Understanding the microbial mechanisms and their energetic metabolisms in these environments is critical to obtain clues for exploring subsurface life on Mars. A wide variety of potential electron donors and acceptors have been discussed in this review (summarized in Figure 14.4), with examples of metabolic pathways, enzymes involved, and microbial populations. Hydrogen and methane are the two sources of electrons that have been seriously considered here and in most other studies as being the most easily available resource in the subsurface, followed by iron and sulfur. As shown in several of the examples, some of these environments have multiple extreme conditions and the microorganisms present have unique adaptations at their disposal to withstand and thrive in those settings. Evaluating the energy released or required by a biochemical reaction is based on the extreme system under consideration. A proper assessment of the energy budgets at appropriate temperature, pressure, and the chemical environment is essential in determining which of such reactions would yield enough energy for microbial growth to possibly occur [14.3]. This information can be used to predict potential life and its activities under Martian conditions.

Future missions to the Red Planet are to be approached with a comprehensive view of what to expect in these environments. As researchers are constantly equipped with cutting-edge data and novel clues (biomarkers, isotopic signature, etc.), we are now better prepared to explore and understand the functioning of the Martian environment. Additionally, future manned expeditions will focus on generating useful sources of energy for human settlement

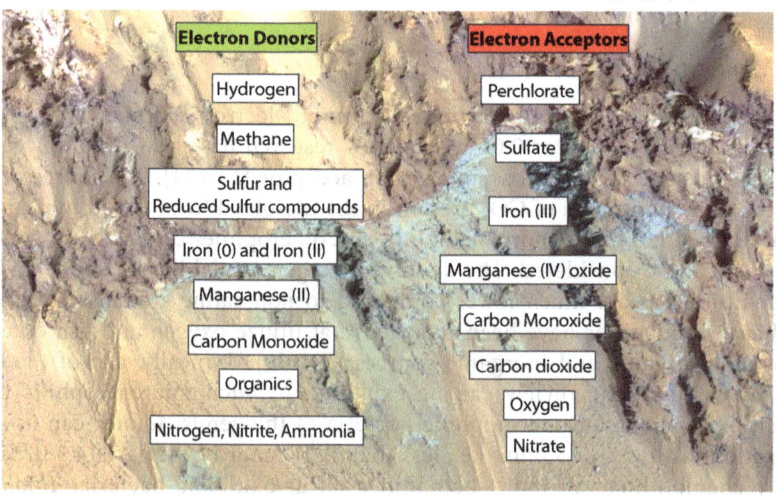

Figure 14.4 A summary of the list of potential electron donors and acceptors available in Martian subsurface. Background is an image of the exposed bedrock in the rim of Hale Crater. (NASA Mars Reconnaissance Orbiter; public domain).

and propelling rockets and engines. For example, hydrogen is an excellent fuel and apart from the geological sources of hydrogen available on Mars, scientists are looking to tap microbially produced hydrogen as well. Hydrogenases, the key enzymes involved in the production of hydrogen, from varied extremophilic microorganisms have already been considered for their capabilities and limitations.

References

14.1. Adams, M.W., The metabolism of hydrogen by extremely thermophilic, sulfur-dependent bacteria. *FEMS Microbiol. Rev.*, 6, 2-3, 219–237, 1990.
14.2. Alain, K., Marteinsson, V.T., Miroshnichenko, M.L., Bonch-Osmolovskaya, E.A., Prieur, D., Birrien, J.L., *Marinitoga piezophila* sp. nov., a rod-shaped, thermo-piezophilic bacterium isolated under high hydrostatic pressure from a deep-sea hydrothermal vent. *Int. J. Syst. Evol. Microbiol.*, 52, 4, 1331–1339, 2002.
14.3. Amend, J.P. and Shock, E.L., Energetics of overall metabolic reactions of thermophilic and hyperthermophilic Archaea and Bacteria. *FEMS Microbiol. Rev.*, 25, 2, 175–243, 2001.
14.4. Amils, R., González-Toril, E., Fernández-Remolar, D., Gómez, F., Aguilera, Á., Rodríguez, N., Malki, M., García-Moyano, A., González-Fairén, A., de la Fuente, V., Sanz, J.L., Extreme environments as Mars terrestrial analogs: The Rio Tinto case. *Planet. Space Sci.*, 55, 3, 370–381, 2007.
14.5. Atreya, S.K., Mahaffy, P.R., Wong, A.S., Methane and related trace species on Mars: Origin, loss, implications for life, and habitability. *Planet. Space Sci.*, 55, 3, 358–369, 2007.
14.6. Azua-Bustos, A., González-Silva, C., Corsini, G., The hyperarid core of the Atacama Desert, an extremely dry and carbon deprived habitat of potential interest for the field of carbon science. *Front. Microbiol.*, 8, #993, 2017.
14.7. Bach, W. and Edwards, K.J., Iron and sulfide oxidation within the basaltic ocean crust: Implications for chemolithoautotrophic microbial biomass production. *Geochim. Cosmochim. Acta*, 67, 20, 3871–3887, 2003.
14.8. Bar-On, Y.M., Phillips, R., Milo, R., The biomass distribution on Earth. *Proc. Natl. Acad. Sci.*, 115, 25, 6506–6511, 2018.
14.9. Berndt, M.E., Allen, D.E., Seyfried, W.E., Jr., Reduction of CO2 during serpentinization of olivine at 300 C and 500 bar. *Geology*, 24, 4, 351–354, 1996.
14.10. Blank, J.G., Green, S.J., Blake, D., Valley, J.W., Kita, N.T., Treiman, A., Dobson, P.F., An alkaline spring system within the Del Puerto Ophiolite (California, USA): A Mars analog site. *Planet. Space Sci.*, 57, 5-6, 533–540, 2009.
14.11. Boston, P.J., Ivanov, M.V., McKay, C.P., On the possibility of chemosynthetic ecosystems in subsurface habitats on Mars. *Icarus*, 95, 2, 300–308, 1992.
14.12. Boxe, C.S., Hand, K.P., Nealson, K.H., Yung, Y.L., Saiz-Lopez, A., An active nitrogen cycle on Mars sufficient to support a subsurface biosphere. *Int. J. Astrobiol.*, 11, 2, 109–115, 2012.
14.13. Bridges, J.C. and Grady, M.M., Evaporite mineral assemblages in the nakhlite (martian) meteorites. *Earth Planet. Sci. Lett.*, 176, 3-4, 267–279, 2000.
14.14. Brisbarre, N., Fardeau, M.L., Cueff, V., Cayol, J.L., Barbier, G., Cilia, V., Ravot, G., Thomas, P., Garcia, J.L., Ollivier, B., *Clostridium caminithermale* sp. nov., a slightly halophilic and moderately thermophilic bacterium isolated from an Atlantic deep-sea hydrothermal chimney. *Int. J. Syst. Evol. Microbiol.*, 53, 4, 1043–1049, 2003.
14.15. Cady, S.L., Farmer, J.D., Grotzinger, J.P., Schopf, J.W., Steele, A., Morphological biosignatures and the search for life on Mars. *Astrobiology*, 3, 2, 351–368, 2003.
14.16. Canion, A., Overholt, W.A., Kostka, J.E., Huettel, M., Lavik, G., Kuypers, M.M., Temperature response of denitrification and anaerobic ammonium oxidation rates and microbial

community structure in Arctic fjord sediments. *Environ. Microbiol.*, 16, 10, 3331–3344, 2014.

14.17. Cavicchioli, R., Extremophiles and the search for extraterrestrial life. *Astrobiology*, 2, 3, 281–292, 2002.

14.18. Chapelle, F.H., O'Neill, K., Bradley, P.M., Methé, B.A., Ciufo, S.A., Knobel, L.L., Lovley, D.R., A hydrogen-based subsurface microbial community dominated by methanogens. *Nature*, 415, 6869, #312, 2002.

14.19. Chastain, B.K. and Kral, T.A., Zero-valent iron on Mars: An alternative energy source for methanogens. *Icarus*, 208, 1, 198–201, 2010.

14.20. Chastain, B.K. and Chevrier, V., Methane clathrate hydrates as a potential source for martian atmospheric methane. *Planet. Space Sci.*, 55, 10, 1246–1256, 2007.

14.21. Christensen, P.R., Bandfield, J.L., Clark, R.N., Edgett, K.S., Hamilton, V.E., Hoefen, T., Kieffer, H.H., Kuzmin, R.O., Lane, M.D., Malin, M.C. *et al.*, Detection of crystalline hematite mineralization on Mars by the Thermal Emission Spectrometer: Evidence for near-surface water. *J. Geophys. Res.: Planets*, 105, E4, 9623–9642, 2000.

14.22. Ciarletti, V., Corbel, C., Plettemeier, D., Cais, P., Clifford, S.M., Hamran, S.E., WISDOM GPR designed for shallow and high-resolution sounding of the Martian subsurface. *Proc. IEEE*, 99, 5, 824–836, 2011.

14.23. Conrad, R., Frenzel, P., Cohen, Y., Methane emission from hypersaline microbial mats: Lack of aerobic methane oxidation activity. *FEMS Microbiol. Ecol.*, 16, 4, 297–305, 1995.

14.24. D'Amico, S., Collins, T., Marx, J.C., Feller, G., Gerday, C., Psychrophilic microorganisms: Challenges for life. *EMBO Rep.*, 7, 4, 385–389, 2006.

14.25. Deppenmeier, U., Müller, V., Gottschalk, G., Pathways of energy conservation in methanogenic archaea. *Arch. Microbiol.*, 165, 3, 149–163, 1996.

14.26. Dick, G.J., Lee, Y.E., Tebo, B.M., Manganese (II)-oxidizing *Bacillus* spores in Guaymas Basin hydrothermal sediments and plumes. *Appl. Environ. Microbiol.*, 72, 5, 3184–3190, 2006.

14.27. Dodsworth, J.A., Hungate, B.A., Hedlund, B.P., Ammonia oxidation, denitrification and dissimilatory nitrate reduction to ammonium in two US Great Basin hot springs with abundant ammonia-oxidizing archaea. *Environ. Microbiol.*, 13, 8, 2371–2386, 2011.

14.28. Dong, Y., Sanford, R.A., Boyanov, M.I., Kemner, K.M., Flynn, T.M., O'Loughlin, E.J., Locke, R.A., Weber, J.R., Egan, S.M., Fouke, B.W., *Tepidibacillus decaturensis* sp. nov., a microaerophilic, moderately thermophilic iron-reducing bacterium isolated from 1.7 km depth groundwater. *Int. J. Syst. Evol. Microbiol.*, 66, 10, 3964–3971, 2016.

14.29. Dong, Y., Kumar, C.G., Chia, N., Kim, P.J., Miller, P.A., Price, N.D., Cann, I.K., Flynn, T.M., Sanford, R.A., Krapac, I.G. *et al.*, *Halomonas sulfidaeris*-dominated microbial community inhabits a 1.8 km-deep subsurface Cambrian Sandstone reservoir. *Environ. Microbiol.*, 16, 6, 1695–1708, 2014.

14.30. Dopson, M. and Johnson, D.B., Biodiversity, metabolism and applications of acidophilic sulfur-metabolizing microorganisms. *Environ. Microbiol.*, 14, 10, 2620–2631, 2012.

14.31. Edwards, K.J., Rogers, D.R., Wirsen, C.O., McCollom, T.M., Isolation and characterization of novel psychrophilic, neutrophilic, Fe-oxidizing, chemolithoautotrophic α-and γ-Proteobacteria from the deep sea. *Appl. Environ. Microbiol.*, 69, 5, 2906–2913, 2003.

14.32. Edwards, K.J., Bond, P.L., Gihring, T.M., Banfield, J.F., An archaeal iron-oxidizing extreme acidophile important in acid mine drainage. *Science*, 287, 5459, 1796–1799, 2000.

14.33. Fang, J., Zhang, L., Bazylinski, D.A., Deep-sea piezosphere and piezophiles: Geomicrobiology and biogeochemistry. *Trends Microbiol.*, 18, 9, 413–422, 2010.

14.34. Farmer, J.D. and Des Marais, D.J., Exploring for a record of ancient Martian life. *J. Geophys. Res.: Planets*, 104, E11, 26977–26995, 1999.

14.35. Feldman, W.C., Prettyman, T.H., Maurice, S., Plaut, J.J., Bish, D.L., Vaniman, D.T., Mellon, M.T., Metzger, A.E., Squyres, S.W., Karunatillake, S. *et al.*, Global distribution of near-surface hydrogen on Mars. *J. Geophys. Res.: Planets*, 109, E9, 1–13, 2004.

14.36. Fernández-Remolar, D., Gómez-Elvira, J., Gómez, F., Sebastian, E., Martıin, J., Manfredi, J.A., Torres, J., González Kesler, C., Amils, R., The Tinto River, an extreme acidic environment under control of iron, as an analog of the Terra Meridiani hematite site of Mars. *Planet. Space Sci.*, 52, 1-3, 239–248, 2004.

14.37. Fischer, E., Martínez, G.M., Elliott, H.M., Rennó, N.O., Experimental evidence for the formation of liquid saline water on Mars. *Geophys. Res. Lett.*, 41, 13, 4456–4462, 2014.

14.38. Fisk, M.R. and Giovannoni, S.J., Sources of nutrients and energy for a deep biosphere on Mars. *J. Geophys. Res.: Planets*, 104, E5, 11805–11815, 1999.

14.39. Frantseva, K., Mueller, M., ten Kate, I.L., van der Tak, F.F., Greenstreet, S., Delivery of organics to Mars through asteroid and comet impacts. *Icarus*, 309, 125–133, 2018.

14.40. Freissinet, C., Glavin, D.P., Mahaffy, P.R., Miller, K.E., Eigenbrode, J.L., Summons, R.E., Brunner, A.E., Buch, A., Szopa, C., Archer, P.D. *et al.*, Organic molecules in the Sheepbed Mudstone, Gale Crater, Mars. *J. Geophys. Res.: Planets*, 120, 3, 495–514, 2015.

14.41. Gilichinsky, D.A., Wilson, G.S., Friedmann, E.I., McKay, C.P., Sletten, R.S., Rivkina, E.M., Vishnivetskaya, T.A., Erokhina, L.G., Ivanushkina, N.E., Kochkina, G.A. *et al.*, Microbial populations in Antarctic permafrost: Biodiversity, state, age, and implication for astrobiology. *Astrobiology*, 7, 2, 275–311, 2007.

14.42. Giuranna, M., Viscardy, S., Daerden, F., Neary, L., Etiope, G., Oehler, D., Formisano, V., Aronica, A., Wolkenberg, P., Aoki, S. *et al.*, Independent confirmation of a methane spike on Mars and a source region east of Gale Crater. *Nat. Geosci.*, 12, 5, 326, 2019.

14.43. Gregory, S.P., Barnett, M.J., Field, L.P., Milodowski, A.E., Subsurface microbial hydrogen cycling: Natural occurrence and implications for industry. *Microorganisms*, 7, 2, #53, 2019.

14.44. Hallberg, K.B., González-Toril, E., Johnson, D.B., *Acidithiobacillus ferrivorans*, sp. nov.; facultatively anaerobic, psychrotolerant iron-, and sulfur-oxidizing acidophiles isolated from metal mine-impacted environments. *Extremophiles*, 14, 1, 9–19, 2010.

14.45. Hannington, M.D., Fe and Mn oxides at seafloor hydrothermal vents. *Catena Suppl.*, 21, 351–370, 1992.

14.46. Hanson, R.S. and Hanson, T.E., Methanotrophic bacteria. *Microbiol. Mol. Biol. Rev.*, 60, 2, 439–471, 1996.

14.47. Hedrich, S. and Johnson, D.B., *Acidithiobacillus ferridurans* sp. nov., an acidophilic iron-, sulfur-and hydrogen-metabolizing chemolithotrophic gammaproteobacterium. *Int. J. Syst. Evol. Microbiol.*, 63, 11, 4018–4025, 2013.

14.48. Herman, D.C. and Frankenberger, W.T., Microbial-mediated reduction of perchlorate in groundwater. *J. Environ. Qual.*, 27, 4, 750–754, 1998.

14.49. Heyer, J., Berger, U., Hardt, M., Dunfield, P.F., *Methylohalobius crimeensis* gen. nov., sp. nov., a moderately halophilic, methanotrophic bacterium isolated from hypersaline lakes of Crimea. *Int. J. Syst. Evol. Microbiol.*, 55, 5, 1817–1826, 2005.

14.50. Hoshino, T. and Inagaki, F., Distribution of anaerobic carbon monoxide dehydrogenase genes in deep subseafloor sediments. *Lett. Appl. Microbiol.*, 64, 5, 355–363, 2017.

14.51. Huber, R., Wilharm, T., Huber, D., Trincone, A., Burggraf, S., König, H., Rachel, R., Rockinger, I., Fricke, H., Stetter, K.O., *Aquifex pyrophilus* gen. nov. sp. nov., represents a novel group of marine hyperthermophilic hydrogen-oxidizing bacteria. *Syst. Appl. Microbiol.*, 15, 3, 340–351, 1992.

14.52. Hynek, B.M., Arvidson, R.E., Phillips, R.J., Geologic setting and origin of Terra Meridiani hematite deposit on Mars. *J. Geophys. Res.: Planets*, 107, E10, 18–1, 2002.

14.53. Ino, K., Hernsdorf, A.W., Konno, U., Kouduka, M., Yanagawa, K., Kato, S., Sunamura, M., Hirota, A., Togo, Y.S., Ito, K. *et al.*, Ecological and genomic profiling of anaerobic methane-oxidizing archaea in a deep granitic environment. *ISME J.*, 12, 1, #31, 2018.
14.54. Islam, T., Torsvik, V., Larsen, Ø., Bodrossy, L., Øvreås, L., Birkeland, N.K., Acid-tolerant moderately thermophilic methanotrophs of the class Gammaproteobacteria isolated from tropical topsoil with methane seeps. *Front. Microbiol.*, 7, #851, 2016.
14.55. Islam, T., Jensen, S., Reigstad, L.J., Larsen, Ø., Birkeland, N.K., Methane oxidation at 55 C and pH 2 by a thermoacidophilic bacterium belonging to the Verrucomicrobia phylum. *Proc. Natl. Acad. Sci.*, 105, 1, 300–304, 2008.
14.56. Islas, S., Velasco, A.M., Becerra, A., Delaye, L., Lazcano, A., Hyperthermophily and the origin and earliest evolution of life. *Int. Microbiol.*, 6, 2, 87–94, 2003.
14.57. Johnson, S.S., Mischna, M.A., Grove, T.L., Zuber, M.T., Sulfur-induced greenhouse warming on early Mars. *J. Geophys. Res.: Planets*, 113, E8, 1–15, 2008.
14.58. Kanso, S., Greene, A.C., Patel, B.K., *Bacillus subterraneus* sp. nov., an iron-and manganese-reducing bacterium from a deep subsurface Australian thermal aquifer. *Int. J. Syst. Evol. Microbiol.*, 52, 3, 869–874, 2002.
14.59. Kendall, M.M. and Boone, D.R., The order Methanosarcinales, in: *The Prokaryotes*, M. Dworkin, S. Falkow, E. Rosenberg, K.H. Schleifer, E. Stackebrandt (Eds.), Springer, New York, NY, USA, 2006.
14.60. King, G.M., Carbon monoxide as a metabolic energy source for extremely halophilic microbes: Implications for microbial activity in Mars regolith. *Proc. Natl. Acad.*, 112, 14, 4465–4470, 2015.
14.61. King, G.M., Contributions of atmospheric CO and hydrogen uptake to microbial dynamics on recent Hawaiian volcanic deposits. *Appl. Environ. Microbiol.*, 69, 7, 4067–4075, 2003.
14.62. Krishnan, K.P., Sinha, R.K., Krishna, K., Nair, S., Singh, S.M., Microbially mediated redox transformations of manganese (II) along with some other trace elements: A study from Antarctic lakes. *Polar Biol.*, 32, 12, 1765–1778, 2009.
14.63. Krukenberg, V., Harding, K., Richter, M., Glöckner, F.O., Gruber-Vodicka, H.R., Adam, B., Berg, J.S., Knittel, K., Tegetmeyer, H.E., Boetius, A., Wegener, G., Candidatus *Desulfofervidus auxilii*, a hydrogenotrophic sulfate-reducing bacterium involved in the thermophilic anaerobic oxidation of methane. *Environ. Microbiol.*, 18, 9, 3073–3091, 2016.
14.64. Krumholz, L.R., Harris, S.H., Tay, S.T., Suflita, J.M., Characterization of two subsurface H2-utilizing bacteria, *Desulfomicrobium hypogeium* sp. nov. and *Acetobacterium psammolithicum* sp. nov., and their ecological roles. *Appl. Environ. Microbiol.*, 65, 6, 2300–2306, 1999.
14.65. Langevin, Y., Poulet, F., Bibring, J.P., Gondet, B., Sulfates in the north polar region of Mars detected by OMEGA/Mars Express. *Science*, 307, 5715, 1584–1586, 2005.
14.66. Lanza, N.L., Fischer, W.W., Wiens, R.C., Grotzinger, J., Ollila, A.M., Cousin, A., Anderson, R.B., Clark, B.C., Gellert, R., Mangold, N. *et al.*, High manganese concentrations in rocks at Gale crater, Mars. *Geophys. Res. Lett.*, 41, 16, 5755–5763, 2014.
14.67. Lefèvre, F., The enigma of methane on Mars, in: *Biosignatures for Astrobiology. Advances in Astrobiology and Biogeophysics*, B. Cavalazzi and F. Westall (Eds.), Springer, Cham, Switzerland, 2019.
14.68. Leshin, L.A., Mahaffy, P.R., Webster, C.R., Cabane, M., Coll, P., Conrad, P.G., Archer, P.D., Jr., Atreya, S.K., Brunner, A.E., Buch, A. *et al.*, Volatile, isotope, and organic analysis of martian fines with the Mars Curiosity rover. *Science*, 341, 6153, 1238937, 2013.
14.69. Litchfield, C.D., Survival strategies for microorganisms in hypersaline environments and their relevance to life on early Mars. *Meteorit. Planet. Sci.*, 33, 4, 813–819, 1998.
14.70. Liu, Y. and Whitman, W.B., Metabolic, phylogenetic, and ecological diversity of the methanogenic archaea. *Ann. N.Y. Acad. Sci.*, 1125, 1, 171–189, 2008.

14.71. Madden, M.E., Leeman, J.R., Root, M.J., Gainey, S., Reduced sulfur–carbon–water systems on Mars may yield shallow methane hydrate reservoirs. *Planet. Space Sci.*, 59, 2-3, 203–206, 2011.
14.72. Magnabosco, C., Lin, L.H., Dong, H., Bomberg, M., Ghiorse, W., Stan-Lotter, H., Pedersen, K., Kieft, T.L., van Heerden, E., Onstott, T.C., The biomass and biodiversity of the continental subsurface. *Nat. Geosci.*, 11, 10, 707, 2018.
14.73. Mahaffy, P.R., Webster, C.R., Atreya, S.K., Franz, H., Wong, M., Conrad, P.G., Harpold, D., Jones, J.J., Leshin, L.A., Manning, H. *et al.*, Abundance and isotopic composition of gases in the Martian atmosphere from the Curiosity rover. *Science*, 341, 6143, 263–266, 2013.
14.74. Mangold, N., Forni, O., Dromart, G., Stack, K., Wiens, R.C., Gasnault, O., Sumner, D.Y., Nachon, M., Meslin, P.Y., Anderson, R.B. *et al.*, Chemical variations in Yellowknife Bay formation sedimentary rocks analyzed by ChemCam on board the Curiosity rover on Mars. *J. Geophys. Res.: Planets*, 120, 3, 452–482, 2015.
14.75. Marion, G.M., Kargel, J.S., Catling, D.C., Modeling ferrous–ferric iron chemistry with application to martian surface geochemistry. *Geochim. Cosmochim. Acta*, 72, 1, 242–266, 2008.
14.76. Martin, W., Baross, J., Kelley, D., Russell, M.J., Hydrothermal vents and the origin of life. *Nat. Rev. Microbiol.*, 6, 11, 805, 2008.
14.77. Martín-Cuadrado, A.B., López-García, P., Alba, J.C., Moreira, D., Monticelli, L., Strittmatter, A., Gottschalk, G., Rodríguez-Valera, F., Metagenomics of the deep Mediterranean, a warm bathypelagic habitat. *PLoS One*, 2, 9, e914, 2007.
14.78. Max, M.D. and Clifford, S.M., The state, potential distribution, and biological implications of methane in the Martian crust. *J. Geophys. Res.: Planets*, 105, E2, 4165–4171, 2000.
14.79. McGlynn, S.E., Energy metabolism during anaerobic methane oxidation in ANME archaea. *Microbes Environ.*, 32, 1, 5–13, 2017.
14.80. McMahon, S., Parnell, J., Blamey, N.J., Evidence for seismogenic hydrogen gas, a potential microbial energy source on Earth and Mars. *Astrobiology*, 16, 9, 690–702, 2016.
14.81. McSween, H.Y., Taylor, G.J., Wyatt, M.B., Elemental composition of the Martian crust. *Science*, 324, 5928, 736–739, 2009.
14.82. Michalski, J.R., Onstott, T.C., Mojzsis, S.J., Mustard, J., Chan, Q.H., Niles, P.B., Johnson, S.S., The Martian subsurface as a potential window into the origin of life. *Nat. Geosci.*, 11, 1, 21, 2018.
14.83. Michoud, G. and Jebbar, M., High hydrostatic pressure adaptive strategies in an obligate piezophile *Pyrococcus yayanosii*. *Sci. Rep.*, 6, #27289, 2016.
14.84. Moores, J.E., Gough, R.V., Martinez, G.M., Meslin, P.Y., Smith, C.L., Atreya, S.K., Mahaffy, P., Newman, C., Webster, C.R., Methane seasonal cycle at Gale Crater on Mars consistent with regolith adsorption and diffusion. *Nat. Geosci.*, 12, 5, 321–325, 2019.
14.85. Mormile, M.R., Biesen, M.A., Gutierrez, M.C., Ventosa, A., Pavlovich, J.B., Onstott, T.C., Fredrickson, J.K., Isolation of *Halobacterium salinarum* retrieved directly from halite brine inclusions. *Environ. Microbiol.*, 5, 11, 1094–1102, 2003.
14.86. Morozova, D., Möhlmann, D., Wagner, D., Survival of methanogenic archaea from Siberian permafrost under simulated Martian thermal conditions. *Orig. Life Evol. Biosph.*, 37, 2, 189–200, 2007.
14.87. Navarro-González, R., Vargas, E., de La Rosa, J., Raga, A.C., McKay, C.P., Reanalysis of the Viking results suggests perchlorate and organics at midlatitudes on Mars. *J. Geophys. Res.: Planets*, 115, E12, 1–11, 2010.
14.88. Nealson, K.H., Inagaki, F., Takai, K., Hydrogen-driven subsurface lithoautotrophic microbial ecosystems (SLiMEs): Do they exist and why should we care? *Trends Microbiol.*, 13, 9, 405–410, 2005.
14.89. Nealson, K.H. and Cox, B.L., Microbial metal-ion reduction and Mars: Extraterrestrial expectations? *Curr. Opin. Microbiol.*, 5, 3, 296–300, 2002.

14.90. Nie, N.X., Dauphas, N., Greenwood, R.C., Iron and oxygen isotope fractionation during iron UV photo-oxidation: Implications for early Earth and Mars. *Earth Planet. Sci. Lett.*, 458, 179–191, 2017.
14.91. Niederberger, T.D., Perreault, N.N., Tille, S., Lollar, B.S., Lacrampe-Couloume, G., Andersen, D., Greer, C.V., Pollard, W., Whyte, L.G., Microbial characterization of a subzero, hypersaline methane seep in the Canadian High Arctic. *ISME J.*, 4, 10, 1326–1329, 2010.
14.92. Nixon, S., Cousins, C.R., Cockell, C., Plausible microbial metabolisms on Mars. *Astron. Geophys.*, 54, 13–16, 2013.
14.93. Norton, C.F. and Grant, W.D., Survival of halobacteria within fluid inclusions in salt crystals. *Microbiology*, 134, 5, 1365–1373, 1988.
14.94. Nyquist, L.E., Bogard, D.D., Shih, C.Y., Greshake, A., Stöffler, D., Eugster, O., Ages and geologic histories of Martian meteorites. *Space Sci. Rev.*, 96, 105–164, 2001.
14.95. Oehler, D.Z. and Etiope, G., Methane seepage on Mars: Where to look and why. *Astrobiology*, 17, 12, 1233–1264, 2017.
14.96. Oren, A., Bardavid, R.E., Mana, L., Perchlorate and halophilic prokaryotes: Implications for possible halophilic life on Mars. *Extremophiles*, 18, 1, 75–80, 2014.
14.97. Ormö, J., Komatsu, G., Chan, M.A., Beitler, B., Parry, W.T., Geological features indicative of processes related to the hematite formation in Meridiani Planum and Aram Chaos, Mars: A comparison with diagenetic hematite deposits in southern Utah, USA. *Icarus*, 171, 2, 295–316, 2004.
14.98. Orosei, R., Lauro, S.E., Pettinelli, E., Cicchetti, A., Coradini, M., Cosciotti, B., Di Paolo, F., Flamini, E., Mattei, E., Pajola, M. *et al.*, Radar evidence of subglacial liquid water on Mars. *Science*, 361, 6401, 490–493, 2018.
14.99. Parro, V., Rodríguez-Manfredi, J.A., Briones, C., Compostizo, C., Herrero, P.L., Vez, E., Sebastián, E., Moreno-Paz, M., Garcia-Villadangos, M., Fernández-Calvo, P. *et al.*, Instrument development to search for biomarkers on Mars: terrestrial acidophile, iron-powered chemolithoautotrophic communities as model systems. *Planet. Space Sci.*, 53, 7, 729–737, 2005.
14.100. Paul, B.G., Ding, H., Bagby, S.C., Kellermann, M.Y., Redmond, M.C., Andersen, G.L., Valentine, D.L., Methane-oxidizing bacteria shunt carbon to microbial mats at a marine hydrocarbon seep. *Front. Microbiol.*, 8, #186, 2017.
14.101. Paul, V.G., Wronkiewicz, D.J., Mormile, M.R., Foster, J.S., Mineralogy and microbial diversity of the microbialites in the hypersaline Storr's Lake, The Bahamas. *Astrobiology*, 16, 4, 282–300, 2016.
14.102. Petersen, J.M., Zielinski, F.U., Pape, T., Seifert, R., Moraru, C., Amann, R., Hourdez, S., Girguis, P.R., Wankel, S.D., Barbe, V. *et al.*, Hydrogen is an energy source for hydrothermal vent symbioses. *Nature*, 476, 7359, 176–180, 2011.
14.103. Pikuta, E.V., Hoover, R.B., Tang, J., Microbial extremophiles at the limits of life. *Crit. Rev. Microbiol.*, 33, 3, 183–209, 2007.
14.104. Price, A., Pearson, V.K., Schwenzer, S.P., Miot, J., Olsson-Francis, K., Nitrate-dependent iron oxidation: A potential Mars metabolism. *Front. Microbiol.*, 9, #513, 2018.
14.105. Pulschen, A.A., Rodrigues, F., Duarte, R.T., Araujo, G.G., Santiago, I.F., Paulino-Lima, I.G., Rosa, C.A., Kato, M.J., Pellizari, V.H., Galante, D., UV-resistant yeasts isolated from a high-altitude volcanic area on the Atacama Desert as eukaryotic models for astrobiology. *Microbiologyopen*, 4, 4, 574–588, 2015.
14.106. Rampelotto, P.H., Resistance of microorganisms to extreme environmental conditions and its contribution to astrobiology. *Sustainability*, 2, 6, 1602–1623, 2010.
14.107. Rapin, W., Meslin, P.Y., Maurice, S., Wiens, R.C., Laporte, D., Chauviré, B., Gasnault, O., Schröder, S., Beck, P., Bender, S.C. *et al.*, Quantification of water content by laser induced breakdown spectroscopy on Mars. *Spectrochim. Acta Part B: At. Spectrosc.*, 130, 82–100, 2017.

14.108. Rittmann, B.E. and McCarty, P.L. (Eds.), *Environmental Biotechnology: Principles and Applications*, Tata McGraw-Hill Education, New York, NY, USA, 2012.

14.109. Roberts, M.F., Organic compatible solutes of halotolerant and halophilic microorganisms. *Saline Syst.*, 1, 1, #5, 2005.

14.110. Roden, E.E., Microbial iron-redox cycling in subsurface environments. *Biochem. Soc. Trans.*, 40, 1249–1256, 2012.

14.111. Schidlowski, M., Stable carbon isotopes: possible clues to early life on Mars. *Adv. Space Res.*, 12, 4, 101–110, 1992.

14.112. Schulte, M., Blake, D., Hoehler, T., McCollom, T., Serpentinization and its implications for life on the early Earth and Mars. *Astrobiology*, 6, 2, 364–376, 2006.

14.113. Schwartzman, D.W. and Lineweaver, C.H., The hyperthermophilic origin of life revisited. *Biochem. Soc. Trans.*, 32, 167–171, 2004.

14.114. Seto, M., Noguchi, K., Van Cappellen, P., Potential for aerobic methanotrophic metabolism on Mars. *Astrobiology*, 19, 10, 1187–1195, 2019. https://doi.org/10.1089/ast.2018.1943.

14.115. Siebach, K.L. and Grotzinger, J.P., Volumetric estimates of ancient water on Mount Sharp based on boxwork deposits, Gale Crater, Mars. *J. Geophys. Res.: Planets*, 119, 1, 189–198, 2014.

14.116. Sinha, N., Nepal, S., Kral, T., Kumar, P., Survivability and growth kinetics of methanogenic archaea at various pHs and pressures: Implications for deep subsurface life on Mars. *Planet. Space Sci.*, 136, 15–24, 2017.

14.117. Simonato, F., Campanaro, S., Lauro, F.M., Vezzi, A., D'Angelo, M., Vitulo, N., Valle, G., Bartlett, D.H., Piezophilic adaptation: a genomic point of view. *J. Biotechnol.*, 126, 1, 11–25, 2006.

14.118. Slobodkin, A.I., Jeanthon, C., L'haridon, S., Nazina, T., Miroshnichenko, M., Bonch-Osmolovskaya, E., Dissimilatory reduction of Fe (III) by thermophilic bacteria and archaea in deep subsurface petroleum reservoirs of Western Siberia. *Curr. Microbiol.*, 39, 2, 99–102, 1999.

14.119. Sorokin, D.Y., Tourova, T.P., Muyzer, G., Kuenen, G.J., *Thiohalospira halophila* gen. nov., sp. nov. and *Thiohalospira alkaliphila* sp. nov., novel obligately chemolithoautotrophic, halophilic, sulfur-oxidizing gammaproteobacteria from hypersaline habitats. *Int. J. Syst. Evol. Microbiol.*, 58, 7, 1685–1692, 2008.

14.120. Sorokin, D.Y., Tourova, T.P., Lysenko, A.M., Mityushina, L.L., Kuenen, J.G., *Thioalkalivibrio thiocyanoxidans* sp. nov. and *Thioalkalivibrio paradoxus* sp. nov., novel alkaliphilic, obligately autotrophic, sulfur-oxidizing bacteria capable of growth on thiocyanate, from soda lakes. *Int. J. Syst. Evol. Microbiol.*, 52, 2, 657–664, 2002.

14.121. Sorokin, D.Y., Jones, B.E., Kuenen, J.G., An obligate methylotrophic, methane-oxidizing *Methylomicrobium* species from a highly alkaline environment. *Extremophiles*, 4, 3, 145–155, 2000.

14.122. Sorokin, D.Y., Tourova, T.P., Kuenen, J.G., A new facultatively autotrophic hydrogen-and sulfur-oxidizing bacterium from an alkaline environment. *Extremophiles*, 4, 4, 237–245, 2000.

14.123. Spietz, R.L., Butterfield, D.A., Buck, N.J., Larson, B.I., Chadwick, W.W., Jr., Walker, S.L., Kelley, D.S., Morris, R.M., Deep-sea volcanic eruptions create unique chemical and biological linkages between the subsurface lithosphere and the oceanic hydrosphere. *Oceanography*, 31, 1, 128–135, 2018.

14.124. Stamenković, V., Ward, L.M., Mischna, M., Fischer, W.W., O_2 solubility in Martian near-surface environments and implications for aerobic life. *Nat. Geosci.*, 11, 12, 905–909, 2018.

14.125. Stern, J.C., Sutter, B., Freissinet, C., Navarro-González, R., McKay, C.P., Archer, P.D., Jr., Buch, A., Brunner, A.E., Coll, P., Eigenbrode, J.L. *et al.*, Evidence for indigenous nitrogen in sedimentary and aeolian deposits from the Curiosity rover investigations at Gale crater, Mars. *Proc. Natl. Acad. Sci.*, 112, 14, 4245–4250, 2015.

14.126. Stevens, A.H., Patel, M.R., Lewis, S.R., Modelled isotopic fractionation and transient diffusive release of methane from potential subsurface sources on Mars. *Icarus*, 281, 240–247, 2017.

14.127. Steven, B., Leveille, R., Pollard, W.H., Whyte, L.G., Microbial ecology and biodiversity in permafrost. *Extremophiles*, 10, 4, 259–267, 2006.

14.128. Stoker, C., Lemke, L., Mandell, H., Mckay, D., George, J., Gomez-Elvira, J., Amils, R., Stevens, T., Miller, D., Mars analog research and technology experiment (MARTE): A simulated Mars drilling mission to search for subsurface life at the Rio Tinto (Spain). *Lunar Planet. Sci. XXXIV*, 1076, 1-2, 2003.

14.129. Tarnas, J.D., Mustard, J.F., Lollar, B.S., Bramble, M.S., Cannon, K.M., Palumbo, A.M., Plesa, A.C., Radiolytic H_2 production on Noachian Mars: Implications for habitability and atmospheric warming. *Earth Planet. Sci. Lett.*, 502, 133–145, 2018.

14.130. Takai, K., Gamo, T., Tsunogai, U., Nakayama, N., Hirayama, H., Nealson, K.H., Horikoshi, K., Geochemical and microbiological evidence for a hydrogen-based, hyperthermophilic subsurface lithoautotrophic microbial ecosystem (HyperSLiME) beneath an active deep-sea hydrothermal field. *Extremophiles*, 8, 4, 269–282, 2004.

14.131. Takai, K., Komatsu, T., Horikoshi, K., *Hydrogenobacter subterraneus* sp. nov., an extremely thermophilic, heterotrophic bacterium unable to grow on hydrogen gas, from deep subsurface geothermal water. *Int. J. Syst. Evol. Microbiol.*, 51, 4, 1425–1435, 2001.

14.132. Templeton, A.S., Staudigel, H., Tebo, B.M., Diverse Mn (II)-oxidizing bacteria isolated from submarine basalts at Loihi Seamount. *Geomicrobiol. J.*, 22, 3-4, 127–139, 2005.

14.133. Tebo, B.M., Bargar, J.R., Clement, B.G., Dick, G.J., Murray, K.J., Parker, D., Verity, R., Webb, S.M., Biogenic manganese oxides: Properties and mechanisms of formation. *Annu. Rev. Earth Planet. Sci.*, 32, 287–328, 2004.

14.134. Timmers, P.H., Welte, C.U., Koehorst, J.J., Plugge, C.M., Jetten, M.S., Stams, A.J., Reverse methanogenesis and respiration in methanotrophic archaea, Archaea, 2017, 1-22, https://doi.org/10.1155/2017/1654237.

14.135. Trent, J.D., Extremophiles in astrobiology: per Ardua ad Astra. *Gravit. Space Biol.*, 13, 2, 5–12, 2000.

14.136. Treude, T., Boetius, A., Knittel, K., Wallmann, K., Jørgensen, B.B., Anaerobic oxidation of methane above gas hydrates at Hydrate Ridge, NE Pacific Ocean. *Mar. Ecol. Prog. Ser.*, 264, 1–14, 2003.

14.137. Trotsenko, Y.A. and Khmelenina, V.N., Aerobic methanotrophic bacteria of cold ecosystems. *FEMS Microbiol. Ecol.*, 53, 1, 15–26, 2005.

14.138. Vaniman, D.T., Bish, D.L., Ming, D.W., Bristow, T.F., Morris, R.V., Blake, D.F., Chipera, S.J., Morrison, S.M., Treiman, A.H., Rampe, E.B. *et al.*, Mineralogy of a mudstone at Yellowknife Bay, Gale crater, Mars. *Science*, 343, 6169, 1243480, 2014.

14.139. Ventosa, A., Nieto, J.J., Oren, A., Biology of moderately halophilic aerobic bacteria. *Microbiol. Mol. Biol. Rev.*, 62, 2, 504–544, 1998.

14.140. Vignais, P.M. and Billoud, B., Occurrence, classification, and biological function of hydrogenases: An overview. *Chem. Rev.*, 107, 10, 4206–4272, 2007.

14.141. Wagner, D., Spieck, E., Bock, E., Pfeiffer, E.M., Microbial life in terrestrial permafrost: Methanogenesis and nitrification in gelisols as potentials for exobiological process, in: *Astrobiology - the Quest for the Conditions of Life*, G. Horneck and C. Baumstark-Khan (Eds.), Springer-Verlag, Berlin, Germany, 2002.

14.142. Wallis, M.K., Wickramasinghe, J.T., Wickramasinghe, N.C., Mars polar cap: A habitat for elementary life. *Int. J. Astrobiol.*, 8, 2, 117–119, 2009.

14.143. Weidler, G.W., Dornmayr-Pfaffenhuemer, M., Gerbl, F.W., Heinen, W., Stan-Lotter, H., Communities of Archaea and Bacteria in a subsurface radioactive thermal spring in the Austrian Central Alps, and evidence of ammonia-oxidizing Crenarchaeota. *Appl. Environ. Microbiol.*, 73, 1, 259–270, 2007.
14.144. Weiss, B.P., Yung, Y.L., Nealson, K.H., Atmospheric energy for subsurface life on Mars? *Proc. Natl. Acad. Sci.*, 97, 4, 1395–1399, 2000.
14.145. Wentworth, S.J., Gibson, E.K., Velbel, M.A., McKay, D.S., Antarctic Dry Valleys and indigenous weathering in Mars meteorites: Implications for water and life on Mars. *Icarus*, 174, 2, 383–395, 2005.
14.146. Westall, F., Life on the early Earth: A sedimentary view. *Science*, 308, 5720, 366–367, 2005.
14.147. Wierzchos, J., Cámara, B., De Los Rios, A., Davila, A.F., Sánchez Almazo, I.M., Artieda, O., Wierzchos, K., Gómez-Silva, B., McKay, C., Ascaso, C., Microbial colonization of Ca-sulfate crusts in the hyperarid core of the Atacama Desert: Implications for the search for life on Mars. *Geobiology*, 9, 1, 44–60, 2011.
14.148. Wierzchos, J., Ascaso, C., McKay, C.P., Endolithic cyanobacteria in halite rocks from the hyperarid core of the Atacama Desert. *Astrobiology*, 6, 3, 415–422, 2006.
14.149. Yamagishi, A., Yokobori, S.I., Yoshimura, Y., Yamashita, M., Hashimoto, H., Kubota, T., Yano, H., Harayama, J., Tabata, M., Kobayashi, K. *et al.*, Japan Astrobiology Mars Project (JAMP): Search for microbes on the Mars surface with special interest in methane-oxidizing bacteria. *Biol. Sci. Space*, 24, 2, 67–82, 2010.
14.150. Yu, X., Amrhein, C., Deshusses, M.A., Matsumoto, M.R., Perchlorate reduction by autotrophic bacteria in the presence of zero-valent iron. *Environ. Sci. Technol.*, 40, 4, 1328–1334, 2006.
14.151. Yung, Y.L. and Chen, P., Methane on Mars. *Astrobiol. Outreach*, 3, 125, 2332–2519, 2015.
14.152. Zeinalipour-Yazdi, C.D., Hargreaves, J.S., Catlow, C.R.A., Nitrogen activation in a Mars–van Krevelen mechanism for ammonia synthesis on Co_3Mo_3N. *J. Phys. Chem. C*, 119, 51, 28368–28376, 2015.

Part IV

THEORY AND HYPOTHESES

15

Origin of Initial Communities of Thermophilic Extremophiles on Earth by Efficient Response to Oscillations in the Environment

Vladimir N. Kompanichenko[1]* and Vladimir F. Levchenko[2]

[1]*Institute for Complex Analysis RAS, Birobidzhan, Russia*
[2]*Institute of Evolutionary Physiology and Biochemistry RAS, St. Petersburg, Russia*

Abstract

A general method is proposed for a primary community emergence on Earth (and other planets) by efficient response to oscillations in the environment. The approach postulates that a minimal "self-sufficient" unit of primordial life is a community, but not a separate microorganism. The conversion of a cluster of prebiotic organic microsystems into the community of primary microorganisms (high-temperature extremophiles) proceeds in hydrothermal environments with multilevel oscillations of physicochemical parameters (temperature, pressure, redox potential, etc.). The transition to life takes place through efficient (intensified) response of the cluster's microsystems to external oscillations, when the common contribution of free energy in the cluster (community) begins to prevail over the common contribution of entropy (in other words, when the "thermodynamic inversion" occurs). Some theoretical confirmations of the origin of life in oscillating (not stable) conditions are given, including an oscillatory mode of many metabolic processes and necessity of external influences (moderate stress) for existence of living beings. Besides, verification of the proposed approach is possible by means of a new way of laboratory experiments: various prebiotic models (microsystems) should be explored in the chamber with the quite high-frequency (split second – 30 minutes) regular oscillations interrupting by irregular fluctuations. A goal of the experiments consists of detection of initial sparks of the living state in the microsystems (for instance, increase of internal energy gradients, strengthening active transport, rise of homochirality degree, etc.). Appearance of such communities on Earth corresponds with the embryonic stage of the evolution of biospheres.

Keywords: Origin of life, prebiotic microsystem, community, hydrothermal system, planet, thermophilic microorganism, oscillation, biosphere

**Corresponding author*: kompanv@yandex.ru
Vladimir N. Kompanichenko: https://www.researchgate.net/profile/Vladimir_Kompanichenko
Vladimir F. Levchenko: lew@lew.spb.org, http://www.iephb.ru/?lab=laboratoriya-modelirovaniya-evolyutsii

Joseph Seckbach and Helga Stan-Lotter (eds.) Extremophiles as Astrobiological Models, (319–328)

15.1 Introduction

Collocations like "the origin of life" and "the origin of biosphere" are often understood as convertible terms. However, their strict sense is different. The origin of life means the appearance of initial living units and/or their communities on early Earth (or other planet), while the biosphere is the life system on a planetary scale. It embraces the whole near-surface geospheres (ocean, land, air) and consists of the hierarchically organized ecosystems. The earliest stage of the biosphere evolution ("pre-biosphere") is indicated as embryosphere [15.18] [15.19] [15.23]. At this stage, the hierarchically organized system of life on Earth was absent. The initial local communities were assumptively separated from each other in space, although they could be connected by means of various (bio)geochemical cycles. The traced evolution of humankind can serve as some analogy or an indirect confirmation of this supposition. A single community or some separated communities of the oldest human beings appeared in Africa. Then, step-by-step, they expanded into other continents and have formed the highly organized human civilization. Taking into consideration the way in which complex life originated, a similar way can be supposed for the simplest life system—a community of primary living units (initial microorganisms).

At the present time, a lot of researchers consider some thermophilic (60–80 °C) or hyper-thermophilic (>80 °C) species of Archaea and Bacteria as the oldest microorganisms, which are at the deepest root of the phylogenetic tree of life [15.1] [15.30] [15.32] [15.34]. These organisms inhabit hot vents at the bottom of the oceans and springs on the continents. It is peculiar that ecological niches of these communities are extremely diversified: they can exist under the widest range of temperature, salinity, pH, redox potential, and other physicochemical parameters. The easiest explanation of this data consists of the origin of primary forms of life on early Earth in hydrothermal environments. Many researchers are actively elaborating the general hydrothermal scenario of the origin of life following its initial version that considered submarine "black smokers" as a cradle of life [15.2]. They emphasize a major role of thermal energy and high gradients between cool oceanic water and hot vents for the origin of life. Unlike them, the authors pay major attention to multilevel (i.e., on micro-macro levels) oscillations of physicochemical parameters peculiar to hydrothermal medium. According to the elaborated approach, life appears by continuous efficient response of prebiotic microsystems to incessant oscillations in the environment [15.10] [15.12] [15.13] [15.14]. The approach (called "inversion"; see also below) implies that initial communities on Earth (as well as on other planets) were thermophilic extremophiles.

A goal of this chapter is to propose the way in which a primary community of initial microorganisms could emerge from a cluster of prebiotic microsystems in highly fluctuating hydrothermal environments.

15.2 Required Conditions for the Origin of Life: Necessity of Rapid-Frequency Oscillations of Parameters

Specificity of a chemical nonequilibrium system, being under external influences, consists of its immediate response to significant changes in the surroundings. Arising (self-assembly) driven by extremely fast chemical reactions induced in response to very short (less than 1 second) pulses of energy were explored, in particular, by Eigen [15.5]; for his work he was awarded the Nobel Prize in Chemistry, along with Norrish and Porter, in 1967. According to this research, a strong impact of temperature, pressure, electric field or another physical factor must initiate the transition of an equilibrium chemical system into the highly nonequilibrium (or far from equilibrium) state. Thereafter the system returns to equilibrium state.

The results of this research have been applied to exploration of prebiotic organic microsystems under oscillating conditions in the context of the origin of life [15.14]. The effect of a single physical impact on a prebiotic organic system can be subdivided into two periods. At first the impact strongly influences ("stresses") the network of chemical reactions (Figure 15.1a), then the system partially compensates the effect of the executed impact, in accordance with Le Chatelier's principle (Figure 15.1b). Under highly fluctuating conditions, new chemical reactions are generated in a prebiotic microsystem as the response (counteraction) to continuous stress. Real chemical equilibrium never is achieved in those nonequilibrium systems if incessant oscillations in the surroundings are maintained. Sometimes under this regime of oscillations and suitable composition of the microsystem the counteraction may temporarily expand and become intensified (efficient) in comparison with the effect of external stress (Figure 15.1c). This is a crucial step (called "thermodynamic inversion") to the origin of life, in accordance with the inversion concept [15.10] [15.11] [15.12] [15.13]. In this way a prebiotic microsystem evolves in the direction of a primary living unit. External stress and the effective response to it are actually in the foundation of life processes. This thesis follows the theory of stress [15.29], which postulates that any particular influence on a living being (mechanical, physical, etc.) immediately initiates the standard set of nonspecific (that is, general) reactions that lead it to a state of stress. After that, the state of stress launches a specific response to the obtained influence.

At the present time it is accepted that three conditions for the origin of life on early Earth are required: availability of organic matter, liquid medium, and source of energy. Following this approach, a fourth required condition should be added: the necessity of multilevel physicochemical oscillations in the maternal environment. The regime of such oscillations

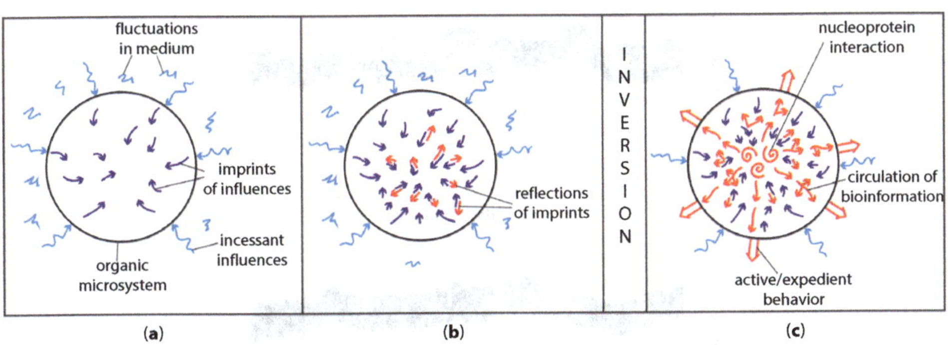

Figure 15.1 Scheme of the conversion of a nonequilibrium prebiotic microsystem into primary form of life under oscillating environment. (a) Arising (self-assembly) of an organic prebiotic microsystem within the oscillating medium; penetration of continuous external influences into the microsystem and transformation of the network of chemical reactions (blue arrows); (b) Initiation of the system's response to continuous external stress, arising of chemical reactions counteracting the stress and partially compensating it; (c) Conversion of the prebiotic microsystem into the initial form of life, when its activity (counteraction) exceeds the pressing from the oscillating environment. The conversion corresponds with the thermodynamic inversion. (Image credit: see [15.14], open access).

includes both fast regular and irregular oscillations. The first oscillations range between 30 minutes (the minimal lifetime of the oldest thermophilic microorganisms, from one division to the next) and a split second (according to [15.5], the time in which induced extremely fast chemical reactions appear). Sudden irregular fluctuations maintain the non-equilibrium state of the maternal environment [15.13] [15.14]. Fluctuation on different levels of the system organization can also be related to "circulations" of different ranks—from large-scale circulations up to "micro-circulations" [15.22] [15.24].

15.3 Parameters of the Environment for the Origin of Life

The proposed fourth required condition for the origin of life allows us to clearly define the most probable cradle of life. The scale of physicochemical fluctuations in aqueous media depends on the scale of the corresponding gradients. The Earth's oceans are characterized by very low gradients of the parameters. Thus, the values of salinity range from 30 to 38 g/l, pH from 8.0 to 8.4. These small gradients determine low amplitudes of physicochemical ocean fluctuations that may occur with slow currents, surf, and even hurricanes. They do not lead to significant changes in temperature, hydrostatic pressure, or hydrochemical composition. Daily oscillations of temperature are characterized with low amplitudes and very long periods (about 24 hours) that greatly prevail during the entire lifetime of the oldest species of thermophilic prokaryotes. Their lifetime can be as short as 30 minutes. So, an ocean cannot be defined as an appropriate medium for the origin of life, taking into consideration its relatively stable conditions. The same concerns lakes and groundwater aquifers with stagnant regime.

Fluctuations in hydrothermal systems are very diverse. For instance, salinity in hydrothermal media varies within the values from <1 to 500 g/l, pH ranges between 1 and 12, and

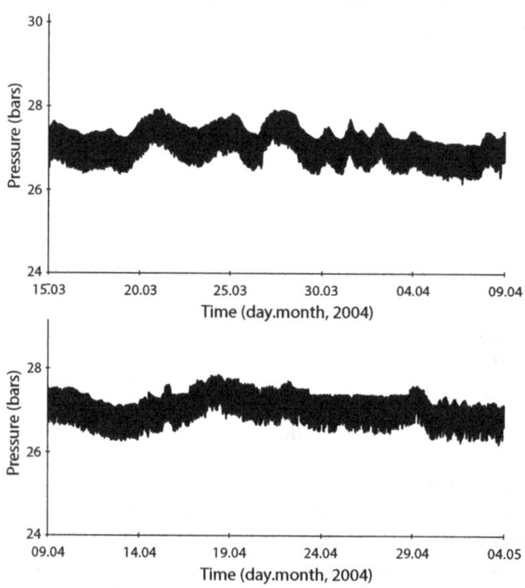

Figure 15.2 Pressure dynamics of water-steam mixture in well No. 30 in 2004 (Mutnovsky hydrothermal system, Kamchatka). X-axis, day/month, Y-axis, pressure (bars). (Image credit: V. Kompanichenko).

temperature within 30–400 °C. Strong fluctuations in present hydrothermal systems are initiated by tectonic activity (including earthquakes and volcanic eruptions). These events possess very high potential for generating multilevel oscillations of physicochemical parameters. Incessant macro- and micro-fluctuations in hydrothermal systems are maintained by continuous counteraction between hydrodynamic pressure of rising fluid and lithostatic pressure of host rocks. This general data can be supported by the direct measurements of pressure and temperature oscillations of water-steam mixture carried out in the Kamchatka geothermal region [15.15] [15.16] [15.17]. An example of pressure dynamics of water-steam mixture is shown on Figure 15.2. The mathematical processing of this data revealed complex dynamics in this record, including irregular abrupt and gradual macro-fluctuations, regular micro-oscillations (their periods are about 20 minutes), and the highest-frequency component consisting of sudden pressure changes and fluctuations with periods lower than 5 minutes.

15.4 Formation of Prebiotic Microsystem Clusters and Their Conversion into Primary Communities of Thermophilic Extremophiles

The dissolved lipid components tend to self-assemble into phase-isolated microsystems, when their concentration in the solution reaches a certain critical level. According to [15.3] [15.26], the microsystems must be capable of forming "protocells" (vesicles with membrane bounded compartments) containing sets of polymers able to be subjected to combinatorial selection. Fresh water of continental hot springs is an appropriate medium to form the vesicles, unlike saline water of submarine hot vents [15.4] [15.26]. Elongation of polymeric chains is possible under changeable conditions—either periodic change of wet-dry conditions in near-surface hydrothermal settings on land [15.3] [15.26] or oscillations of temperature in aqueous medium [15.28]. In particular, proteinoid chains might elongate under high-temperature conditions [15.7] [15.8]. One more important factor of prebiotic chemical evolution consists of continuous maintenance of nonequilibrium conditions which establish a kinetic trap; otherwise the polymers will be degraded [15.3]. As a result of an advanced chemical evolution described in well-grounded scenarios [15.3] [15.26] [15.31], "protocells" and more complex "artificial cells" (which are able to self-replicate) can be formed. These researchers along with others sometimes define such complicated microsystems as "almost life." But none of them obtained actually living units (microorganisms) characterized by extremely fast biochemical reactions, expedient behavior, and close cooperation between the microorganisms in a population. Such scenarios of advanced prebiotic evolution do not explain how real microorganisms became active (aggressive) with respect to the environment tending to expand. So, at the present time there exists a great gap between "almost life" and "actual life" observed in the biosphere. The way the conversion of prebiotic microsystems (protocells, artificial cells) into actually living units (microorganisms) might occur is described below on the basis of the inversion hypothesis [15.13]. This approach demands further experimental verification, of course.

A great number of organic microsystems (integrated into clusters) simultaneously appear during self-assembly in spacious hydrothermal environments. They are influenced by incessant oscillations of physicochemical parameters, as shown in Figure 15.1a.

The oscillations maintain nonequilibrium (far from equilibrium) conditions and heterogeneity within the cluster. Besides which, the oscillations lead to continuous interaction between the microsystems and recombination of macromolecules in them ([15.13], Chapter 7). The cluster contains organics of different complexity: assemblies (three-dimensional microstructures consisting of macromolecules), macromolecules (oligomers and polymers), and simple molecules (Figure 15.3).

Natural hydrothermal systems are characterized by gradual or abrupt temperature and pressure descent in the direction of the surface. Correspondingly, any cluster of assembled prebiotic microsystems exists in the anisotropic space caused by the vertical pressure-temperature gradients. Also, the temperature in subsurface cracks usually changes in the horizontal direction too—from the central stream (maximal) to host rocks (minimal). That means, each microsystem in the cluster takes the unique position characterized by individual values of temperature, pressure and chemical parameters in the given place or space. The individual niche of a prebiotic microsystem within the cluster defines its unique peculiarities in composition, morphology, and internal structure. In this context, the cluster under the described fluctuating conditions can be considered as a highly heterogeneous system, which possesses a huge diversity.

Heterogeneity and cooperative events (processes) are fundamental properties of a chemical system being under far from equilibrium conditions [15.6] [15.9] [15.12] [15.27] [15.28]. If such conditions are continuously maintained in a hydrothermal environment, the cluster is inevitably divided into local zones rich and poor in free energy (the low-entropy and high-entropy zones correspondingly). Cooperation between oscillating

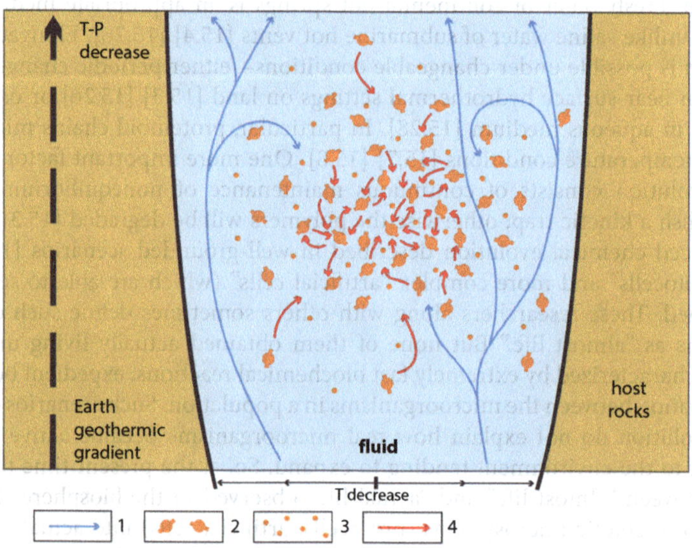

Figure 15.3 Arising of embryo of the initial community in hydrothermal fluid migrating through a crack to the surface. 1 – laminar and convective fluid currents; 2 – oscillating organic microsystems interacting with each other; 3 – various organic molecules (polymers and monomers); 4 – direction of free energy transfer within the cluster as a result of exchange between the prebiotic microsystems. (Image credit: V. Kompanichenko).

prebiotic microsystems within the low-entropy zones is provided by means of incessant exchange and internal recombination of molecules (Figure 15.3) (see also [15.24]). Free energy surplus in the local low-entropy zones allows the oscillating microsystems to execute efficient (intensified) response to incessant influences from the outside world. In this way the microsystems may undergo thermodynamic inversion, when the excessive "over-entropy" free energy appears in them (Figure 15.1c). From this point on, oscillating microsystems become primary forms of life (according to [15.13]). Their further existence and evolution are sustained by cooperation within the local low-entropy network, including integration through initial biogeochemical cycles. In fact, the network itself represents the embryo of community of thermophilic extremophiles. Some theoretical aspects of thermodynamic inversion are considered in Levchenko's book *Thermodynamic Inversion: Origin of Living Systems* ([15.13], Chapters 3 and 7). It is assumed that the emergence of such communities in hydrothermal environments on the early Earth corresponds with the embryonic stage of the biosphere evolution considered by Levchenko [15.19] [15.21] [15.22] [15.23].

15.5 Theoretical and Experimental Verification of the Proposed Approach

The described scenario of the origin of life is based on four required conditions: 1) aqueous medium, 2) availability of organic matter, 3) presence of energy source, and 4) multilevel physicochemical oscillations in the medium. Other concepts of life's origin do not consider the fourth condition as required. That means these concepts admit the origin of life in a medium, where significant oscillations of physicochemical parameters are absent, for instance in ocean (beyond submarine hot vents) or ice layers. Some theoretical arguments that emphasize the necessity of oscillating medium for the origin of life are given below.

1) It is known that an oscillatory mode is peculiar to many metabolic processes in living beings. This fact can be easily explained taking into consideration the fourth condition as a requirement. In this context the metabolic oscillations are simply the consequence of the initial physicochemical oscillations in the maternal medium for the origin of life. As a response to incessant external influences, the induced oscillating reactions were launched in prebiotic microsystems. With the following thermodynamic inversion, they were inherited and involved in the network of metabolic processes. It is difficult to find a reasonable answer to the question of how the oscillating metabolism could emerge in a stable cradle of life.
2) The inversion scenario implies that the conversion of prebiotic microsystems into primary living units occurred through their efficient response to external influences, or stress of moderate level. This thesis is actually stated in the theory of stress: "stress (but not distress) is necessary for life; absence of stress means death" [15.29]. It is unclear how an efficient response to stress could arise in primary microorganisms within a stable environment.

3) According to the concept by Zavarzin [15.33], biogeochemical cycles of the primary biosphere might be launched only by different groups of microorganisms. He shows that the contemporary prokaryotic communities catalyze biogeochemical cycles, all the reactions conjugated into a common system. This notion emphasizes the active role of microorganisms that sustain biogeochemical cycles; it is verified by a lot of data on microbial ecology. Zavarzin's data support our understanding of a primary community as a heterogeneous, but cooperative (integrated) system [15.13] [15.23] [15.25].

A way in which laboratory experiments could check the proposed theoretical scenario of life's origin was also recently offered [15.14]. A goal of such modified laboratory simulations consists of exploration of various prebiotic models (microsystems) in highly fluctuating environment. Appearance of the response to external changes and its development within the organic matrix should be explored. It is important to reveal initial sparks of the living state in various organic microsystems that are in the oscillating medium (for instance, increase of the internal energy gradients, strengthening of active transport, rise of homochirality degree, etc.).

15.6 Conclusion

The main points of this work are as follows:

1) The transformation of prebiotic microsystems into initial forms of life may take place through their efficient response to physicochemical oscillations in hydrothermal environment. It is expected that the primary microorganisms should be thermophilic extremophiles. Other concepts are implied such as that initial microorganisms form at the expense of step-by-step chemical complication of organic molecules and microsystems. The offered principal scheme should be relevant for the origin of life on any planet because it is based on the universal thermodynamic criteria considered in Kompanichenko's book ([15.13], Chapter 2).
2) According to our approach, suitability of hydrothermal environments for the origin of life is mainly related to the presence of physicochemical oscillations of parameters, while most other researchers emphasize the major role of thermal energy and high gradients between oceanic water and hot vent. We consider multilevel oscillations (and corresponding circulations) as one more required condition for the emergence of life (in addition to the availability of organic matter, aqueous medium, and source of energy).
3) Minimal "self-sufficient" unit of the initial life is a community of simplest microorganisms, not a single microorganism. This thesis is considered in detail in some other publications [15.13] [15.20] [15.21] [15.22] [15.23] [15.33]. Arising of the described original communities of the simplest microorganisms (along with the launched local biogeochemical cycles) corresponds with the embryonic stage of the biosphere evolution.

References

15.1. Colman, D.R., Jay, Z.R., Inskeep, W.P., Jennings, R.d.M., Maas, K.R., Rusch, D.B., Takacs-Vesbach, C.D., Novel, deep-branching heterotrophic bacterial populations recovered from thermal spring metagenomes. *Front. Microbiol.*, 7, 304, 2016.

15.2. Corliss, J.B., Baross, J.A., Hoffman, S.E., An hypothesis concerning the relationship between submarine hot springs and the origin of life on the Earth. *Oceanol. Acta*, S 4, 59–69, 1981.

15.3. Damer, B.F. and Deamer, D.W., Coupled phases and combinatorial selection in fluctuating hydrothermal pools: A scenario to guide experimental approaches to the origin of cellular life. *Life*, 5, 1, 872–887, 2015.

15.4. Deamer, D.W., Damer, B.F., Kompanichenko, V.N., Hydrothermal chemistry and the origin of cellular life. *Astrobiology*, 19, 12, 1523–1537, 2019.

15.5. Eigen, M., Selforganization of matter and the evolution of biological macromolecules. *Naturwissenschaften*, 58, 465–523, 1971.

15.6. Feistel, R. and Ebeling, W., *Physics of Self-Organization and Evolution*, Verlag GmbH & Co, Wiley, VCH, 2011.

15.7. Fox, S. and Dose, K., *Molecular Evolution and the Origin of Life*, Dekker, New York, 1975.

15.8. Fox, S., Balin, P., Pappelis, A., Yu, B., Experimental retracement of terrestrial origin of an excitable cell: Was it predictable?, in: *Chemical Evolution: Physics of the Origin and Evolution of Life*, J. Chela-Flores and F. Raulin (Eds.), pp. 21–32, Kluwer, Netherlands, 1996.

15.9. Haken, H., *Synergetics*, Springer-Verlag, Berlin, New York, 1978.

15.10. Kompanichenko, V.N., Three stages of the origin-of-life process: Bifurcation, stabilization and inversion. *Int. J. Astrobiol.*, 7, 1, 27–46, 2008.

15.11. Kompanichenko, V.N., Changeable hydrothermal media as a potential cradle of life on a planet. *Planet. Space Sci.*, 57, 468–476, 2009.

15.12. Kompanichenko, V.N., Inversion concept of the origin of life. *Orig. Life Evol. Biosph.*, 42, 2–3, 153–178, 2012.

15.13. Kompanichenko, V.N., *Thermodynamic Inversion: Origin of Living Systems*, Springer International Publishing, Cham, Switzerland, 2017.

15.14. Kompanichenko, V.N., The rise of a habitable planet: Four required conditions for the origin of life in the universe. *Geosciences*, 9, 92, 2019.

15.15. Kompanichenko, V.N., Exploring the Kamchatka geothermal region in the context of life's beginning. *Life*, 9, 2, 41, 2019.

15.16. Kompanichenko, V.N., Poturay, V.A., Shlufman, K.V., Hydrothermal systems of Kamchatka as the model for prebiotic environment. *Orig. Life Evol. Biosph.*, 45, 1-2, 93–103, 2015.

15.17. Kompanichenko, V.N. and Shlufman, K.V., The amplitude-frequency function of pressure variations: Steam-water mixture in the Verkhne-Mutnovskii hydrothermal system. *J. Volcanol. Seismol.*, 7, 5, 338–344, 2013.

15.18. Levchenko, V.F., The hypothesis of embryosphere as a consequence of principles of modeling. *Proc. Russ. Acad. Sci.*, 2, 317–320, (in Russian), 1993.

15.19. Levchenko, V.F., The seed of life. *Int. J. Comp. Anticip. Syst. (Belgium)*, 13, 62–76, 2001.

15.20. Levchenko, V.F., Evolution and origin of life: Some general approaches, in: *Astrobiology in Russia. Proceedings of International Astrobiology Conference*, Russian Astrobiology Center, NASA, St. Petersburg, pp. 7–21, 2002.

15.21. Levchenko, V.F., Origin of life and biosphere is indivisible process (in Russian), in: *Charles Darwin and Modern Biology*, pp. 338–347, Nestor-Istorya Publishing, St. Petersburg, 2010.

15.22. Levchenko, V.F., The origin of biosphere. *Int. J. Comp. Anticip. Syst. (Belgium)*, 25, 40–48, 2010.

15.23. Levchenko, V.F., *Three Stages of Evolution of Life on Earth*, Lambert Academic Publishing, Germany (in Russian), 2011.
15.24. Levchenko, V., Kazansky, A., Sabirov, M., Semenova, E., Early biosphere: Origin and evolution, in: *The Biosphere*, N. Ishwaran, Rijeka (Croatia) (Ed.), pp. 4–32, InTechOpen, 2012.
15.25. Levchenko, V.F. and Kotolupov, V.A., Levels of organization of living systems: Cooperons. *J. Evol. Biochem. Physiol.*, 46, 6, 530–538, 2010.
15.26. Milshteyn, D., Damer, B., Havig, J., Deamer, D., Amphiphilic compounds assemble into membranous vesicles in hydrothermal hot spring water but not in seawater. *Life*, 8, 2, 11, 2018.
15.27. Nicolis, G. and Prigogine, I., *Self-Organization in Nonequilibrium Systems*, Wiley, New York, 1977.
15.28. Prigogine, I. and Stengers, I., *Order Out of Chaos*, Bantam, New York, 1984.
15.29. Selye, H., *Stress without Distress*, J. B. Lippincott & Company, Philadelphia, New York, 1974.
15.30. Stetter, K.O., Hyperthermophiles in the history of life. *Philos. Trans. R. Soc. B*, 361, 1837–1841, 2006.
15.31. Sugawara, T., Kurihara, K., Suzuki, K., Constructive approach toward protocells, in: *Engineering of Chemical Complexity. World Scientific Review*, Mikhailov, A. (Ed.), pp. 1–17, 2012.
15.32. Woese, C.R., Microbial evolution. *Microbiol. Rev.*, 51, 221–270, 1987.
15.33. Zavarzin, G.A., Does evolution make the essence of biology? *Her. Russ. Acad. Sci.*, 76, 3, 292–302, 2006.
15.34. Xue, H., Tong, K.-L., Marck, C., Grosjean, H., Wong, J.T.-F., Transfer RNA paralogs: Evidence for genetic code-amino acid biosynthesis coevolution and an archaean root of life. *Gene*, 310, 59–66, 2003.

16

Extremophiles and Horizontal Gene Transfer: Clues to the Emergence of Life

Sohan Jheeta

NoR CEL, Leeds, UK

Abstract

Since time immemorial, one of the fundamental questions that preoccupies the human mind is, how did all this life around us come into existence? It is not surprising that there is no quick and easy answer to this question, as we do not even know the size of the puzzle that lays in front of us. Therefore, the approach that needs to be adopted is a multidisciplinary understanding of biology, chemistry, physics, geology and their subsidiary scientific subjects (e.g., genetics, microbial biochemistry and geobiology) as well as reporting on ancient microorganisms preserved in sediments and volcanic lava deposits. Moreover, we should also ask what part did extremophiles play in the emergence of life. In this chapter, certain front runner hypotheses, ideas and the latest methodologies are brought into focus. In addition, the reader is guided through a possible scenario posed for chemical evolution, as elucidated by element (e.g., $^{13}C/^{12}C$ ratio) fractionation techniques, and introduced to a construct of the last universal common ancestor (LUCA), which gave rise to the three domains of life; namely, Archaea, Bacteria and Eukarya. In doing so, a possible role played by viruses is explored—in particular that of RNA viruses. Also brought to the fore, are the latest techniques (e.g., the use of 16S rRNA genes) which are assisting scientists to elucidate the routes to the emergence of life.

Keywords: RNA, DNA, HGT, T-Luca, extremophile, origin of life, prebiotic, phylogenetic

16.1 Introduction

Cellular life forms on the Earth emerged from that very last entity that ever lived, namely the last universal common ancestor (LUCA), before the emergence of the two domains of life and then subsequently a third domain, which was a more complex chimera of the first two. The concept of LUCA is principally a hypothetical construct, as there is no direct microfossil rock evidence for its existence. However, its presence has been inferred from the fact that all contemporary cellular life forms possess various types of RNAs, including rRNA, mRNA and tRNA; other nucleotides (e.g., nicotinamide adenine dinucleotide, NAD; and coenzyme A, CoA) and small molecules, such as pyruvic acid ($CH_3COCOOH$), which

Email: sohan@sohanjheeta.com
Sohan Jheeta: www.sohanjheeta.com, https://www.researchgate.net/profile/Sohan_Jheeta, www.nor-cel.com, www.linkedin.com/in/sohan-jheeta-5a083894

Joseph Seckbach and Helga Stan-Lotter (eds.) Extremophiles as Astrobiological Models, (329–358)

are all thought to be relics of the bygone epoch when LUCA flourished [16.56] [16.139] [16.140]. The reason that RNAs are thought to be the relics of early living entities [16.58] [16.118] is that these molecules have at least five attributes, namely they can: 1) act as a repository for precursor chemical information (cf. mRNA and genomes of RNA viruses); 2) carry out self-replication via template-directed propagation; 3) self-process, e.g., carry out activities displayed by group I introns [16.12]; 4) suppress gene expression via RNA interference [16.68]; and 5) act as RNA-based catalysts, namely ribozymes [16.1] [16.13]. Albeit ribozymes are not as ultraefficient as proteinaceous catalysts (i.e., enzymes) and are also somewhat limited in scope, they are still sufficiently diverse, specific and quick enough (cf. clays as below) to pave the way for the emergence of life. Of all the non-coding RNAs (ncRNAs) tRNA is thought to be the very first quasi-living entity and it is this molecule that formed the crosslink between RNAs and amino acids, bringing about the emergence of a protein world by enabling catalysis of peptide bonds [-C(O)-N(H)-] synthesis from amino acids by acting as a carrier. However, the invention of the very first enzyme did not need to be ultraefficient, it just had to be rudimentary, with enough leeway to carry out slightly more efficient reactions when compared to the limited scope of ribozymes or any other catalysts available, e.g., clay surface-mediated reactions [16.52] [16.59] [16.115]. These collective operational activities of RNAs are generally encompassed under the banner of the RNA world concept of genetic information hypothesis [16.23] [16.35]. The essence of this chapter is to follow the possible timeline from the emergence of transitional-LUCAs (T-LUCAs) through to LUCAs and then eventually on to cellular life forms. The exchange and transfer of information as well as the significant roles of extremophiles and viruses will be explored along the way. In-depth prebiotic chemistry will not form part of this chapter.

16.2 T-LUCAs, LUCAs and Progenotes

Before continuing with the origin of cellular life forms, it is only wise to touch upon the concept of "progenotes." This concept was first put forward by Woese and Fox [16.139]. The general idea being that progenotes, like T-LUCAs, were also genotypically and phenotypically rapidly evolving entities. Progenotes predate LUCA by a long, extensive evolutionary history, which represents a phase when the molecular machineries for replication, transcription and translation were still in the process of evolving the relationship between genotype and phenotype [16.58] [16.139]. In this context, T-LUCAs and progenotes could be construed as being virtually similar entities which co-existed together, at various developmental stages, in communities.

16.3 Prebiotic World and T-LUCA

The existence of an entity like the LUCA was actually originally predicted by Charles Darwin, as shown in a page from his notebook (Figure 16.1a), although he never actually used the LUCA acronym. He expounded on this in his famous 1859 book entitled *Theory of Evolution by Natural Selection* [16.20], where he extrapolated that each and every organism that ever lived, or is living on Earth, is descended and thus related to all the organisms that preceded it and, as a result, by winding the evolutionary clock back, one would eventually arrive at LUCA—a brilliantly inspired deduction. Although LUCA is thought to be the last common ancestor

(a) (b)

Figure 16.1 (a) Transmutation of species as depicted on page 36 of Darwin's notebook B dated 1837. (b) Lost City white smokers. (Credit: Internet ref: National Science Foundation, Washington, USA [16.49]).

from which all cellular life forms emerged, this may not be the full picture. It is highly feasible that during the RNA world there could have been a myriad of quasi-entities which could reasonably be described collectively as pre-LUCAs [16.41] [16.58] [16.74]. These entities were probably rapidly evolving and so in fact it may be inaccurate to refer to them as "pre-LUCAs"; a more precise name is T-LUCA, as that is technically what they were. These entities probably lived within the vicinity of one another [16.95], as they would have been formed during the prebiotic chemistry world in the same locale by processes as yet to be elucidated fully. Nevertheless, what T-LUCA needed in order to develop was some sort of "bag" made from organic molecules, within which life's chemical reactions could take place; in the absence of such an organic bag, life's reactants would be too far apart in an open watery environment to be able to encounter one another. There are two potential scenarios by which such an organic bag could have come into existence. First, some scientists believe that the nearest candidates were tiny, millimeter-sized clay bubbles, such as those found at the warm alkaline hydrothermal white smokers (Figure 16.1b) at the Lost City [16.65] [16.66], which are located about 1 km from black smokers at the bottom of oceans [16.55]. Clay bubbles are thought to be important because it is postulated, under the metabolism first hypothesis, that the very first simple organic membranes may have been formed on the interior surface walls of interlinked NiS-FeS mesh within clay bubbles [16.72] [16.91]. The clay surrounding these organic bubble-shaped membranes could then eventually dissolve away, releasing the potential vesicles which then contained the chemicals of life [16.19] [16.26] [16.40].

The second possibility is that some scientists posit that 10-carbon-atoms-long n-monocarboxylic acid chains (e.g., decanoic acid) were delivered onto the early Earth by carbonaceous chondritic meteorites (e.g., Murchison and Tagish Lake) and that these chains were sufficiently long enough to form slightly advanced "soap-like" bubbles which were double layered vesicles, as in Figure 16.2a [16.51]; compare the levels of sophistication with those of the contemporary biomembranes of Archaea with ether (R-O-R′) linkage and, Bacteria [16.2] [16.21] [16.38] [16.92] and Eukarya with ester (R-CO-R′) linkage as in Figure 16.2b [16.47] and Table 16.1. These vesicles, although crude and simple, eventually encapsulated life's chemicals probably during the hydration and dehydration cycles occurring during the ebb and flow of tides on the early shores [16.81] [16.125]. Once the encapsulation of life's chemicals had taken place, T-LUCAs had arrived [16.79].

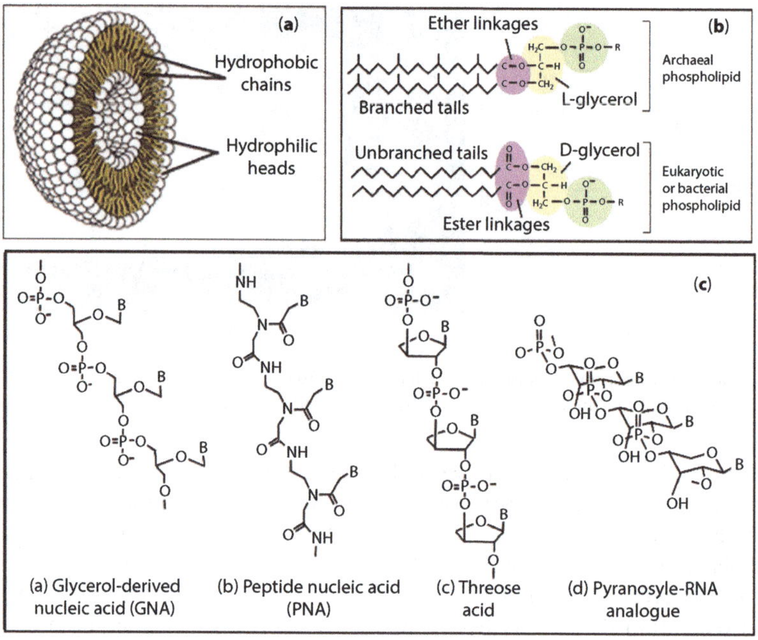

Figure 16.2 (a) Depicts a typical bilayer vesicle formed when (for example) decanoic acid is mixed with water. (b) Showing that membranes in Archaea use ether (R-O-R′) bonds and, membranes in Bacteria and Eukarya use ester bonds (R-CO-R′). As a result, contemporary biomembranes are extremely complex and sophisticated. (c) Shows, by way of examples, various possible precursor informational molecules; Note that B = nitrogenous base (Image credits: (Image credits: (a) Internet ref: Villarreal [16.51]; (b) Khan Academy [16.47]); (c) S. Jheeta).

Next, the T-LUCAs would have needed to co-operate with one another and exchange chemicals in order to mature. The exchange of chemicals between the early rudimentary membranes played a pivotal role in the transfer of chemicals from one T-LUCA to another. These transfers could have been achieved in three possible ways during the maturation of T-LUCA. First, the transfer could have been initiated via high-intensity primordial lightning, which is comparable to that observed within the plumes of modern-day volcanic eruptions and is also much more intense than desert lightning [16.4] [16.60], noting that the Earth's primordial atmosphere was an ash-laden, electrified and turbulent environment. Thus, primordial lightning generated high energy electric fields which would have formed "hydro-holes" or "hydro-channels" within the membranes of the T-LUCAs, through which the exchanges/transfers could have been achieved [16.59] [16.73]. In the second scenario, Brownian motion could have caused T-LUCAs to collide with and stick to one another, potentially resulting in the fusion of the pre-cell contents of two or more T-LUCAs, leading to either a more mature entity or the demise of the resultant T-LUCA. Thirdly, any destroyed T-LUCAs may have become leaky and spilled their contents into the immediate surrounding environment and this material could then have been taken up by other maturing T-LUCAs during the tidal hydration/dehydration cycles mentioned previously, adding to their inventory of chemical compounds; the net effect being that the T-LUCA was evolving and maturing continuously.

Whatever the mechanism, the compounds that were being exchanged and transferred were, in essence, information carrying precursor molecules, which may have included polyribonucleotides (RNA), glycerol-derived nucleic acid (GNA), peptide nucleotides (PNA), threose nucleic acid (TNA), and/or pyranosyl-RNA [16.96] [16.109], as in Figure 16.2c.

These precursor informational chemicals would also have been evolving, which resulted in RNA outperforming PNA, TNA, GNA or pyranosyl-RNA. Does this "selection" of the RNA molecule mark the beginning of Darwinian evolution? Maybe it is indicative of it, as ultimately RNA was and still is an efficient information carrier, even able to transport rudimentary "genes" during those early days. In addition, RNA informational transfers could be equated to rudimentary horizontal "gene" transfers (HGT), because any chemical information that RNA transferred would have been conserved. Consider Spiegelman's "little monster" serial dilution experiment, which began with studies on an RNA virus, Qβ bacteriophage with a genome size of ~4217 nucleotides. In essence, this phage was a positive-sense, single-stranded RNA ((+)ssRNA) virus, meaning that, as a +sense, it could be used directly by ribosomes for protein making, which was thought to be an exclusive preserve of transcript mRNA expression; DNA's intervention was superfluous to requirements, thus the DNA-dependent RNA polymerase, which carried out the transcription of RNA from a DNA template was effectively made redundant. The protein expressed by the Qβ bacteriophage was a Qβ RNA replicase, an enzyme which carried out RNA template-directed replication of Qβ bacteriophage's genome, a previously unknown mechanism.

In 1965 Spiegelman [16.120] demonstrated that in a solution containing 4800 nucleotides long RNAs, plus Qβ RNA replicase purified from Qβ bacteriophage together with free RNA nucleotides and salt (12 mM $MgCl_2$), the long RNAs actually replicated. The investigation also highlighted that along with the replication of RNA strands they also generated shorter RNA lengths during each successive generation; and that as the lengths of these RNA strands became smaller, the replication process speeded up. This was achieved by transferring small samples from the first test tube into the next tube containing free nucleotides and salt; i.e., a microbiological technique known as serial dilution. After 74 successive serial dilutions, which equates to 74 generations, the RNA length reduced to 218 nucleotides. The essence of this demonstration and other similar experiments is that (a) the RNA molecules appear to be evolving [16.73] [16.85] [16.123]; (b) it lends some credence to the RNA world hypothesis [16.23] [16.53] and (c) T-LUCA was most likely to be an RNA-based entity. This could be considered to be true in light of the fact that there are infectious RNA molecules, specifically viroids and satellites, which could be construed to be important too during the chemical evolution of life epoch [16.22] [16.94] [16.133]. In short, T-LUCAs were edging toward the emergence of fully fledged LUCAs.

16.4 Emergence of LUCA

Both T-LUCAs and LUCAs were probably "alive and kicking" between 4.3 and 4.0 billion (10^9) years ago, when the Earth was only relatively young (~350 million years old), as posited by ^{13}C and ^{12}C isotope ratio fractionation studies (see below). The environments in which they emerged could only be described as harsh and inhospitable places. This was due to many factors, one being that tides occurred every 6 hours, since the Moon was much closer to the Earth and so tidal waves would have been frequent and extremely high, pounding the shores of the early island land masses. In addition, numerous islands of active volcanoes spewed out heated magma and a plentiful supply of baking hot ash as well as the following compounds: sulfur monoxide (SO), sulfur dioxide (SO_2), sulfur trioxide (SO_3) and carbonyl sulfide (OCS); carbon monoxide (CO) and carbon dioxide (CO_2); ammonia (NH_3), methane (CH_4) and dihydrogen (H_2) gases; water vapors with a mix of sulfurous (H_2SO_3) and sulfuric (H_2SO_4) acids; hydrochloric (HCl) and carbonic acid (H_2CO_3)—and as a result there would also have been heavy acidic rain precipitation [16.60] [16.63]. Radiogenic isotopes (e.g., ^{40}K, ^{235}U, ^{238}U, ^{232}Th) heated the land mass still further; intense lightning [16.4] [16.60] and heavy bombardment by impactors such as meteorites, comets, asteroids and even planetesimal collisions (cf. the Moon formation) generated yet more heat, and finally meteoritic shock waves melted the rocks, superheating both any water and the primordial atmosphere [16.15]. The same violent impactors further delivered both organic molecules and water approximately 4.4 billion years ago onto the surface of the Earth, as testified by the highly stable and resilient mineral, zircon, found in ancient sediments dated to be 4.2 billion years old [16.37] as well as the isotopic distribution ratio of deuterium to hydrogen (D/H) in the Solar System; the presence of the numerous craters on the Moon and Mercury stands as testament to the heavy violent bombardment phase [16.61]. It is during this turbulent period that prebiotic chemical evolution was also edging towards the RNA world, with the eventual emergence of T-LUCAs indicating that the Earth was already pregnant with life.

LUCAs appeared during the late bombardment epoch, perhaps *circa* 4.3 to 4.1 billion years ago. Fractionation studies of ^{13}C and ^{12}C isotopes within ancient rocks found at sites in desolate and bleak places such as Burgess Shale, Canada; Isua, Greenland and the Warrawoona/Pilbara region in Western Australia, have placed the time of emergence of LUCA and/or cellular life at 4.1 billion years ago [16.6] [16.24]. The sensitivity of the fractionation techniques relies upon finding crack-free specimens of zircon mineral crystals which completely encase graphite inclusions. Graphite (technically coal) may be the embodiment of signals arising from the mass demise of entities that inhabited that particular locale. The encased graphite should yield evidence, if any, of enrichment of ^{12}C with respect to the signals from heavier carbon isotopes (^{13}C) present in the ambient environment, including in the atmosphere and hydrosphere.

The logical reasoning behind the fractionation techniques' evidence for the existence of an interface between LUCA and/or emerging cellular life is that carboxylase enzymes (e.g., Ribulose-1,5-bisphosphate carboxylase/oxygenase, RuBisCO) could select lighter, stable ^{12}C isotopes over the heavier ^{13}C ones during the carbon assimilation by ancient microorganisms; this implies that a complex enzyme like RuBisCO may also have been present in these early entities. In other words, ^{13}C isotopes were actively discriminated against by autotrophs (highly probably chemolithoautotrophs) by not being incorporated into the organic matter. The $^{13}C/^{12}C$ ratio, measured in parts per thousand, will thus have a negative value for graphite containing remains of assimilated ^{12}C (Eq. 16.1). Conversely, if the values for $^{13}C/^{12}C$ ratios are either zero or positive it implies that ^{13}C enrichment, as yet an unknown process, is a result of a non-living source [16.37]. The scientists carrying out these studies say that this is the evidence for very early carbon-assimilating entities [16.6].

$$\delta^{13}C = \left[\frac{\left({}^{13}C/{}^{12}C\right) Sample}{\left({}^{13}C/{}^{12}C\right) Standard} - 1 \right] X\,1000(0/00) \qquad \text{(Eq. 16.1)}$$

Similar investigations using sulfur assimilation ratios (i.e., $\delta^{36}S/\delta^{34}S$) produced comparable results, showing that the methodology is credible [16.70].

16.5 Chemical Composition of LUCA

Regarding LUCA's chemical make-up, Weiss in 2016 carried out large-scale studies referred to as "phylogenetic bracketing," on 6.1 million protein coding genes and reported that LUCA had 355 genes [16.70] [16.134]—compare this to van Tonder's estimation of the core genes of *Helicobacter pylori* of 244 [16.129]—and was a chemolithoautotrophic, thermophilic strict anaerobe, which was able to fix both CO_2 and N_2 and was also H_2-dependent for energy generation (Eq. 16.2, Eq. 16.3, Eq. 16.4 and Eq. 16.5 in Table 16.2). Its biochemistry was achieved via the ancient reductive acetyl CoA (Wood–Ljungdahl) pathway probably belonging to hydrogenotrophic methanogens and/or acetogen groups [16.127]. On a biochemical level, these studies indicated that LUCA was more likely to have had a free-floating circular DNA-based repository genetic code rather than an RNA one and, therefore, it's also likely to have had a DNA-based replication system. It was able to carry out aminoacylation, transacylation and elongation—i.e., it possessed a version of a ribosome [16.84]. This means it was capable of protein synthesis and also that building block homochirality had been established; chirality being a type of stereochemistry, of which there are two stereoisomers, namely either left (levo, L) or right (dextro, D) handedness, with the rest being defined as achiral (e.g., glycine). Thus, D-sugars in the backbones of both RNA and DNA were being utilized, and L-amino acids were being incorporated into polypeptides during their synthesis.

Was LUCA an RNA entity or was it a transitional RNA/DNA based entity? Whereas Weiss *et al.* [16.134] leaned more towards LUCA being DNA based, Leipe *et al.* [16.84] thought that LUCA possessed both RNA and DNA, but unlike contemporary life forms, LUCA lacked principal enzymes, namely DNA polymerases, which meant that it did not replicate with DNA polymerases as in the newly emerging Archaea and Bacteria; instead LUCA's genome replicated via reverse transcription using a reverse transcriptase enzyme to make a complementary DNA. This enzyme is a feature of all RNA viruses (e.g., retroviruses) and to date over 1000 different kinds of reverse transcriptases in bacteria have been identified, denoting the probable importance of these enzymes [16.94]. Krishnamurthy [16.74] indicated that at some point in the story of life there was, in the absence of separating and editing mechanisms in the RNA world, a simultaneous presence of both RNA and DNA molecules with the preexisting alternative building structures, i.e., XNA (where X = unknown nitrogenous base). In this milieu there was a gradual chemical evolution from a heterogeneous molecular environment to an homogeneous one during the oligomeric and/or polymeric transitions which led to the eventual emergence of the homogeneous backbones of an "RNA:DNA" hybrid world with the eventual emergence of the DNA/protein world. The net result being that it was a "chimeric oligonucleotide system" from which life emerged rather than a "pure" RNA world [16.34] [16.58] [16.74]. The DNA in Krishnamurthy's posit may have been uracil-DNA viruses [16.80], which are still found in nature today; uracil-DNA is a nucleic acid that has deoxyribose sugar as a backbone and a uracil base instead of thymine. Woese *et al.* [16.140] and Diener [16.22] both thought that LUCA was an RNA-based organism living in the RNA world [16.35]. Forterre [16.28] ventured that LUCA was an RNA cell and that the first stage in the evolution towards a DNA world is the emergence of a protein world. RNA invented proteins, which in turn invented DNA (RNA → proteins → DNA). This has a certain credence in that DNA can be made, as pointed out above, by using a reverse transcriptase enzyme, where a thymine base is inserted instead of uracil during the transcription of an RNA strand. The difference between uracil and thymine is that the latter has a methyl group ($-CH_3$) at position 5 in the pyrimidine ring. So, the answer to the question posed above is as follows: since T-LUCAs were evolving from simple to complex entities, this meant chemical evolution at fairly rudimentary levels was being selected, replaced or being deselected entirely, resulting in improvements being made with advancing complexity. Each and every LUCA would have been at different evolutionary stages in their lives; to define a fully-fledged LUCA as a particular stage is not possible. Of course, scientists have attempted to give it various attributes but that is merely wishful thinking; therefore, both T-LUCAs and LUCAs are evolving entities just as contemporary cellular life forms have been evolving ever since their emergence. Out of the myriad entities in the LUCA world, there must have been at least two resilient LUCA extremophiles with a hardy chemistry emerging from inhospitable environments as pointed out previously. These gave birth to their equally extremophilic and even more hardy and resilient progeny, namely Archaea and Bacteria, which were far more superior and dominant, their DNA genome being highly conserved and tweaked. Both the T-LUCAs and LUCAs had little chance of competing and thus, in the long run, lost out to the more ruthless cellular predators; they might even have ended up as nutritious food for the newly emerging life forms—much like the praying mantis, whereupon after copulation the female mercilessly devours the male. LUCA's demise was such that there are not even rock fossil remains to be found anywhere, only chemical fossils (e.g., RNAs) to indicate that they were here first. Cellular life is king.

16.6 Emergence of Cellular Life Forms

Taking it as read that cellular life forms emerged from LUCA, once the transition of LUCAs had begun, cellular life arrived in the differing entities of Archaea, Bacteria and, subsequently, Eukarya, giving credence to the theory that there were at least two types of LUCAs from which contemporary life evolved [16.38]. Archaea and Bacteria had a completely different evolution as shown by the biochemical composition of their membranes; Archaea used isoprene ether molecules as their backbone and Bacteria settled for fatty acid ester bonded membranes (Figure 16.2b); in addition, Table 16.1 also compares and contrasts other aspects of the three domains. Archaea and Bacteria are simpler entities when compared to Eukarya, which is highly sophisticated. By the time Eukarya appeared, *circa* 2.0 billion years ago [16.102] [16.113], there had been at least two major advancements, the first being the invagination of the cell membrane, which led to the formation of organelles—for example, a boundary of nuclear membrane housing the genetic material (primarily chromosome(s) and reservoir of nucleoli of RNAs) and thereby protecting it against attacks from nucleolytic enzymes, namely deoxyribonucleases (DNases) and exonucleases; and from invading viral genomes (genetic information encoded in DNA or RNA), so as to keep the chromosomal DNA unadulterated as far as possible. This is in stark contrast to the Archaea and Bacteria domains which are anucleated; these two domains collectively are classified as prokaryotes [16.119].

The second advancement that occurred was a "marriage" between an archaeal host cell (probably Lokiarchaeota) and a bacterial energy generator intercellular resident, probably

Table 16.1 Various attributes of Archaea, Bacteria and Eukarya compared and contrasted.

Trait	Archaea	Bacteria	Eukarya
*Cell wall	None	Peptidoglycan	cellulose, hemicellulose, pectin and lignin
Carbon linkage of lipids	Ether *(R-O-R′)	Ester *(R-CO-R′)	Ester *(R-CO-R′)
Phosphate backbone of lipids	*sn*-glycerol-1-phosphate	*sn*-glycerol-3-phosphate	*sn*-glycerol-3-phosphate
*Cell lipid membrane enzymes	Geranylgerylglyceryl phosphate synthase	Glycerol-3-phosphate dehydrogenase	Glycerol-3-phosphate dehydrogenase
Metabolism	Bacterial like	Bacterial	Eukaryotic
Nucleus	No	No	Yes
Organelles	No	No	Yes
Splicosomal introns	No	No	Yes
Telomeres	No	No	Yes
Chromosome shape	Circular	Mostly circular	Linear
DNA replication	Eukaryotic-like	Bacterial	Eukaryotic
Transcription	Eukaryotic-like	Bacterial	Eukaryotic
Translation	Eukaryotic-like	Bacterial	Eukaryotic
Spore production	No	Yes	Yes
*Ribosomal genes	16S rRNA	16S rRNA	18S rRNA
*Life style: reproduction	Asexually, binary fission, fragmentation and budding	Both sexually and asexually	Both sexually and asexually

Modified from Tan [16.126].
*Author S. Jheeta added the extra information.

an ancient species of Alphaproteobacteria [16.119] [16.128] [16.138] [16.142]; this is where a smaller bacterial cell with the ability to generate rapid, high levels of useable chemical energy via oxidative phosphorylation (e.g., adenosine triphosphate, ATP) during aerobic respiration resided within a larger archaeal cell which afforded it protection. The smaller cell came to be called a mitochondrion and the union resulted in a symbiotic relationship; the larger host cell affording protection for the smaller resident cell, which in turn returned the favor by supplying the host cell with a plentiful supply of energy. To decrease the workload and increase the energy demands from the mitochondrion, the vast majority of its genes from circular chromosomes were transferred to the chromosomes of the larger, eukaryotic cell. Such unions became permanent and irreversible for the bulk of eukaryotes [16.99], with very few exceptions; for example, some species of *Monocercomonoides* disposed of their mitochondria altogether [16.62], suggesting that mitochondria are not totally necessary for survival of such eukaryotes, although their life style would be sluggish.

16.7 Evidence for the Earliest Cellular Life Forms

Initially the evidence for the existence of early microorganisms came from ancient preserved sediments and volcanic lava deposits from the various remote sites mentioned previously. Upon investigation, these samples showed the presence of optically discernible morphology within extinct prokaryote-like life forms, including cyanobacteria, preserved in approximately 3.5 billion years old cherts from Western Australia, as shown in Figure 16.3 [16.110] [16.111] [16.112]. To identify these morphologies the investigator used combined imaging of microfossils using laser Raman spectroscopy and chemical fractionation techniques. On the other hand, there is less evidence for eukaryotes, as their cells do not preserve very well (unlike prokaryotes which have cell walls and are therefore more resilient) plus they are difficult to find prior to the Paleoproterozoic period ~2 billion years ago [16.71] [16.102] [16.113].

Schopf's studies have been criticized by Brasier *et al.* [16.10] and Dalton [16.18], both citing that what Schopf and company observed were only morphosized rocks due to heating (e.g., by plate tectonic processes) and that there was no way such fossils could be preserved since the geological time frame is far too long. The debate concerning the palaeontological evidence for prokaryotes being present as early as 3.5 billion years ago is now pretty well settled, as shown in studies by Wacey *et al.* [16.131]; one of the principal authors on the Wacey paper was Brasier.

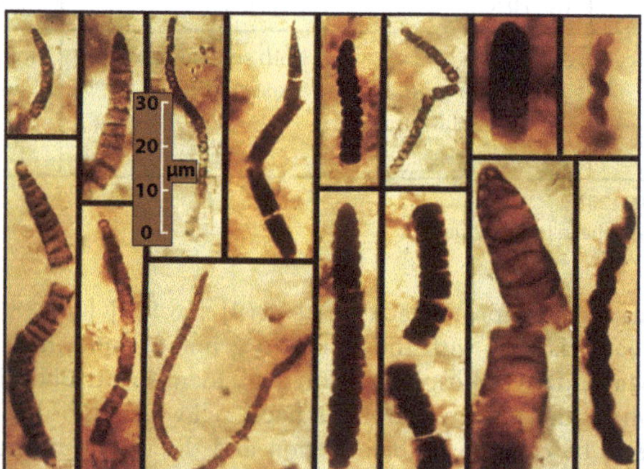

Figure 16.3 The depicted scale and typical results obtained by Schopf. (Credit: JW Schopf).

Table 16.2 Summary of an alternate way of grouping extremophiles based on their domain categorization as well as their energy requirement and acquisition of carbon, rather than their habitat. The author has focussed on the energy and nutrient requirements of archaea and bacteria.

Groups	Some examples of prokaryotes	Further notes
Photolithoautotrophs **Energy:** sunlight **donor:** water photolysis **C source:** CO_2	**Bacteria:** Cyanobacteria	Some cyanobacteria are free-living, carrying out oxygenic photosynthesis to obtain their energy needs. They can also fix CO_2 as a source of carbon. Other cyanobacteria are present as endosymbiotic chloroplast organelles within red algae, green algae, and land plants. NADH is used in anabolic reactions. $H_2O \xrightarrow{h\nu} 2e^- + [2H^+] + \frac{1}{2}O_2$ (Eq. 16.2) $NAD^+ + [2H^+] + 2e^- \rightarrow NADH + H^+$ (Eq. 16.3)
Chemolithoautotrophs **Energy:** inorganic bond breaking ***e⁻/H⁺* donor:** inorganic **C source:** CO_2	**Archaea:** *Ferroglobus, Methanobacteria* or *Pyrolobus* **Bacteria:** Nitrifying bacteria, sulfur-oxidizing bacteria, iron-oxidizing bacteria, Knallgas-bacteria	Most methanogens reduce CO_2 (e- acceptor) using H_2 (e- donor) to generate CH_4. The methanogen hydrogenase catalyses H_2 to generate highly reactive ions, (further in-depth details [16.127]). H_2 can be obtained from serpentinization reactions of olivine, or it may be produced by other inhabitants in the locale. $H_2 \leftrightarrow 2e^- + 2H^+ \quad E_0' = -414\,mV$ (Eq. 16.4) $CO_2 + 8e^- + 8H^+ \rightarrow 2H_2O + CH_4$ $\Delta G^{o\prime} = -131\ kJ\ mol^{-1}$ (Eq. 16.5) Being autotrophs, methanogens can also carry out CO_2 fixation. $4e^-$ and $4H^+$ can also be redirected to fix C via the reductive acetyl-CoA pathway. $CO_2 + 4e^- + 4H^+ \rightarrow CH_2O + H_2O$ (Eq. 16.6) Formaldehyde (CH_2O) can be used to make hexose sugars.
Chemoorgano autotrophs **Energy:** organic bond breaking ***e⁻/H⁺* donor:** organic **C source:** CO_2	**Archaea:** *Methanosarcinales* *Pyrococcus* *Sulfolobus*	Only members of *Methanosarcinales* can liberate methane as well as fix carbon during the fermentation of acetate—see Eq. 16.6 above $CH_3COO^- + H^+ \rightarrow CH_4 + CO_2$ $\Delta G^{o\prime} = -36\ kJ\ mol^{-1}$ (Eq. 16.7)

(Continued)

Table 16.2 Summary of an alternate way of grouping extremophiles based on their domain categorization as well as their energy requirement and acquisition of carbon, rather than their habitat. The author has focussed on the energy and nutrient requirements of archaea and bacteria. (*Continued*)

Photolithoheterotrophs **Energy:** sunlight ***e⁻/H⁺* donor:** inorganic **C source:** organic	Bacteria: Purple non-S bacteria	These microorganisms can use sunlight but are anoxygenic as they can use hydrogen (Eq. 16.4), sulfide (Eq. 16.8) or organic compounds as electron donors. Photosynthesis is carried out using type 1 photosynthesis via bacteriochlorophyll and carotenoids. They are unable to use CO_2, instead they utilise carbohydrates, fatty acids, and alcohols as their sole carbon source. $S^{2-} \rightarrow 2e^- + S^0$ (Eq. 16.8)
Photoorganoheterotrophs **Energy:** sunlight ***e⁻/H⁺* donor:** organic **C source:** organic	**Bacteria:** Heliobacteria, Rhodobacter, some green non-sulfur bacteria	These microorganisms contain bacteriochlorophyll and use anoxygenic photosynthesis to derive their energy requirements. They also use light wavelengths that cannot be deployed by photolithoautotrophs.
Chemolithoheterotrophs **Energy:** inorganic bond breaking ***e⁻/H⁺* donor:** inorganic **C source:** organic	Bacteria: Iron and sulphur-oxidising and N-fixing bacteria	Vast majority of the Kingdoms of Animalia, Bacteria, Fungi, and Protista are known to be chemoheterotrophs, principally relying on the primary producers.
Chemorganoheterotrophs **Energy:** organic bond breaking ***e⁻/H⁺* donor:** organic **C source:** organic	**Bacteria:** *Bacillus, Shigella, Salmonella, E. coli* *Yersinia pestis, Oceanithermus profundus* and *Actinobacteria*	

Prokaryotes can either harness light energy (photoautotrophs) from the Sun, or harness chemical bond energy (chemoautotrophs) from inorganic compounds. On the balance of probability most scientists nowadays believe that chemolithoautotrophs emerged first. Why these and not photoautotrophs? Because chemolithoautotrophs derive their energy from breaking energetic inorganic chemical bonds such as those of dihydrogen (H_2) molecules found in anaerobic places such as marshland, sediments and alkaline hydrothermal vents. If such reactions drive the chemical reactions, this would give some credence to the metabolism first hypothesis and there is mounting evidence in favor of this hypothesis and chemolithoautotrophy, as promulgated by Russell *et al.* [16.106] [16.107] [16.108], Martin *et al.* [16.91], Sojo *et al.* [16.117], Lane *et al.* [16.76] and Koonin and Martin [16.72]; readers may wish to read the following comprehensive hydrothermal vent hypothesis paper entitled: "A Hydrothermal-Sedimentary Context for the Origin of Life" [16.135]. In addition, these scientists state that it is practically impossible to abiotically make porphyrins, a group of heterocyclic organic compounds of which a principal example is chlorophyll [16.30] [16.58]. There is also the question of the early Sun and primordial atmosphere. The Sun's energy output capacity was only 75% of that of today [16.64] and, in the absence of ozone (O_3), the primordial atmosphere extended far beyond what is observed nowadays. This means that an exceptionally small amount of visible light reached the surface of the Earth and so the light-harnessing abilities of chlorophyll, bacteriochlorophyll and associated pigments would have been next to none. Whether chemolithoautotrophs or photoautotrophs emerged first, is not an overly relevant point here, however. Table 16.2 and section 16.9 brings into focus that photosynthetic apparatuses are observed in a small selection of species of the Bacterial domain but are almost totally absent in that of Archaea because these derive their energy from breaking inorganic chemical bonds and not by utilizing light; ergo archaea emerged before photoautotrophs. What is more significant is that these very early emerging cellular life forms lived in environments that were dire and uncongenial; so, they had to have been obligate anaerobic extremophiles such as hydrogenotrophic methanogens and/or hyperthermophiles (e.g., species of *Thermococcus*; [16.127]).

16.8 The Hypotheses: Genetic First vs. Metabolism First

Although there are at least two dozen hypotheses purporting to explain the origin of life, in essence they can all be grouped into two camps; namely, genetic first and metabolism first. The former promulgates that information in the form of chemicals such as RNA came first and thus forms the basis of the RNA world hypothesis. The latter promulgates that pathways (e.g., glycolysis), cycles (for instance, Krebs cycle) and hypercycles—interactions between two or more cycles—arose first, with genetics emerging subsequently. There is, however, a blurring of lines between the two hypotheses in that they are both intertwined with one another as follows: all proteins need specific chemical instructions for them to be made, in a prescribed order. These specific instructions are encoded in an mRNA and with the help from tRNA and rRNA, amino acids are stitched together as per the mRNA's instructions. However, the synthesis of nucleotides and transcription of DNA is dependent on the proteins—it seems this is a perpetual conundrum of the chicken and egg cycle. In actual fact, there is no conundrum, as both of these two diametrically conflicting models cannot be completely separated, since they are entangled with one another and there cannot really be a question of which came first when explaining the origin of life. A similar sort of perspective is taken by both Krishnamurthy [16.74] and Forterre [16.27] [16.28]. The former researcher advocates that both RNA and DNA arose simultaneously during the same milieu and within the same time frame. The latter investigator envisages that proteins came before DNA and so they invented DNA (recall: RNA → proteins → DNA). The author sides with the idea that tRNA forged a crosslink in the form of an activated amino acid and tRNA linkage [16.103]. In the absence of any real, hardcore rock fossil samples and/or improved, discerning chemical isotope fractionation evidence dating from approximately 4.0 billion years ago, it is difficult to give a fully definitive answer about how life began *de novo*. Even if life could be made in a "test tube," it may not be using the same algorithm as when life emerged on the Earth; there could be a million other ways to make life. One thing is undeniable and that is that life arose by a gradual complexification of chemistry—life is an extremely complex chemistry at its zenith. This author is not entirely concerned with exactly how life began on Earth, as this can never be fully deciphered, but rather, whether or not life could be made afresh in a laboratory.

16.9 Extremophiles

What are extremophiles? They are hardy microorganisms, the vast majority of them belonging to the domain: Archaea, which are considered to be the most ancient microorganisms on Earth, even when compared to those of the Bacteria domain; this was realized when the ribosomal RNA genes (in particular 16S ones), membranes and cell walls (Table 16.1) as well as biochemistries from both Bacteria and Eukarya were compared and contrasted.

There are various categories and subdivisions of these extremely resilient microorganisms which can all tolerate extreme environments. For example, psychrophiles, as in Alpha- and Betaproteobacteria, can survive hundreds of meters below the polar ice cap at -20 °C, as well as in permafrost [16.9] [16.55]; thermophiles (*Thermococcus litoralis*) are comfortable in hot and boiling springs with temperatures in excess of 80 °C [16.97] (Figures 16.4a), and hyperthermophiles (*Pyrolobus fumarii* and *Escherichia coli* strain MG1655) make their home in the vicinity of black smokers (Figure 16.4b) on the ocean seabeds where the temperatures can exceed 400 °C and the pressures can be as high as 2.0 GPa [16.114]. All these can be compared to mesophiles which probably account for 95% of microbes, which have a temperature range between 20 to 45 °C. The muddy seabed also harbors a range of obligate anaerobes (e.g., *Clostridium perfringens*) which are only comfortable in oxygen-free environments as they are instantly killed when exposed to any amount of oxygen.

In addition, endoliths (as in certain species of cyanobacteria) and cryptoendoliths relish the idea of living within rock pores (Figure 16.5a) and between aggregate grains (*Bacillus marismortui* and *Virgibacillus pantothenticus*) [16.130], and thrive in fissures, aquifers, tar pits and faults filled with groundwater in the Earth's deep subsurface [16.69] [16.137].

Polyextremophiles can endure a range of different environments simultaneously; for example, *Sulfolobus acidocaldarius*, an archaeon, flourishes both at a low pH of 3 as well as at a high temperature of 80 °C, thus making it both an acidophile and thermophile respectively [16.106]. Conversely, *Methanosalsum zhilinae* [16.67] and *Methanothermococcus thermolithotrophicus* [16.136] are alkalophiles and flourish between pH 9.8–10. *Deinococcus radiodurans* has been found to survive quite comfortably in the vicinity of nuclear reactors [16.5] [16.43] with radiation levels high enough to incinerate humans. The world's desert sands (e.g., Atacama and Sahara) are subjected to long periods of drought, with annual rainfalls of less than 1 mm; air and sand is heated to scorching temperatures in excess of 45 °C daily; and, extended durations of high levels of UV radiation impinges upon their surfaces because there is only a very thin layer of ozone above [16.17], making it unbearable

(a)

(b)

Figure 16.4 (a) Hot spring at Wai-O-Tapu park, New Zealand. (b) Example of black smokers at a hydrothermal vent (Image credits: (a) S. Jheeta; (b) Tunnicliffe Laboratory, University of Victoria, Canada [16.50]).

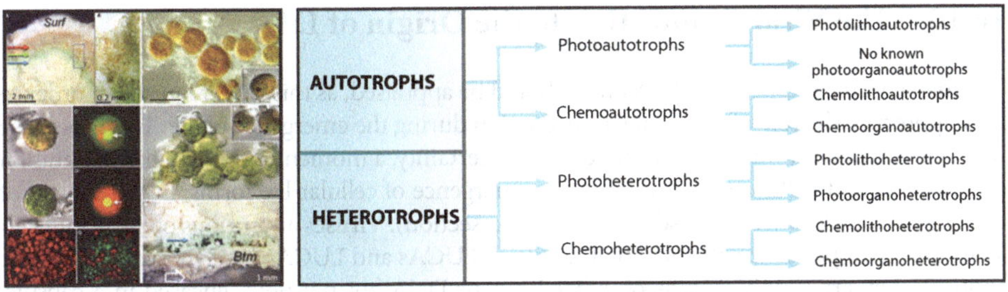

(a) (b)

Figure 16.5 (a) Images showing examples of various endoliths; for example, the various colored bands in the top left and bottom right images are actually microscopic inhabitants making their living by residing between mineral grains. (b) Classification of organisms on the basis of energy and carbon procurements. Also see examples of microorganisms in Table 16.2 (Image credits: (a) Wierzchos *et al.* [16.137]; (b) S. Jheeta).

for the vast majority of organisms; yet still, some fauna and flora survive in these bleak desert terrains, including resilient microorganisms, such as species of *Bacillus* and their spores, as well as chlorophototrophic and eukaryotic algae, and prokaryotic cyanobacteria [16.137]. Further afield, scientists have recovered bacteria from the outer edges of the atmosphere at an altitude of 100 km, where deep space begins [16.132]. In fact, it has been shown in the exobiological experiments (BIOPAN and EXPOSE; Horneck [16.42]) executed on board the International Space Station, at an altitude of 408 km and during a period of over 6 months, that spores of bacteria (*Bacillus subtilis* and *Chroococcidiopsis*) and of lichen (*Xanthoria elegans*) could survive being exposed to the harsh environment of space, which includes extremely low temperatures and pressures, high solar radiation and lethal cosmic rays [16.89] [16.93] [16.98]. Table 16.2 shows a summary of the lifestyle of a broad range of extremophiles as well as their source of energy requirements and carbon acquisition facets.

Brief notes on various groups (Figure 16.5b for classification) follow:

- Photoautotrophs: There is only one division in this category – the photolithoautotrophs as there is no known photoorganoautotroph.
- Chemoautotrophs: This is divided in to two categories. Chemolithoautotrophs can use inorganic compounds such as dihydrogen (H_2) to generate required energy by reducing CO_2 and, in addition, CO_2 is assimilated simultaneously into bio-organic compounds (Table 16.2). In the other case, chemoorganoautotrophs derive the substrate-level phosphorylated ATP from fermentation of compounds such as acetate (CH_3COO^-) and/or methanol (CH_4OH) with the liberation of CH_4. C is fixed at the same time.
- Photoheterotrophs: both categories (photolithoheterotrophs and photoorganoheterotrophs) have bacteriochlorophyll. The former uses both inorganic and organic compounds (e.g., alcohols) instead of CO_2, in the process generating energy and assimilating bio-organic compounds.
- Chemoheterotrophs: Chemolithoheterotrophs and Chemoorganoheterotrophs together make up a conglomerate group on planet Earth, which acquires carbon from other living entities and could generate their energy requirement from both inorganic and organic sources.

16.10 The Viral Connection to the Origin of Life

The role of viruses (especially RNA ones) should be appraised, as this is prerequisite in order to fully appreciate the role of horizontal transduction during the emergence of life. This is because RNA viruses played, with a significant degree of certainty, a momentous role in the maturation of T-LUCAs and LUCAs on the road to the emergence of cellular life forms, in the context of an HGT process known as transduction (see next section). Viruses are curious obligate intercellular molecular parasitic entities in that, unlike T-LUCAs and LUCAs (and subsequent cellular entities), they are simply lengths of either RNA or DNA but not both, encased in a capsule, namely capsid; they do not harbor any other aspects of cellular entities (e.g., organelles such as ribosomes, vacuoles, lysosomes, Golgi apparatus and endoplasmic reticula as well as mitochondria and chloroplasts) whatsoever. They are merely "sets" of genetic informational vehicles, not unlike transposons and retrotransposons, except that they have their own capsid capsule to travel in. Viroids and satellites, on the other hand, are similar to transposons and retrotransposons and are thus naked, meaning that they don't have any capsules; however, like viruses they are also intercellular molecular parasites [16.22] [16.133]. Upon infection, all intercellular molecular viral parasites sequester a host cell's protein-making apparatus for making their virion capsules, as well as replicating the newly injected viral genome strand(s); the newly made strands are then housed in a newly synthesized virion before being ejected from the host cell; during the sequestering of the host cell apparatuses the infected cell may be destroyed.

The origin of viruses is still shrouded in mystery and very little is known about their beginnings except to say that they existed alongside LUCAs [16.27] [16.28] [16.29]. Claverie suggested that viruses could have predated LUCAs based on biochemical/genetic analysis and further added that these initial viruses were composed solely of ncRNAs, with DNA and DNA replication machineries emerging later [16.16] [16.74] [16.84]. Forterre's assertion is based on concepts that some prokaryotes share homology with viral capsid proteins and/or ATPase proteins for packaging and so, the RNA viruses were present during the RNA world epoch [16.27] [16.28] [16.29]. DNA superseded RNA as it was actually invented by and from RNA. If one compares uracil/thymine nitrogenous bases and ribose/deoxyribose sugars in RNA/DNA respectively, one will find that RNA predated DNA; more importantly the transition of RNA to DNA meant that uracil → thymine was a necessary step as DNA is less error prone and more stable when compared to RNA; in fact, it is the absence of an oxygen atom at the C2 position within the pentose ring structure which allows the DNA to form a more stable structure. There is evidence, as pointed out elsewhere in this chapter, of an intermediate of "DNA" viruses that contain uracil-DNA. This implies that viruses existed at the time of LUCAs, a fact upon which most of the scientific community agree. One of the abiding features of RNA viruses is that since they do not contain ribosomes, this mean they cannot be represented on the phylogenetic tree of life, as will become apparent later in this chapter.

16.11 Horizontal Gene Transfer (HGT)

What is HGT and why is it significant? A broad definition of HGT is that it is a mechanism whereby there is either a transfer or exchange of mobile genetic elements (MGEs) between extant donor and recipient cells, or an active take-up of strands of "naked" nucleic acids from the environment by competent cells, followed in both instances by incorporation, recombination or replication of any received genetic material into the recipient cell's genome, with the recipient cell becoming transformed in the process. Compare this to vertical gene transfer which is broadly the entire transfer of genomes, as well as the protoplasm with its RNA content from parent to offspring via the process known as reproduction. MGEs include transposons, retrotransposons, genomic islands, phage genomes, DNA plasmids, transfer of RNAs within sperms, transfer of RNA by parasitic plants (e.g., species of the genus *Cuscuta* RNAs → *Sambucus* species) and naked RNAs (e.g., viroids and viral satellites) as well as DNAs in the environment from long dead entities. HGT is necessary because, especially in microorganisms, it can bring about instantaneous modification of phenotypic changes (e.g., repair of photosynthetic apparatus [16.116]) and/or genome modification, as well as assist in gene duplication and gene pool augmentation, plus it also aids in the acquisition of new genotypic and/or phenotypic traits. In higher organisms (e.g., mammals) MGEs may be achieved via the germline. So, HGT is the exchange of MGEs with the net result being that the extant recipient cell may not always resemble either the donor cell or its former self. This is because a recipient cell may receive many genes from many different individual donor cells within the same genus, as well as from different genera and even different phyla, viruses, or even strands of nucleic acids from demised species within the environment, thus potentially giving rise to a new strain or even a species of living entity, thereby bringing about biodiversity; for example, 10% of the genetic make-up of the different strains of lactic acid bacteria is due to phage mediated genetic diversity [16.7] [16.88]. However, it should be noted that such received genes may not all be conserved; i.e., passed on to the next generation; for example, cyanobacterium which has a dysfunctional photosynthetic apparatus receives a functional gene from another cyanobacterium via the mechanism of transduction but the acquired gene may not be propagated [16.86] [16.116].

T-LUCAs were, as pointed out earlier, heavily engaged in the processes of HGT, albeit at that time the exchange of material occurring, highly probably via the hydro-holes, was the result of primordial high-intensity electric lightning [16.4] [16.57]. This level of HGT simplicity was necessary because T-LUCAs' cell membranes were, in all essence, very simple in comparison with contemporary cellular life's cells; they were probably only decanoic acid vesicles. On the other hand, a LUCA's membrane was sufficiently developed to become more discerning, compared to that of a T-LUCA, bearing in mind that LUCAs, in essence, were already almost fully developed cellular entities and so it is not unreasonable to assume that a LUCA's membrane would have been sufficiently capable of significantly more sophisticated HGT [16.32] [16.58]. It should be noted that the biomembranes of cellular life forms are highly developed with the presence of glycoproteins, globular proteins, surface proteins, glycolipids, cholesterol and carbohydrate molecules. Also present are the biosensors required for detecting factors in the external environment, for example, attack from naked nucleic acids and/or viral particles, as well as the cell receptors necessary for docking of various molecules as in acetylcholine, a neurotransmitter. This level of advancement in HGTs demonstrates their transition from simple processes (e.g. those in LUCAs) to the complex ones required in cellular life forms. Since the geological time frame for the emergence of life was only 300 million years duration [16.82], there must have been as yet undiscovered processes which brought about an exponential rise in both the levels and complexity of HGT mechanisms in order to transform T-LUCAs into LUCAs and then to the two domains of cellular life.

In the absence of HGT it is highly probable that life on Earth would have been exceedingly slow to evolve and would also lack a large-scale biodiversity. In short, it is primarily the ubiquitous nature of HGT's "quick fix" way of acquiring novel genotypic and phenotypic traits which include new physiological, biochemical, structural, self-protection and neutralizing (e.g., from toxins) attributes, which led to the emergence of such a vast array of biodiversity of life on Earth. The role of HGT cannot be underestimated as it played a major part from its initial beginnings during the T-LUCA epoch right up to present day.

16.12 Mechanisms of HGT

The more widely recognized and well-understood modes of HGT include bacterial transformation, as well as bacterial conjugation and transduction by viruses (e.g., bacteriophages), and these are thought to be the result of Darwinian evolution [16.57]. To fully comprehend how a living entity acquires physiological properties and metabolic traits which it did not previously have, means having knowledge of the methods of the exchange/transfer of MGEs. There are five, in theory, mechanisms by which an extant entity can acquire new genotypes or phenotypes and these are briefly outlined below. Should the reader require more detailed information, further reading of Francino's book, *Horizontal Gene Transfer in Microorganisms*, is recommended [16.31].

Bacterial transformation [16.39] (Figure 16.6a) is an active process whereby a competent cell takes up "naked" strands of nucleic acids of generally shortish length from the environment; these strands may be appropriated from living or dead cell(s) or even from long extinct organisms (e.g., mammoths [16.101]). The MGEs are taken up due to expression by specific genes located on the main chromosome of the recipient cell, resulting in the production of particular proteins and other complexes within the cell membranes, e.g., via type II secretion or type IV pili mechanisms [16.3] [16.14] [16.87]. This mechanism was first demonstrated by Fredrick Griffith in 1928, where he showed that, in brief, "smooth" strains of virulent *Streptococcus pneumoniae* killed live mice when injected into them, because the bacterial cells are protected from the mouse's immune system due to the presence of smooth capsules around the cells; when injected

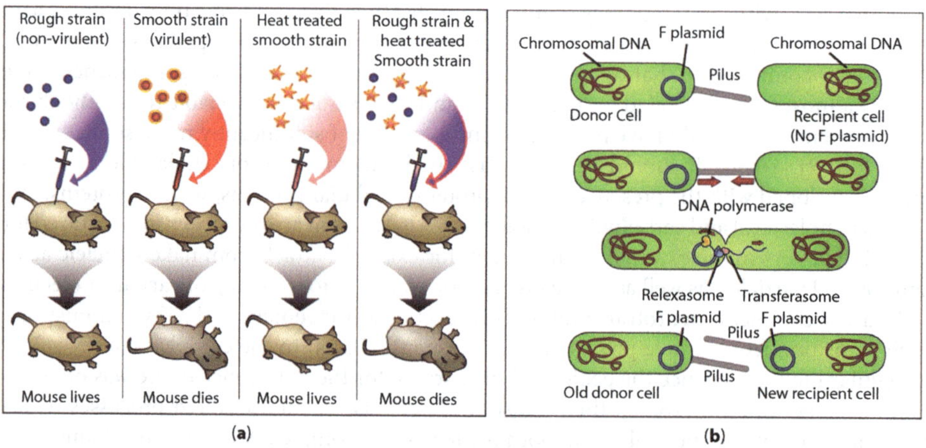

Figure 16.6 (a) Schematic illustration of the process of transformation. (b) Schematic showing the process of conjugation, which begins with the formation of pilus and ends with the eventual transfer of a copy of plasmid to the recipient cell. (Image credits: (a) Internet ref: Griffith's experiment [16.45]; (b) Internet ref: bacterial conjugation [16.44]).

with either "rough" or heat-treated smooth strains, the mice were unaffected. However, when a mixture of both rough and heat-treated smooth strains was injected, the mice died—the conclusion being that the transforming factor (i.e., DNA) from the heat-treated, now dead smooth strains, transformed the rough strains by conveying them with "smooth strain" properties which killed the mice [16.39].

Bacterial conjugation (Figure 16.6b) is also an active process exhibited by the majority of prokaryotes during which a whole plasmid or a small section of plasmid from one donor cell is actively transferred unidirectionally to another competent recipient cell via pili; the recipient cell remaining passive throughout the process [16.83]. For example, the bacterium *Staphylococcus aureus* became resistant to the overuse of the antibiotic, methicillin, because the plasmid which conferred the resistance was obtained, via pili, from bacterial cells that were already immune—such strains are now referred to as MRSA (internet ref: methicillin-resistant *Staphylococcus aureus* [16.48]). During this mode of HGT, a cell may receive genes from different individual donor cells within the same genus, as well as from different genera and even different phyla. Conjugation is widespread amongst species of both Gram-positive and Gram-negative bacteria as well as the Crenarchaeota phylum belonging to the domain of Archaea [16.25].

Transduction is a process whereby viral capsids are used to transfer MGEs from one extant cell to another living cell, without both cells ever coming into contact with one another [16.64] [16.121]; such transfers can occur between different species, genera [16.100] and even between different classes; for example, phage SN-T can transfer MGEs between the Archaea species of Alpha-proteobacter, Beta-proteobacter and Gamma-proteobacter [16.8] [16.54]. It has also been shown that the modification of photosynthetic apparatus in marine phytoplankton is widespread including in cyanobacteria, as was demonstrated by the recent identification of photosystem I (PS I) and photosystem II (PS II) genes in viruses (namely, cyanophages) of marine cyanobacteria (e.g., PS I of the genera *Synechococcus* and *Prochlorococcus*). In this case, cyanophages seemingly help to optimize photosynthesis by providing additional photosystem genes with different properties [16.86] [16.102]. In today's world, the importance of the transduction mechanism cannot be underestimated, viruses being the source of the largest pool of genetic diversity. It is estimated that there are 10^{30} viruses in the biosphere of the oceans alone [16.104] [16.124]. To put this number into some sort of context, the different species belonging to the three domains of life total only 10^8 [16.46]; although this estimation is thought to be woefully low as we can only cultivate less than 5% of prokaryotic species in a laboratory, the rest are simply too fastidious. However, phylogenetic studies have now enabled techniques such as polymerase chain reactions (PCR) to tentatively identify indirectly the ribosomal RNA genes of these fastidious microorganisms (even though they can't be cultivated under laboratory conditions). This technique is a process whereby ribosomal genes (i.e., the DNA segment which codes for rRNAs, including 16S rRNA), being highly conserved, can be amplified using polymerase (typically Taq polymerase) so as to make detection of fastidious prokaryotes possible. Such tests can be carried out on any niche environment—animal intestines, marshlands, desert and tundra subsoils as well as ocean planktons, one of the largest of all biospheres, etc. However, the results have revealed that the total numbers of microbial species identified is still low, meaning that it is unlikely to be anywhere near as high as the 10^{30} level for virions. There are an estimated 10^{29} viral infections occurring per day worldwide, generating yet more virions. Of this number, even if we were to assume that only one virion in a trillion (10^{12}) is involved in successful transduction, the combined number of transfers of MGEs by transduction probably exceeds what could be achieved by bacterial transformation and bacterial conjugation put together. Basically, both prokaryotes and eukaryotes live in a viral world.

The next two mechanisms of HGT are recent discoveries and, as such, very little is known about their *modus operandi* at present:

- *Gene Transfer Agents* (*GTAs*) are believed to be ancient modes of MGEs transfer systems involving small, diverse and phage-like particles which are found in a wide range of prokaryotes [16.78]. The mechanism of transfer is very similar to transduction except that GTAs, unlike the phages, do not lead to the lysis of the host cell [16.77]. More importantly, they transfer small (~4.5 kb) bits of MGEs to the recipient cell. This system has been widely studied in the purple non-sulfur bacterium, *Rhodobacter capsulatus*, which belongs to the class Alpha-proteobacteria.
- *Membrane Vesicle Transfers* (*MVTs*) are executed via small cellular membranous vesicles emanating from the surface of a donor cell in which MGEs are enclosed; these MGEs are transferred to living recipient cells [16.36] [16.75] [16.90]. In 2012, Gaudin and co-workers showed that the species of Thermococcales, namely *T. gammatolerans* and *T. kodakarensis* are able to form vesicles simply by budding off from cells, a process akin to the ectosomes in eukaryotes. In addition to facilitating the transfer of MGEs, these vesicles also have an additional function in that they afford protection against thermo-degradation of the plC70 plasmid during the process of transfer between donor and recipient cells in *T. kodakarensis*. In addition, readers may wish to read an excellent paper entitled "Origin of Life: LUCA and Extracellular Membrane Vesicles" [16.33].

16.13 Clues to the Origins of Life and a Phylogenetic Tree

It is undoubtedly clear that cellular extremophiles emerged from resilient LUCAs, in particular hyperthermophilic ones [16.11], which were thought to be the vanguard of the origin of cellular forms of life. Figure 16.7a, the phylogenetic tree of life, depicts that both Archaea and Bacteria bifurcated from a single tree trunk at about the same time, but if one LUCA did, in fact, give rise to the two domains of life, how credible was this bifurcation? In 2002, Cavalier-Smith suggests that Archaea transmuted from Bacteria, which emerged first; others (e.g., Woese *et al.* [16.140]) posit, that in general, Archaea appears to be more primitive than Bacteria and therefore emerged prior to it [16.11]; see Tables 16.1 and 16.2. Moreover, on the balance of probability there is no particular reason why any number of resilient LUCAs could not have pushed through to evolve into any number of cellular life forms at about the same geological-time-frame, but eventually only Archaea and Bacteria survived as final "products," as testified by microfossil rocks [16.110] [16.111] [16.112] and carbon isotope fractionation evidence [16.2] [16.24] and so Williams *et al.* [16.138] concluded the emergence of two domains, namely Archaea and Bacteria. Whether the phylogenetic tree of life is represented with one or two trunks is not overly central to the argument, in fact there is even a possibility of a third option, which is referred to as a "bush," such as depicted in Figure 16.7b. This author believes that a "tangled bush" is a more appropriate description for the

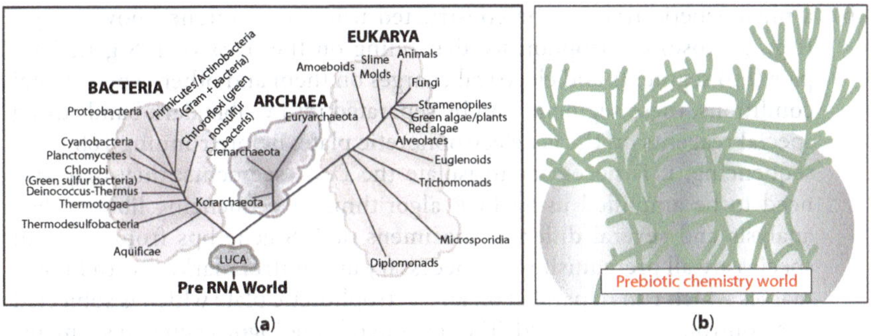

Figure 16.7 (a) The phylogenetic tree of life is a theoretical construct to explain the emergence of the three domains of life, namely Archaea, Bacteria and Eukarya, and also indicates their relatedness; the major one being that Archaea is more related to Eukarya than it is to Bacteria. (b) Communities of T-LUCAs emerged from prebiotic chemistry which eventually gave rise to a community of LUCAs which were almost fully functioning cellular life forms. Then emerged the various strands, out of which Archaea and Bacteria survived; Figure 16.7b shows a typical representation of a tangled bush; the cross-linkage within the bush is brought about due to extensive horizontal gene transfers. (Image credits: (a & b) S. Jheeta).

phylogenetic tree. However, the story of the phylogenetic tree of life does not end here, as more will be explained in relation to the "unrooted" tree of life, later in this chapter.

The study of the hardy prokaryotes (Table 16.2) which thrived in the extreme environments on early Earth, could elucidate the routes of the emergence of cellular life forms from prebiotic chemistry via LUCAs; these hardy prokaryotes make good candidates for phylogenetic studies. Nevertheless, this does not mean that eukaryotes are of no interest; on the contrary, they too are relevant; for example, Table 16.1 compares and contrasts various attributes of all three domains of life and puts these attributes into some sort of perspective. It should be noted that viruses are exempted from phylogenetic studies because they lack ribosomes.

In order to construct a phylogenetic tree of life, three aspects of genetics are of paramount importance and all three requirements are met by ribosomes, because these particular structures are present in all cellular life forms and have remained both structurally and functionally unchanged since at least the LUCA epoch. These requirements are as follows:

a) The conserved genes to be studied need to be universally present in all cellular life forms. Ribosomes are a cell's "factories" for synthesizing proteins which also demand a specific order of amino acids bond linkage (i.e., a peptide bond), eventually resulting in proteins with the same level of structural and functional integrities as that of mRNA specificity. The rest of the genome may change radically, due to the rampant introduction of new MGEs via HGT. The components of the ribosome which are highly conserved are the 16S rRNA gene regions (where S in 16S is a Svedberg unit of sedimentation) in prokaryotes; since the ribosomes of eukaryotes are slightly heavier, their conserved gene regions are called 18S rRNAs. In both cases the rRNA gene regions disclose the proximity of evolutionary relationship in the form of genotypic rather than phenotypic features;

b) a phylogenetic tree can be constructed using a technique known as gel (e.g., agarose) electrophoresis; depending on the mass of 16S gene base pair (bp) segments and electrical charges on them and other experimental conditions, these segments can be separated into visible dark bands using special reagents on the gel electrophoretic plate. Before running 16S DNA sequencing, it is necessary to isolate the DNA segments; they may also need to be amplified using PCR algorithms. A sequencing library, data analysis and several different specimens of 16S gene bps from different species are all prerequisite; also necessary are control markers with known masses which run in parallel on an electrophoretic plate which is subjected to a controlled electric field. The separated homologous gene bp segments are now compared and contrasted for amassed differences in sequences in different species (Figure 16.8a) in relation to the control marker masses. The amassed differences in the genotypic features can then be identified as newly emerging species or relatedness; and

c) the rate of mutations needs to be within acceptable levels and must be due to natural causes. Although any changes in the 16S or 18S rRNA gene regions of the DNA are highly conserved, yet the conservation is still not 100% efficient, meaning changes do occur which are extremely slow and rare, taking millions of years rather than rapid changes, as experienced by the rest of the genome in general, occurring due to the rampant nature of HGT and mutation.

Overall, such phylogenetic studies give information primarily about the relatedness of all cellular life forms on Earth. Still, such studies are not foolproof because relative abundance may be unreliable as it may overinflate biodiversity evaluation due to not being able to differentiate between species that are too closely related (Figure 16.8b).

Furthermore, other phylogenetic analysis of species across the three domains involving metagenetic data and comparative studies—systems and synthetic biology together with molecular palaeontology and even systems chemistry advances—increasingly revealed that different species crossbreed regularly. Differing rates of evolution within both clades and

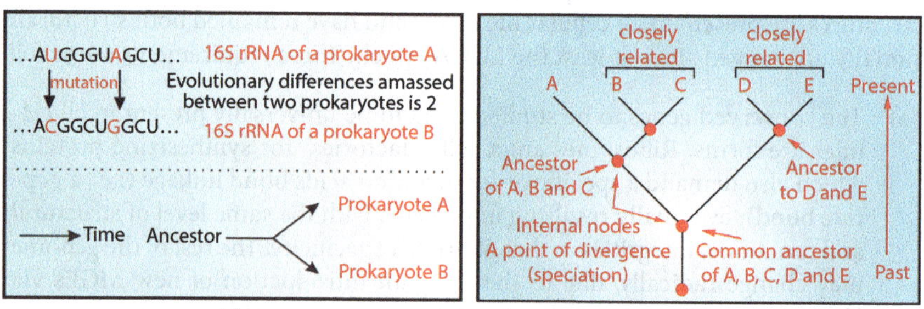

(a) (b)

Figure 16.8 (a) Shows that the evolutionary-time-frame is 2 units from the ancestor; one unit for species A and the other for species B, also shown is a map of the phylogenetic tree bifurcation from when the two prokaryotes parted from their ancestor [16.141]. (b) Interpretation as follows: the level of relativity of organisms, e.g., B and C, are more closely related to each other than to A, and, even less to D and E; A and E are only distantly related. (Image credit: S. Jheeta).

clusters, together with abrupt molecular radiation, and even convergent evolution, affect speciation. With this level of knowledge of promiscuity due to HGTs, it is now becoming clear that, as yet, there may be a third type of tree of life, namely an "unrooted tree of life". The advocates of this line of enquiry point out that it is not possible to trace any cellular life forms back to a LUCA, which may be seen as a "mirage". Further, if the trunk in the "mono-tree of life" is removed at the point where the Archaea and Bacteria bifurcated, it becomes a "node" (Figure 16.8b) and the need for LUCA construct effectively becomes meaningless. So, which one of the trees represent the starting point in the evolutionary history of the 3 domains of life? As far as the unrooted tree of life goes, it is not overly germane point, as far as LUCA is concerned. What is relevant is that the examination of 16S rRNA or of 18S rRNA genes indicates that all life forms are intricately intertwined and that other overwhelming evidence suggests that Archaea and Bacteria did emerge initially from LUCA-like organism(s), with Eukarya being a chimera of these two domains. Either the mono or bush of life trees are possible when compared and contrasted with the unrooted one. However, the author believes that, perhaps, the bush of life is a truer representative, due to levels of promiscuity displayed amongst the living entities which, perhaps, emerged from various LUCAs initially, but only two made to the fore.

16.14 Conclusion

It has been proven that all known rRNAs within all past and present cells are related to one another down the evolutionary timeline, all the way from the present day. In fact, all the phylogenetic studies carried out by whatever algorithm reveal that the last universal common ancestor of all life on Earth forms a central focal point at the base of the phylogenetic tree/bush of life from which Archaea and Bacteria bifurcated. At some point during its evolution LUCA was an RNA organism, i.e., a T-LUCA, but how it chemically evolved from its humble beginning during the prebiotic era is yet to be fully deciphered. However, it can be stated as a matter of fact that extremophilic LUCAs gave rise to extremophilic prokaryotes followed by eukaryotes. It is by investigating the DNA base sequence of the 16S or 18S rRNAs genes of both prokaryotes and eukaryotes that we can say that there were, indeed, the living entities at the base of the phylogenetic tree/bush of life which are our ancestral beings!

Acknowledgment

Special thanks to Professor J. William Schopf, University of California, Los Angeles, for the use of his image in Figure 16.3.

References

16.1. Altman, S., Structural biology. *Nature*, 7, 827–828, 2000.

16.2. Apel, C.L., Deamer, D.W., Mautner, M.N., Self-assembled vesicles of monocarboxylic acids and alcohols:conditions for stability and for the encapsulation of biopolymers. *Biochim. Biophys. Acta*, 1559, 1, 1–9, 2002.

16.3. Averhoff, B. and Friedrich, A., Type IV pili-related natural transformation systems: DNA transport in mesophilic and thermophilic bacteria. *Arch. Microbiol.*, 180, 6, 385–393, 2003.
16.4. Basiuk, V.A. and Navarro-Gonzalez, R., Possible role of volcanic ash-gas clouds in the Earth's prebiotic chemistry. *Orig. Life Evol. Biosph.*, 26, 173–194, 1996.
16.5. Bauche, C. and Laval, J., Repair of oxidized bases in the extremely radiation-resistant bacterium Deinococcus radiodurans. *J. Bacteriol.*, 181, 1, 262–269, 1999.
16.6. Bell, E.A., Boehnke, P., Harrison, T.M., Mao, W.L., Potentially biogenic carbon preserved in a 4.1 billion-year-old zircon. *Proc. Natl. Acad. Sci. U.S.A.*, 112, 47, 14518–14521, 2015.
16.7. Berger, B., Pridmore, R.D., Barretto, C., Delmas-Julien, F., Schreiber, K., Arigoni, F., Brussow, H., Similarity and differences in the Lactobacillus acidophilus group identified by polyphasic analysis and comparative genomics. *J. Bacteriol.*, 189, 1311–1321, 2007.
16.8. Beumer, A. and Robinson, J.B., A broad-host-range, generalized transducing phage (SN-T) acquires 16S rRNA genes from different genera of bacteria. *Appl. Environ. Microbiol.*, 71, 8301–8304, 2005.
16.9. Bowman, J.P., McCammon, S.A., Brown, M.V., Nichols, D.S., McMeekin, T.A., Diversity and association of psychrophilic bacteria in Antarctic sea ice. *Appl. Environ. Microbiol.*, 63, 8, 3068–3078, 1997.
16.10. Brasier, M.D., Green, O.R., Jeplhcoat, A.P., Kleppe, A.K., Von Kranendonk, M.J., Lindsey, J.F., Steele, A., Grassineau, N.V., Letters to nature: Questioning the evidence for Earth's oldest fossils. *Nature*, 416, 6876, 76–81, 2002.
16.11. Cavalier-Smith, T., The neomuran origin of archaebacteria, the negibacterial root of the universal tree and bacterial mega-classification. *Int. J. Syst. Evol. Microbiol.*, 52, 7–76, 2002.
16.12. Cech, T.R., Self-splicing of group I introns. *Annu. Rev. Biochem.*, 59, 543–68, 1990.
16.13. Cech, T.R., Zaug, A.J., Grabowski, P.J., *In vitro* splicing of the ribosomal RNA precursor of Tetrahymena: Involvement of a guanosine nucleotides in the excision of the intervening sequence. *Cell*, 27, 487–496, 1981.
16.14. Chen, I. and Dubnau, D., DNA uptake during bacterial transformation. *Nat. Rev. Microbiol.*, 2, 241–249, 2004.
16.15. Chyba, C.F. and Sagan, C., Endogenous production, exogenous delivery and impact-shock synthesis of organic molecules: An inventory for the origins of life. *Nature*, 355, 125–132, 1992.
16.16. Claverie, J.M., Viruses take center stage in cellular evolution. *Genome Biol.*, 7, 110, 2006.
16.17. Cordero, R.R., Damiani, A., Jorquera, J., Sepulveda, E., Ultraviolet radiation in the Atacama Desert. *Antonie van Leeuwenhoek*, 111, 8, 1301–1313, 2018.
16.18. Dalton, R., Squaring up over ancient life. *Nature*, 417, 782–784, 2002.
16.19. Damer, B. and Deamer, D., Coupled phases and combinatorial selection in fluctuating hydrothermal pools: A scenario to guide experimental approaches to the origin of cellular life. *Life*, 5, 872–887, 2015.
16.20. Darwin, C., *On the Origin of Species by Means of Natural Selection, or the Preservation of Favoured Races in the Struggle for Life*, 1st ed., John Murray, London, 1859.
16.21. Deamer, D., The role of lipid membranes in life's origin. *Life*, 7, 5, 2017.
16.22. Diener, T.O., Potato spindle tuber 'virus': IV. A replicating, low molecular weight RNA. *Virology*, 45, 2, 411–428, 1971.
16.23. Diener, T.O., Circular RNAs: Relics of precellular evolution? *Proc. Natl. Acad. Sci. U.S.A.*, 86, 9370–9374, 1989.
16.24. Dodd, M.S., Papineau, D., Grenne, T., Slack, J.F., Rittner, M., Pirajno, F., O'Neil, J., Little, C.T.S., Evidence for early life in Earth's oldest hydrothermal vent precipitates. *Nature*, 543, 60–64, 2017.
16.25. Dreiseikelmann, B., Translocation of DNA across bacterial membranes. *Microbiol. Rev.*, 58, 3, 293–316, 1994.

16.26. Dworkin, J.P., Deamer, D.W., Sandford, S.A., Allamandola, L.J., Self-assembling amphiphilic molecules: Synthesis in the simulated interstellar/precometary ices. *Proc. Natl. Acad. Sci. U.S.A.*, 98, 3, 815–819, 2001.
16.27. Forterre, P., The two ages of the RNA world, and the transition to the DNA world: A story of viruses and cells. *Biochimie*, 87, 793–803, 2005.
16.28. Forterre, P., Three RNA cells for ribosomal lineages and three DNA viruses to replicate their genomes: A hypothesis for the origin of cellular domain. *Proc. Natl. Acad. Sci. U.S.A.*, 103, 3669–3674, 2006.
16.29. Forterre, P., Defining life: The virus viewpoint. *Orig. Life Evol. Biosph.*, 40, 151–160, 2010.
16.30. Fox, S. and Strasdeit, H., A possible prebiotic origin on volcanic islands of oligopyrrole-type photopigments and electron transfer cofactors. *Astrobiology*, 13, 578–595, 2013.
16.31. Francino, M.P., *Horizontal Gene Transfer in Microorganisms*, Caister Academic Press, Norfolk, UK, 2012.
16.32. Galperin, M.Y., Diversity of structure and function of response regulator output domains. *Curr. Opin. Microbiol.*, 13, 150–159, 2010.
16.33. Gaudin, M., Gauliard, E., Schouten, S., Houel-Renault, L., Lenormand, P., Marguet, E., Forterre, P., Hyperthermophilic archaea produce membrane vesicles that can transfer DNA. *Environ. Microbiol. Rep.*, 5, 1, 1–8, 2012.
16.34. Gavette, J.V., Stoop, M., Hud, N.V., Krishnamurthy, R., RNA-DNA chimeras in the context of an RNA world transition to an RNA/DNA-world. *Angew. Chem. Int.*, 55, 13204–13209, 2016.
16.35. Gilbert, W., Origin of life: The RNA world. *Nature*, 319, 618, 1986.
16.36. Gill, S. and Forterre, P., Origin of life: LUCA and extracellular membrane vesicles (EMVs). *Int. J. Astrobiol.*, 15, 1, 7–15, 2016.
16.37. Gilmour, I., Wright, I., Wright, J., *Origin of Earth and Life*, p. 57, The Open University, UK, 2003.
16.38. Gordon, R. and Sharov, A.A., (Eds) *Habitability of the Universe before Earth*, pp. 321–341, Elsevier, Academic Press, USA, 2017.
16.39. Griffith, F., The significance of pneumococcal types. *J. Hyg.*, 27, 2, 113–159, 1928.
16.40. Hanczyc, M.M., Fujikawa, S.M., Szostak, J.W., Experimental models of primitive cellular compartments: Encapsulation, growth, and division. *Science*, 302, 618–622, 2003.
16.41. Holmes, E.C., What does virus evolution tell us about virus origins? *J. Virol.*, 85, 11, 5247–5251, 2011.
16.42. Horneck, G., Klaus, M.D., Mancinelli, R.L., Space microbiology. *Microbiol. Mol. Biol. Rev.*, 74, 121–156, 2010.
16.43. Huyghe, P., Conan the bacterium. *Sciences*, 38, 16–19, 1998.
16.44. Internet ref: Bacterial conjugation, https://simple.m.wikipedia.org/wiki/Bacterial_conjugation.
16.45. Internet ref: Griffith's experiment, https://en.wikipedia.org/wiki/Griffith%27s_experiment.
16.46. Internet ref: Goldenberg, S. (23/08/2011) Planet Earth is home to 8.7 million species, scientists estimate. *Guardian Newspaper*, http://www.theguardian.com/environment/2011/aug/23/species-earth-estimate-scientists (accessed confirmed 21/06/2019).
16.47. Internet ref: Khan academy, https://ka-perseus-images.s3.amazonaws.com/9d44a501fb30b8301640773ed409226119208fdf.png.
16.48. Internet ref: Methicillin resistant *Stapylococcus aureus*, http://en.wikipedia.org/wiki/Methicillin-resistant_Staphylococcus_aureus.
16.49. Internet ref: National Science Foundation, Washington, USA, http://www.nsf.gov/od/lpa/news/press/01/pr0156.htm.
16.50. Internet ref: Tunnicliffe Laboratory, University of Victoria, Canada, https://tunnicliffelab.wordpress.com/.

16.51. Internet ref: Villarreal, M.R., https://en.wikipedia.org/wiki/Lipid_bilayer#/media/File: Phospholipids_aqueous_solution_structures.svg.
16.52. Iqubal, M.A., Sharma, R., Jheeta, S., Kamaluddin, Thermal condensation of glycine and alanine on metal ferrite surface: primitive peptide bond formation scenario. *Life*, 7, 15, 2017.
16.53. Jandura, A. and Krause, H.M., Opinion: The new RNA world: Growing evidence for long noncoding RNA functionality. *Trends Genet.*, 33, 10, 665–676, 2017.
16.54. Jensen, E.C., Schrader, H.S., Rieland, B., Thompson, T.L., Lee, K.W., Nickerson, K.W., Kokjohn, T.A., Prevalence of broad-host-range lytic bacteriophages of *Sphaerotilus natans*, *Escherichia coli*, and *Pseudomonas aeruginosa*. *Appl. Environ. Microbiol.*, 64, 2, 575–580, 1998.
16.55. Jheeta, S., Final frontiers: The hunt for life elsewhere in the universe. *Astrophys. Space Sci.*, 348, 1–10, 2013.
16.56. Jheeta, S., Horizontal gene transfer and its part in the reorganisation of genetics during the LUCA epoch. *Life*, 3, 518–523, 2013.
16.57. Jheeta, S., The routes of emergence of life from LUCA during the RNA and viral world: A conspectus. *Life*, 5, 1445–1453, 2015.
16.58. Jheeta, S., The landscape of the emergence of life. *Life*, 7, 27, 1–12, 2017.
16.59. Jheeta, S. and Joshi, P.C., Prebiotic RNA synthesis by montmorillonite catalysis. *Life*, 4, 318–330, 2014.
16.60. Johnson, A.P., Cleaves, H.J., Dworkin, J.P., Glavin, D.P., Lazcano, A., Bada, J.L., The Miller volcanic spark discharge experiment. *Science*, 322, 404, 2008.
16.61. Jørgensen, U.G., Appel, P.W.U., Hatsukawa, Y., Frei, R., Oshima, M., Toh, Y., Kimura, A., The Earth–Moon system during the late heavy bombardment period – Geochemical support for impacts dominated by comets. *Icarus*, 204, 368–380, 2009.
16.62. Karnkowska, A., Vacek, V., Zubacova, Z., Treitli, S.C., Petrzelkova, R., Eme, L., Novak, L., Zarsky, V., Barlow, L.D., Herman, E.K. *et al.*, A eukaryote without a mitochondrial organelle. *Curr. Biol.*, 26, 1274–1284, 2016.
16.63. Kasting, J.F., Earth's early atmosphere. *Science*, 259, 5097, 920–926, 1993.
16.64. Kellenberger, E., The genetic control of the shape of a virus. *Sci. Am.*, 215, 32–39, 1966.
16.65. Kelley, D.S., Karson, J.A., Blackman, D.K., Früh-Green, G.L., Butterfield, D.A., Lilley, M.D., Olson, E.J., Schrenk, M.O., Roe, K.K., Lebon, G.T., Rivizzigno, P., An off-axis hydrothermal vent field near the Mid-Atlantic Ridge at 308 N. *Nature*, 412, 145–149, 2001.
16.66. Kelley, D.S., Karson, J.A., Früh-Green, G.L., Yoerger, D.R., Shank, T.M., Butterfield, D.A., Hayes, J.M., Schrenk, M.O., Olson, E.J., Proskurowski, G. *et al.*, A serpentinite-hosted ecosystem: The Lost City hydrothermal field. *Science*, 307, 1428–1434, 2005.
16.67. Kendall, M.M. and Boone, D.R., Cultivation of methanogens from shallow marine sediments at Hydrate Ridge, Oregon. *Archaea*, 2, 31–38, 2006.
16.68. Ketting, R.F., Fischer, S.E.J., Bernstein, E., Sijen, T., Hannon, G.J., Plasterk, R.H.A., Dicer functions in RNA interference and in synthesis of small RNA involved in developmental timing in *C. elegans*. *Genes Dev.*, 15, 2654–2659, 2001.
16.69. Kim, J.-S. and Crowley, D.E., Microbial diversity in natural asphalts of the rancho La Brea tar pits. *Appl. Environ. Microbiol.*, 73, 4579–4591, 2007.
16.70. Knoll, A.H., *Life of a Young Planet*, p. 8, Princeton University Press, USA, 2003.
16.71. Knoll, A.H., Javaux, E.J., Hewitt, D., Cohen, P., Eukaryotic organisms in Proterozoic oceans. *Philos. Trans. R. Soc. B*, 361, 1470, 1023–1038, 2006.
16.72. Koonin, E.V. and Martin, W., On the origin of genomes and cells within inorganic compartments. *Trends Genet.*, 21, 12, 647–654, 2005.
16.73. Kotnik, T., Lightning-triggered electroporation and electrofusion as possible contributors to natural horizontal gene transfer. *Phys. Life Rev.*, 13, 351–370, 2013.

16.74. Krishnamurthy, R., The emergence of RNA. *Isr. J. Chem.*, 55, 837–850, 2015.
16.75. Kulp, A. and Kuehn, M.J., Biological functions and biogenesis of secreted bacterial outer membrane vesicles. *Annu. Rev. Microbiol.*, 64, 163–184, 2010.
16.76. Lane, N., Allen, J.F., Martin, W., How did LUCA make a living? Chemiosmosis in the origin of life. *Bioessays*, 32, 271–280, 2010.
16.77. Lang, A.S. and Beatty, T., Importance of widespread gene transfer agent genes in a-proteobacteria. *Trends Microbiol.*, 15, 2, 54–60, 2006.
16.78. Lang, A.S., Zhaxybayeva, O., Beatty, J.T., Gene transfer agents: Phage-like elements of genetic exchange. *Nat. Rev.: Microbiol.*, 10, 7, 472–482, 2013.
16.79. Lanier, K.A. and Williams, L.D., The origin of life: Models and data. *J. Mol. Evol.*, 84, 85–92, 2017.
16.80. Lasken, R.S., Schuster, D.M., Rashtchian, A., Archaebacterial DNA polymerases tightly bind uracil-containing DNA. *J. Biol. Chem.*, 271, 30, 17692–17696, 1996.
16.81. Lathe, R., Tidal chain reaction and the origin of replicating biopolymers. *Int. J. Astrobiol.*, 4, 1, 19–31, 2005.
16.82. Lazcano, A. and Miller, S.L., How long did it take for life to begin and evolve to cyanobacteria? *J. Mol. Evol.*, 39, 546–554, 1994.
16.83. Lederberg, J. and Tatum, E.L., Gene recombination in E. coli. *Nature*, 158, 558, 1946.
16.84. Leipe, D.D., Aravind, L., Koonin, E.V., Did DNA replication evolve twice independently? *Nucleic Acid Res.*, 27, 17, 3389–401, 1999.
16.85. Lincoln, T.A. and Joyce, G.F., Replication of an RNA enzyme. *Science*, 323, 5918, 1229–1232, 2009.
16.86. Lindell, D., Jaffe, J.D., Johnson, Z.I., Church, G.M., Chisholm, S.W., Photosynthesis genes in marine viruses yield proteins during host infection. *Nature*, 438, 3, 86–79, 2005.
16.87. Lorenz, M.G. and Wackernagel, W., Bacterial gene transfer by natural genetic transformation in the environment. *Microbiol. Rev.*, 58, 3, 563–602, 1994.
16.88. Villarreal, L.P. and Witzany, G., The DNA habitat and its RNA inhabitants: At the dawn of RNA sociology. *Genomics Insights*, 6, 1–12, 2013.
16.89. Mancinelli, R.L., The effect of the space environment on the survival of *Halorubrum chaoviator* and *Synechococcus* (Nägeli): Data from the space experiment OSMO on EXPOSE-R. *Int. J. Astrobiol.*, 14, 123–128, 2014.
16.90. Manning, A.J. and Kuehn, M.J., Functional advantages conferred by extracellular prokaryotic membrane vesicles. *J. Mol. Microbiol. Biotechnol.*, 23, 1-2, 131–141, 2013.
16.91. Martin, W., Baross, J., Kelley, D., Russell, M.J., Hydrothermal vents and the origin of life. *Nat. Rev. Microbiol.*, 6, 805–814, 2008.
16.92. McCollom, T.M. and Ritter, G., Lipid synthesis under hydrothermal conditions by Fischer-Tropsch-type reactions. *Orig. Life Evol. Biosph.*, 29, 153–166, 1999.
16.93. Meeßen, J., Wuthenow., P., Schille, P., Rabbow, E., de Vera, J.-P.P., Ott, S., Resistance of the lichen Buellia frigida to simulated space conditions during the preflight tests for BIOMEX—Viability assay and morphological stability. *Astrobiology*, 15, 8, 601–615, 2015.
16.94. Moelling, K., *Viruses - More Friends than Foes*, p. 61, World Scientific Publishing Company Private Ltd, Singapore, 2017.
16.95. Mulkidjanian, A.Y., Bychkov, A.Y., Dibrova, D.V., Galperin, M.Y., Koonin, E.V., Origin of first cells at terrestrial, anoxic geothermal fields. *Proc. Natl. Acad. Sci. U.S.A.*, 109, E821–E830, 2012.
16.96. Nelson, K.E., Levy, M., Miller, S.L., Peptide nucleic acids rather than RNA may have been the first genetic molecule. *Proc. Natl. Acad. Sci. U.S.A.*, 97, 8, 3868–3871, 2000.
16.97. Neuner, A., Jannasch, H.W., Belkin, S., Stetter, K.O., *Thermococcus litoralis* sp. nov.: A new species of extremely thermophilic marine archaebacteria. *Arch. Microbiol.*, 153, 2, 205–207, 1990.

16.98. Nicholson, W.L., Munakata, N., Horneck, G., Melosh, H.J., Setlow, P., Resistance of Bacillus endospores to extreme terrestrial and extra-terrestrial environments. *Microbiol. Mol. Biol. Rev.*, 64, 548–572, 2000.
16.99. Nowack, E.C.M., Vogel, H., Groth, M., Grossman, A.R., Melkonian, M., Glöckner, G., Endosymbiotic gene transfer and transcriptional regulation of transferred genes in *Paulinella chromatophora. Mol. Biol. Evol.*, 28, 1, 407–422, 2011.
16.100. Nzula, S., Vandamme, P., Govan, J.R.W., Sensitivity of the *Burkholderia cepacia* complex and *Pseudomonas aeruginosa* to transducing bacteriophages. *FEMS Immunol. Med. Microbiol.*, 28, 307–312, 2000.
16.101. Overballe-Petersen, S. and Willerslev, E., Horizontal transfer of short and degraded DNA has evolutionary implications for microbes and eukaryotic sexual reproduction. *Bioessays*, 36, 1005–1010, 2014.
16.102. Payne, J.L., Boyer, A.G., Brown, J.H., Finnegan, S., Kowalewski, M., Krause, R.A., Jr., Lyons, S.K., McClain, C.R., McShea, D.W., Novack-Gottshall, P.M. *et al.*, Two-phase increase in the maximum size of life over 3.5 billion years reflects biological innovation an environmental opportunity. *Proc. Natl. Acad. Sci. U.S.A.*, 106, 24–27, 2009.
16.103. Prosdocimia, F., Jheeta, S., Farias, S.T., Conceptual challenges for the emergence of the biological system: Cell theory and self-replication. *Med. Hypotheses*, 119, 79–83, 2018.
16.104. Rohwer, F., Prangishvili, D., Lindell, D., Roles of viruses in the environment. *Environ. Microbiol.*, 11, 11, 2771–2774, 2009.
16.105. Rothschild, L.J. and Mancinelli, R.L., Life in extreme environments. *Nature*, 409, 1092–101, 2001.
16.106. Russell, M.J. and Hall, A.J., The emergence of life from iron monosulphide bubbles at a submarine hydrothermal redox and pH front. *J. Geol. Soc. London*, 154, 377–402, 1997.
16.107. Russell, M.J., Hall, A.J., Martin, W., Serpentinization as a source of energy at the origin of life. *Geobiology*, 8, 5, 355–371, 2010.
16.108. Russell, M.J., Hall, A.J., Turner, D., In vitro growth of iron sulphide chimneys: Possible culture chambers for origin-of-life experiments. *Terra Nova*, 1, 3, 238–241, 1989.
16.109. Schoning, K.-U., Scholz, P., Guntha, S., Wu, X., Krishnamurthy, R., Eschenmoser, A., Chemical aetiology of nucleic acid structure: A threofuranosyl-(3' 2') oligonucleotide system. *Science*, 290, 5404, 134–1351, 2006.
16.110. Schopf, J.W., Microfossils of the early archaean apex chert: New evidence of the antiquity of life. *Science*, 260, 640–646, 1993.
16.111. Schopf, J.W., Kitajima, K., Spicuzza, M.J., Kudryavtsev, A.B., Valley, J.W., SIMS analyses of the oldest known assemblage of microfossils document their taxon-correlated carbon isotope compositions. *Proc. Natl. Acad. Sci. U.S.A.*, 115, 1, 53–58, 2018.
16.112. Schopf, J.W., Kudryavtsev, A.B., Agresti, D.G., Wdowiak, T.J., Czaja, A.D., Laser-Raman imagery of Earth's earliest fossils. *Nature*, 416, 6876, 73–6, 2002.
16.113. Seilacher, A., Bose, P.K., Pflüger, F., Triploblastic animals more than 1 billion years ago: Trace fossil evidence from India. *Science*, 282, 80–83, 1998.
16.114. Sharma, A., Scott, J.H., Cody, G.D., Fogel, M.L., Hazen, R.M., Hemley, R.J., Huntress, W.T., Microbial activity at gigapascal pressures. *Science*, 295, 5559, 1514–1516, 2002.
16.115. Sharma, R., Iqubal, M.A., Jheeta, S., Adsorption and oxidation of aromatic amines on metal(II) hexacyanocobaltate(III) complexes: Implication for oligomerization of exotic aromatic compounds. *Inorganics*, 5, 18, 2017.
16.116. Sharon, I., Alperovitch, A., Rohwer, F., Haynes, M., Glaser, F., Atamna-Ismaeel, N., Pinter, R.Y., Partensky, F., Koonin, E.V., Wolf, Y.I. *et al.*, All-in-one: Photosystem-I gene cassettes in marine viruses. *Nature*, 461, 7261, 258–262, 2009.

16.117. Sojo, V., Herschy, B., Whicher, A., Camprubí, E., Lane, N., The origin of life in alkaline hydrothermal vents. *Astrobiology*, 16, 2, 181–97, 2016.

16.118. Sorci, L., Martynowski, D., Rodionova, D.A., Eyobo, Y., Zogaj, X., Klose, K.E., Nikolaev, E.V., Magni, G., Zhang, H., Osterman, A.L., Nicotinamide mononucleotide synthetase is the key enzyme or an alternative route of NAD biosynthesis in Francisella tularensis? *Proc. Natl. Acad. Sci. U.S.A.*, 106, 9, 3083–3088, 2009. Epub 2009 Feb 9.

16.119. Spang, A., Saw, J.H., Jørgensen, S.L., Zaremba-Niedzwiedzka, K., Martijn, J., Lind, A.E., van Eijk, R., Schleper, C., Guy, L., Ettema, T.J.G., Complex archaea that bridge the gap between prokaryotes and eukaryotes. *Nature*, 521, 7551, 173–179, 2015.

16.120. Spiegelman, S., Haruna, I., Holland, I.B., Braudreau, G., Mills, D., The synthesis of a self-propagating and infectious nucleic acid with a purified enzyme. *Biochemistry*, 54, 919–927, 1965.

16.121. Stanier, R.Y., Doudoroff, M., Adelberg, E.A., *General Microbiology*, pp. 514–558, Macmillan Publishers Ltd, 1973.

16.122. Stanier, R.Y. and van Niel, C.B., The concept of a bacterium. *Arch. Mikrobiol.*, 42, 17–35, 1962.

16.123. Sumper, M. and Luce, R., Evidence for *de novo* production of self-replicating and environmentally adapted RNA structures by bacteriophage Q3 replicase. *Proc. Natl. Acad. Sci. U.S.A.*, 72, 1, 162–166, 1975.

16.124. Suttle, C.A., Marine viruses—Major players in the global ecosystem. *Nat. Rev. Microbiol.*, 5, 801–812, 2007.

16.125. Szostak, J.W., Bartel, D.P., Luisi, P.L., Synthesizing life. *Nature*, 409, 387–390, 2001.

16.126. Tan, C. and Tomkins, J., Information processing differences between Archaea and Eukarya – implications for homologs and the myths of eukaryogenesis. *Answ. Res. J.*, 8, 121–141, 2015.

16.127. Thauer, R.K., Kaster, A.-K., Goenrich, M., Schick, M., Hiromoto, T., Shima, S., Hydrogenases from methanogenic archaea, nickel, a novel cofactor, and H_2 storage. *Annu. Rev. Biochem.*, 79, 507–36, 2010.

16.128. Thrash, J.C., Boyd, A., Huggett, M.J., Grote, J., Carini, P., Yoder, R.J., Robbertse, B., Spatafora, J.W., Rappé, M.S., Giovannoni, S.J., Phylogenomic evidence for a common ancestor of mitochondria and the SAR11 clade. *Sci. Rep.*, 1, 13, 2011.

16.129. van Tonder, A.J., Mistry, S., Bray, J.E., Hill, D.M.C., Cody, A.J., Farmer, C.L., Klugman, P.K., von Gottberg, A., Bentley, S.D., Parkhill, J. *et al.*, Defining the estimated core genome of bacterial populations using a Bayesian decision model. *PLoS Comput. Biol.*, 10, 8, e100378, 2014.

16.130. Vreeland, R.H., Rosenzweig, W.D., Powers, D.W., Isolation of a 250 million-year-old halotolerant bacterium from a primary salt crystal. *Nature*, 407, 897–900, 2000.

16.131. Wacey, D., Kilburn, M.R., Saunders, M., Cliff, J., Brasier, M.D., Microfossils of sulphur-metabolizing cells in 3.4-billion-year-old rocks of Western Australia. *Nat. Geosci.*, 4, 698–702, 2011.

16.132. Wainwright, M., Wickramasinghe, N.C., Narlikar, J.V., Rajaratnam, P., Microorganisms cultured from stratospheric air samples obtained at 41 km. *FEMS Microbiol. Lett.*, 218, 161–165, 2003.

16.133. Wang, M.-B., Bian, X.-Y., Wu, L.-M., Liu, L.-X., Smith, N.A., Isenegger, D., Wu, R.-M., Masuta, C., Vance, V.B., Watson, *et al.*, On the role of RNA silencing in the pathogenicity and evolution of viroids and viral satellites. *Proc. Natl. Acad. Sci. U.S.A.*, 101, 9, 3275–3280, 2004.

16.134. Weiss, M.C., Sousa, F.L., Mrnjavac, N., Neukirchen, S., Roettger, M., Nelson-Sathi, S., Martin, W.F., The physiology and habitat of the last universal common ancestor. *Nat. Microbiol.*, 1, #16116, 2016.

16.135. Westall, F., Hickman-Lewis, K., Hinman, N., Gautret, P., Campbell, K.A., Bréhéret, J.G., Foucher, F., Hubert, J., Sorieul, S., Dass, *et al.*, A Hydrothermal-sedimentary context for the origin of life. *Astrobiology*, 18, 3, 259–293, 2018.
16.136. Whitman, W.B. and Jeanthon, C., Methanococcales, in: *The Prokaryotes*, M.S. Dworkin Falkow, E. Rosenberg, K.H. Schleifer, E. Stackebrandt (Eds.), Springer, New York, NY, 2006.
16.137. Wierzchos, J., DiRuggiero, J., Vítek, P., Octavio, A., Souza-Egipsy, V., Skaloud, P., Tisza, M., Davila, A., Vílchez, C., Garbayo, I., Ascaso, C., Adaptation strategies of lane endolithic chlorophototrophs to survive the hyperarid and extreme solar radiation environment of the Atacama Desert. *Front. Microbiol.*, 6, 934, 2015.
16.138. Williams, T.A., Foster, P.G., Cox, C.J., Embley, T.M., An archaeal origin of eukaryotes supports only two primary domains of life. *Nature*, 504, 231–236, 2013.
16.139. Woese, C.R. and Fox, G.E., Phylogenetic structure of the prokaryotic domain: The primary kingdoms. *Proc. Natl. Acad. Sci. U.S.A.*, 74, 11, 5088–90, 1977.
16.140. Woese, C.R., Kandler, O., Wheelis, M.L., Towards a natural system of organisms: Proposal for the domains Archaea, Bacteria, and Eucarya. *Proc. Natl. Acad. Sci. U.S.A.*, 87, 4576–4579, 1990.
16.141. Yarus, M., *Life from an RNA World*, p. 15, Publisher: Harvard University Press, USA, 2010.
16.142. Zaremba-Niedzwiedzka, K., Caceres, E.F., Saw, J.H., Bäckström, D., Juzokaite, L., Vancaester, E., Seitz, K.W., Anantharaman, K., Starnawski, P., Kjeldsen, K.U. *et al.*, Asgard archaea illuminate the origin of eukaryotic cellular complexity. *Nature*, 541, 353–358, 2017.

17

What Do the DPANN Archaea and the CPR Bacteria Tell Us about the Last Universal Common Ancestors?

Charles H. Lineweaver

Research School of Astronomy and Astrophysics, Research School of Earth Sciences, Australian National University, Canberra, Australia

Abstract

The recent discovery of CPR bacteria pushes the common ancestor of bacteria deeper into the past. The recent discovery of DPANN archaea pushes the common ancestor of archaea deeper into the past. Together, these can push the last universal common ancestor (LUCA) of all extant cellular life deeper into the past. CPR and DPANN seem to be missing some features thought to be core features of all life. The small genomes, cells sizes and episymbiotic nature of CPR and DPANN may represent an ancestral state. I review efforts to include viruses in the phylogenetic tree of life and postulate the usefulness of different LUCAs: the LUCA of all extant cellular life, the LUCA of all cellular life (including extinct cellular life) and the LUCA of all life (including viruses). As we find out more about CPR, DPANN, mimiviruses, and ssRNA viruses, we are presented with organisms with less autonomy and an increasingly long list of "missing" biosynthetic pathways. Rather than excluding these less autonomous organisms from the category we call life, they probably give us our best view of our earliest life-like ancestors. Multiple LUCAs that include our less autonomous, life-like ancestors can help displace the dominant essentialist view of life with its imaginary well-defined boundary between life and non-life.

Keywords: Origin of life, LUCA, DPANN, CPR, viruses

17.1 Introduction

Over the past few decades, genomic discoveries have changed the deepest divergences in the phylogenetic tree of life. Until recently, it was thought that all eukaryotes are equally distant from all archaea. However, recent analyses suggest that eukaryotes are more closely related to Asgard archaea than to other archaea [17.33] [17.35]. Our closest relatives within the Asgard superphylum seem to be the Heimdallarchaeota [17.27] [17.35]. This embedding of the eukaryotes within archaea reduces the number of domains of life from three to two (Figure 17.1).

Email: charley.lineweaver@anu.edu.au
Charles H. Lineweaver: http://www.mso.anu.edu.au/~charley/, https://scholar.google.com/citations?user=-4KUT64YAAAAJ&hl=en, https://www.researchgate.net/profile/Charley_Lineweaver

Joseph Seckbach and Helga Stan-Lotter (eds.) Extremophiles as Astrobiological Models, (359–368)

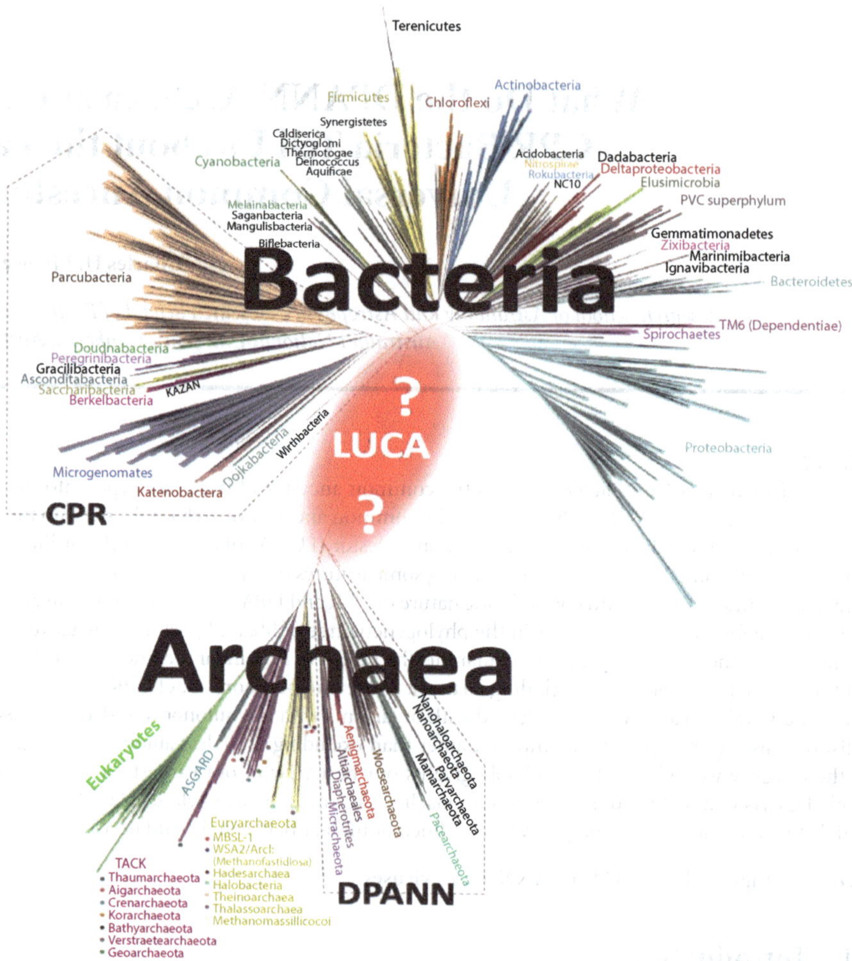

Figure 17.1 The tree of life. Humans and other eukaryotes are represented by the green sliver in the lower left embedded within the ASGARD Archaea. The red oval in the center is an estimate of the position of LUCA (Last Universal Common Ancestor of all extant cellular life). The newly discovered "Candidate Phyla Radiation" (CPR) is enclosed by the dashed line in the upper left. The newly discovered DPANN is enclosed by the dashed line in the lower right. Both CPR and DPANN are deeply rooted. The vast majority of organisms in CPR and DPANN have been identified with metagenomics and have not been cultured (see for example Figure. 1 in [17.13]). (Image credit: modified from [17.8]).

The recent discovery of CPR and DPANN has modified an even deeper part of the tree of life and has arguably increased the known microbial diversity of Earth by a factor of five (Figure 17.1, [17.8]). The discoveries of CPR and DPANN can be attributed to two advances: (i) recognition of the importance of extreme environments, including the subsurface biosphere which had previously been neglected due to the difficulty of sampling, and (ii) metagenomic whole genome sampling and the computer assemblage of nearly complete genomes [17.4]. The relatively new field of metagenomics (cultivation-independent genomics) is not just 16S rRNA amplicon sequencing but genome-resolved metagenomics which can tell us about the presence or absence of various pathways and metabolisms. 16S rRNA sequencing was unable to do this.

17.2 The Discovery of DPANN and CPR

The acronym DPANN stands for the clads Diapherotrites, Parvarchaeota, Aenigmarchaeota, Nanoarchaeota and Nanohaloarchaea, and is a superphylum of archaea first proposed by Rinke *et al.* [17.26]. Baker *et al.* [17.4] reported ARMAN (Archaeal Richmond Mine acidophilic nanoorganisms), also known as Micrarchaeota and Parvarchaeota. Woesearchaeota and Pacearchaeota were discovered later and proposed to be members of the DPANN [17.7]. Another phylum, Altiarchaeota [17.25], was placed into this superphylum [17.28] [17.33], but see also [17.1]. Candidate phyla radiation (CPR) may be comparable in breadth to all other bacteria [17.6] [17.13].

17.3 Common Features of CPR and DPANN

Most phylogenies suggest that DPANN archaea form a monophyletic group at the base of the archaeal tree [17.7] [17.28] [17.33] [17.11], but see also [17.2] [17.24]. DPANN and CPR seem to be the most deeply rooted archaea and bacteria respectively [17.13] [17.8] [17.9]. Unifying general features of CPR and DPANN are their small cell sizes, small genomes, frequent episymbiotic associations with other bacteria and archaea, and limited metabolic repertoires [17.8] [17.9] [17.11]. Many lack some basic biosynthetic pathways for nucleotides, amino acids, and lipids [17.20] [17.8]. Some CPR lineages are missing what were considered to be universal ribosomal proteins [17.6].

Although only a few CPR and DPANN have been co-cultured with their host, most are predicted to be symbionts with symbiotic lifestyles. Their dependence on other microbes seems to be a characteristic taxonomic feature. Their mutual dependence has been described as cooperative biogeochemical cycling or metabolic handoffs [17.12] [17.3]. Their complex interdependencies and obligate symbioses within a microbial community probably explain the difficulty of growing many of these organisms in pure cultures. Some CPR and DPANN are episymbionts and are associated with the surfaces of host cells, rather than being contained within host cells as endosymbionts [17.8].

The question in the title of this chapter (What Do the DPANN Archaea and the CPR Bacteria Tell Us about the Last Universal Common Ancestors?) can now be reformulated as: Is the dependence of CPR and DPANN on other microbes an ancestral feature characteristic of very early life, or is it the result of a loss (often associated with parasitism) of previously existing capabilities? Were CPR and DPANN previously more complete, autonomous and free-living? Or does the lack (in many of them) of some basic biosynthetic pathways for nucleotides, amino acids, and lipids, suggest that our earliest ancestors were more dependent upon each other and less autonomous. Castelle and Banfield [17.8] write, "Likely, the evolutionary trajectories of CPR and DPANN have been shaped by combination of gene loss, lateral transfer, and convergent evolution."

Lateral transfer was a principal driving force in early cellular evolution [17.34]. However, it is much less certain that gene loss or convergent evolution played a significant role.

In which case, the interconnections between the metabolisms of microbial community members (metabolic handoffs) could be an ancestral feature of early life. Parasitic loss should not be systematically associated with deeply rooted organisms close to the root, as seems to be the case with CPR and DPANN.

17.4 LUCA and the Deep-Rootedness of CPR and DPANN

In Figure 17.2, the branch length to an extant organism is the distance from LUCA to the end of the branch. However, since we only have a very approximate idea of where LUCA is located (somewhere in the red oval), we cannot say which branch lengths are the shortest. Potential candidates for having the shortest branch lengths (and therefore being the best extant representative of LUCA) are Wirthbacteria within CPR and Riflebacteria just outside the CPR region [17.6]. Amongst the DPANN archaea, Nanohaloarchaeota and Nanoarchaeota are good candidates for having the shortest branch lengths.

Although the branchings near the root of the tree are not as well supported as the more recent divergences, in this tree CPR is a monophyletic group. That is, all CPR are equally

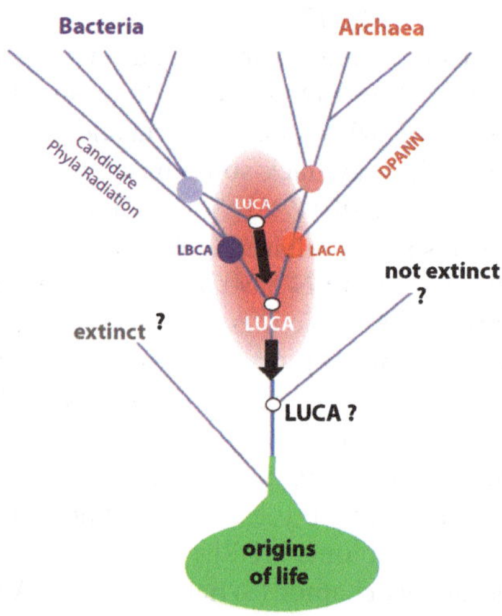

Figure 17.2 If all extant cellular life on Earth were known to us, the last universal common ancestor (LUCA) of all extant cellular life would be fixed in time and would have fixed properties. It would have lived about 3 or 4 billion years ago and its properties could be inferred by comparing the most distantly related extant cellular life forms (e.g., Weiss *et al.* [17.31]). However, all extant cellular life on Earth is not known. Discoveries are still being made. The discovery of CPR [17.13] [17.8] has pushed the last bacterial common ancestor (LBCA) back in time from the light blue to the dark blue dot. The discovery of a new group of archaea known as DPANN [17.4] [17.26] [17.7] [17.28] has pushed the last archaeal common ancestor (LACA) back in time from the light red to the dark red dot [17.33]. The existence of these two deeper ancestors can then push the last universal common ancestor (LUCA) of all extant cellular life deeper into the past: the LUCA label in small white font can become the LUCA in larger white font. If metagenomics reveals an even deeper branch (such as the branch labeled "not extinct?") then LUCA recedes even deeper into the past and becomes a better representative of the origins of life. (Figure modified from Lineweaver & Chopra [17.17]).

distant from other bacteria, and there is a common ancestor of all CPR that is not an ancestor of other bacteria. This is not true of DPANN. If we insist on using exclusively monophyletic names to characterize clads, then apes are a kind of monkey, tetrapods are a kind of fish and vertebrates are a kind of invertebrate. Since the root of archaea is all within the new DPANN grouping, one could even say that archaea are a kind of DPANN—but the convention seems to be to extend the definition of archaea to include DPANN.

17.5 Short Branches, Deep Branches and Multiple LUCAs

The idea of multiple LUCAs seems problematic. How can there be more than one LUCA? Figure 17.2 introduces and helps explain the potential epistemological recession of LUCA deeper and deeper into the past as we discover more deeply rooted life forms. Our understanding of the microbial diversity on Earth is incomplete. The discovery of CPR and DPANN is an example of the on-going discovery of organisms that are deeper branching and have shorter branches in the tree of life.

There is an important distinction to be made between deep branches and short branches. Extant organisms with short branches are more representative of LUCA than extant organisms with long branches. However, branch length depends on where LUCA is.

The "depth" of a branch depends on the depth of its divergence from another branch.

In other words, branches don't have depths, divergences do. We can say that CPR and DPANN are deeply rooted in the sense that their divergences from the lineages that lead to the other bacteria and other archaea (respectively) are deep and close to, or within the red oval. Figure 17.3 illustrates the distinction between short and deep branches.

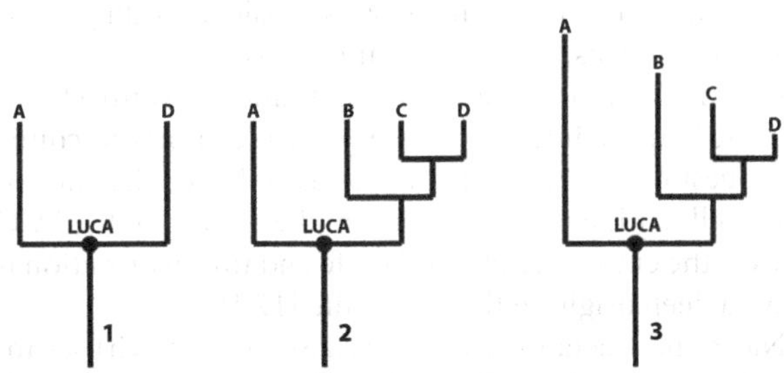

Figure 17.3 The Deep Branch Fallacy. In tree 1, A and D are equally representative of their common ancestor LUCA. If we sequence more life forms and add B and C, to create tree 2, it is a common fallacy to then say "A is a deeper branch" or, "A is more primitive and therefore more representative of LUCA." But A's or D's relationship to LUCA does not change simply by adding close relatives of D to the tree. We could have added a few close relatives of A to the tree and gotten the opposite result. I call this the deep branch fallacy. This fallacy is particularly nefarious in the context of human evolution where the San people are often said to represent the deepest branch and therefore the most primitive branch of human evolution. If we modify tree 2 to include information on branch length we could get tree 3, where the shortest branched organism (D) is the best representative of LUCA. There is no objective "deepest" branch in a tree; the deepest branch depends on which branch has been most densely sampled. From D's perspective, A is the deepest branch (= most distant relative). From A's perspective, B, C and D are equally "deep." If we add 10 divergences along the branch that leads to A, the branch that leads to B, C and D becomes the deepest. (Image credit: the author).

An important goal of phylogenetic trees is to identify the last universal common ancestor of all cellular life, and to find its metabolism [17.31] [17.32]. This estimate of LUCA's metabolism then becomes our best proxy for the first metabolisms of life. LUCA is normally understood as the last universal common ancestor of all known extant cellular life. Thus, as we learn more—as we more comprehensively sample extant life—LUCA can recede deeper into the past, as shown in Figure 17.2. If we were able to sample the DNA of billion-year-old fossils, then it would be profitable to remove the word "extant" and define a new LUCA of all cellular life (including extinct cellular life). This would move LUCA much deeper into the past since most cellular life has gone extinct. LUCA should not be confused with the origin of life. We have good reasons to believe that many branches diverged from the lineage that led to LUCA, and that these branches have gone extinct. If they hadn't gone extinct LUCA would be even deeper and closer to the origin of life.

17.6 Viruses: LUCA without 'Cellular'

Viruses may be about an order of magnitude more numerous than bacteria [17.5]. About one third of biologists think that viruses are not life. Another third think they are life. The remaining third think the question is not worth discussing: "It is an irrelevant semantic point whether we choose to label viruses as alive" [17.10]. See Szostak [17.29] for a discussion of why attempts to define life do not help us understand the origin of life.

There are no viruses in Figure 17.1. Some biologist think that viruses do not belong in the tree of life [17.21] [17.18] while others are convinced they do (e.g., [17.19] [17.14]). Still others think that not only do viruses belong in the tree of life, they are the key to understanding the origins of life [17.30] [17.15] [17.23].

According to Koonin and Dolja [17.15], "Viruses were defined as one of the two principal types of organisms in the biosphere, namely, as capsid-encoding organisms in contrast to ribosome-encoding organisms, i.e., all cellular life forms."

Phylogenetic trees that include viruses can be found in Ward [17.30] and Nasir and Caetano-Anolles [17.23]. These interesting and plausible efforts to connect viruses to the rest of life call into question efforts to define life as only cellular life, and suggest a new LUCA of viruses and cellular life that is different and earlier than the LUCA of cellular life. Viruses are spread over the entire tree of cellular life and this distribution is what one would expect if viruses have a deep origin in the tree of life [17.16].

The analysis of Nasir and Caetano-Anolles [17.23] connects viruses to the cellular phylogenetic tree and is based on Folding Super Families of genes that are shared by archaea, bacteria, eukaryotes and viruses. They find that basal viral groups are minus-single-strand RNA, retrotranscribing viruses and double-stranded RNA viral families with segmented genomes (see their Figure 6b).

The discovery of giant viruses with genomes and physical sizes comparable to cellular organisms, and with remnants of protein translation machinery and virus-specific parasites (virophages) have raised important questions about the origin of viruses. The viruses most closely related to cellular life are "Megavirales," which includes mimiviruses and Hypoviridae. This evidence advocates for their inclusion in global phylogenomic studies and their consideration as a distinct and ancient form of life [17.22].

The analysis of Nasir and Caetano-Anolles [17.23] suggests a long history of widespread episodes of horizontal transfer of genetic information between viruses and cellular life and that viruses have an ancient origin. If viruses are ancient, then current controversies over whether they are alive or not [17.16] is good evidence that there is no well-defined boundary between life and non-life.

References

17.1. Adam, P.S., Borrel, G., Brochier-Armanet, C., Gribaldo, S., The growing tree of Archaea: New perspectives on their diversity, evolution and ecology. *ISME J.*, 11, 2407–2425, 2017.

17.2. Aouad, M., Taib, N., Oudart, A., Lecocq, M., Gouy, M., Brochier-Armanet, C., Extreme halophilic archaea derive from two distinct methanogen class II lineages. *Mol. Phylogenet. Evol.*, 127, 46–54, 2018.

17.3. Anantharaman, K., Brown, C.T., Hug, L.A., Sharon, I., Castelle, C.J., Probst, A.J., Thomas, B.C., Singh, A., Wilkins, M.J., Karaoz, U. *et al.*, Thousands of microbial genomes shed light on interconnected biogeochemical processes in an aquifer system. *Nat. Commun.*, 7, 13219, 2016.

17.4. Baker, B.J., Tyson, G.W., Webb, R.I., Flanagan, J., Hugenholtz, P., Allen, E.E., Banfield, J.F., Lineages of acidophilic archaea revealed by community genomic analysis. *Science*, 314, 5807, 1933–1935, 2006.

17.5. Bamford, D.H., What does structure tell us about virus evolution? *Curr. Opin. Struct. Biol.*, 15, 655–663, 2005.

17.6. Brown, C.T., Hug, L.A., Thomas, B.C., Sharon, I., Castelle, C.J., Singh, A., Wilkins, M.J., Wrighton, K.C., Williams, K.H., Banfield, J.F., Unusual biology across a group comprising more than 15% of domain Bacteria. *Nature*, 523, 208–211, 2015.

17.7. Castelle, C.J., Wrighton, K.C., Thomas, B.C., Hug, L.A., Brown, C.T., Wilkins, M.J., Frischkorn, K.R., Tringe, S.G., Singh, A., Markillie, L.M. *et al.*, Genomic expansion of domain archaea highlights roles for organisms from new phyla in anaerobic carbon cycling. *Curr. Biol.*, 25, 6, 690–701, 2015.

17.8. Castelle, C.J. and Banfield, J.F., Major new microbial groups expand diversity and alter our understanding of the tree of life. *Cell*, 172, 1181–1197, 2018.

17.9. Castelle, C.J., Brown, C.T., Anantharaman, K., Probst, A.J., Huang, R.H., Banfield, J.F., Biosynthetic capacity, metabolic variety and unusual biology in CPR and DPANN radiations. *Nat. Rev. Microbiol.*, 16, 629–645, 2018.

17.10. Dawkins, R. and Wong, Y., *The Ancestor's Tale: A Pilgrimage to the Dawn of Life*, Weidenfeld and Nicholson, London, 2016.

17.11. Dombrowski, N., Lee, J.-H., Williams, T.A., Offre, P., Spang, A., Genomic diversity, lifestyles and evolutionary origins of DPANN archaea. *FEMS Microbiol. Lett.*, 366, 2, fnz008, 2019.

17.12. Hug, L.A., Thomas, B.C., Sharon, I., Brown, C.T., Sharma, R., Hettich, R.L., Wilkins, M.J., Williams, K.H., Singh, A., Banfield, J.F., Critical biogeochemical functions in the subsurface are associated with bacteria from new phyla and little studied lineages. *Environ. Microbiol.*, 18, 1, 159–173, 2016.

17.13. Hug, L.A., Baker, B.J., Anantharaman, K., Brown, C.T., Probst, A.J., Castelle, C.J., Butterfield, C.N., Hernsdorf, A.W., Amano, Y., Ise, K. *et al.*, A new view of the tree of life. *Nat. Microbiol.*, 1, 5, 16048, 2016.

17.14. Koonin, E.V., Senkevich, T.G., Dolja, V.V., Compelling reasons why viruses are relevant for the origin of cells. *Nat. Rev. Microbiol.*, 7, 615, 2009.

17.15. Koonin, E.V. and Dolja, V.V., Virus world as an evolutionary network of viruses and capsid-less selfish elements. *Microbiol. Mol. Biol. Rev.*, 78, 2, 278–303, 2014.

17.16. Lineweaver, C.H., We have not detected extraterrestrial life, or have we?, in: *Life as We Know It: Cellular Origins and Life in Extreme Habitats and Astrobiology*, J. Seckbach and M. Walsh (Eds.), p. 445, Springer Life Sciences, Dordrecht, The Netherlands, 2006.

17.17. Lineweaver, C.H. and Chopra, A., The biological overview effect: Our place in nature. *J. Big Hist.*, 3, 3, 109–122, 2019.

17.18. Lopez-Garcia, P. and Moreira, D., Yet viruses cannot be included in the tree of life. *Nat. Rev. Microbiol.*, 7, 615–617, 2009.

17.19. Ludmir, E.B. and Enquist, L.W., Viral genomes are part of the phylogenetic tree of life. *Nat. Rev. Microbiol.*, 7, 615, 2009.

17.20. Luef, B., Frischkorn, K.R., Wrighton, K.C., Holman, H.-Y.N., Birarda, G., Thomas, B.C., Singh, A., Williams, K.H., Siegerist, C.E., Tringe, S.G. *et al.*, Diverse uncultivated ultra-small bacterial cells in groundwater. *Nat. Commun.*, 6, 6372, 2015.

17.21. Moneira, D. and Lopez-Garcia, P., Ten reasons to exclude viruses from the tree of life. *Nat. Rev. Microbiol.*, 7, 4, 306–11, 2009.

17.22. Nasir., A., Kim, K.M., Caetano-Anolles, G., Giant viruses coexisted with the cellular ancestors and represent a distinct supergroup along with superkingdoms Archaea, Bacteria and Eukarya. *BMC Evol. Biol.*, 12, 156, 2012.

17.23. Nasir, A. and Caetano-Anolles, G., A phylogenomic data-driven exploration of viral origins and evolution. *Sci. Adv.*, 1, 8, e1500527, 2015.

17.24. Petitjean, C., Deschamps, P., Lopez-Garcia, P., Moreira, D., Rooting the domain archaea by phylogenomic analysis supports the foundation of the new kingdom Proteoarchaeota. *Genome Biol. Evol.*, 7, 191–204, 2015.

17.25. Probst, A.J., Ladd, B., Jarett, J.K., Geller-McGrath, D.E., Sieber, C.M.K., Emerson, J.B., Anantharaman, K., Thomas, B.C., Malmstrom, R.R., Stieglmeier, M. *et al.*, Differential depth distribution of microbial function and putative symbionts through sediment-hosted aquifers in the deep terrestrial subsurface. *Nat. Microbiol.*, 3, 328–336, 2018.

17.26. Rinke, C., Schwientek, P., Sczyrba, A., Ivanova, N.N., Anderson, I.J., Cheng, J.F., Darling, A., Malfatti, S., Swan, B.K., Gies, E.A. *et al.*, Insights into the phylogeny and coding potential of microbial dark matter. *Nature*, 499, 7459, 431–437, 2013.

17.27. Spang, A., Saw, J.H., Jørgensen, S.L., Zaremba-Niedzwiedzka, K., Martijn, J., Lind, A.E., van Eijk, R., Schleper, C., Guy, L., Ettema, T.J.G., Complex archaea that bridge the gap between prokaryotes and eukaryotes. *Nature*, 521, 173–179, 2015.

17.28. Spang, A., Caceres, E.F., Ettema, T.J.G., Genomic exploration of the diversity, ecology, and evolution of the archaeal domain of life. *Science*, 357, 6351, eaaf3883, 2017.

17.29. Szostak, J.W., Attempts to define life do not help to understand the origin of life. *J. Biomol. Struct. Dyn.*, 29, 4, 599–600, 2012, http://dx.doi.org/10.1080/073911012010524998.

17.30. Ward, P.D., *Life as We Don't Know It: The NASA Search for (and Synthesis of) Alien Life*, Viking Books, New York, 2005.

17.31. Weiss, M.C., Sousa, F.L., Mrnjavac, N., Neukirchen, S., Roettger, M., Nelson-Sathi, S., Martin, W.F., The physiology and habitat of the last universal common ancestor. *Nat. Microbiol.*, 1, 16116, 2016.

17.32. Weiss, M.C., Preiner, M., Xavier, J.C., Zimorski, V., Martin, W.F., The last universal common ancestor between ancient Earth chemistry and the onset of genetics. *PLoS Genet.*, 14, 8, e1007518, 2018.

17.33. Williams, T.A., Szollosi, G.J., Spang, A., Foster, P.G., Heaps, S.E., Boussau, B., Ettema, T.J.G., Embley, T.M., Integrative modeling of gene and genome evolution roots the archaeal tree of life. *Proc. Natl. Acad. Sci. U.S.A.*, 114, E4602–E4611, 2017.
17.34. Woese, C.R., On the evolution of cells. *Proc. Natl. Acad. Sci. U.S.A.*, 99, 13, 8742–8747, 2002.
17.35. Zaremba-Niedzwiedzka, K., Caceres, E.F., Saw, J.H., Bäckström, D., Juzokaite, L., Vancaester, E., Seitz, K.W., Anantharaman, K., Starnawski, P., Kjeldsen, K.U. *et al.*, Asgard archaea illuminate the origin of eukaryotic cellular complexity. *Nature*, 541, 353–358, 2017.

18

Can Biogeochemistry Give Reliable Biomarkers in the Solar System?

Julian Chela-Flores[1,2]

[1]*The Abdus Salam International Centre for Theoretical Physics, Trieste, Italy*
[2]*Institute of Advanced Studies (IDEA), Caracas, Bolivarian Republic of Venezuela*

Abstract
In this chapter we suggest that possible habitats for extremophiles in the Solar System can be tested with sulfur isotopes as biomarkers. This search is feasible with miniaturized mass spectrometry that has already reached a significant degree of development. This instrument is suitable for the forthcoming exploration of the Moon, Mars and the icy surface of the Jovian moon Europa. We suggest that these biomarkers may be relevant on the icy moons of giant and ice planets. We conclude by raising the question, "Are there environments for extremophiles on exoplanets?"

Keywords: Solar System, astrobiology, biosignatures, habitability, mass spectrometry, icy moons, ESA's JUICE mission, NASA's Ice Clipper mission

Abbreviations

BOSS	= biofilm organisms surfing space
CDT	= Canyon Diablo meteorite.
Delta-34 parameter ($\delta^{34}S$)	= Abbreviation used for the sulfur isotope fractionation: $\delta^{34}S = [^{34}S/^{32}S_{sa} / (^{34}S/^{32}S)_{CDT} - 1] \times 10^3$ [‰] Its value is close to zero when the sample coincides with the value of the standard CDT, which is a troilite (FeS) found in an Arizona crater. The delta^{34}S parameter allows a comparison of a sample (sa) with the standard CDT. The sulfur isotopes are the dominant sulfur isotope (^{32}S) and the next in abundance (^{34}S).
DNA	= deoxyribonucleic acid
ESA	= European Space Agency
Ga BP, Gyr BP	= alternative abbreviations for a billion years before the present

Email: chelaf@ictp.it
Julian Chela-Flores: http://en.wikipedia.org/wiki/Julian_Chela-Flores

Joseph Seckbach and Helga Stan-Lotter (eds.) Extremophiles as Astrobiological Models, (369–384)

HK	= S. Hegde and L. Kaltenegger
ICTP	= The Abdus Salam International Centre for Theoretical Physics
ISS	= International Space Station
JUICE	= Jupiter Icy Moons Explorer, a Jovian system mission
LR	= Viking mission labeled release experiment
Ma BP, Myrs BP	= alternatives for a million years before the present
MASPEX	= mass spectrometer for Planetary Exploration/Europa
MS	= mass spectrometer
MUSE	= mission to the Uranian system
NASA	= National Aeronautics and Space Administration
PEP	= particle environment package
RNA	= ribonucleic acid
rRNA	= small subunit ribosomal ribonucleic acid
ODINUS	= origins, dynamics, and interiors of Neptunian and Uranian systems
OSS	= outer Solar System
SUDA	= Surface Dust Mass Analyzer

18.1 Evidence of Life in the Solar System

We put special emphasis on the well-tested techniques of stable isotope geochemistry, especially those of biogeochemistry. We take advantage of the phenomenon of the relative abundance of different isotopes of an element, which are known to vary substantially when driven by living microorganisms. There are confirmed ancient fossils of sulfate-reducing bacteria in the early Earth, when minerals arose from sedimentation of dissolved marine sulfate. Upon evaporation distinctive minerals are formed—true signals testifying life's antiquity. Those ancient fossils show measurable isotopic variations. In our search for the origin of life in the Solar System we go from the early Earth to low orbits, the Moon and Mars [18.12].

The investigation does not end with the Red Planet, but to the contrary it continues all the way to the icy worlds of the outer Solar System, beginning with the moons of the giant planets. We continue to the outer Solar System and anticipate possible biogeochemical biosignatures that are retrievable from the icy moons of the icy giants: Uranus and Neptune [18.11]. We review relevant instrumentation for testing the presence of life itself, whether extinct, or extant, in some of the above-mentioned sites. We conclude with raising the possibility of testing the universality of biochemistry, namely whether those biological signals we are searching for in the Solar System can be applied to other solar systems.

18.2 Extremophiles on Earth

Some aspects of the microbiology of the polar regions are relevant for astrobiology [18.59], such as, on the surface of Europa sulfur patches are accumulating on glacial ice lying over saline springs that are rich in sulfide and sulfate [18.27].

Besides which, the biogenic sulfur-related traffic through the icy surfaces of the lakes is well understood in the Dry Valley lakes lying on the western shore of McMurdo Sound [18.9]. There is another major Europan analog in North America: a valley that has been given the name of "Borup Fiord Pass," where minerals accumulate on glacial ice. This site is a geologically relevant feature that lies on Ellesmere Island in northern Canada. The valley lies over saline springs that are rich in sulfide and sulfate that make their way to the icy patchy surface. The biosignatures that have been studied in this environment are not related to the sulfur isotopes, as we have done in our above-mentioned work. Rather, what has been studied in Ellesmere Island is the process of biomineralization [18.26].

This interesting biogenic process takes place when biominerals are generated as a result of interactions of microbial life and its environment. In chemosynthesis this is a well-understood process [18.36].

These sites have been of interest to astrobiology [18.21]. Such environments harbor unique microbial ecosystems that could orient us on how to plan our experiments on Europa. Lake Joyce is of special interest to NASA, as it is ice covered all the year round: Its icy surface is 6 meters deep. Yet, even the few percent of light that penetrates through the ice is enough to support an algal ecosystem. Many of the structures on the lake bottom look like what we find in the Archean rock record from about 3 Gyr before the present (BP), because its waters harbor carbonate structures known as microbialites. These unique structures are formed with layers of cyanobacteria.

We are interested in finding out how these organisms are able to grow in the dark, cold waters of Lake Joyce: In these environments extremophiles that are trapped in microbial mats may also be living under the Taylor Glacier in the Taylor Valley. These microbes probably lived in the ocean at one time, but when the floor of the Dry Valleys rose more than a million years ago, this glacier was covered by seawater and trapped the microorganisms in pockets of water.

Lake Vostok is the largest of about 80 subglacial lakes in Antarctica [18.62]. Its surface is approximately 14,000 km^2 and its volume is 1,800 km^3. Indeed, this Ontario-sized lake in East Antarctica is also deep, with a maximum depth of 670 m. On the other hand, from the point of view of microbiology, the habitat provided by Lake Vostok presents us an analogue for the Europa environment, as well as the underground lake underneath the Martian south pole [18.49]. The ice above the lake has been cored to a depth of over 3,600 m, stopping just over 100 m over the surface of the lake itself.

This work has revealed great diversity of single-celled organisms. Besides which, it appears that in the lake water temperatures do not drop too far below zero centigrade, with the possibility of geothermal heating raising the temperatures above this level. Data from work deep in the ice core can be extrapolated to the lake itself. We infer that Lake Vostok may support a microbial population, in spite of the fact that a large volume of water has been isolated from the atmosphere for over one million years.

Bacterial density is found to be two- to seven-fold higher in accretion ice than in the overlying glacial ice. Phylogenetic analysis of the amplified small subunit ribosomal ribonucleic acid (rRNA) gene sequences in this accretion ice has revealed the presence of several microbes: alphaproteobacteria, betaproteobacteria and gammaproteobacteria [18.15]. It has been estimated that the youngest water is at least 400,000 years old. It is a window into life forms and climates of primordial eras. The zone of ice layer up to 3,309 m (referred to as I), and the layer between 3,310 to 3,509 m (zone II) provide detailed information about the paleoclimate record spanning the last 420,000 years.

The basal portion of the ice core from 3,539 to 3,623 m has many features differing from overlying glacial ice and its geochemical composition indicates that it represents actual lake water that has accreted (i.e., frozen) to the underneath of ice sheet. In spite of extremely cold air temperatures above the ice (an average of -55 °C), liquid water is stable in the lake owing to the combined effect of geothermal heating, together with the insulating properties of the overlying ice sheet [18.61].

18.3 Extremophiles in Low Orbits Around the Earth

The existence of extremophiles has raised the question in astrobiology whether microbes could survive in extraterrestrial environments. Several experiments have been undertaken in order to ascertain whether analogous microorganisms could be searched for either in on Mars at present or in the past.

The International Space Station (ISS) is a habitable artificial satellite in low Earth orbit (with an altitude between 330 and 435 km). It has been constructed between 1998 and 2011, and ISS is expected to operate until 2028. The ISS serves as a microgravity space environment, where experiments in microbiology have taken place. We are particularly interested in those experiments involving microbes that could survive in other planets or their satellites. Amongst the extremophiles, Archaea is one of the three main domains of life on Earth. Some archaeans have been studied in the ISS: *Halorubrum chaoviatoris* [18.38]; *Halobacterium noricense* and *Halococcus dombrowskii* [18.65]; *Methanobacterium* (peptidoglycans are missing from their cell membranes) and *Methanosarcina barkeri* (produces methane anaerobically) [18.43]. In addition, it is also remarkable to have learnt that the nematode *Caenorhabditis elegans* tolerates hyper-accelerations up to 400,000*g* [18.1].

18.4 Have There Been Extremophiles on the Moon?

On Earth, most of the sulfur is found in its crust. The crust is about 1% of the Earth. Besides, around 3% of the crust is sulfur. This geologic distribution is expected to be the same on the Moon. On the other hand, the oldest known Earth rock might have been retrieved from a sample returned from the Moon by Apollo 14 [18.5]. We are then faced with the suggestive hypothesis of the existence of a habitability window on the Moon [18.58]. Evidently, if this were the case, any living organisms would have been extremophiles.

Certain aspects of lunar geology and hydrology have made it possible to entertain the habitability hypothesis:

- Outgassing, and associated peak in lunar atmospheric pressure were the result of mare basalt eruptions 3.5 billion years ago [18.46].
- The products of lunar volcanism indicate that the lunar interior also contains more water than was once appreciated, and that the lunar mantle may even be comparably water-rich as Earth's upper mantle [18.32].

Since the time of the Apollo missions, we have known that sulfur plays an important role, due to its relative abundance amongst the Moon's biogenic elements. This provides an option for the emergence of sulfur-metabolizing extremophiles. The total sulfur abundance measurements of lunar soils have shown an average of total sulfur content of 850 ppm [18.24]. Apollo 11 and 17 have detected the greatest amount of sulfur in mare basalts (mean values of 2033 and 1860 ppm, respectively). It is significant that the sulfur content in mare basalts is almost 10 times greater than terrestrial basalts.

Our present knowledge of lunar samples suggests that the best place to collect sulfur on the Moon is from mare soils and rocks. Although sulfur is relatively rare, it does rank eleventh in weight abundance among the elements in average lunar mare rocks. Gibson and Moore [18.25] found that the high-Ti mare basalts, in particular, have high sulfur content in the range of 0.16% to 0.27% by weight. These authors also make the important point that lunar basalts actually have more sulfur than terrestrial basalts, which seldom have more than 0.15% [18.71]. Appropriate instrumentation would be in terms of miniaturized instruments that are already available [18.68] [18.72] [18.73] [18.74].

Given a sulfur abundance comparable with the terrestrial case, it is not to be excluded that sulfate metabolizers could have emerged early on the lunar landscape. In fact, microscopic sulfides in approximately 3.47-Gyr-old barites from North Pole, Australia, clearly indicate microbial sulfate reduction on Earth [18.48] [18.60].

It is reasonable then to reconsider the suggestion of a habitability window on the Moon by extremophiles (the SMC hypothesis). In order to test it, we could search for evidence in the lunar paleoregolith layers trapped between lava flows of a time contemporary with the possible early atmosphere.

18.5 Have There Been Extremophiles on Mars?

The suggestive term "Second Genesis" was introduced to denote the emergence of life in places other than Earth [18.39]. This possibility can be searched both with perseverance and with suitable instrumentation [18.28] [18.30]. This is a fundamental phenomenon of astrobiology that could be detected on several locations inside our solar system (cf. Section 18.6), including Europa [18.29] [18.52]. The special case of Mars was discussed recently [18.12]. In the short term such detection is more likely to be successful on thousands of exoplanets in our galaxy [18.18], and in other galaxies where they are being detected [18.20].

Regarding the search for the origin of life on Mars, a Viking mission was a pioneering effort [18.64]. Especially notable was the labeled release (LR) experiment. It showed that the Martian soil responded for metabolic activity [18.35]. This result was based on the positive LR result that was based on the absence of an oxidative agent in the Martian soil. A second Viking experiment found no trace of organic material. A possible rationalization of this conundrum was suggested with the Phoenix mission results, which included the discovery of perchlorate (a chemical compound containing the ion, ClO^-). The missing information of the existence of this surficial salt on Mars in 1976 may have been a factor in hiding the presence of organics on Mars [18.45]. But the reactivity of perchlorate could nevertheless be a nonbiological explanation of the LR results [18.40].

For the above reasons, a search for biosignatures that are independent of the presence of organics is a reasonable objective. Indeed, in a recent publication a non-organic biosignature—sulfur—has been proposed for the future exploration of Europa in the 2020s [18.12]. On the Red Planet the additional advantage is that the search would be independent of either the present harsh environmental conditions, or even excessive depths that may eventually be reachable with future rovers.

On Mars it has been evident for a long time that the southern hemisphere is especially significant and mostly pristine (i.e., unlike the terrestrial case, where traces of the Late Heavy Bombardment, some 3.8 Gyrs BP still remain intact). It corresponds to the Early Archean period on terrestrial stratigraphy and to the Early Noachian on Mars.

Although we are not certain of the presence of life on Mars now, or earlier, the southern hemisphere may still retain analogous conditions to those that led to the origin of life on Earth, where they have by now totally disappeared, due to the harsh effects of metamorphism and plate tectonics.

In any case, during the early geological history of both terrestrial planets, only Mars has left a window for testing at about the same time when we are certain that life emerged on Earth. While there are numerous factors that militate in favor of present habitability of the Red Planet, there is no evidence yet that ancient habitability could be detectable with biosignatures of reliable microbial fossils.

Useful information will follow from careful analysis of the changes that putative early sulfur metabolizing bacteria may have made in isotopic ratios, both on Earth and on Mars. The detection of biosignatures awaits the challenge of improved instrumentation. With so much valuable and appealing potential information to be retrieved, future rovers should carry such instrumentation in their payloads. This is particularly important, as at earlier times the role of sulfur gradually became more relevant during the Noachian. In the early part of the period (4.6 to 3.7 Gyrs before the present, BP), there must have been a reduced amount of sulfur compounds. Later on, due to volcanism, likely candidates for the dense atmospheric components (towards the later Noachian/Hesperian transition) were SO_2 and H_2; consequently, the resulting high atmospheric pressure may have inhibited SO_2 outgassing, while allowing H_2O and CO_2 [18.23].

18.6 Europa is a Likely Location for an Extremophilic Ecosystem

To date, organics have not been detected on Europa. Furthermore, the radiation environment of Europa is so strong that any organics in existence, regardless of whether they are biogenic in origin, would have very short lifetimes. For these reasons, we have argued [18.10] [18.14] that it is unlikely that forthcoming missions with Europa or Ganymede flybys will ever carry instruments capable of detecting biosynthetic molecules unique to known cell biology, namely, amino acids, lipids, polysaccharides, and nucleic acids (DNA, RNA).

Alternatively, the ESA JUICE mission [18.29] and NASA's Europa Clipper [18.52] will carry instruments that could potentially investigate S isotopes in sufficient detail to detect a sulfur, rather than a cellular biomolecule biomarker. Mass spectrometers (MS) have unique capabilities that can be applied to studies of habitability and, perhaps, more importantly, MS can be applied to life detection on icy worlds. Miniaturized MS have already reached a significant degree of development, but we expect some feasible improvements in the foreseeable future [18.68] [18.72] [18.73] [18.74].

The NASA mission has the capability to measure biogenic stable S-isotope fractionation: The MAss SPectrometer for Planetary EXploration/Europa (MASPEX) was designed to determine the composition of the surface and subsurface ocean by measuring Europa's tenuous atmosphere and any surface material ejected into space. Another relevant instrument of the NASA mission is the SUrface Dust Mass Analyzer (SUDA) will measure the composition of small solid particles ejected from Europa and provide an opportunity to directly sample the surface and potential plumes on low-altitude flybys. Significantly, the latter instrument is capable of identifying biosignatures in the ice matrix of ejecta.

Pappalardo *et al.* [18.50] reviewed the evidence for the existence of an active ocean on Europa, a hypothesis under consideration for some time [18.54] and supported by Na/K measurements in the extended atmosphere of Europa [18.8] [18.53]. While fractionated, S-bearing compounds may be found consistent with biological activity, additional biomarkers would be needed to determine whether their source on Europa is due to a Second Genesis on Europa.

Possible candidates for biota on the icy worlds are extremophilic microorganisms, which account for a major portion of life on Earth. It is worth noting that variations in microbial sulfate reduction, which were not previously detected as a product of non-living processes, have been discovered in pure culture experiments [18.63], which revealed a large S mass-dependent fractionation of $\delta^{34}S$ of up to -70‰. Given the fast progress in miniaturized MS instrumentation, it should be possible to measure $\delta^{34}S$ up to the ‰ level in the near future [18.72] [18.73] [18.74].

This biogenic phenomenon is exclusively observed in a handful of natural environments on Earth, which raises the question as to whether an Europan biota would be composed of sulfate reducers and disproportionators that could give rise to measurable geochemical biomarkers. Such a possibility could be tested with the MS approved for the JUICE and Europa Clipper missions. Geochemical biomarkers on the icy surface of Europa could arise, for example, from mineralogy that is indicative of habitable environments. Alternatively, the source of geochemical biomarkers could arise from isotopic fractionation of biological or metabolic processes. At the University of Bern, many instruments capable of addressing these possibilities with an impressive flight hardware development heritage [18.56] [18.57] are part of the JUICE payload. These instruments include the particle environment package (PEP) with a neutral and ion gas mass spectrometer [18.42]. The accuracy necessary for testing the presence of biogenicity with S isotope on Europa would be achievable with the PEP instrumentation.

A lander has also been approved with a payload that includes a MS, as was the case for previous spacecraft missions such as Cassini, Phoenix, and Curiosity [18.50]. At present, however, it is not completely clear whether the effects of radiation on the detection efficiency of these instruments can be reduced on the Europa icy surface [18.69].

Europa's exo-atmosphere is normally considered to be an extension of its surface [18.19]. If its chemical elements are endogenic, the ultimate source must be regions that have a young surface, where the upwelling of subsurface material may occur [18.67]. Tests of biogenicity are feasible on exo-atmospheric particles that arise from surficial patches.

Such tests raise the possibility that the chemical elements of the exo-atmosphere may be from the subsurface ocean. Given the available accuracy of the instruments, anomalous S isotope ratios may provide the first evidence of life in the outer Solar System.

18.7 Are There Other Environments for Extremophiles in the Solar System?

18.7.1 Titan

The atmosphere of Titan is made up mainly of nitrogen and methane with small amounts of hydrogen cyanide, cyanoacetylene, and other organic compounds. From the Earth, Titan has the appearance of an orange haze, whose major components have been studied with various means, going back to the old data from the Voyager 1 flyby. With the Cassini mission we are beginning to grasp the nature of possible geophysical characteristics of Titan's diverse landscapes [18.37] and especially its hydrosphere. In fact, a dark, lake-like feature was observed in a series of images showing an outline of a surface darker than anything else in its surroundings: it has shore-like boundaries.

The Cassini spacecraft has not found signs of acetylene on Titan. A separate study found evidence that hydrogen is disappearing near the moon's surface. The discoveries support a theory that Titan's microbes could survive by breathing hydrogen gas and eating acetylene, producing methane as a result. A non-biological explanation is difficult to conceive of, since a strong catalyst would be needed to remove the hydrogen and acetylene falling from Titan's atmosphere under the very cold surface temperatures on Titan. Indeed, what is observed is exactly what we would expect if life on Titan is present that uses hydrogen and acetylene in its metabolic pathway and produces methane as a result.

By using a Cassini spectrometer, it was found that there was a lack of acetylene [18.17]. Both acetylene and benzene are expected to be produced when sunlight strikes the methane gas in Titan's atmosphere, and then falls to the surface.

Abundant benzene was found on the surface, but no acetylene, even though it is predicted to be the more abundant of the two compounds in the atmosphere. The lack of acetylene, as well as a previously found deficit of ethane, could have been due to a smaller than expected production of these compounds.

We should underline that low acetylene and low ethane, plus low hydrogen, could make a good case for astrobiological studies. Several researchers have suggested that hydrocarbons on Titan could be the basis for life, playing the same role that liquid water does on Earth.

Just as organisms on Earth combine molecular oxygen with organic compounds to get energy, organisms on Titan might react molecular hydrogen with organic materials such as acetylene. In other words, the suggestion is that on Titan methane-based organisms called methanogens could consume hydrogen, acetylene, and ethane. Hence, results of the Huygens probe could indicate the presence of such life by anomalous depletions of acetylene and ethane as well as hydrogen at the surface [18.41]. Because the combination of acetylene and molecular hydrogen produces methane, it would also solve a major mystery on Titan.

Methane is easily broken down by sunlight and destroyed; yet, somehow the moon maintains a plentiful supply of it. Organisms inducing reactions of hydrogen and acetylene might be the answer. Each of the observations separately can be explained by abiotic processes, but the coupling of the two together is intriguing, and really calls for more detailed *in-situ* measurements on a future mission that could test more sensitively for evidence of life, especially in Titan's lakes and seas.

18.7.2 Enceladus

Saturn's icy moon Enceladus displays evidence of active ice volcanism. In the year 2005 the Cassini spacecraft found a cloud of water vapor over the moon's south pole. Images have revealed many jets, some of them reaching up to some 500 km from the surface, with particle velocities from 1.25 ± 0.1 km/s [18.51] to a maximum of 3.40 km/s [18.66].

Warm fractures were detected, where evaporating ice is probably supplying the observed vapor. This is remarkable as Enceladus is the smallest body known to us with active cryovolcanism. The geologically young surface of this satellite is reminiscent of Jupiter's moon Europa. Surprisingly, Enceladus retains an atmosphere that suggests this moon is sufficiently active geologically to reintroduce the water vapor, as some of it may be escaping into interplanetary space due to its low gravity.

Could microbial life exist on Enceladus, where no sunlight reaches, photosynthesis is impossible and no oxygen is available? To answer this question, we need look no further than our own planet to find examples of the types of exotic ecosystems that could make life possible on Saturn's geyser moon ([18.31] and references therein). The answer appears to be, yes, it could be possible since in recent years, several extremophiles have been found on Earth that thrive in places where the Sun does not shine and oxygen is not present, because no photosynthesis takes place.

Microbes have been discovered that survive on the energy from the chemical interaction between different kinds of minerals, and others that live off the energy from the radioactive decay in rocks.

The Enceladus atmosphere was first detected as Cassini flew within 175 kilometers of Enceladus. Data collected during that flyby confirm an extended and dynamic atmosphere. Water vapor was detected by the sophisticated instrumentation on board Cassini: the ion and neutral mass spectrometer and the ultraviolet imaging spectrograph.

On the other hand, the mass spectrometer found the water vapor to be about 65% of the atmosphere, while the fraction of molecular hydrogen is approximately 20%. The remaining 15% is mostly CO_2 and a combination of molecular nitrogen and CO. The evidence points to the feasibility of life in Enceladus. But, "How would life get started?"

A problem in answering that question is that we do not know how life originated on Earth, nor have we been able to reproduce the origin of life in the laboratory. There are several theories for the way life originated on Earth. Do they apply to Enceladus? Two of the theories for the origin of life do seem to apply to Enceladus, the "primordial soup" theory, and the deep-sea vent theory.

18.7.3 Triton, Icy Moons, Pluto and Charon

Out of all these moons, Triton has been of special interest since the first flyby of the Voyager mission. Indeed, we may assume that the formation of Triton would have been in the solar nebula, unlike the case of Titan, which may have been formed in the disk of material that surrounded Saturn. Its density is the same as that of the two other giant moons of the outer Solar System, the Galilean satellites Ganymede and Callisto.

At present, there is interest in developing astrobiology missions to search for evidence of life beyond the gas giants. As we learn more about other planetary bodies in our solar system, it has been recognized that the surfaces of other planets, such as Neptune's moon Triton, also serve as candidates for hosting geochemical biomarkers [18.22] [18.47]. Hussmann *et al.* [18.34] considered models of the internal structure of Triton and other moons of the ice giants. These models predict a Europa-like rocky core, a subsurface ocean, and—significantly—an outer icy surface for Triton, Ariel, Iapetus, Umbriel, Titania, Oberon, and Pluto's moon Charon [18.55]. Triton's rotation around Neptune is in a direction opposite to the planet's rotation ("retrograde motion"), an unusual case for the Solar System, adding further support to the hypothesis of the captured object.

Keeping in mind the geochemical considerations that apply in particular to Europa, it is critical to discuss at this juncture what would be required to facilitate the search for geochemical biomarkers by way of the elemental isotopic content of the icy surfaces and exo-atmospheres of more distant unexplored moons in the Solar System [18.13].

Specific missions to the ice giants have not, to date, been endorsed, although proposals for future missions to them [18.44] are currently under discussion. These proposals include origins, dynamics, and interiors of Neptunian and Uranian systems (ODINUS) [18.70], OSS [18.16], an orbital mission to Uranus [18.2], and a mission to the Uranian system (MUSE) [18.7]. If these mission concepts are accepted, we could find out whether worlds in the realm of ice giants are inhabited.

18.8 Are There Environments for Extremophiles on Exoplanets?

Biogeochemical biosignatures on the surface of exoplanets are not candidates for discovering extraterrestrial life. Alternative biosignatures are possible given the correlation between geomicrobiology and observational astronomy that is possible by extending the red edge of terrestrial chlorophyll—observable in the infrared from space telescopes—to microbial extremophilic life [18.33]. It is possible to push the frontier of habitability of extremophiles from the Solar System to other planetary systems now known after the work of the Kepler Mission [18.4].

A major experiment in the ESA EXPOSE facility for astrobiology experiments in low Earth orbit, has been mounted outside the International Space Station (ISS). The experiment is dedicated to astrobiology. It has been called Biofilm Organisms Surfing Space (BOSS) [18.3]. Its objective is to test whether microorganisms are more tolerant to Martian conditions compared to their planktonic counterparts. Interestingly, from the point of view of this chapter two of the organisms that were exposed are extremophilic.

Firstly, they have used *Deinococcus geothermalis*. This species is extremophilic, produces orange-pigmented colonies and is extremely gamma radiation-resistant. Secondly, *Chroococcidiopsis* has been investigated. The reason why *Chroococcidiopsis* has been selected for the colonizing test on Mars is based on the following qualities: growth on rocks and tolerance to survive the high energy cosmic radiations [18.6].

Hegde and Kaltenegger [18.33] attempted to use some experimental techniques that could suggest the first steps that ought to be taken when direct observation of the surface of exoplanets eventually becomes available. This would be a preliminary and significant step towards finding extremophilic life on exoplanets. Their starting point is the remark that on Earth chlorophyll in vegetation reflects near-infrared light, a phenomenon that occurs in organisms that have chlorophyll, for they reflect a "red edge" that is visible with orbital satellites capable of observing the infrared light.

If our future telescopes were able to distinguish surficial details on exoplanets a red edge would be a candidate for a biomarker, as would other colors that are known to be associated with extremophiles on Earth.

References

18.1. Alves, T., de Souza, J., Campos Pereira, T., *Caenorhabditis elegans* tolerates hyperaccelerations up to 400,000 x g. *Astrobiology*, 18, 825–833, 2018.

18.2. Arridge, C.S., Achilleos, N., Agarwal, J., Agnor, C.B., Ambrosi, R., André, N., Badman, S.V., Baines, K., Banfield, D., Barthélémy, M. *et al.*, The science case for an orbital mission to Uranus: Exploring the origins and evolution of ice giant planets. *Planet. Space Sci.*, 104, Part A, 122–140, 2014.

18.3. Baqué, M., de Vera, J.-P., Rettberg, P., Billi, D., The BOSS and BIOMEX space experiments on the EXPOSE-R2 mission: Endurance of the desert cyanobacterium *Chroococcidiopsis* under simulated space vacuum, Martian atmosphere, UVC radiation and temperature extremes. *Acta Astronaut.*, 91, 180–186, 2013.

18.4. Batalha, N.M., Rowe, J.F., Bryson, S.T., Barclay, T., Burke, C.J., Caldwell, D.A., Christiansen, J.L., Mullally, F., Thompson, S.E., Brown, T.M. *et al.*, Planetary candidates observed by Kepler, III: Analysis of the First 16 months of data. *Astrophys. J. Suppl.*, 204, 2, #24, 21 pp. 2013.

18.5. Bellucci, J.J., Nemchin, A.A., Grange, M., Robinson, K.L., Collins, G., Whitehouse, M.J., Snape, J.F., Norman, M.D., Kring, D.A., Terrestrial-like zircon in a clast from an Apollo 14 breccia. *Earth Planet. Sci. Lett.*, 510, 173–185, 2019.

18.6. Billi, D., Friedmann, H., Caiola, O., Ionizing-radiation resistance in the desiccation-tolerant cyanobacterium *Chroococcidiopsis. Appl. Environ. Microbiol.*, 66, 1489, 2000.

18.7. Bocanegra-Bahamón, T., Bracken, C., Costa Sitjà, M., Dirkx, D., Gerth, I., Konstantinidis, K., Labrianidis, C., Laneuville, M., Luntzer, A., MacArthur, J.L. *et al.*, MUSE—Mission to the uranian system: Unveiling the evolution and formation of ice giants. *Adv. Space Res.*, 55, 2190–2216, 2015.

18.8. Cassidy, T.A., Johnson, R.E., Tucker, O.J., Trace constituents of Europa's atmosphere. *Icarus*, 201, 182–190, 2009.

18.9. Chela-Flores, J., The sulphur dilemma: Are there biosignatures on Europa's icy and patchy surface? *Int. J. Astrobiol.*, 5, 17–22, 2006.

18.10. Chela-Flores, J., Instrumentation for the search of habitable ecosystems in the future exploration of Europa and Ganymede. *Int. J. Astrobiol.*, 9, 101–108, 2010.

18.11. Chela-Flores, J., Forum article: Instrumentation for testing whether the icy moons of the gas and ice giants are inhabited. *Astrobiology*, 17, 958–961, 2017.

18.12. Chela-Flores, J., Testing S isotopes as biomarkers for Mars. *Int. J. Astrobiol.*, 18, 436–439, 2019.

18.13. Chela-Flores, J., Cicuttin, A., Crespo, M.L., Tuniz, C., Biogeochemical fingerprints of life: Earlier analogies with polar ecosystems suggest feasible instrumentation for probing the Galilean moons. *Int. J. Astrobiol.*, 14, 3, 427–434, 2015.

18.14. Chela-Flores, J. and Kumar, N., Returning to Europa: Can traces of surficial life be detected? *Int. J. Astrobiol.*, 7, 263–269, 2008.

18.15. Christner, B.C., Roysto-Bishop, G., Foreman, C.M., Arnold, B.R., Tranter, M., Welh, K.A., Lyons, W.B., Tspain, A.I., Studinger, M., Priscu, J.C., Limnological conditions in subglacial Lake Vostok. Antarctica. *Limnol. Occeanogr.*, 51, 2485-2501, 2006.

18.16. Christophe, B., Spilker, L.J., Anderson, J.D., Andre, N., Asmar, S.W., Aurnou, J., Banfield, D., Barucci, A., Bertolami, O., Bingham, R. *et al.*, OSS (Outer Solar System): A fundamental and planetary physics mission to Neptune, Triton and the Kuiper Belt. *Exp. Astron.*, 34, 203–242, 2012.

18.17. Clark, R.N., Curchin, J.M., Barnes, J.W., Jaumann, R., Soderblom, L., Cruikshank, D.P., Brown, R.H., Rodriguez, S., Lunine, J., Stephan, K. *et al.*, Detection and mapping of hydrocarbon deposits on Titan. *J. Geophys. Res.*, 115, E10005, 2010.

18.18. Crossfield, I.J.M., Petigura, E., Schlieder, J., Howard, A.W., Fulton, B.J., Aller, K.M., Ciardi, D.R., Lepine, S., Barclay, T., de Pater, I. *et al.*, A nearby M star with three transiting super-Earths discovered by K2. *Astrophys J.*, 804, 10, 2015, arXiv:1501.03798 [astro-ph].

18.19. Coustenis, A., Tokano, T., Burger, M.H., Cassidy, T.A., Lopes, R.M., Lorenz, R.D., Retherford, K.D., Schubert, G., Atmospheric/exospheric characteristics of icy satellites. *Space Sci. Rev.*, 153, 155–184, 2010.

18.20. Dai, X. and Guerras, E., Probing extragalactic planets using quasar microlensing. *Astrophys. J. Lett.*, 853, L27, 2018.

18.21. Doran, P.T., Lyons, W.B., McKnight, D.M., *Life in Antarctic Deserts and Other Cold Dry Environments*, Cambridge University Press, United Kingdom, 2010.

18.22. Gaeman, J., Hier-Majumdera, S., Roberts, J.H., Sustainability of a subsurface ocean within Triton's interior. *Icarus*, 220, 339–347, 2012.

18.23. Gaillard, F., Michalski, J., Berger, G., McLennan, S.M., Scaillet, B., Geochemical reservoirs and timing of sulfur cycling on Mars. *Space Sci. Rev.*, 174, 251–300, 2013.

18.24. Gibson, E.K. and Chang, S., The Moon biogenic elements, in: *Exobiology in Solar System Exploration*, Carle, G.C., Schwartz, D.E., Huntington, J.L. (Eds.), 1992, NASA SP-512: 28-43.

18.25. Gibson, E.K., Jr. and Moore, G., Sulfur abundances and distributions in the valley of Taurus-Littrow, in: *Proceedings of the Lunar Science Conference, 5th, Houston, Texas*, Pergamon Press, New York, pp. 1823–1827, 1973.

18.26. Gleeson, D., Pappalardo, R.T., Grasby, S.E., Anderson, M.S., Beauchamp, B., Castano, R., Chien, S., Doggett, T., Mandrake, L., Wagstaff, K., Characterization of a sulfur-rich, Arctic spring site and field analog of Europa using hyperspectral data. *Remote Sens. Environ.*, 114, 1297–1311, 2010.

18.27. Gleeson, D.F., Pappalardo, R.T., Anderson, M.S., Grasby, S.E., Mielke, R.E., Wright, K.E., Templeton, A.S., Biosignature detection at an Arctic analog to Europa. *Astrobiology*, 12, 1–16, 2012.

18.28. Goesmann, F., Brinckerhoff, W.B., Raulin, F., Goetz, W., Danell, R.M., Getty, S.A., Siljestrom, S., Mißbach, H., Steininger, H., Arevalo, R.D., Jr. *et al.*, The Mars organic molecule analyzer (MOMA) instrument: Characterization of organic material in martian sediments. *Astrobiology*, 17, 655–685, 2017.

18.29. Grasset, O., Dougherty, M.K., Coustenis, A., Bunce, E.J., Erd, C., Titov, D.V., Blanc, M., Coates, A., Drossart, P., Fletcher, L. *et al.*, JUpiter ICy moons Explorer (JUICE): An ESA mission to orbit Ganymede and to characterise the Jupiter system. *Planet. Space Sci.*, 78, 1–21, 2013.

18.30. Grotzinger, J.P., Crisp, J., Vasavada, A.R., Anderson, R.C., Baker, C.J., Barry, R., Blake, D.F., Conrad, P., Edgett, K.S., Ferdowski, B. *et al.*, Mars Science Laboratory mission and science investigation. *Space Sci. Rev.*, 170, 5–56, 2012.

18.31. Guzman, M., Lorenz, R., Hurley, D., Farrell, W., Spencer, J., Hansen, C., Hurford, T., Ibea, J., Carlson P. and McKay, C.P., Collecting amino acids in the Enceladus plume. *Int. J. Astrobiol.*, 8, 1, 47–59, 2018.

18.32. Hauri, E.H., Saal, A.E., Nakajima, M., Anand, M., Rutherford, M.J., Van Orman, J.A., LeVoyer, M., Origin and evolution of water in the Moon's interior. *Annu. Rev. Earth Planet. Sci.*, 45, 89–111, 2017.

18.33. Hegde, S. and Kaltenegger, L., Colors of extreme exo-Earth environments. *Astrobiology*, 13, 47–56, 2013.
18.34. Hussmann, H., Sohl, F., Spohn, T., Subsurface oceans and deep interiors of medium-sized outer planet satellites and large trans-Neptunian objects. *Icarus*, 185, 258–273, 2006.
18.35. Klein, H.P., Horowitz, N.H., Levin, G.V., Oyama, V.I., Lederberg, J., Rich, A., Hubbard, J.S., Hobby, G.L., Straat, P.A., Berdahl, B.J. *et al.*, The Viking biological investigation: Preliminary results. *Science*, 194, 99–105, 1976.
18.36. Konhauser, K., *Introduction to Geomicrobiology*, Blackwell Publishing, Oxford, 2007.
18.37. Lunine, J.I., Elachi, C., Wall, S.D., Janssen, M.A., Allison, M.D., Anderson, Y., Boehmer, R., Callahan, P., Encrenaz, P., Flamini, E. *et al.*, Titan's diverse landscapes as evidenced by Cassini RADAR's third and fourth looks at Titan. *Icarus*, 195, 415–433, 2008.
18.38. Mancinelli, R.L., The effect of the space environment on the survival of Halorubrum chaoviator and Synechococcus (Nägeli): Data from the Space Experiment OSMO on EXPOSE-R. *Int. J. Astrobiol.*, 14, 1, 123–128, 2015.
18.39. McKay, C.P., The search for a second genesis of life in our solar system, in: *The First Steps of Life in the Universe*, J. Chela-Flores, T. Owen, F. Raulin (Eds.), pp. 269–267, Kluwer Academic Publishers, Dordrecht, The Netherlands, 2001.
18.40. McKay, C.P., Loving Mars: A Review of To Mars With Love by Patricia Ann Straat. *Astrobiology*, 19, 4, 628–629, 2019.
18.41. McKay, C.P. and Smith, H.D., Possibilities for methanogenic life in liquid methane on the surface of Titan. *Icarus*, 178, 274–276, 2005.
18.42. Meyer, S., Tulej, M., Wurz, P., Mass spectrometry of planetary exospheres at high relative velocity: Direct comparison of open- and closed-source measurements. *Geosci. Instrum. Methods Data Syst.*, 6, 1–8, 2017.
18.43. Morozova, D., Möhlmann, D., Wagner, D., Survival of methanogenic archaea from Siberian permafrost under simulated Martian thermal conditions. *Orig. Life Evol. Biosph.*, 37, 2, 189–200, 2006.
18.44. Hofstadter, M., Murphy, N., Reh, K., *NASA Study: Ice Giant Mission Status and Planning CAPS Meeting, Irvine, CA*, 2015.
18.45. Navarro-Gonzáles, R., Vargas, E., de la Rosa, J., Raga, A.C., McKay, C.P., Reanalysis of the Viking results suggests perchlorate and organics at midlatitudes on Mars. *J. Geophys. Res.: Planets*, 115, E12010, 2010.
18.46. Needham, D.H. and Kring, D.A., Lunar volcanism produced a transient atmosphere around the ancient Moon. *Earth Planet. Sci. Lett.*, 478, 175–178, 2017.
18.47. Nimmo, F. and Spencer, J., Powering Triton's recent geological activity by obliquity tides: Implications for Pluto geology. *Icarus*, 246, 2–10, 2014.
18.48. Ono, S., Photochemistry of sulfur dioxide and the origin of mass-independent isotope fractionation in Earth's atmosphere. *Annu. Rev. Earth Planet. Sci.*, 45, 301–329, 2017.
18.49. Orosei, R., Lauro, S.E., Pettinelli, E., Cicchetti, A., Coradini, M., Cosciotti, B., Di Paolo, F., Flamini, E., Mattei, E., Pajola, M. *et al.*, Radar evidence of subglacial liquid water on Mars. *Science*, 361, 6401, 490–493, 2018.
18.50. Pappalardo, R., Pappalardo, R.T., Vance, S., Bagenal, F., Bills, B.G., Blaney, D.L., Blankenship, D.D., Brinckerhoff, W.B., Connerney, J.E., Hand, K.P. *et al.*, Science potential from a Europa lander. *Astrobiology*, 13, 740–773, 2013.
18.51. Perry, M.E., Teolis, B.D., Grimes, J., Miller, G.P., Hurley, D.M., Waite, J.H., Jr., Perryman, R.S., McNutt, R.L., Jr., Direct measurement of the velocity of the Enceladus vapor plumes. *47th Lunar and Planetary Science Conference*, The Woodlands, Texas, p. #2846, 2016.
18.52. Phillips, C.B. and Pappalardo, R.T., Europa Clipper mission concept: Exploring Jupiter's ocean moon. *Eos Trans. Am. Geophys. Union*, 95, 20, 165–167, 2014.

18.53. Plainaki, C., Cassidy, T.A., Shematovich, V.I., Milillo, A., Wurz, P., Vorburger, A., Roth, L., Galli, A., Rubin, M., Blöcker, A. *et al.*, Towards a global unified model of Europa's tenuous atmosphere. *Space Sci. Rev.*, 214, 40, 2018.
18.54. Reynolds, R.T., Squyres, S.W., Colburn, D.S., McKay, C.P., On the habitability of Europa. *Icarus*, 56, 246–254, 1983.
18.55. Rhoden, A.R., Henning, W., Hurford, T.A., Hamilton, D.P., The interior and orbital evolution of Charon as preserved in its geologic record. *Icarus*, 246, 11–20, 2015.
18.56. Riedo, A., Bieler, A., Neuland, M., Tulej, M., Wurz, P., Performance evaluation of a miniature laser ablation time-of-flight mass spectrometer designed for *in situ* investigations in planetary space research. *J. Mass Spectrom.*, 48, 1–15, 2013.
18.57. Riedo, A., Neuland, M., Meyer, S., Tulej, M., Wurz, P., Coupling of LMS with fs-laser ablation ion source: Elemental and isotope composition measurements. *J. Anal. At. Spectrom.*, 28, 1256–1269, 2013.
18.58. Schulze-Makuch, D. and Crawford, I.A., Was there an early habitability window for Earth's Moon? *Astrobiology*, 18, 1–4, 2018.
18.59. Seckbach, J. and Chela-Flores, J., Microorganisms in extreme environments with a view to astrobiology in the outer Solar System. *Paper presented at the Conference: Testing Habitability of Unexplored Environments the Outer Solar System. Instruments, Methods, and Missions for Astrobiology XVII*, San Diego, USA, 2015.
18.60. Shen, Y., Buick, R., Canfield, D.E., Isotopic evidence for microbial sulphate reduction in the early Archaean era. *Nature*, 410, 77–81, 2001.
18.61. Siegert, M.J., Tranter, M., Ellis-Evans, J.C., Priscu, J.C., Lyons, W.B., The hydrochemistry of Lake Vostok and the potential for life in Antarctic subglacial lakes. *Hydrol. Processes*, 17, 795–814, 2003.
18.62. Siegert, M.J., Carter, S., Tabacco, I., Popov S. and Blankenship, D.D., A revised inventory of Antarctic subglacial lake. *Antarct. Sci.*, 17, 453–460, 2005.
18.63. Sim, M.S., Bosak, T., Ono, S., Large sulfur isotope fractionation does not require disproportionation. *Science*, 333, 74–77, 2011.
18.64. Soffen, G.A. and Snyder, C.W., First Viking mission to Mars. *Science*, 193, 759–766, 1976.
18.65. Stan-Lotter, H., Radax, C., Gruber, C., Legat, A., Pfaffenhuemer, M., Wieland, H., Leuko, S., Weidler, G., Kömle, N., Kargl, G., Astrobiology with haloarchaea from Permo-Triassic rock salt. *Int. J. Astrobiol.*, 1, 4, 271–284, 2002.
18.66. Teolis, B.D., Perry, M.E., Hansen, C.J., Waite, J.H., Porco, C.C., Spencer, J.R., Howett, C.J.A., Enceladus plume structure and time variability: Comparison of Cassini observations. *Astrobiology*, 17, 9, 926–940, 2017.
18.67. Tobie, G., Giese, B., Hurford, T.A., Lopes, R.M., Nimmo, F., Postberg, F., Retherford, K.D., Schmidt, J., Spencer, J.R., Tokano, T., Turtle, E.P., Surface, subsurface and atmosphere exchanges on the satellites of the outer Solar System. *Space Sci. Rev.*, 153, 375–410, 2010.
18.68. Tulej, M., Neubeck, A., Ivarsson, M., Riedo, A., Neuland, M.B., Meyer, S., Wurz, P., Chemical composition of micrometer-sized filaments in an aragonite host by a miniature laser ablation/ionization mass spectrometer. *Astrobiology*, 15, 1–14, 2015.
18.69. Tulej, M., Meyer, S., Lüthi, M., Lasi, D., Galli, A., Piazza, D., Desorgher, L., Reggiani, D., Hajdas, W., Karlsson, S. *et al.*, Experimental investigation of the radiation shielding efficiency of a MCP detector in the radiation environment near Jupiter's Moon Europa: Nuclear instruments and methods. *Phys. Res. B*, 383, 21–37, 2016.
18.70. Turrini, D., Politi, R., Peron, R., Grassi, D., Plainaki, C., Barbieri, M., Lucchesi, D.M., Magni, G., Altieri, F., Cottini, V. *et al.*, The comparative exploration of the ice giant planets with twin spacecraft: Unveiling the history of our Solar System. *Planet. Space Sci.*, 104, 93–107, 2014.

18.71. Vaniman, D., Pettit, D., Heiken, G., Uses of lunar sulfur, in: *Proceedings of the Second Conference of Lunar Bases and Space Activities held in Houston, TX, April 5-7, 1988*, pp. 429–435, 1992.
18.72. Wiesendanger, R., Tulej, M., Riedo, A., Frey, S., Shea, H., Wurz, P., Improved detection sensitivity for heavy trace elements using a miniature laser ablation ionisation mass spectrometer. *J. Anal. At. Spectrom.*, 32, 11, 2182–2188, 2017.
18.73. Wiesendanger, R., Tulej, M., Grimaudo, V., Cedeño-López, A., Lukmanov, R., Riedo, A., Wurz, P., A method for improvement of mass resolution and isotope accuracy for laser ablation time-of-flight mass spectrometers. *J. Chemom.*, 33, 1, e3081, 2018.
18.74. Wiesendanger, R., Wacey, D., Tulej, M., Neubeck, A., Ivarsson, I., Grimaudo, V., Moreno-Garcia, P., Cedeno-Lopez, A., Riedo, A., Wurz, P., Chemical and optical identification of micrometer-sized 1.9 billion-year-old fossils by combining a miniature laser ablation ionization mass spectrometry system with an optical microscope. *Astrobiology*, 18, 1071–1080, 2018.

Index

Printed in the USA/Agawam, MA
September 16, 2026

910661.014